Basic Engineering

Why is knowledge of mathematics important in engineering?

A career in any engineering or scientific field will require both basic and advanced mathematics. Without mathematics to determine principles, calculate dimensions and limits, explore variations, prove concepts, and so on, there would be no mobile telephones, televisions, stereo systems, video games, microwave ovens, computers, or virtually anything electronic. There would be no bridges, tunnels, roads, skyscrapers, automobiles, ships, planes, rockets or most things mechanical. There would be no metals beyond the common ones, such as iron and copper, no plastics, no synthetics. In fact, society would most certainly be less advanced without the use of mathematics throughout the centuries and into the future.

Electrical engineers require mathematics to design, develop, test, or supervise the manufacturing and installation of electrical equipment, components, or systems for commercial, industrial, military, or scientific use.

Mechanical engineers require mathematics to perform engineering duties in planning and designing tools, engines, machines, and other mechanically functioning equipment; they oversee installation, operation, maintenance, and repair of such equipment as centralised heat, gas, water, and steam systems.

Aerospace engineers require mathematics to perform a variety of engineering work in designing, constructing, and testing aircraft, missiles, and spacecraft; they conduct basic and applied research to evaluate adaptability of materials and equipment to aircraft design and manufacture and recommend improvements in testing equipment and techniques.

Nuclear engineers require mathematics to conduct research on nuclear engineering problems or apply principles and theory of nuclear science to problems concerned with release, control, and utilisation of nuclear energy and nuclear waste disposal.

Petroleum engineers require mathematics to devise methods to improve oil and gas well production and determine the need for new or modified tool designs; they oversee drilling and offer technical advice to achieve economical and satisfactory progress.

Industrial engineers require mathematics to design, develop, test, and evaluate integrated systems for managing industrial production processes, including human work factors, quality control, inventory control, logistics and material flow, cost analysis, and production co-ordination.

Environmental engineers require mathematics to design, plan, or perform engineering duties in the prevention, control, and remediation of environmental health hazards, using various engineering disciplines; their work may include waste treatment, site remediation, or pollution control technology.

Civil engineers require mathematics at all levels in civil engineering – structural engineering, hydraulics and geotechnical engineering are all fields that employ mathematical tools such as differential equations, tensor analysis, field theory, numerical methods and operations research.

Knowledge of mathematics is therefore needed by each of the engineering disciplines listed above.

It is intended that this text – *Basic Engineering Mathematics* – will provide a step by step approach to learning all the early, fundamental mathematics needed for your future engineering studies.

To Sue

Basic Engineering Mathematics

Sixth Edition

John Bird, BSc (Hons), CEng, CMath, CSci, FIMA, FIET, FCollT

Routledge
Taylor & Francis Group

LONDON AND NEW YORK

Sixth edition published 2014
by Routledge
2 Park Square, Milton Park, Abingdon, Oxon, OX14 4RN

and by Routledge
711 Third Avenue, New York, NY 10017

Routledge is an imprint of the Taylor & Francis Group, an informa business

First edition published by Newnes 1999
Fifth edition published by Newnes 2010

British Library Cataloguing-in-Publication Data
A catalogue record for this book is available from the British Library.

Library of Congress Cataloging-in-Publication Data
Bird, J. O.
Basic engineering mathematics / John Bird. – Sixth edition.
pages cm
1. Engineering mathematics. I. Title.
TA330.B513 2014
620.001'51–dc23

ISBN13: 978-0-415-66278-9 (pbk)
ISBN13: 978-1-315-85884-5 (ebk)

Typeset in Times by
Servis Filmsetting Ltd, Stockport, Cheshire

Contents

Preface

Basic Engineering Mathematics 6th Edition introduces and then consolidates basic mathematical principles and promotes awareness of mathematical concepts for students needing a broad base for further vocational studies. In this sixth edition, new material has been added to some of the chapters, together with around 40 extra practical problems interspersed throughout the text. The four chapters only available on the website in the previous edition have been included in this edition. In addition, some multiple choice questions have been included to add interest to the learning.

The text covers:

(i) **Basic mathematics** for a wide range of introductory/access/foundation mathematics courses

(ii) **'Mathematics for Engineering Technicians'** for BTEC First NQF Level 2; *chapters 1 to 12, 16 to 18, 20, 21, 23, and 25 to 27 are needed for this module*.

(iii) The mandatory **'Mathematics for Technicians'** for BTEC National Certificate and National Diploma in Engineering, NQF Level 3; *chapters 7 to 10, 14 to 17, 19, 20 to 23, 25 to 27, 31, 32, 34 and 35 are needed for this module. In addition, chapters 1 to 6, 11 and 12 are helpful revision for this module*.

(iv) **GCSE revision**, and for similar mathematics courses in English-speaking countries worldwide.

Basic Engineering Mathematics 6th Edition provides a lead into *Engineering Mathematics 7th Edition*.

Each topic considered in the text is presented in a way that assumes in the reader little previous knowledge of that topic.

Theory is introduced in each chapter by a brief outline of essential theory, definitions, formulae, laws and procedures. However, these are kept to a minimum, for problem solving is extensively used to establish and exemplify the theory. It is intended that readers will gain real understanding through seeing problems solved and then solving similar problems themselves.

This textbook contains some **750 worked problems**, followed by over **1600 further problems** (all with answers – at the end of the book). The further problems are contained within **161 Practice Exercises**; each Practice Exercise follows on directly from the relevant section of work. Fully worked solutions to all 1600 problems have been made freely available to all via the website – see below. **420 line diagrams** enhance the understanding of the theory. Where at all possible the problems mirror potential practical situations found in engineering and science.

At regular intervals throughout the text are **15 Revision Tests** to check understanding. For example, Revision Test 1 covers material contained in chapters 1 and 2, Revision Test 2 covers the material contained in chapters 3 to 5, and so on. These Revision Tests do not have answers given since it is envisaged that lecturers/instructors could set the Tests for students to attempt as part of their course structure. Lecturers/instructors may obtain a complimentary set of solutions of the Revision Tests in an **Instructor's Manual** available from the publishers via the internet – see below.

At the end of the book a list of relevant **formulae** contained within the text is included for convenience of reference.

'Learning by example' is at the heart of *Basic Engineering Mathematics 6th Edition*.

JOHN BIRD
Defence College of Technical Training,
HMS Sultan,
formerly of University of Portsmouth
and Highbury College, Portsmouth

John Bird is the former Head of Applied Electronics in the Faculty of Technology at Highbury College, Portsmouth, UK. More recently, he has combined freelance lecturing at the University of Portsmouth, with Examiner responsibilities for Advanced Mathematics with City and Guilds, and examining for the International Baccalaureate Organization. He is the author of some 125 textbooks on engineering and mathematical subjects, with worldwide sales of 1 million copies. He is currently a Senior Training Provider at the Defence School of Marine Engineering in the Defence College of Technical Training at HMS *Sultan*, Gosport, Hampshire, UK.

Free Web downloads

For students

1. **Full solutions** to the 1600 questions contained in the 161 Practice Exercises

2. Download **Multiple choice questions and answer sheet**

3. **List of Essential Formulae**

4. **Famous Engineers/Scientists** – From time to time in the text, 16 famous mathematicians/engineers are referred to and emphasised with an asterisk*. Background information on each of these is available via the website. Mathematicians/engineers involved are: **Boyle, Celsius, Charles, Descartes, Faraday, Henry, Hertz, Hooke, Kirchhoff, Leibniz, Napier, Newton, Ohm, Pythagoras, Simpson and Young.**

For instructors/lecturers

1. **Full solutions** to the 1600 questions contained in the 161 Practice Exercises

2. **Full solutions** and marking scheme to each of the **15 Revision Tests** – named as **Instructors Manual**

3. **Revision Tests** – available to run off to be given to students

4. Download **Multiple choice questions and answer sheet**

5. **List of Essential Formulae**

6. **Illustrations** – all 420 available on Power-Point

7. **Famous Engineers/Scientists** – 16 are mentioned in the text, as listed above

Acknowledgements

The publisher wishes to thank CASIO Electronic Co. Ltd, London for permission to reproduce the image of the Casio fx-991ES calculator on page 24.

The publisher also wishes to thank the AA Media Ltd for permission to reproduce the map of Portsmouth on page 149.

Chapter 1

Basic arithmetic

Why it is important to understand: **Basic arithmetic**

Being numerate, i.e. having an ability to add, subtract, multiply and divide whole numbers with some confidence, goes a long way towards helping you become competent at mathematics. Of course electronic calculators are a marvellous aid to the quite complicated calculations often required in engineering; however, having a feel for numbers 'in our head' can be invaluable when estimating. Do not spend too much time on this chapter because we deal with the calculator later; however, try to have some idea how to do quick calculations in the absence of a calculator. You will feel more confident in dealing with numbers and calculations if you can do this.

At the end of this chapter, you should be able to:

- understand positive and negative integers
- add and subtract integers
- multiply and divide two integers
- multiply numbers up to 12×12 by rote
- determine the highest common factor from a set of numbers
- determine the lowest common multiple from a set of numbers
- appreciate the order of operation when evaluating expressions
- understand the use of brackets in expressions
- evaluate expressions containing $+, -, \times, \div$ and brackets

1.1 Introduction

Whole numbers are simply the numbers 0, 1, 2, 3, 4, 5, ... (and so on). **Integers** are like whole numbers, but they also include negative numbers. $+3, +5$ and $+72$ are examples of positive integers; $-13, -6$ and -51 are examples of negative integers. Between positive and negative integers is the number 0 which is neither positive nor negative.

The four basic arithmetic operators are add $(+)$, subtract $(-)$, multiply $(\times)$ and divide $(\div)$.

It is assumed that adding, subtracting, multiplying and dividing reasonably small numbers can be achieved without a calculator. However, if revision of this area is needed then some worked problems are included in the following sections.

When **unlike signs** occur together in a calculation, the overall sign is **negative**. For example,

$$3 + (-4) = 3 + -4 = 3 - 4 = -1$$

and

$$(+5) \times (-2) = -10$$

Basic Engineering Mathematics. 978-0-415-66278-9, © 2014 John Bird. Published by Taylor & Francis. All rights reserved.

Like signs together give an overall **positive sign**. For example,

$$3 - (-4) = 3 - -4 = 3 + 4 = \mathbf{7}$$

and

$$(-6) \times (-4) = \mathbf{+24}$$

1.2 Revision of addition and subtraction

You can probably already add two or more numbers together and subtract one number from another. However, if you need revision then the following worked problems should be helpful.

Problem 1. Determine $735 + 167$

$$
\begin{array}{r}
\mathbf{H\,T\,U} \\
7\ 3\ 5 \\
+\ 1\ 6\ 7 \\
\hline
9\ 0\ 2 \\
\hline
1\ 1 \\
\end{array}
$$

(i) $5 + 7 = 12$. Place 2 in units (U) column. Carry 1 in the tens (T) column.

(ii) $3 + 6 + 1$ (carried) $= 10$. Place the 0 in the tens column. Carry the 1 in the hundreds (H) column.

(iii) $7 + 1 + 1$ (carried) $= 9$. Place the 9 in the hundreds column.

Hence, $\mathbf{735 + 167 = 902}$

Problem 2. Determine $632 - 369$

$$
\begin{array}{r}
\mathbf{H\,T\,U} \\
6\ 3\ 2 \\
-\ 3\ 6\ 9 \\
\hline
2\ 6\ 3 \\
\hline
\end{array}
$$

(i) $2 - 9$ is not possible; therefore change one ten into ten units (leaving 2 in the tens column). In the units column, this gives us $12 - 9 = 3$

(ii) Place 3 in the units column.

(iii) $2 - 6$ is not possible; therefore change one hundred into ten tens (leaving 5 in the hundreds column). In the tens column, this gives us $12 - 6 = 6$

(iv) Place the 6 in the tens column.

(v) $5 - 3 = 2$

(vi) Place the 2 in the hundreds column.

Hence, $\mathbf{632 - 369 = 263}$

Problem 3. Add $27, -74, 81$ and -19

This problem is written as $27 - 74 + 81 - 19$.

Adding the positive integers:	27
	81
Sum of positive integers is	108
Adding the negative integers:	74
	19
Sum of negative integers is	93
Taking the sum of the negative integers from the sum of the positive integers gives	108
	−93
	15

Thus, $\mathbf{27 - 74 + 81 - 19 = 15}$

Problem 4. Subtract -74 from 377

This problem is written as $377 - -74$. Like signs together give an overall positive sign, hence

$$
377 - -74 = 377 + 74 \qquad
\begin{array}{r}
3\ 7\ 7 \\
+\ \ 7\ 4 \\
\hline
4\ 5\ 1 \\
\hline
\end{array}
$$

Thus, $\mathbf{377 - -74 = 451}$

Problem 5. Subtract 243 from 126

The problem is $126 - 243$. When the second number is larger than the first, take the smaller number from the larger and make the result negative. Thus,

$$
126 - 243 = -(243 - 126) \qquad
\begin{array}{r}
2\ 4\ 3 \\
-\ 1\ 2\ 6 \\
\hline
1\ 1\ 7 \\
\hline
\end{array}
$$

Thus, $\mathbf{126 - 243 = -117}$

Problem 6. Subtract 318 from −269

The problem is −269 − 318. The sum of the negative integers is

$$\begin{array}{r} 269 \\ +318 \\ \hline 587 \end{array}$$

Thus, **−269 − 318 = −587**

Now try the following Practice Exercise

Practice Exercise 1   Further problems on addition and subtraction (answers on page 422)

In Problems 1−15, determine the values of the expressions given, without using a calculator.

1. $67\,\text{kg} - 82\,\text{kg} + 34\,\text{kg}$

2. $73\,\text{m} - 57\,\text{m}$

3. $851\,\text{mm} - 372\,\text{mm}$

4. $124 - 273 + 481 - 398$

5. £927 − £114 + £182 − £183 − £247

6. $647 - 872$

7. $2417 - 487 + 2424 - 1778 - 4712$

8. $-38419 - 2177 + 2440 - 799 + 2834$

9. £2715 − £18250 + £11471 − £1509 + £113274

10. $47 + (-74) - (-23)$

11. $813 - (-674)$

12. $3151 - (-2763)$

13. $4872\,\text{g} - 4683\,\text{g}$

14. $-23148 - 47724$

15. $53774 - 38441$

16. Calculate the diameter d and dimensions A and B for the template shown in Fig. 1.1. All dimensions are in millimetres.

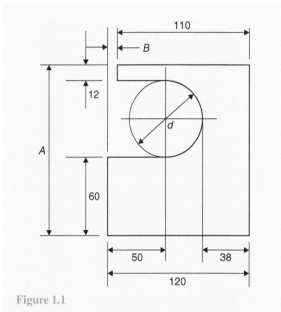

Figure 1.1

1.3 Revision of multiplication and division

You can probably already multiply two numbers together and divide one number by another. However, if you need a revision then the following worked problems should be helpful.

Problem 7. Determine 86 × 7

$$\begin{array}{r} \textbf{H T U} \\ 8\ 6 \\ \times \quad 7 \\ \hline 6\ 0\ 2 \\ \hline 4 \end{array}$$

(i) $7 \times 6 = 42$. Place the 2 in the units (U) column and 'carry' the 4 into the tens (T) column.

(ii) $7 \times 8 = 56; 56 + 4 \text{ (carried)} = 60$. Place the 0 in the tens column and the 6 in the hundreds (H) column.

Hence, $86 \times 7 = 602$

A good grasp of **multiplication tables** is needed when multiplying such numbers; a reminder of the multiplication table up to 12 × 12 is shown below. Confidence with handling numbers will be greatly improved if this table is memorised.

Multiplication table

×	2	3	4	5	6	7	8	9	10	11	12
2	4	6	8	10	12	14	16	18	20	22	24
3	6	9	12	15	18	21	24	27	30	33	36
4	8	12	16	20	24	28	32	36	40	44	48
5	10	15	20	25	30	35	40	45	50	55	60
6	12	18	24	30	36	42	48	54	60	66	72
7	14	21	28	35	42	49	56	63	70	77	84
8	16	24	32	40	48	56	64	72	80	88	96
9	18	27	36	45	54	63	72	81	90	99	108
10	20	30	40	50	60	70	80	90	100	110	120
11	22	33	44	55	66	77	88	99	110	121	132
12	24	36	48	60	72	84	96	108	120	132	144

Problem 8. Determine 764×38

$$
\begin{array}{r}
764 \\
\times \; 38 \\
\hline
6112 \\
22920 \\
\hline
29032 \\
\hline
\end{array}
$$

(i) $8 \times 4 = 32$. Place the 2 in the units column and carry 3 into the tens column.

(ii) $8 \times 6 = 48; 48 + 3$ (carried) $= 51$. Place the 1 in the tens column and carry the 5 into the hundreds column.

(iii) $8 \times 7 = 56; 56 + 5$ (carried) $= 61$. Place 1 in the hundreds column and 6 in the thousands column.

(iv) Place 0 in the units column under the 2

(v) $3 \times 4 = 12$. Place the 2 in the tens column and carry 1 into the hundreds column.

(vi) $3 \times 6 = 18; 18 + 1$ (carried) $= 19$. Place the 9 in the hundreds column and carry the 1 into the thousands column.

(vii) $3 \times 7 = 21; 21 + 1$ (carried) $= 22$. Place 2 in the thousands column and 2 in the ten thousands column.

(viii) $6112 + 22920 = 29032$

Hence, **$764 \times 38 = 29032$**

Again, knowing multiplication tables is rather important when multiplying such numbers.

It is appreciated, of course, that such a multiplication can, and probably will, be performed using a **calculator**. However, there are times when a calculator may not be available and it is then useful to be able to calculate the 'long way'.

Problem 9. Multiply 178 by -46

When the numbers have different signs, the result will be negative. (With this in mind, the problem can now be solved by multiplying 178 by 46). Following the procedure of Problem 8 gives

$$
\begin{array}{r}
178 \\
\times \; 46 \\
\hline
1068 \\
7120 \\
\hline
8188 \\
\hline
\end{array}
$$

Thus, $178 \times 46 = 8188$ and **$178 \times (-46) = -8188$**

Problem 10. Determine $1834 \div 7$

$$
\begin{array}{r}
262 \\
7{\overline{)1834}}
\end{array}
$$

(i)   7 into 18 goes 2, remainder 4. Place the 2 above the 8 of 1834 and carry the 4 remainder to the next digit on the right, making it 43

(ii)   7 into 43 goes 6, remainder 1. Place the 6 above the 3 of 1834 and carry the 1 remainder to the next digit on the right, making it 14

(iii)   7 into 14 goes 2, remainder 0. Place 2 above the 4 of 1834

Hence, $1834 \div 7 = 1834/7 = \dfrac{1834}{7} = 262$

The method shown is called **short division**.

Problem 11. Determine $5796 \div 12$

$$
\begin{array}{r}
483 \\
12\overline{)5796} \\
\underline{48} \\
99 \\
\underline{96} \\
36 \\
\underline{36} \\
00
\end{array}
$$

(i)   12 into 5 won't go. 12 into 57 goes 4; place 4 above the 7 of 5796

(ii) $4 \times 12 = 48$; place the 48 below the 57 of 5796

(iii) $57 - 48 = 9$

(iv)   Bring down the 9 of 5796 to give 99

(v)   12 into 99 goes 8; place 8 above the 9 of 5796

(vi) $8 \times 12 = 96$; place 96 below the 99

(vii) $99 - 96 = 3$

(viii)   Bring down the 6 of 5796 to give 36

(ix)   12 into 36 goes 3 exactly.

(x)   Place the 3 above the final 6

(xi) $3 \times 12 = 36$; Place the 36 below the 36

(xii) $36 - 36 = 0$

Hence, $5796 \div 12 = 5796/12 = \dfrac{5796}{12} = 483$

The method shown is called **long division**.

Now try the following Practice Exercise

Practice Exercise 2   Further problems on multiplication and division (answers on page 422)

Determine the values of the expressions given in Problems 1 to 9, without using a calculator.

1.   (a) 78×6 (b) 124×7

2.   (a) £261×7 (b) £462×9

3.   (a) $783 \,\text{kg} \times 11$ (b) $73 \,\text{kg} \times 8$

4.   (a) $27 \,\text{mm} \times 13$ (b) $77 \,\text{mm} \times 12$

5.   (a) 448×23 (b) $143 \times (-31)$

6.   (a) $288 \,\text{m} \div 6$ (b) $979 \,\text{m} \div 11$

7.   (a) $\dfrac{1813}{7}$ (b) $\dfrac{896}{16}$

8.   (a) $\dfrac{21424}{13}$ (b) $15900 \div -15$

9.   (a) $\dfrac{88737}{11}$ (b) $46858 \div 14$

10.   A screw has a mass of 15 grams. Calculate, in kilograms, the mass of 1200 such screws $(1 \,\text{kg} = 1000 \,\text{g})$.

11.   Holes are drilled 35.7 mm apart in a metal plate. If a row of 26 holes is drilled, determine the distance, in centimetres, between the centres of the first and last holes.

12.   A builder needs to clear a site of bricks and top soil. The total weight to be removed is 696 tonnes. Trucks can carry a maximum load of 24 tonnes. Determine the number of truck loads needed to clear the site.

1.4   Highest common factors and lowest common multiples

When two or more numbers are multiplied together, the individual numbers are called **factors**. Thus, a factor is a number which divides into another number exactly. The **highest common factor (HCF)** is the largest number which divides into two or more numbers exactly.

For example, consider the numbers 12 and 15

The factors of 12 are 1, 2, 3, 4, 6 and 12 (i.e. all the numbers that divide into 12).

The factors of 15 are 1, 3, 5 and 15 (i.e. all the numbers that divide into 15).

1 and 3 are the only **common factors**; i.e. numbers which are factors of **both** 12 and 15

Hence, **the HCF of 12 and 15 is 3** since 3 is the highest number which divides into **both** 12 and 15

A **multiple** is a number which contains another number an exact number of times. The smallest number which is exactly divisible by each of two or more numbers is called the **lowest common multiple (LCM)**.

For example, the multiples of 12 are 12, 24, 36, 48, 60, 72, ... and the multiples of 15 are 15, 30, 45, 60, 75, ...

60 is a common multiple (i.e. a multiple of **both** 12 and 15) and there are no lower common multiples.

Hence, **the LCM of 12 and 15 is 60** since 60 is the lowest number that both 12 and 15 divide into.

Here are some further problems involving the determination of HCFs and LCMs.

Problem 12. Determine the HCF of the numbers 12, 30 and 42

Probably the simplest way of determining an HCF is to express each number in terms of its lowest factors. This is achieved by repeatedly dividing by the prime numbers 2, 3, 5, 7, 11, 13, ... (where possible) in turn. Thus,

$$12 = \boxed{2} \times 2 \times \boxed{3}$$
$$30 = \boxed{2} \qquad \times \boxed{3} \times 5$$
$$42 = \boxed{2} \qquad \times \boxed{3} \times 7$$

The factors which are common to each of the numbers are 2 in column 1 and 3 in column 3, shown by the broken lines. Hence, **the HCF is 2 × 3**; i.e. **6**. That is, 6 is the largest number which will divide into 12, 30 and 42.

Problem 13. Determine the HCF of the numbers 30, 105, 210 and 1155

Using the method shown in Problem 12:

$$30 = 2 \times \boxed{3} \times \boxed{5}$$
$$105 = \boxed{3} \times \boxed{5} \times 7$$
$$210 = 2 \times \boxed{3} \times \boxed{5} \times 7$$
$$1155 = \boxed{3} \times \boxed{5} \times 7 \times 11$$

The factors which are common to each of the numbers are 3 in column 2 and 5 in column 3. Hence, **the HCF is 3 × 5 = 15**

Problem 14. Determine the LCM of the numbers 12, 42 and 90

The LCM is obtained by finding the lowest factors of each of the numbers, as shown in Problems 12 and 13 above, and then selecting the largest group of any of the factors present. Thus,

$$12 = \boxed{2 \times 2} \times 3$$
$$42 = 2 \qquad \times 3 \qquad\qquad \times \boxed{7}$$
$$90 = 2 \qquad \times \boxed{3 \times 3} \times \boxed{5}$$

The largest group of any of the factors present is shown by the broken lines and is 2 × 2 in 12, 3 × 3 in 90, 5 in 90 and 7 in 42

Hence, **the LCM is 2 × 2 × 3 × 3 × 5 × 7 = 1260** and is the smallest number which 12, 42 and 90 will all divide into exactly.

Problem 15. Determine the LCM of the numbers 150, 210, 735 and 1365

Using the method shown in Problem 14 above:

$$150 = \boxed{2} \times \boxed{3} \times \boxed{5 \times 5}$$
$$210 = 2 \times 3 \times 5 \qquad \times 7$$
$$735 = \qquad 3 \times 5 \qquad \times \boxed{7 \times 7}$$
$$1365 = \qquad 3 \times 5 \qquad \times 7 \qquad \times \boxed{13}$$

Hence, **the LCM is 2 × 3 × 5 × 5 × 7 × 7 × 13**
$$= 95550$$

Now try the following Practice Exercise

Practice Exercise 3   Further problems on highest common factors and lowest common multiples (answers on page 422)

Find (a) the HCF and (b) the LCM of the following groups of numbers.

1. 8, 12   2. 60, 72

3. 50, 70   4. 270, 900

5. 6, 10, 14   6. 12, 30, 45

7. 10, 15, 70, 105   8. 90, 105, 300

9. 196, 210, 462, 910   10. 196, 350, 770

1.5   Order of operation and brackets

1.5.1   Order of operation

Sometimes addition, subtraction, multiplication, division, powers and brackets may all be involved in a calculation. For example,

$$5 - 3 \times 4 + 24 \div (3 + 5) - 3^2$$

This is an extreme example but will demonstrate the order that is necessary when evaluating.

When we read, we read from left to right. However, with mathematics there is a definite order of precedence which we need to adhere to. The order is as follows:

Brackets
Order (or p**O**wer)
Division
Multiplication
Addition
Subtraction

Notice that the first letters of each word spell **BODMAS**, a handy aide-mémoire. **O**rder means p**O**wer. For example, $4^2 = 4 \times 4 = 16$

$5 - 3 \times 4 + 24 \div (3 + 5) - 3^2$ is   evaluated   as follows:

$5 - 3 \times 4 + 24 \div (3 + 5) - 3^2$

$= 5 - 3 \times 4 + 24 \div 8 - 3^2$ (**B**racket is removed and
$3 + 5$ replaced with 8)

$= 5 - 3 \times 4 + 24 \div 8 - 9$ (**O**rder means p**O**wer; in
this case, $3^2 = 3 \times 3 = 9$)

$= 5 - 3 \times 4 + 3 - 9$ (**D**ivision: $24 \div 8 = 3$)

$= 5 - 12 + 3 - 9$ (**M**ultiplication: $-3 \times 4 = -12$)

$= 8 - 12 - 9$ (**A**ddition: $5 + 3 = 8$)

$= -13$ (**S**ubtraction: $8 - 12 - 9 = -13$)

In practice, **it does not matter if multiplication is performed before division or if subtraction is performed before addition**. What is important is that **the process of multiplication and division must be completed before addition and subtraction**.

1.5.2   Brackets and operators

The basic laws governing the **use of brackets and operators** are shown by the following examples.

(a) $2 + 3 = 3 + 2$; i.e. the order of numbers when adding does not matter.

(b) $2 \times 3 = 3 \times 2$; i.e. the order of numbers when multiplying does not matter.

(c) $2 + (3 + 4) = (2 + 3) + 4$; i.e. the use of brackets when adding does not affect the result.

(d) $2 \times (3 \times 4) = (2 \times 3) \times 4$; i.e. the use of brackets when multiplying does not affect the result.

(e) $2 \times (3 + 4) = 2(3 + 4) = 2 \times 3 + 2 \times 4$;   i.e.   a number placed outside of a bracket indicates that the whole contents of the bracket must be multiplied by that number.

(f) $(2 + 3)(4 + 5) = (5)(9) = 5 \times 9 = 45$; i.e. adjacent brackets indicate multiplication.

(g) $2[3 + (4 \times 5)] = 2[3 + 20] = 2 \times 23 = 46$;   i.e. when an expression contains inner and outer brackets, **the inner brackets are removed first**.

Here are some further problems in which BODMAS needs to be used.

Problem 16.   Find the value of $6 + 4 \div (5 - 3)$

The order of precedence of operations is remembered by the word BODMAS. Thus,

$$6 + 4 \div (5 - 3) = 6 + 4 \div 2 \qquad \text{(Brackets)}$$
$$= 6 + 2 \qquad \text{(Division)}$$
$$= 8 \qquad \text{(Addition)}$$

Problem 17.   Determine the value of
$13 - 2 \times 3 + 14 \div (2 + 5)$

$$13 - 2 \times 3 + 14 \div (2 + 5) = 13 - 2 \times 3 + 14 \div 7 \quad \text{(B)}$$
$$= 13 - 2 \times 3 + 2 \quad \text{(D)}$$
$$= 13 - 6 + 2 \quad \text{(M)}$$
$$= 15 - 6 \quad \text{(A)}$$
$$= 9 \quad \text{(S)}$$

Problem 18.   Evaluate
$16 \div (2+6) + 18[3 + (4 \times 6) - 21]$

$16 \div (2+6) + 18[3 + (4 \times 6) - 21]$

$= 16 \div (2+6) + 18[3 + 24 - 21]$ (B: inner bracket is determined first)

$= 16 \div 8 + 18 \times 6$ (B)

$= 2 + 18 \times 6$ (D)

$= 2 + 108$ (M)

$= \mathbf{110}$ (A)

Note that a number outside of a bracket multiplies all that is inside the brackets. In this case,

$18[3 + 24 - 21] = 18[6]$, which means $18 \times 6 = 108$

Problem 19.   Find the value of
$23 - 4(2 \times 7) + \dfrac{(144 \div 4)}{(14 - 8)}$

$23 - 4(2 \times 7) + \dfrac{(144 \div 4)}{(14 - 8)} = 23 - 4 \times 14 + \dfrac{36}{6}$ (B)

$= 23 - 4 \times 14 + 6$ (D)

$= 23 - 56 + 6$ (M)

$= 29 - 56$ (A)

$= \mathbf{-27}$ (S)

Problem 20.   Evaluate
$\dfrac{3 + \sqrt{(5^2 - 3^2)} + 2^3}{1 + (4 \times 6) \div (3 \times 4)} + \dfrac{15 \div 3 + 2 \times 7 - 1}{3 \times \sqrt{4} + 8 - 3^2 + 1}$

$\dfrac{3 + \sqrt{(5^2 - 3^2)} + 2^3}{1 + (4 \times 6) \div (3 \times 4)} + \dfrac{15 \div 3 + 2 \times 7 - 1}{3 \times \sqrt{4} + 8 - 3^2 + 1}$

$= \dfrac{3 + 4 + 8}{1 + 24 \div 12} + \dfrac{15 \div 3 + 2 \times 7 - 1}{3 \times 2 + 8 - 9 + 1}$

$= \dfrac{3 + 4 + 8}{1 + 2} + \dfrac{5 + 2 \times 7 - 1}{3 \times 2 + 8 - 9 + 1}$

$= \dfrac{15}{3} + \dfrac{5 + 14 - 1}{6 + 8 - 9 + 1}$

$= 5 + \dfrac{18}{6}$

$= 5 + 3 = \mathbf{8}$

Now try the following Practice Exercise

Practice Exercise 4   Further problems on order of precedence and brackets (answers on page 422)

Evaluate the following expressions.

1. $14 + 3 \times 15$

2. $17 - 12 \div 4$

3. $86 + 24 \div (14 - 2)$

4. $7(23 - 18) \div (12 - 5)$

5. $63 - 8(14 \div 2) + 26$

6. $\dfrac{40}{5} - 42 \div 6 + (3 \times 7)$

7. $\dfrac{(50 - 14)}{3} + 7(16 - 9) - 7$

8. $\dfrac{(7 - 3)(1 - 6)}{4(11 - 6) \div (3 - 8)}$

9. $\dfrac{(3 + 9 \times 6) \div 3 - 2 \div 2}{3 \times 6 + (4 - 9) - 3^2 + 5}$

10. $\dfrac{(4 \times 3^2 + 24) \div 5 + 9 \times 3}{2 \times 3^2 - 15 \div 3} +$
$\dfrac{2 + 27 \div 3 + 12 \div 2 - 3^2}{5 + (13 - 2 \times 5) - 4}$

11. $\dfrac{1 + \sqrt{25} + 3 \times 2 - 8 \div 2}{3 \times 4 - \sqrt{(3^2 + 4^2)} + 1} -$
$\dfrac{(4 \times 2 + 7 \times 2) \div 11}{\sqrt{9} + 12 \div 2 - 2^3}$

For fully worked solutions to each of the problems in Practice Exercises 5 to 7 in this chapter, go to the website:
www.routledge.com/cw/bird

Chapter 2

Fractions

Why it is important to understand: Fractions

Engineers use fractions all the time, examples including stress to strain ratios in mechanical engineering, chemical concentration ratios and reaction rates, and ratios in electrical equations to solve for current and voltage. Fractions are also used everywhere in science, from radioactive decay rates to statistical analysis. Calculators are able to handle calculations with fractions. However, there will be times when a quick calculation involving addition, subtraction, multiplication and division of fractions is needed. Again, do not spend too much time on this chapter because we deal with the calculator later; however, try to have some idea how to do quick calculations in the absence of a calculator. You will feel more confident to deal with fractions and calculations if you can do this.

At the end of this chapter, you should be able to:

- understand the terminology numerator, denominator, proper and improper fractions and mixed numbers
- add and subtract fractions
- multiply and divide two fractions
- appreciate the order of operation when evaluating expressions involving fractions

2.1 Introduction

A mark of 9 out of 14 in an examination may be written as $\dfrac{9}{14}$ or 9/14. $\dfrac{9}{14}$ is an example of a fraction. The number above the line, i.e. 9, is called the **numerator**. The number below the line, i.e. 14, is called the **denominator**.

When the value of the numerator is less than the value of the denominator, the fraction is called a **proper fraction**. $\dfrac{9}{14}$ is an example of a proper fraction.

When the value of the numerator is greater than the value of the denominator, the fraction is called an **improper fraction**. $\dfrac{5}{2}$ is an example of an improper fraction.

A **mixed number** is a combination of a whole number and a fraction. $2\dfrac{1}{2}$ is an example of a mixed number. In fact, $\dfrac{5}{2} = 2\dfrac{1}{2}$

There are a number of everyday examples in which fractions are readily referred to. For example, three people equally sharing a bar of chocolate would have $\dfrac{1}{3}$ each. A supermarket advertises $\dfrac{1}{5}$ off a six-pack of beer; if the beer normally costs £2 then it will now cost £1.60. $\dfrac{3}{4}$ of the employees of a company are women; if the company has 48 employees, then 36 are women.

Calculators are able to handle calculations with fractions. However, to understand a little more about fractions we will in this chapter show how to add, subtract, multiply and divide with fractions without the use of a calculator.

Problem 1. Change the following improper fractions into mixed numbers:

$$\text{(a) } \frac{9}{2} \quad \text{(b) } \frac{13}{4} \quad \text{(c) } \frac{28}{5}$$

(a) $\frac{9}{2}$ means 9 halves and $\frac{9}{2} = 9 \div 2$, and $9 \div 2 = 4$ and 1 half, i.e.

$$\frac{9}{2} = 4\frac{1}{2}$$

(b) $\frac{13}{4}$ means 13 quarters and $\frac{13}{4} = 13 \div 4$, and $13 \div 4 = 3$ and 1 quarter, i.e.

$$\frac{13}{4} = 3\frac{1}{4}$$

(c) $\frac{28}{5}$ means 28 fifths and $\frac{28}{5} = 28 \div 5$, and $28 \div 5 = 5$ and 3 fifths, i.e.

$$\frac{28}{5} = 5\frac{3}{5}$$

Problem 2. Change the following mixed numbers into improper fractions:

$$\text{(a) } 5\frac{3}{4} \quad \text{(b) } 1\frac{7}{9} \quad \text{(c) } 2\frac{3}{7}$$

(a) $5\frac{3}{4}$ means $5 + \frac{3}{4}$. 5 contains $5 \times 4 = 20$ quarters.

Thus, $5\frac{3}{4}$ contains $20 + 3 = 23$ quarters, i.e.

$$5\frac{3}{4} = \frac{23}{4}$$

The quick way to change $5\frac{3}{4}$ into an improper fraction is $\frac{4 \times 5 + 3}{4} = \frac{23}{4}$

(b) $1\frac{7}{9} = \frac{9 \times 1 + 7}{9} = \frac{16}{9}$

(c) $2\frac{3}{7} = \frac{7 \times 2 + 3}{7} = \frac{17}{7}$

Problem 3. In a school there are 180 students of which 72 are girls. Express this as a fraction in its simplest form

---

The fraction of girls is $\frac{72}{180}$

Dividing both the numerator and denominator by the lowest prime number, i.e. 2, gives

$$\frac{72}{180} = \frac{36}{90}$$

Dividing both the numerator and denominator again by 2 gives

$$\frac{72}{180} = \frac{36}{90} = \frac{18}{45}$$

2 will not divide into both 18 and 45, so dividing both the numerator and denominator by the next prime number, i.e. 3, gives

$$\frac{72}{180} = \frac{36}{90} = \frac{18}{45} = \frac{6}{15}$$

Dividing both the numerator and denominator again by 3 gives

$$\frac{72}{180} = \frac{36}{90} = \frac{18}{45} = \frac{6}{15} = \frac{2}{5}$$

So $\frac{72}{180} = \frac{2}{5}$ **in its simplest form**.

Thus, $\frac{2}{5}$ **of the students are girls.**

2.2 Adding and subtracting fractions

When the denominators of two (or more) fractions to be added are the same, the fractions can be added 'on sight'.

For example, $\frac{2}{9} + \frac{5}{9} = \frac{7}{9}$ and $\frac{3}{8} + \frac{1}{8} = \frac{4}{8}$

In the latter example, dividing both the 4 and the 8 by 4 gives $\frac{4}{8} = \frac{1}{2}$, which is the simplified answer. This is called **cancelling**.

Addition and subtraction of fractions is demonstrated in the following worked examples.

Problem 4. Simplify $\frac{1}{3} + \frac{1}{2}$

(i) Make the denominators the same for each fraction. The lowest number that both denominators divide into is called the **lowest common multiple** or **LCM** (see Chapter 1, page 5). In this example, the LCM of 3 and 2 is 6

(ii) 3 divides into 6 twice. Multiplying both numerator and denominator of $\frac{1}{3}$ by 2 gives

$$\frac{1}{3} = \frac{2}{6}$$

(iii) 2 divides into 6, 3 times. Multiplying both numerator and denominator of $\frac{1}{2}$ by 3 gives

$$\frac{1}{2} = \frac{3}{6}$$

(iv) Hence,

$$\frac{1}{3} + \frac{1}{2} = \frac{2}{6} + \frac{3}{6} = \mathbf{\frac{5}{6}}$$

Problem 5. Simplify $\dfrac{3}{4} - \dfrac{7}{16}$

(i) Make the denominators the same for each fraction. The lowest common multiple (LCM) of 4 and 16 is 16

(ii) 4 divides into 16, 4 times. Multiplying both numerator and denominator of $\frac{3}{4}$ by 4 gives

$$\frac{3}{4} = \frac{12}{16}$$

(iii) $\dfrac{7}{16}$ already has a denominator of 16

(iv) Hence,

$$\frac{3}{4} - \frac{7}{16} = \frac{12}{16} - \frac{7}{16} = \mathbf{\frac{5}{16}}$$

Problem 6. Simplify $4\dfrac{2}{3} - 1\dfrac{1}{6}$

$4\dfrac{2}{3} - 1\dfrac{1}{6}$ is the same as $\left(4\dfrac{2}{3}\right) - \left(1\dfrac{1}{6}\right)$ which is the same as $\left(4 + \dfrac{2}{3}\right) - \left(1 + \dfrac{1}{6}\right)$ which is the same as

$4 + \dfrac{2}{3} - 1 - \dfrac{1}{6}$ which is the same as $3 + \dfrac{2}{3} - \dfrac{1}{6}$ which is the same as $3 + \dfrac{4}{6} - \dfrac{1}{6} = 3 + \dfrac{3}{6} = 3 + \dfrac{1}{2}$

Thus, $4\dfrac{2}{3} - 1\dfrac{1}{6} = \mathbf{3\dfrac{1}{2}}$

Problem 7. Evaluate $7\dfrac{1}{8} - 5\dfrac{3}{7}$

$$7\frac{1}{8} - 5\frac{3}{7} = \left(7 + \frac{1}{8}\right) - \left(5 + \frac{3}{7}\right) = 7 + \frac{1}{8} - 5 - \frac{3}{7}$$

$$= 2 + \frac{1}{8} - \frac{3}{7} = 2 + \frac{7 \times 1 - 8 \times 3}{56}$$

$$= 2 + \frac{7 - 24}{56} = 2 + \frac{-17}{56} = 2 - \frac{17}{56}$$

$$= \frac{112}{56} - \frac{17}{56} = \frac{112 - 17}{56} = \frac{95}{56} = \mathbf{1\frac{39}{56}}$$

Problem 8. Determine the value of $4\dfrac{5}{8} - 3\dfrac{1}{4} + 1\dfrac{2}{5}$

$$4\frac{5}{8} - 3\frac{1}{4} + 1\frac{2}{5} = (4 - 3 + 1) + \left(\frac{5}{8} - \frac{1}{4} + \frac{2}{5}\right)$$

$$= 2 + \frac{5 \times 5 - 10 \times 1 + 8 \times 2}{40}$$

$$= 2 + \frac{25 - 10 + 16}{40}$$

$$= 2 + \frac{31}{40} = \mathbf{2\frac{31}{40}}$$

Now try the following Practice Exercise

Practice Exercise 5   Introduction to fractions (answers on page 422)

1. Change the improper fraction $\dfrac{15}{7}$ into a mixed number.

2. Change the improper fraction $\dfrac{37}{5}$ into a mixed number.

3. Change the mixed number $2\dfrac{4}{9}$ into an improper fraction.

4. Change the mixed number $8\dfrac{7}{8}$ into an improper fraction.

5. A box contains 165 paper clips. 60 clips are removed from the box. Express this as a fraction in its simplest form.

6. Order the following fractions from the smallest to the largest.

$$\frac{4}{9}, \frac{5}{8}, \frac{3}{7}, \frac{1}{2}, \frac{3}{5}$$

7. A training college has 375 students of which 120 are girls. Express this as a fraction in its simplest form.

Evaluate, in fraction form, the expressions given in Problems 8 to 20.

8. $\dfrac{1}{3} + \dfrac{2}{5}$ 9. $\dfrac{5}{6} - \dfrac{4}{15}$

10. $\dfrac{1}{2} + \dfrac{2}{5}$ 11. $\dfrac{7}{16} - \dfrac{1}{4}$

12. $\dfrac{2}{7} + \dfrac{3}{11}$ 13. $\dfrac{2}{9} - \dfrac{1}{7} + \dfrac{2}{3}$

14. $3\dfrac{2}{5} - 2\dfrac{1}{3}$ 15. $\dfrac{7}{27} - \dfrac{2}{3} + \dfrac{5}{9}$

16. $5\dfrac{3}{13} + 3\dfrac{3}{4}$ 17. $4\dfrac{5}{8} - 3\dfrac{2}{5}$

18. $10\dfrac{3}{7} - 8\dfrac{2}{3}$ 19. $3\dfrac{1}{4} - 4\dfrac{4}{5} + 1\dfrac{5}{6}$

20. $5\dfrac{3}{4} - 1\dfrac{2}{5} - 3\dfrac{1}{2}$

2.3 Multiplication and division of fractions

2.3.1 Multiplication

To multiply two or more fractions together, the numerators are first multiplied to give a single number and this becomes the new numerator of the combined fraction. The denominators are then multiplied together to give the new denominator of the combined fraction.

For example, $\dfrac{2}{3} \times \dfrac{4}{7} = \dfrac{2 \times 4}{3 \times 7} = \dfrac{\mathbf{8}}{\mathbf{21}}$

Problem 9. Simplify $7 \times \dfrac{2}{5}$

$$7 \times \frac{2}{5} = \frac{7}{1} \times \frac{2}{5} = \frac{7 \times 2}{1 \times 5} = \frac{14}{5} = \mathbf{2}\frac{\mathbf{4}}{\mathbf{5}}$$

Problem 10. Find the value of $\dfrac{3}{7} \times \dfrac{14}{15}$

Dividing numerator and denominator by 3 gives

$$\frac{3}{7} \times \frac{14}{15} = \frac{1}{7} \times \frac{14}{5} = \frac{1 \times 14}{7 \times 5}$$

Dividing numerator and denominator by 7 gives

$$\frac{1 \times 14}{7 \times 5} = \frac{1 \times 2}{1 \times 5} = \frac{\mathbf{2}}{\mathbf{5}}$$

This process of dividing both the numerator and denominator of a fraction by the same factor(s) is called **cancelling**.

Problem 11. Simplify $\dfrac{3}{5} \times \dfrac{4}{9}$

$$\frac{3}{5} \times \frac{4}{9} = \frac{1}{5} \times \frac{4}{3} \text{ by cancelling}$$

$$= \frac{\mathbf{4}}{\mathbf{15}}$$

Problem 12. Evaluate $1\dfrac{3}{5} \times 2\dfrac{1}{3} \times 3\dfrac{3}{7}$

Mixed numbers **must** be expressed as improper fractions before multiplication can be performed. Thus,

$$1\frac{3}{5} \times 2\frac{1}{3} \times 3\frac{3}{7} = \left(\frac{5}{5} + \frac{3}{5}\right) \times \left(\frac{6}{3} + \frac{1}{3}\right) \times \left(\frac{21}{7} + \frac{3}{7}\right)$$

$$= \frac{8}{5} \times \frac{7}{3} \times \frac{24}{7} = \frac{8 \times 1 \times 8}{5 \times 1 \times 1} = \frac{64}{5}$$

$$= \mathbf{12}\frac{\mathbf{4}}{\mathbf{5}}$$

Problem 13. Simplify $3\dfrac{1}{5} \times 1\dfrac{2}{3} \times 2\dfrac{3}{4}$

The mixed numbers need to be changed to improper fractions before multiplication can be performed.

$$3\frac{1}{5} \times 1\frac{2}{3} \times 2\frac{3}{4} = \frac{16}{5} \times \frac{5}{3} \times \frac{11}{4}$$

$$= \frac{4}{1} \times \frac{1}{3} \times \frac{11}{1} \text{ by cancelling}$$

$$= \frac{4 \times 1 \times 11}{1 \times 3 \times 1} = \frac{44}{3} = 14\frac{2}{3}$$

2.3.2 Division

The simple rule for division is **change the division sign into a multiplication sign and invert the second fraction**.

For example, $\quad \dfrac{2}{3} \div \dfrac{3}{4} = \dfrac{2}{3} \times \dfrac{4}{3} = \dfrac{8}{9}$

Problem 14. Simplify $\dfrac{3}{7} \div \dfrac{8}{21}$

$$\frac{3}{7} \div \frac{8}{21} = \frac{3}{7} \times \frac{21}{8} = \frac{3}{1} \times \frac{3}{8} \text{ by cancelling}$$

$$= \frac{3 \times 3}{1 \times 8} = \frac{9}{8} = 1\frac{1}{8}$$

Problem 15. Find the value of $5\dfrac{3}{5} \div 7\dfrac{1}{3}$

The mixed numbers must be expressed as improper fractions. Thus,

$$5\frac{3}{5} \div 7\frac{1}{3} = \frac{28}{5} \div \frac{22}{3} = \frac{28}{5} \times \frac{3}{22} = \frac{14}{5} \times \frac{3}{11} = \frac{42}{55}$$

Problem 16. Simplify $3\dfrac{2}{3} \times 1\dfrac{3}{4} \div 2\dfrac{3}{4}$

Mixed numbers must be expressed as improper fractions before multiplication and division can be performed:

$$3\frac{2}{3} \times 1\frac{3}{4} \div 2\frac{3}{4} = \frac{11}{3} \times \frac{7}{4} \div \frac{11}{4} = \frac{11}{3} \times \frac{7}{4} \times \frac{4}{11}$$

$$= \frac{1 \times 7 \times 1}{3 \times 1 \times 1} \text{ by cancelling}$$

$$= \frac{7}{3} = 2\frac{1}{3}$$

Now try the following Practice Exercise

Practice Exercise 6   Multiplying and dividing fractions (answers on page 422)

Evaluate the following.

1. $\dfrac{2}{5} \times \dfrac{4}{7}$
2. $5 \times \dfrac{4}{9}$
3. $\dfrac{3}{4} \times \dfrac{8}{11}$
4. $\dfrac{3}{4} \times \dfrac{5}{9}$
5. $\dfrac{17}{35} \times \dfrac{15}{68}$
6. $\dfrac{3}{5} \times \dfrac{7}{9} \times 1\dfrac{2}{7}$
7. $\dfrac{13}{17} \times 4\dfrac{7}{11} \times 3\dfrac{4}{39}$
8. $\dfrac{1}{4} \times \dfrac{3}{11} \times 1\dfrac{5}{39}$
9. $\dfrac{2}{9} \div \dfrac{4}{27}$
10. $\dfrac{3}{8} \div \dfrac{45}{64}$
11. $\dfrac{3}{8} \div \dfrac{5}{32}$
12. $\dfrac{3}{4} \div 1\dfrac{4}{5}$
13. $2\dfrac{1}{4} \times 1\dfrac{2}{3}$
14. $1\dfrac{1}{3} \div 2\dfrac{5}{9}$
15. $2\dfrac{4}{5} \div \dfrac{7}{10}$
16. $2\dfrac{3}{4} \div 3\dfrac{2}{3}$
17. $\dfrac{1}{9} \times \dfrac{3}{4} \times 1\dfrac{1}{3}$
18. $3\dfrac{1}{4} \times 1\dfrac{3}{5} \div \dfrac{2}{5}$

19. A ship's crew numbers 105, of which $\dfrac{1}{7}$ are women. Of the men, $\dfrac{1}{6}$ are officers. How many male officers are on board?

20. If a storage tank is holding 450 litres when it is three-quarters full, how much will it contain when it is two-thirds full?

21. Three people, P, Q and R, contribute to a fund. P provides 3/5 of the total, Q provides 2/3 of the remainder and R provides £8. Determine (a) the total of the fund and (b) the contributions of P and Q.

22. A tank contains 24,000 litres of oil. Initially, $\dfrac{7}{10}$ of the contents are removed, then $\dfrac{3}{5}$ of the remainder is removed. How much oil is left in the tank?

2.4 Order of operation with fractions

As stated in Chapter 1, sometimes addition, subtraction, multiplication, division, powers and brackets can all be involved in a calculation. A definite order of precedence must be adhered to. The order is:

Brackets

Order (or p**O**wer)

Division

Multiplication

Addition

Subtraction

This is demonstrated in the following worked problems.

Problem 17. Simplify $\dfrac{7}{20} - \dfrac{3}{8} \times \dfrac{4}{5}$

$$\frac{7}{20} - \frac{3}{8} \times \frac{4}{5} = \frac{7}{20} - \frac{3 \times 1}{2 \times 5} \text{ by cancelling}$$

$$= \frac{7}{20} - \frac{3}{10} \qquad \text{(M)}$$

$$= \frac{7}{20} - \frac{6}{20}$$

$$= \frac{1}{20} \qquad \text{(S)}$$

Problem 18. Simplify $\dfrac{1}{4} - 2\dfrac{1}{5} \times \dfrac{5}{8} + \dfrac{9}{10}$

$$\frac{1}{4} - 2\frac{1}{5} \times \frac{5}{8} + \frac{9}{10} = \frac{1}{4} - \frac{11}{5} \times \frac{5}{8} + \frac{9}{10}$$

$$= \frac{1}{4} - \frac{11}{1} \times \frac{1}{8} + \frac{9}{10} \text{ by cancelling}$$

$$= \frac{1}{4} - \frac{11}{8} + \frac{9}{10} \qquad \text{(M)}$$

$$= \frac{1 \times 10}{4 \times 10} - \frac{11 \times 5}{8 \times 5} + \frac{9 \times 4}{10 \times 4}$$

(since the LCM of 4, 8 and 10 is 40)

$$= \frac{10}{40} - \frac{55}{40} + \frac{36}{40}$$

$$= \frac{10 - 55 + 36}{40} \qquad \text{(A/S)}$$

$$= -\frac{9}{40}$$

Problem 19. Simplify

$$2\frac{1}{2} - \left(\frac{2}{5} + \frac{3}{4}\right) \div \left(\frac{5}{8} \times \frac{2}{3}\right)$$

$$2\frac{1}{2} - \left(\frac{2}{5} + \frac{3}{4}\right) \div \left(\frac{5}{8} \times \frac{2}{3}\right)$$

$$= \frac{5}{2} - \left(\frac{2 \times 4}{5 \times 4} + \frac{3 \times 5}{4 \times 5}\right) \div \left(\frac{5}{8} \times \frac{2}{3}\right) \qquad \text{(B)}$$

$$= \frac{5}{2} - \left(\frac{8}{20} + \frac{15}{20}\right) \div \left(\frac{5}{8} \times \frac{2}{3}\right) \qquad \text{(B)}$$

$$= \frac{5}{2} - \frac{23}{20} \div \left(\frac{5}{4} \times \frac{1}{3}\right) \text{ by cancelling} \qquad \text{(B)}$$

$$= \frac{5}{2} - \frac{23}{20} \div \frac{5}{12} \qquad \text{(B)}$$

$$= \frac{5}{2} - \frac{23}{20} \times \frac{12}{5} \qquad \text{(D)}$$

$$= \frac{5}{2} - \frac{23}{5} \times \frac{3}{5} \text{ by cancelling}$$

$$= \frac{5}{2} - \frac{69}{25} \qquad \text{(M)}$$

$$= \frac{5 \times 25}{2 \times 25} - \frac{69 \times 2}{25 \times 2} \qquad \text{(S)}$$

$$= \frac{125}{50} - \frac{138}{50} \qquad \text{(S)}$$

$$= -\frac{13}{50}$$

Problem 20. Evaluate

$$\frac{1}{3} \text{ of } \left(5\frac{1}{2} - 3\frac{3}{4}\right) + 3\frac{1}{5} \div \frac{4}{5} - \frac{1}{2}$$

$$\frac{1}{3} \text{ of } \left(5\frac{1}{2} - 3\frac{3}{4}\right) + 3\frac{1}{5} \div \frac{4}{5} - \frac{1}{2}$$

$$= \frac{1}{3} \text{ of } 1\frac{3}{4} + 3\frac{1}{5} \div \frac{4}{5} - \frac{1}{2} \qquad \text{(B)}$$

$$= \frac{1}{3} \times \frac{7}{4} + \frac{16}{5} \div \frac{4}{5} - \frac{1}{2} \qquad \text{(O)}$$

(Note that the 'of' is replaced with a multiplication sign.)

$$= \frac{1}{3} \times \frac{7}{4} + \frac{16}{5} \times \frac{5}{4} - \frac{1}{2} \qquad \text{(D)}$$

$$= \frac{1}{3} \times \frac{7}{4} + \frac{4}{1} \times \frac{1}{1} - \frac{1}{2} \text{ by cancelling}$$

$$= \frac{7}{12} + \frac{4}{1} - \frac{1}{2} \qquad \text{(M)}$$

$$= \frac{7}{12} + \frac{48}{12} - \frac{6}{12} \qquad \text{(A/S)}$$

$$= \frac{49}{12}$$

$$= \mathbf{4\frac{1}{12}}$$

Now try the following Practice Exercise

Practice Exercise 7   Order of operation with fractions (answers on page 422)

Evaluate the following.

1. $2\frac{1}{2} - \frac{3}{5} \times \frac{20}{27}$

2. $\frac{1}{3} - \frac{3}{4} \times \frac{16}{27}$

3. $\frac{1}{2} + \frac{3}{5} \div \frac{9}{15} - \frac{1}{3}$

4. $\frac{1}{5} + 2\frac{2}{3} \div \frac{5}{9} - \frac{1}{4}$

5. $\frac{4}{5} \times \frac{1}{2} - \frac{1}{6} \div \frac{2}{5} + \frac{2}{3}$

6. $\frac{3}{5} - \left(\frac{2}{3} - \frac{1}{2}\right) \div \left(\frac{5}{6} \times \frac{3}{2}\right)$

7. $\frac{1}{2}$ of $\left(4\frac{2}{5} - 3\frac{7}{10}\right) + \left(3\frac{1}{3} \div \frac{2}{3}\right) - \frac{2}{5}$

8. $\dfrac{6\frac{2}{3} \times 1\frac{2}{5} - \frac{1}{3}}{6\frac{3}{4} \div 1\frac{1}{2}}$

9. $1\frac{1}{3} \times 2\frac{1}{5} \div \frac{2}{5}$

10. $\frac{1}{4} \times \frac{2}{5} - \frac{1}{5} \div \frac{2}{3} + \frac{4}{15}$

11. $\dfrac{\frac{2}{3} + 3\frac{1}{5} \times 2\frac{1}{2} + 1\frac{1}{3}}{8\frac{1}{3} \div 3\frac{1}{3}}$

12. $\frac{1}{13}$ of $\left(2\frac{9}{10} - 1\frac{3}{5}\right) + \left(2\frac{1}{3} \div \frac{2}{3}\right) - \frac{3}{4}$

For fully worked solutions to each of the problems in Practice Exercises 5 to 7 in this chapter, go to the website:

www.routledge.com/cw/bird

This assignment covers the material contained in Chapters 1 and 2. *The marks available are shown in brackets at the end of each question.*

1. Evaluate
   $1009\,\text{cm} - 356\,\text{cm} - 742\,\text{cm} + 94\,\text{cm}$. (3)

2. Determine £284 × 9 (3)

3. Evaluate
   (a) $-11239 - (-4732) + 9639$
   (b) -164×-12
   (c) 367×-19 (6)

4. Calculate (a) $\$153 \div 9$ (b) $1397\,\text{g} \div 11$ (4)

5. A small component has a mass of 27 grams. Calculate the mass, in kilograms, of 750 such components. (3)

6. Find (a) the highest common factor and (b) the lowest common multiple of the following numbers: 15 40 75 120 (7)

Evaluate the expressions in questions 7 to 12.

7. $7 + 20 \div (9 - 5)$ (3)

8. $147 - 21(24 \div 3) + 31$ (3)

9. $40 \div (1 + 4) + 7[8 + (3 \times 8) - 27]$ (5)

10. $\dfrac{(7-3)(2-5)}{3(9-5) \div (2-6)}$ (3)

11. $\dfrac{(7 + 4 \times 5) \div 3 + 6 \div 2}{2 \times 4 + (5 - 8) - 2^2 + 3}$ (5)

12. $\dfrac{(4^2 \times 5 - 8) \div 3 + 9 \times 8}{4 \times 3^2 - 20 \div 5}$ (5)

13. Simplify
   (a) $\dfrac{3}{4} - \dfrac{7}{15}$
   (b) $1\dfrac{5}{8} - 2\dfrac{1}{3} + 3\dfrac{5}{6}$ (8)

14. A training college has 480 students of which 150 are girls. Express this as a fraction in its simplest form. (2)

15. A tank contains 18 000 litres of oil. Initially, $\dfrac{7}{10}$ of the contents are removed, then $\dfrac{2}{5}$ of the remainder is removed. How much oil is left in the tank? (4)

16. Evaluate
   (a) $1\dfrac{7}{9} \times \dfrac{3}{8} \times 3\dfrac{3}{5}$
   (b) $6\dfrac{2}{3} \div 1\dfrac{1}{3}$
   (c) $1\dfrac{1}{3} \times 2\dfrac{1}{5} \div \dfrac{2}{5}$ (10)

17. Calculate
   (a) $\dfrac{1}{4} \times \dfrac{2}{5} - \dfrac{1}{5} \div \dfrac{2}{3} + \dfrac{4}{15}$
   (b) $\dfrac{\dfrac{2}{3} + 3\dfrac{1}{5} \times 2\dfrac{1}{2} + 1\dfrac{1}{3}}{8\dfrac{1}{3} \div 3\dfrac{1}{3}}$ (8)

18. Simplify $\left\{\dfrac{1}{13} \text{ of } \left(2\dfrac{9}{10} - 1\dfrac{3}{5}\right)\right\} + \left(2\dfrac{1}{3} \div \dfrac{2}{3}\right) - \dfrac{3}{4}$ (8)

For lecturers/instructors/teachers, fully worked solutions to each of the problems in Revision Test 1, together with a full marking scheme, are available at the website:
www.routledge.com/cw/bird

Chapter 3

Decimals

Why it is important to understand: Decimals

Engineers and scientists use decimal numbers all the time in calculations. Calculators are able to handle calculations with decimals; however, there will be times when a quick calculation involving addition, subtraction, multiplication and division of decimals is needed. Again, do not spend too much time on this chapter because we deal with the calculator later; however, try to have some idea how to do quick calculations involving decimal numbers in the absence of a calculator. You will feel more confident to deal with decimal numbers in calculations if you can do this.

At the end of this chapter, you should be able to:

- convert a decimal number to a fraction and vice-versa
- understand and use significant figures and decimal places in calculations
- add and subtract decimal numbers
- multiply and divide decimal numbers

3.1 Introduction

The decimal system of numbers is based on the digits 0 to 9.

There are a number of everyday occurrences in which we use decimal numbers. For example, a radio is, say, tuned to 107.5 MHz FM; 107.5 is an example of a decimal number.

In a shop, a pair of trainers cost, say, £57.95; 57.95 is another example of a decimal number. 57.95 is a decimal fraction, where a decimal point separates the integer, i.e. 57, from the fractional part, i.e. 0.95

57.95 actually means (5 × 10) + (7 × 1)

$$+ \left(9 \times \frac{1}{10}\right) + \left(5 \times \frac{1}{100}\right)$$

3.2 Converting decimals to fractions and vice-versa

Converting decimals to fractions and vice-versa is demonstrated below with worked examples.

Problem 1. Convert 0.375 to a proper fraction in its simplest form

(i) 0.375 may be written as $\dfrac{0.375 \times 1000}{1000}$ i.e.
$$0.375 = \frac{375}{1000}$$

(ii) Dividing both numerator and denominator by 5 gives $\dfrac{375}{1000} = \dfrac{75}{200}$

(iii) Dividing both numerator and denominator by 5 again gives $\dfrac{75}{200} = \dfrac{15}{40}$

(iv) Dividing both numerator and denominator by 5 again gives $\dfrac{15}{40} = \dfrac{3}{8}$

Since both 3 and 8 are only divisible by 1, we cannot 'cancel' any further, so $\dfrac{3}{8}$ is the 'simplest form' of the fraction.

Hence, **the decimal fraction $0.375 = \dfrac{3}{8}$ as a proper fraction**.

Problem 2. Convert 3.4375 to a mixed number

Initially, the whole number 3 is ignored.

(i) 0.4375 may be written as $\dfrac{0.4375 \times 10000}{10000}$ i.e. $0.4375 = \dfrac{4375}{10000}$

(ii) Dividing both numerator and denominator by 25 gives $\dfrac{4375}{10000} = \dfrac{175}{400}$

(iii) Dividing both numerator and denominator by 5 gives $\dfrac{175}{400} = \dfrac{35}{80}$

(iv) Dividing both numerator and denominator by 5 again gives $\dfrac{35}{80} = \dfrac{7}{16}$

Since both 5 and 16 are only divisible by 1, we cannot 'cancel' any further, so $\dfrac{7}{16}$ is the 'lowest form' of the fraction.

(v) Hence, $0.4375 = \dfrac{7}{16}$

Thus, **the decimal fraction $3.4375 = 3\dfrac{7}{16}$ as a mixed number**.

Problem 3. Express $\dfrac{7}{8}$ as a decimal fraction

To convert a proper fraction to a decimal fraction, the numerator is divided by the denominator.

$$
\begin{array}{r}
0.8\,7\,5 \\
8\overline{)7.0\,0\,0}
\end{array}
$$

(i) 8 into 7 will not go. Place the 0 above the 7

(ii) Place the decimal point above the decimal point of 7.000

(iii) 8 into 70 goes 8, remainder 6. Place the 8 above the first zero after the decimal point and carry the 6 remainder to the next digit on the right, making it 60

(iv) 8 into 60 goes 7, remainder 4. Place the 7 above the next zero and carry the 4 remainder to the next digit on the right, making it 40

(v) 8 into 40 goes 5, remainder 0. Place 5 above the next zero.

Hence, **the proper fraction $\dfrac{7}{8} = 0.875$ as a decimal fraction**.

Problem 4. Express $5\dfrac{13}{16}$ as a decimal fraction

For mixed numbers it is only necessary to convert the proper fraction part of the mixed number to a decimal fraction.

$$
\begin{array}{r}
0.8\,1\,2\,5 \\
16\overline{)13.0\,0\,0\,0}
\end{array}
$$

(i) 16 into 13 will not go. Place the 0 above the 3

(ii) Place the decimal point above the decimal point of 13.0000

(iii) 16 into 130 goes 8, remainder 2. Place the 8 above the first zero after the decimal point and carry the 2 remainder to the next digit on the right, making it 20

(iv) 16 into 20 goes 1, remainder 4. Place the 1 above the next zero and carry the 4 remainder to the next digit on the right, making it 40

(v) 16 into 40 goes 2, remainder 8. Place the 2 above the next zero and carry the 8 remainder to the next digit on the right, making it 80

(vi) 16 into 80 goes 5, remainder 0. Place the 5 above the next zero.

(vii) Hence, $\dfrac{13}{16} = 0.8125$

Thus, **the mixed number $5\dfrac{13}{16} = 5.8125$ as a decimal fraction**.

Now try the following Practice Exercise

Practice Exercise 8   Converting decimals to fractions and vice-versa (answers on page 423)

1. Convert 0.65 to a proper fraction.

2. Convert 0.036 to a proper fraction.

3. Convert 0.175 to a proper fraction.

4. Convert 0.048 to a proper fraction.

5. Convert the following to proper fractions.
   (a) 0.66   (b) 0.84   (c) 0.0125
   (d) 0.282   (e) 0.024

6. Convert 4.525 to a mixed number.

7. Convert 23.44 to a mixed number.

8. Convert 10.015 to a mixed number.

9. Convert 6.4375 to a mixed number.

10. Convert the following to mixed numbers.
    (a) 1.82   (b) 4.275   (c) 14.125
    (d) 15.35   (e) 16.2125

11. Express $\dfrac{5}{8}$ as a decimal fraction.

12. Express $6\dfrac{11}{16}$ as a decimal fraction.

13. Express $\dfrac{7}{32}$ as a decimal fraction.

14. Express $11\dfrac{3}{16}$ as a decimal fraction.

15. Express $\dfrac{9}{32}$ as a decimal fraction.

3.3 Significant figures and decimal places

A number which can be expressed exactly as a decimal fraction is called a **terminating decimal**. For example,

$$3\dfrac{3}{16} = 3.1825 \text{ is a terminating decimal}$$

A number which cannot be expressed exactly as a decimal fraction is called a **non-terminating decimal**. For example,

$$1\dfrac{5}{7} = 1.7142857\ldots \text{ is a non-terminating decimal}$$

The answer to a non-terminating decimal may be expressed in two ways, depending on the accuracy required:

(a) correct to a number of **significant figures**, or

(b) correct to a number of **decimal places** i.e. the number of figures after the decimal point.

The last digit in the answer is unaltered if the next digit on the right is in the group of numbers 0, 1, 2, 3 or 4. For example,

$$1.714285\ldots = \mathbf{1.714} \text{ correct to 4 significant figures}$$
$$= \mathbf{1.714} \text{ correct to 3 decimal places}$$

since the next digit on the right in this example is 2
The last digit in the answer is increased by 1 if the next digit on the right is in the group of numbers 5, 6, 7, 8 or 9. For example,

$$1.7142857\ldots = \mathbf{1.7143} \text{ correct to 5 significant figures}$$
$$= \mathbf{1.7143} \text{ correct to 4 decimal places}$$

since the next digit on the right in this example is 8

Problem 5. Express 15.36815 correct to
(a) 2 decimal places,   (b) 3 significant figures,
(c) 3 decimal places,   (d) 6 significant figures

(a) $15.36815 = \mathbf{15.37}$ correct to 2 decimal places.

(b) $15.36815 = \mathbf{15.4}$ correct to 3 significant figures.

(c) $15.36815 = \mathbf{15.368}$ correct to 3 decimal places.

(d) $15.36815 = \mathbf{15.3682}$ correct to 6 significant figures.

Problem 6. Express 0.004369 correct to
(a) 4 decimal places,   (b) 3 significant figures

(a) $0.004369 = \mathbf{0.0044}$ correct to 4 decimal places.

(b) $0.004369 = \mathbf{0.00437}$ correct to 3 significant figures.

Note that the zeros to the right of the decimal point do not count as significant figures.

Now try the following Practice Exercise

Practice Exercise 9 Significant figures and decimal places (answers on page 423)

1. Express 14.1794 correct to 2 decimal places.

2. Express 2.7846 correct to 4 significant figures.

3. Express 65.3792 correct to 2 decimal places.

4. Express 43.2746 correct to 4 significant figures.

5. Express 1.2973 correct to 3 decimal places.

6. Express 0.0005279 correct to 3 significant figures.

3.4 Adding and subtracting decimal numbers

When adding or subtracting decimal numbers, care needs to be taken to ensure that the decimal points are beneath each other. This is demonstrated in the following worked examples.

Problem 7. Evaluate $46.8 + 3.06 + 2.4 + 0.09$ and give the answer correct to 3 significant figures

The decimal points are placed under each other as shown. Each column is added, starting from the right.

$$\begin{array}{r} 46.8 \\ 3.06 \\ 2.4 \\ +\underline{0.09} \\ \underline{52.35} \\ 11\ 1 \end{array}$$

(i) $6 + 9 = 15$. Place 5 in the hundredths column. Carry 1 in the tenths column.

(ii) $8 + 0 + 4 + 0 + 1$ (carried) $= 13$. Place the 3 in the tenths column. Carry the 1 into the units column.

(iii) $6 + 3 + 2 + 0 + 1$ (carried) $= 12$. Place the 2 in the units column. Carry the 1 into the tens column.

(iv) $4 + 1$(carried) $= 5$. Place the 5 in the hundreds column.

Hence,

$$46.8 + 3.06 + 2.4 + 0.09 = 52.35$$

$$= 52.4, \textbf{correct to 3 significant figures}$$

Problem 8. Evaluate $64.46 - 28.77$ and give the answer correct to 1 decimal place

As with addition, the decimal points are placed under each other as shown.

$$\begin{array}{r} 64.46 \\ -\underline{28.77} \\ \underline{35.69} \end{array}$$

(i) $6 - 7$ is not possible; therefore 'borrow' 1 from the tenths column. This gives $16 - 7 = 9$. Place the 9 in the hundredths column.

(ii) $3 - 7$ is not possible; therefore 'borrow' 1 from the units column. This gives $13 - 7 = 6$. Place the 6 in the tenths column.

(iii) $3 - 8$ is not possible; therefore 'borrow' from the hundreds column. This gives $13 - 8 = 5$. Place the 5 in the units column.

(iv) $5 - 2 = 3$. Place the 3 in the hundreds column.

Hence,

$$64.46 - 28.77 = 35.69$$

$$= 35.7 \textbf{ correct to 1 decimal place}$$

Problem 9. Evaluate $312.64 - 59.826 - 79.66 + 38.5$ and give the answer correct to 4 significant figures

The sum of the positive decimal fractions $= 312.64 + 38.5 = 351.14$

The sum of the negative decimal fractions $= 59.826 + 79.66 = 139.486$

Taking the sum of the negative decimal fractions from the sum of the positive decimal fractions gives

$$\begin{array}{r} 351.140 \\ -\underline{139.486} \\ \underline{211.654} \end{array}$$

Hence, $351.140 - 139.486 = \mathbf{211.654} = \mathbf{211.7}$, **correct to 4 significant figures**.

Now try the following Practice Exercise

Practice Exercise 10   Adding and subtracting decimal numbers (answers on page 423)

Determine the following without using a calculator.

1. Evaluate $37.69 + 42.6$, correct to 3 significant figures.

2. Evaluate $378.1 - 48.85$, correct to 1 decimal place.

3. Evaluate $68.92 + 34.84 - 31.223$, correct to 4 significant figures.

4. Evaluate $67.841 - 249.55 + 56.883$, correct to 2 decimal places.

5. Evaluate $483.24 - 120.44 - 67.49$, correct to 4 significant figures.

6. Evaluate $738.22 - 349.38 - 427.336 + 56.779$, correct to 1 decimal place.

7. Determine the dimension marked x in the length of the shaft shown in Fig. 3.1. The dimensions are in millimetres.

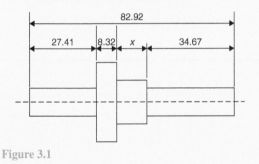

Figure 3.1

3.5 Multiplying and dividing decimal numbers

When multiplying decimal fractions:

(a) the numbers are multiplied as if they were integers, and

(b) the position of the decimal point in the answer is such that there are as many digits to the right of it as the sum of the digits to the right of the decimal points of the two numbers being multiplied together.

This is demonstrated in the following worked examples.

Problem 10. Evaluate 37.6×5.4

$$
\begin{array}{r}
376 \\
\times 54 \\
\hline
1504 \\
18800 \\
\hline
20304 \\
\hline
\end{array}
$$

(i) $376 \times 54 = 20304$

(ii) As there are $1 + 1 = 2$ digits to the right of the decimal points of the two numbers being multiplied together, $37.\underline{6} \times 5.\underline{4}$, then

$$37.6 \times 5.4 = 203.04$$

Problem 11. Evaluate $44.25 \div 1.2$, correct to (a) 3 significant figures, (b) 2 decimal places

$$44.25 \div 1.2 = \frac{44.25}{1.2}$$

The denominator is multiplied by 10 to change it into an integer. The numerator is also multiplied by 10 to keep the fraction the same. Thus,

$$\frac{44.25}{1.2} = \frac{44.25 \times 10}{1.2 \times 10} = \frac{442.5}{12}$$

The long division is similar to the long division of integers and the steps are as shown.

$$
\begin{array}{r}
36.875 \\
12 \overline{)442.500} \\
\underline{36} \\
82 \\
\underline{72} \\
105 \\
\underline{96} \\
90 \\
\underline{84} \\
60 \\
\underline{60} \\
0 \\
\end{array}
$$

(i) 12 into 44 goes 3; place the 3 above the second 4 of 442.500

(ii) $3 \times 12 = 36$; place the 36 below the 44 of 442.500

(iii) $44 - 36 = 8$

(iv) Bring down the 2 to give 82

(v) 12 into 82 goes 6; place the 6 above the 2 of 442.500

(vi) $6 \times 12 = 72$; place the 72 below the 82

(vii) $82 - 72 = 10$

(viii) Bring down the 5 to give 105

(ix) 12 into 105 goes 8; place the 8 above the 5 of 442.500

(x) $8 \times 12 = 96$; place the 96 below the 105

(xi) $105 - 96 = 9$

(xii) Bring down the 0 to give 90

(xiii) 12 into 90 goes 7; place the 7 above the first zero of 442.500

(xiv) $7 \times 12 = 84$; place the 84 below the 90

(xv) $90 - 84 = 6$

(xvi) Bring down the 0 to give 60

(xvii) 12 into 60 gives 5 exactly; place the 5 above the second zero of 442.500

(xviii) Hence, $44.25 \div 1.2 = \dfrac{442.5}{12} = \mathbf{36.875}$

So,

(a) $44.25 \div 1.2 = \mathbf{36.9}$, **correct to 3 significant figures**.

(b) $44.25 \div 1.2 = \mathbf{36.88}$, **correct to 2 decimal places**.

Problem 12. Express $7\dfrac{2}{3}$ as a decimal fraction, correct to 4 significant figures

Dividing 2 by 3 gives $\dfrac{2}{3} = 0.666666\ldots$

and $7\dfrac{2}{3} = 7.666666\ldots$

Hence, $7\dfrac{2}{3} = \mathbf{7.667}$ **correct to 4 significant figures**.
Note that $7.6666\ldots$ is called **7.6 recurring** and is written as $\mathbf{7.\dot{6}}$

Now try the following Practice Exercise

Practice Exercise 11   Multiplying and dividing decimal numbers (answers on page 423)

In Problems 1 to 8, evaluate without using a calculator.

1. Evaluate 3.57×1.4

2. Evaluate 67.92×0.7

3. Evaluate 167.4×2.3

4. Evaluate 342.6×1.7

5. Evaluate $548.28 \div 1.2$

6. Evaluate $478.3 \div 1.1$, correct to 5 significant figures.

7. Evaluate $563.48 \div 0.9$, correct to 4 significant figures.

8. Evaluate $2387.4 \div 1.5$

In Problems 9 to 14, express as decimal fractions to the accuracy stated.

9. $\dfrac{4}{9}$, correct to 3 significant figures.

10. $\dfrac{17}{27}$, correct to 5 decimal places.

11. $1\dfrac{9}{16}$, correct to 4 significant figures.

12. $53\dfrac{5}{11}$, correct to 3 decimal places.

13. $13\dfrac{31}{37}$, correct to 2 decimal places.

14. $8\dfrac{9}{13}$, correct to 3 significant figures.

15. Evaluate $421.8 \div 17$, (a) correct to 4 significant figures and (b) correct to 3 decimal places.

16. Evaluate $\dfrac{0.0147}{2.3}$, (a) correct to 5 decimal places and (b) correct to 2 significant figures.

17. Evaluate (a) $\dfrac{12.\dot{6}}{1.5}$ (b) $5.\dot{2} \times 12$

18. A tank contains 1800 litres of oil. How many tins containing 0.75 litres can be filled from this tank?

For fully worked solutions to each of the problems in Practice Exercises 8 to 11 in this chapter, go to the website:

Chapter 4

Using a calculator

Why it is important to understand: Using a calculator

The availability of electronic pocket calculators, at prices which all can afford, has had a considerable impact on engineering education. Engineers and student engineers now use calculators all the time since calculators are able to handle a very wide range of calculations. You will feel more confident to deal with all aspects of engineering studies if you are able to correctly use a calculator accurately.

At the end of this chapter, you should be able to:

- use a calculator to add, subtract, multiply and divide decimal numbers
- use a calculator to evaluate square, cube, reciprocal, power, root and $\times 10^x$ functions
- use a calculator to evaluate expressions containing fractions and trigonometric functions
- use a calculator to evaluate expressions containing π and e^x functions
- evaluate formulae, given values

4.1 Introduction

In engineering, calculations often need to be performed. For simple numbers it is useful to be able to use mental arithmetic. However, when numbers are larger an electronic calculator needs to be used.

There are several calculators on the market, many of which will be satisfactory for our needs. It is essential to have a **scientific notation calculator** which will have all the necessary functions needed and more.

This chapter assumes you have a **CASIO fx-991ES PLUS calculator**, or similar, as shown in Fig. 4.1.

Besides straightforward addition, subtraction, multiplication and division, which you will already be able to do, we will check that you can use squares, cubes, powers, reciprocals, roots, fractions and trigonometric functions (the latter in preparation for Chapter 21). There are several other functions on the calculator which we do not need to concern ourselves with at this level.

4.2 Adding, subtracting, multiplying and dividing

Initially, after switching on, press **Mode**.

Of the three possibilities, use **Comp**, which is achieved by pressing **1**.

Next, press **Shift** followed by **Setup** and, of the eight possibilities, use **Mth IO**, which is achieved by pressing **1**.

By all means experiment with the other menu options – refer to your 'User's guide'.

All calculators have **+**, **−**, **×** and **÷ functions** and these functions will, no doubt, already have been used in calculations.

Problem 1. Evaluate $364.7 \div 57.5$ correct to 3 decimal places

(i) Type in 364.7

Figure 4.1 A Casio fx-991ES PLUS Calculator

(ii) Press ÷

(iii) Type in 57.5

(iv) Press = and the fraction $\dfrac{3647}{575}$ appears.

(v) Press the $S \Leftrightarrow D$ function and the decimal answer 6.34260869... appears.

Alternatively, after step (iii) press Shift and = and the decimal will appear.

Hence, **364.7 ÷ 57.5 = 6.343 correct to 3 decimal places**.

Problem 2. Evaluate $\dfrac{12.47 \times 31.59}{70.45 \times 0.052}$ correct to 4 significant figures

(i) Type in 12.47

(ii) Press ×

(iii) Type in 31.59

(iv) Press ÷

(v) The denominator must have brackets; i.e. press (

(vi) Type in 70.45 × 0.052 and complete the bracket; i.e.)

(vii) Press = and the answer 107.530518... appears.

Hence, $\dfrac{\mathbf{12.47 \times 31.59}}{\mathbf{70.45 \times 0.052}} = \mathbf{107.5}$ **correct to 4 significant figures**.

Now try the following Practice Exercise

Practice Exercise 12   Addition, subtraction, multiplication and division using a calculator (answers on page 423)

1. Evaluate $378.37 - 298.651 + 45.64 - 94.562$

2. Evaluate 25.63×465.34 correct to 5 significant figures.

3. Evaluate $562.6 \div 41.3$ correct to 2 decimal places.

4. Evaluate $\dfrac{17.35 \times 34.27}{41.53 \div 3.76}$ correct to 3 decimal places.

5. Evaluate $27.48 + 13.72 \times 4.15$ correct to 4 significant figures.

6. Evaluate $\dfrac{(4.527 + 3.63)}{(452.51 \div 34.75)} + 0.468$ correct to 5 significant figures.

7. Evaluate $52.34 - \dfrac{(912.5 \div 41.46)}{(24.6 - 13.652)}$ correct to 3 decimal places.

8. Evaluate $\dfrac{52.14 \times 0.347 \times 11.23}{19.73 \div 3.54}$ correct to 4 significant figures.

9. Evaluate $\dfrac{451.2}{24.57} - \dfrac{363.8}{46.79}$ correct to 4 significant figures.

10. Evaluate $\dfrac{45.6 - 7.35 \times 3.61}{4.672 - 3.125}$ correct to 3 decimal places.

4.3 Further calculator functions

4.3.1 Square and cube functions

Locate the x^2 and x^3 functions on your calculator and then check the following worked examples.

Problem 3. Evaluate 2.4^2

(i) Type in 2.4

(ii) Press x^2 and 2.4^2 appears on the screen.

(iii) Press = and the answer $\dfrac{144}{25}$ appears.

(iv) Press the $S \Leftrightarrow D$ function and the fraction changes to a decimal 5.76

Alternatively, after step (ii) press Shift and = . Thus, $\mathbf{2.4^2 = 5.76}$

Problem 4. Evaluate 0.17^2 in engineering form

(i) Type in 0.17

(ii) Press x^2 and 0.17^2 appears on the screen.

(iii) Press Shift and = and the answer 0.0289 appears.

(iv) Press the ENG function and the answer changes to 28.9×10^{-3}, which is **engineering form**.

Hence, $\mathbf{0.17^2 = 28.9 \times 10^{-3}}$ **in engineering form**. The ENG function is extremely important in engineering calculations.

Problem 5. Change 348620 into engineering form

(i) Type in 348620

(ii) Press = then ENG.

Hence, $\mathbf{348620 = 348.62 \times 10^3}$ **in engineering form**.

Problem 6. Change 0.0000538 into engineering form

(i) Type in 0.0000538

(ii) Press = then ENG.

Hence, $\mathbf{0.0000538 = 53.8 \times 10^{-6}}$ **in engineering form**.

Problem 7. Evaluate 1.4^3

(i) Type in 1.4

(ii) Press x^3 and 1.4^3 appears on the screen.

(iii) Press = and the answer $\dfrac{343}{125}$ appears.

(iv) Press the $S \Leftrightarrow D$ function and the fraction changes to a decimal: 2.744

Thus, $\mathbf{1.4^3 = 2.744}$

Now try the following Practice Exercise

Practice Exercise 13   Square and cube functions (answers on page 423)

1. Evaluate 3.5^2

2. Evaluate 0.19^2

3. Evaluate 6.85^2 correct to 3 decimal places.

4. Evaluate $(0.036)^2$ in engineering form.

5. Evaluate 1.563^2 correct to 5 significant figures.

6. Evaluate 1.3^3

7. Evaluate 3.14^3 correct to 4 significant figures.

8. Evaluate $(0.38)^3$ correct to 4 decimal places.

9. Evaluate $(6.03)^3$ correct to 2 decimal places.

10. Evaluate $(0.018)^3$ in engineering form.

4.3.2 Reciprocal and power functions

The reciprocal of 2 is $\dfrac{1}{2}$, the reciprocal of 9 is $\dfrac{1}{9}$ and the reciprocal of x is $\dfrac{1}{x}$, which from indices may be written as x^{-1}. Locate the reciprocal, i.e. x^{-1} on the calculator. Also, locate the power function, i.e. $x^{\square}$, on your calculator and then check the following worked examples.

Problem 8. Evaluate $\dfrac{1}{3.2}$

(i)   Type in 3.2

(ii) Press x^{-1} and 3.2^{-1} appears on the screen.

(iii) Press $=$ and the answer $\dfrac{5}{16}$ appears.

(iv) Press the $S \Leftrightarrow D$ function and the fraction changes to a decimal: 0.3125

Thus, $\dfrac{1}{3.2} = \mathbf{0.3125}$

Problem 9. Evaluate 1.5^5 correct to 4 significant figures

(i)   Type in 1.5

(ii) Press $x^{\square}$ and $1.5^{\square}$ appears on the screen.

(iii) Press 5 and 1.5^5 appears on the screen.

(iv) Press Shift and $=$ and the answer 7.59375 appears.

Thus, $\mathbf{1.5^5 = 7.594}$ **correct to 4 significant figures**.

Problem 10. Evaluate $2.4^6 - 1.9^4$ correct to 3 decimal places

(i)   Type in 2.4

(ii) Press $x^{\square}$ and $2.4^{\square}$ appears on the screen.

(iii) Press 6 and 2.4^6 appears on the screen.

(iv) The cursor now needs to be moved; this is achieved by using the cursor key (the large blue circular function in the top centre of the calculator). Press $\rightarrow$

(v)   Press $-$

(vi) Type in 1.9, press $x^{\square}$, then press 4

(vii) Press $=$ and the answer 178.07087... appears.

Thus, $\mathbf{2.4^6 - 1.9^4 = 178.071}$ **correct to 3 decimal places**.

Now try the following Practice Exercise

Practice Exercise 14   Reciprocal and power functions (answers on page 423)

1. Evaluate $\dfrac{1}{1.75}$ correct to 3 decimal places.

2. Evaluate $\dfrac{1}{0.0250}$

3. Evaluate $\dfrac{1}{7.43}$ correct to 5 significant figures.

4. Evaluate $\dfrac{1}{0.00725}$ correct to 1 decimal place.

5. Evaluate $\dfrac{1}{0.065} - \dfrac{1}{2.341}$ correct to 4 significant figures.

6. Evaluate 2.1^4

7. Evaluate $(0.22)^5$ correct to 5 significant figures in engineering form.

8. Evaluate $(1.012)^7$ correct to 4 decimal places.

9. Evaluate $(0.05)^6$ in engineering form.

10. Evaluate $1.1^3 + 2.9^4 - 4.4^2$ correct to 4 significant figures.

4.3.3 Root and $\times 10^x$ functions

Locate the square root function $\sqrt{\square}$ and the $\sqrt[\square]{\square}$ function (which is a Shift function located above the $x^{\square}$ function) on your calculator. Also, locate the $\times 10^x$ function and then check the following worked examples.

Problem 11. Evaluate $\sqrt{361}$

(i)   Press the $\sqrt{\square}$ function.

(ii) Type in 361 and $\sqrt{361}$ appears on the screen.

(iii) Press $=$ and the answer 19 appears.

Thus, $\sqrt{361} = \mathbf{19}$

Problem 12. Evaluate $\sqrt[4]{81}$

(i)   Press the $\sqrt[\square]{\square}$ function.

(ii) Type in 4 and $\sqrt[4]{\square}$ appears on the screen.

(iii) Press $\rightarrow$ to move the cursor and then type in 81 and $\sqrt[4]{81}$ appears on the screen.

(iv) Press $=$ and the answer 3 appears.

Thus, $\sqrt[4]{81} = \mathbf{3}$

Problem 13. Evaluate $6 \times 10^5 \times 2 \times 10^{-7}$

(i)   Type in 6

(ii)   Press the $\times 10^x$ function (note, you do not have to use $\times$)

(iii) Type in 5

(iv)   Press $\times$

(v)   Type in 2

(vi)   Press the $\times 10^x$ function.

(vii) Type in -7

(viii) Press $=$ and the answer $\dfrac{3}{25}$ appears.

(ix)   Press the $S \Leftrightarrow D$ function and the fraction changes to a decimal: 0.12

Thus, $\mathbf{6 \times 10^5 \times 2 \times 10^{-7} = 0.12}$

Now try the following Practice Exercise

1.   Evaluate $\sqrt{4.76}$ correct to 3 decimal places.

2.   Evaluate $\sqrt{123.7}$ correct to 5 significant figures.

3.   Evaluate $\sqrt{34528}$ correct to 2 decimal places.

4.   Evaluate $\sqrt{0.69}$ correct to 4 significant figures.

5.   Evaluate $\sqrt{0.025}$ correct to 4 decimal places.

6.   Evaluate $\sqrt[3]{17}$ correct to 3 decimal places.

7.   Evaluate $\sqrt[4]{773}$ correct to 4 significant figures.

8.   Evaluate $\sqrt[5]{3.12}$ correct to 4 decimal places.

9.   Evaluate $\sqrt[3]{0.028}$ correct to 5 significant figures.

10. Evaluate $\sqrt[6]{2451} - \sqrt[4]{46}$ correct to 3 decimal places.

Express the answers to questions 11 to 15 in engineering form.

11. Evaluate $5 \times 10^{-3} \times 7 \times 10^8$

12. Evaluate $\dfrac{3 \times 10^{-4}}{8 \times 10^{-9}}$

13. Evaluate $\dfrac{6 \times 10^3 \times 14 \times 10^{-4}}{2 \times 10^6}$

14. Evaluate $\dfrac{56.43 \times 10^{-3} \times 3 \times 10^4}{8.349 \times 10^3}$ correct to 3 decimal places.

15. Evaluate $\dfrac{99 \times 10^5 \times 6.7 \times 10^{-3}}{36.2 \times 10^{-4}}$ correct to 4 significant figures.

4.3.4 Fractions

Locate the $\dfrac{\square}{\square}$ and $\square\dfrac{\square}{\square}$ functions on your calculator (the latter function is a Shift function found above the $\dfrac{\square}{\square}$ function) and then check the following worked examples.

Problem 14. Evaluate $\dfrac{1}{4} + \dfrac{2}{3}$

(i)   Press the $\dfrac{\square}{\square}$ function.

(ii)   Type in 1

(iii) Press $\downarrow$ on the cursor key and type in 4

(iv) $\dfrac{1}{4}$ appears on the screen.

(v)   Press $\rightarrow$ on the cursor key and type in $+$

(vi)   Press the $\dfrac{\square}{\square}$ function.

(vii) Type in 2

(viii) Press $\downarrow$ on the cursor key and type in 3

(ix)   Press $\rightarrow$ on the cursor key.

(x)   Press $=$ and the answer $\dfrac{11}{12}$ appears.

(xi)   Press the $S \Leftrightarrow D$ function and the fraction changes to a decimal $0.9166666\ldots$

Thus, $\dfrac{1}{4} + \dfrac{2}{3} = \dfrac{11}{12} = \mathbf{0.9167}$ as a decimal, correct to 4 decimal places.

It is also possible to deal with **mixed numbers** on the calculator. Press Shift then the $\dfrac{\square}{\square}$ function and $\square\dfrac{\square}{\square}$ appears.

Problem 15. Evaluate $5\frac{1}{5} - 3\frac{3}{4}$

(i) Press Shift then the $\frac{\square}{\square}$ function and $\square\frac{\square}{\square}$ appears on the screen.

(ii) Type in 5 then $\rightarrow$ on the cursor key.

(iii) Type in 1 and $\downarrow$ on the cursor key.

(iv) Type in 5 and $5\frac{1}{5}$ appears on the screen.

(v) Press $\rightarrow$ on the cursor key.

(vi) Type in $-$ and then press Shift then the $\frac{\square}{\square}$ function and $5\frac{1}{5} - \square\frac{\square}{\square}$ appears on the screen.

(vii) Type in 3 then $\rightarrow$ on the cursor key.

(viii) Type in 3 and $\downarrow$ on the cursor key.

(ix) Type in 4 and $5\frac{1}{5} - 3\frac{3}{4}$ appears on the screen.

(x) Press $\rightarrow$ on the cursor key.

(xi) Press $=$ and the answer $\frac{29}{20}$ appears.

(xii) Press $S \Leftrightarrow D$ function and the fraction changes to a decimal 1.45

Thus, $5\frac{1}{5} - 3\frac{3}{4} = \frac{29}{20} = 1\frac{9}{20} = 1.45$ as a decimal.

Now try the following Practice Exercise

Practice Exercise 16   Fractions (answers on page 423)

1. Evaluate $\frac{4}{5} - \frac{1}{3}$ as a decimal, correct to 4 decimal places.

2. Evaluate $\frac{2}{3} - \frac{1}{6} + \frac{3}{7}$ as a fraction.

3. Evaluate $2\frac{5}{6} + 1\frac{5}{8}$ as a decimal, correct to 4 significant figures.

4. Evaluate $5\frac{6}{7} - 3\frac{1}{8}$ as a decimal, correct to 4 significant figures.

5. Evaluate $\frac{1}{3} - \frac{3}{4} \times \frac{8}{21}$ as a fraction.

6. Evaluate $\frac{3}{8} + \frac{5}{6} - \frac{1}{2}$ as a decimal, correct to 4 decimal places.

7. Evaluate $\frac{3}{4} \times \frac{4}{5} - \frac{2}{3} \div \frac{4}{9}$ as a fraction.

8. Evaluate $8\frac{8}{9} \div 2\frac{2}{3}$ as a mixed number.

9. Evaluate $3\frac{1}{5} \times 1\frac{1}{3} - 1\frac{7}{10}$ as a decimal, correct to 3 decimal places.

10. Evaluate $\dfrac{\left(4\frac{1}{5} - 1\frac{2}{3}\right)}{\left(3\frac{1}{4} \times 2\frac{3}{5}\right)} - \frac{2}{9}$ as a decimal, correct to 3 significant figures.

4.3.5 Trigonometric functions

Trigonometric ratios will be covered in Chapter 21. However, very briefly, there are three functions on your calculator that are involved with trigonometry. They are:

   sin which is an abbreviation of **sine**
   cos which is an abbreviation of **cosine**, and
   tan which is an abbreviation of **tangent**

Exactly what these mean will be explained in Chapter 21.

There are two main ways that angles are measured, i.e. in **degrees** or in **radians**. Pressing Shift, Setup and 3 shows degrees, and Shift, Setup and 4 shows radians.

Press 3 and your calculator will be in **degrees mode**, indicated by a small D appearing at the top of the screen. Press 4 and your calculator will be in **radian mode**, indicated by a small R appearing at the top of the screen.

Locate the sin, cos and tan functions on your calculator and then check the following worked examples.

Problem 16. Evaluate $\sin 38°$

(i) Make sure your calculator is in degrees mode.

(ii) Press sin function and sin(appears on the screen.

(iii) Type in 38 and close the bracket with) and sin (38) appears on the screen.

(iv) Press $=$ and the answer 0.615661475... appears.

Thus, **$\sin 38° = 0.6157$, correct to 4 decimal places**.

Problem 17. Evaluate $5.3 \tan (2.23 \text{ rad})$

(i) Make sure your calculator is in radian mode by pressing Shift then Setup then 4 (a small R appears at the top of the screen).

(ii) Type in 5.3 then press tan function and 5.3 tan(appears on the screen.

(iii) Type in 2.23 and close the bracket with) and 5.3 tan (2.23) appears on the screen.

(iv) Press = and the answer −6.84021262… appears.

Thus, **5.3 tan (2.23 rad) = −6.8402, correct to 4 decimal places**.

___

Now try the following Practice Exercise

Evaluate the following, each correct to 4 decimal places.

1. Evaluate $\sin 67°$

2. Evaluate $\cos 43°$

3. Evaluate $\tan 71°$

4. Evaluate $\sin 15.78°$

5. Evaluate $\cos 63.74°$

6. Evaluate $\tan 39.55° - \sin 52.53°$

7. Evaluate $\sin(0.437 \text{ rad})$

8. Evaluate $\cos(1.42 \text{ rad})$

9. Evaluate $\tan(5.673 \text{ rad})$

10. Evaluate $\dfrac{(\sin 42.6°)(\tan 83.2°)}{\cos 13.8°}$

4.3.6 π and e^x functions

Press Shift and then press the $\times 10^x$ function key and π appears on the screen. Either press Shift and = (or = and $S \Leftrightarrow D$) and the value of π appears in decimal form as 3.14159265…

Press Shift and then press the ln function key and $e^\square$ appears on the screen. Enter 1 and then press = and $e^1 = e = 2.71828182…$

Now check the following worked examples involving π and e^x functions.

Problem 18.   Evaluate 3.57π

(i) Enter 3.57

(ii) Press Shift and the $\times 10^x$ key and 3.57π appears on the screen.

(iii) Either press Shift and = (or = and $S \Leftrightarrow D$) and the value of 3.57π appears in decimal as 11.2154857…

Hence, **3.57 π = 11.22 correct to 4 significant figures**.

Problem 19.   Evaluate $e^{2.37}$

(i) Press Shift and then press the ln function key and $e^\square$ appears on the screen.

(ii) Enter 2.37 and $e^{2.37}$ appears on the screen.

(iii) Press Shift and = (or = and $S \Leftrightarrow D$) and the value of $e^{2.37}$ appears in decimal as 10.6973922…

Hence, **$e^{2.37} = 10.70$ correct to 4 significant figures**.

___

Now try the following Practice Exercise

Evaluate the following, each correct to 4 significant figures.

1. 1.59π

2. $2.7(\pi - 1)$

3. $\pi^2\left(\sqrt{13} - 1\right)$

4. $3e^\pi$

5. $8.5e^{-2.5}$

6. $3e^{2.9} - 1.6$

7. $3e^{(2\pi - 1)}$

8. $2\pi e^{\frac{\pi}{3}}$

9. $\sqrt{\left[\dfrac{5.52\pi}{2e^{-2} \times \sqrt{26.73}}\right]}$

10. $\sqrt{\left[\dfrac{e^{\left(2 - \sqrt{3}\right)}}{\pi \times \sqrt{8.57}}\right]}$

4.4   Evaluation of formulae

The statement $y = mx + c$ is called a **formula** for y in terms of m, x and c.

y, m, x and c are called **symbols** or **variables**.

When given values of m, x and c we can evaluate y.

There are a large number of formulae used in engineering and in this section we will insert numbers in place of symbols to evaluate engineering quantities.

Just four examples of important formulae are:

1. A straight line graph is of the form $y = mx + c$ (see Chapter 17).

2. Ohm's law* states that $V = I \times R$.

3. Velocity is expressed as $v = u + at$.

4. Force is expressed as $F = m \times a$.

Here are some practical examples. Check with your calculator that you agree with the working and answers.

Problem 20. In an electrical circuit the voltage V is given by Ohm's law, i.e. $V = IR$. Find, correct to 4 significant figures, the voltage when $I = 5.36\,$A and $R = 14.76\,\Omega$

$$V = IR = (5.36)(14.76)$$

Hence, **voltage $V = 79.11\,$V, correct to 4 significant figures**.

Problem 21. The surface area A of a hollow cone is given by $A = \pi rl$. Determine, correct to 1

*Who was Ohm? – **Georg Simon Ohm** (16 March 1789–6 July 1854) was a Bavarian physicist and mathematician who discovered what came to be known as Ohm's law – the direct proportionality between voltage and the resultant electric current. To find out more go to www.routledge.com/cw/bird

decimal place, the surface area when $r = 3.0\,$cm and $l = 8.5\,$cm

$$A = \pi rl = \pi (3.0)(8.5)\,\text{cm}^2$$

Hence, **surface area $A = 80.1\,\text{cm}^2$, correct to 1 decimal place**.

Problem 22. Velocity v is given by $v = u + at$. If $u = 9.54\,$m/s, $a = 3.67\,$m/s^2 and $t = 7.82\,$s, find v, correct to 3 significant figures

$$v = u + at = 9.54 + 3.67 \times 7.82$$
$$= 9.54 + 28.6994 = 38.2394$$

Hence, **velocity $v = 38.2\,$m/s, correct to 3 significant figures**.

Problem 23. The area, A, of a circle is given by $A = \pi r^2$. Determine the area correct to 2 decimal places, given radius $r = 5.23\,$m

$$A = \pi r^2 = \pi (5.23)^2 = \pi (27.3529)$$

Hence, **area, $A = 85.93\,\text{m}^2$, correct to 2 decimal places**.

Problem 24. Density $= \dfrac{\text{mass}}{\text{volume}}$. Find the density when the mass is $6.45\,$kg and the volume is $300 \times 10^{-6}\,\text{m}^3$

$$\textbf{Density} = \frac{\text{mass}}{\text{volume}} = \frac{6.45\,\text{kg}}{300 \times 10^{-6}\,\text{m}^3} = \textbf{21500\,kg/m}^3$$

Problem 25. The power, P watts, dissipated in an electrical circuit is given by the formula $P = \dfrac{V^2}{R}$. Evaluate the power, correct to 4 significant figures, given that $V = 230\,$V and $R = 35.63\,\Omega$

$$P = \frac{V^2}{R} = \frac{(230)^2}{35.63} = \frac{52900}{35.63} = 1484.70390\ldots$$

Press ENG and $1.48470390\ldots \times 10^3$ appears on the screen.

Hence, **power, $P = 1485\,$W or $1.485\,$kW correct to 4 significant figures**.

Now try the following Practice Exercise

Practice Exercise 19   Evaluation of formulae (answers on page 423)

1. The area A of a rectangle is given by the formula $A = lb$. Evaluate the area when $l = 12.4$ cm and $b = 5.37$ cm.

2. The circumference C of a circle is given by the formula $C = 2\pi r$. Determine the circumference given $r = 8.40$ mm.

3. A formula used in connection with gases is $R = \dfrac{PV}{T}$. Evaluate R when $P = 1500$, $V = 5$ and $T = 200$

4. The velocity of a body is given by $v = u + at$. The initial velocity u is measured and found to be 12 m/s. If the acceleration a is 9.81 m/s^2 calculate the final velocity, v, when time t is 15 seconds.

5. Calculate the current I in an electrical circuit, where $I = V/R$ amperes when the voltage V is measured and found to be 7.2 V and the resistance R is $17.7\,\Omega$.

6. Find the distance s, given that $s = \dfrac{1}{2}gt^2$ when time $t = 0.032$ seconds and acceleration due to gravity $g = 9.81$ m/s^2. Give the answer in millimetres.

7. The energy stored in a capacitor is given by $E = \dfrac{1}{2}CV^2$ joules. Determine the energy when capacitance $C = 5 \times 10^{-6}$ farads and voltage $V = 240$ V.

8. Find the area A of a triangle, given $A = \dfrac{1}{2}bh$, when the base length b is 23.42 m and the height h is 53.7 m.

9. Resistance R_2 is given by $R_2 = R_1(1 + \alpha t)$. Find R_2, correct to 4 significant figures, when $R_1 = 220$, $\alpha = 0.00027$ and $t = 75.6$

10. Density $= \dfrac{\text{mass}}{\text{volume}}$. Find the density when the mass is 2.462 kg and the volume is 173 cm^3. Give the answer in units of kg/m^3. (Note that 1 cm$^3 = 10^{-6}$ m^3)

11. Velocity $=$ frequency $\times$ wavelength. Find the velocity when the frequency is 1825 Hz and the wavelength is 0.154 m.

12. Evaluate resistance R_T, given
$$\frac{1}{R_T} = \frac{1}{R_1} + \frac{1}{R_2} + \frac{1}{R_3} \text{ when } R_1 = 5.5\,\Omega,$$
$R_2 = 7.42\,\Omega$ and $R_3 = 12.6\,\Omega$.

Here are some further practical examples. Again, check with your calculator that you agree with the working and answers.

Problem 26. The volume V cm^3 of a right circular cone is given by $V = \dfrac{1}{3}\pi r^2 h$. Given that radius $r = 2.45$ cm and height $h = 18.7$ cm, find the volume, correct to 4 significant figures

$$V = \frac{1}{3}\pi r^2 h = \frac{1}{3}\pi (2.45)^2 (18.7)$$

$$= \frac{1}{3} \times \pi \times 2.45^2 \times 18.7$$

$$= 117.544521\ldots$$

Hence, **volume, $V = 117.5$ cm^3, correct to 4 significant figures**.

Problem 27. Force F newtons is given by the formula $F = \dfrac{Gm_1m_2}{d^2}$, where m_1 and m_2 are masses, d their distance apart and G is a constant. Find the value of the force given that $G = 6.67 \times 10^{-11}$, $m_1 = 7.36$, $m_2 = 15.5$ and $d = 22.6$. Express the answer in standard form, correct to 3 significant figures

$$F = \frac{Gm_1m_2}{d^2} = \frac{(6.67 \times 10^{-11})(7.36)(15.5)}{(22.6)^2}$$

$$= \frac{(6.67)(7.36)(15.5)}{(10^{11})(510.76)} = \frac{1.490}{10^{11}}$$

Hence, **force $F = 1.49 \times 10^{-11}$ newtons, correct to 3 significant figures**.

Problem 28. The time of swing, t seconds, of a simple pendulum is given by $t = 2\pi\sqrt{\dfrac{l}{g}}$

Determine the time, correct to 3 decimal places, given that $l = 12.9$ and $g = 9.81$

$$t = 2\pi\sqrt{\frac{l}{g}} = (2\pi)\sqrt{\frac{12.9}{9.81}} = 7.20510343\ldots$$

Hence, **time $t = 7.205$ seconds, correct to 3 decimal places**.

Problem 29. Resistance, $R\,\Omega$, varies with temperature according to the formula $R = R_0(1 + \alpha t)$. Evaluate R, correct to 3 significant figures, given $R_0 = 14.59$, $\alpha = 0.0043$ and $t = 80$

$$R = R_0(1 + \alpha t) = 14.59[1 + (0.0043)(80)]$$
$$= 14.59(1 + 0.344)$$
$$= 14.59(1.344)$$

Hence, **resistance, $R = 19.6\,\Omega$, correct to 3 significant figures**.

Problem 30. The current, I amperes, in an a.c. circuit is given by $I = \dfrac{V}{\sqrt{(R^2 + X^2)}}$. Evaluate the current, correct to 2 decimal places, when $V = 250$ V, $R = 25.0\,\Omega$ and $X = 18.0\,\Omega$.

$$I = \frac{V}{\sqrt{(R^2 + X^2)}} = \frac{250}{\sqrt{(25.0^2 + 18.0^2)}} = 8.11534341\ldots$$

Hence, **current, $I = 8.12$ A, correct to 2 decimal places**.

Now try the following Practice Exercise

Practice Exercise 20   Evaluation of formulae (answers on page 423)

1. Find the total cost of 37 calculators costing £12.65 each and 19 drawing sets costing £6.38 each.

2. Power $= \dfrac{\text{force} \times \text{distance}}{\text{time}}$. Find the power when a force of 3760 N raises an object a distance of 4.73 m in 35 s.

3. The potential difference, V volts, available at battery terminals is given by $V = E - Ir$. Evaluate V when $E = 5.62$, $I = 0.70$ and $r = 4.30$

4. Given force $F = \dfrac{1}{2}m(v^2 - u^2)$, find F when $m = 18.3$, $v = 12.7$ and $u = 8.24$

5. The current I amperes flowing in a number of cells is given by $I = \dfrac{nE}{R + nr}$. Evaluate the current when $n = 36$, $E = 2.20$, $R = 2.80$ and $r = 0.50$

6. The time, t seconds, of oscillation for a simple pendulum is given by $t = 2\pi\sqrt{\dfrac{l}{g}}$. Determine the time when $l = 54.32$ and $g = 9.81$

7. Energy, E joules, is given by the formula $E = \dfrac{1}{2}LI^2$. Evaluate the energy when $L = 5.5$ and $I = 1.2$

8. The current I amperes in an a.c. circuit is given by $I = \dfrac{V}{\sqrt{(R^2 + X^2)}}$. Evaluate the current when $V = 250$, $R = 11.0$ and $X = 16.2$

9. Distance s metres is given by the formula $s = ut + \dfrac{1}{2}at^2$. If $u = 9.50$, $t = 4.60$ and $a = -2.50$, evaluate the distance.

10. The area, A, of any triangle is given by $A = \sqrt{[s(s-a)(s-b)(s-c)]}$ where $s = \dfrac{a+b+c}{2}$. Evaluate the area, given $a = 3.60$ cm, $b = 4.00$ cm and $c = 5.20$ cm.

11. Given that $a = 0.290$, $b = 14.86$, $c = 0.042$, $d = 31.8$ and $e = 0.650$, evaluate v given that $v = \sqrt{\left(\dfrac{ab}{c} - \dfrac{d}{e}\right)}$

12. Deduce the following information from the train timetable shown in Table 4.1.

   (a) At what time should a man catch a train at Fratton to enable him to be in London Waterloo by 14.23 h?

   (b) A girl leaves Cosham at 12.39 h and travels to Woking. How long does the journey take? And, if the distance between Cosham and Woking is 55 miles, calculate the average speed of the train.

   (c) A man living at Havant has a meeting in London at 15.30 h. It takes around 25 minutes on the underground to reach his destination from London Waterloo. What train should he catch from Havant to comfortably make the meeting?

   (d) Nine trains leave Portsmouth harbour between 12.18 h and 13.15 h. Which train should be taken for the shortest journey time?

For fully worked solutions to each of the problems in Practice Exercises 12 to 20 in this chapter, go to the website:

www.routledge.com/cw/bird

Table 4.1 Train timetable from Portsmouth Harbour to London Waterloo

Saturdays

OUTWARD Train Alterations		Time S04	Time S03	Time S08	Time S02	Time S03	Time S04	Time S04	Time S01	Time S02
Portsmouth Harbour	dep	12:18SW	12:22GW	12:22GW	12:45SW	12:45SW	12:45SW	12:54SW	13:12SN	13:15SW
Portsmouth & Southsea	arr	12:21	12:25	12:25	12:48	12:48	12:48	12:57	13:15	13:18
Portsmouth & Southsea	dep	12:24	12:27	12:27	12:50	12:50	12:50	12:59	13:16	13:20
Fratton	arr	12:27	12:30	12:30	12:53	12:53	12:53	13:02	13:19	13:23
Fratton	dep	12:28	12:31	12:31	12:54	12:54	12:54	13:03	13:20	13:24
Hilsea	arr	12:32						13:07		
Hilsea	dep	12:32						13:07		
Cosham	arr		12:38	12:38				13:12		
Cosham	dep		12:39	12:39				13:12		
Bedhampton	arr	12:37								
Bedhampton	dep	12:37								
Havant	arr	12:39			13:03	13:03	13:02		13:29	13:33
Havant	dep	12:40			13:04	13:04	13:04		13:30	13:34
Rowlands Castle	arr	12:46								
Rowlands Castle	dep	12:46								
Chichester	arr								13:40	
Chichester	dep								13:41	
Barnham	arr								13:48	
Barnham	dep								13:49	
Horsham	arr								14:16	
Horsham	dep								14:20	
Crawley	arr								14:28	
Crawley	dep								14:29	
Three Bridges	arr								14:32	
Three Bridges	dep								14:33	
Gatwick Airport	arr								14:37	
Gatwick Airport	dep								14:38	
Horley	arr								14:41	
Horley	dep								14:41	
Redhill	arr								14:47	
Redhill	dep								14:48	
East Croydon	arr								15:00	
East Croydon	dep								15:00	
Petersfield	arr	12:56			13:17	13:17	13:17			13:47
Petersfield	dep	12:57			13:18	13:18	13:18			13:48
Liss	arr	13:02								
Liss	dep	13:02								
Liphook	arr	13:09								
Liphook	dep	13:09								
Haslemere	arr	13:14C			13:31	13:31	13:30C			14:01
Haslemere	dep	13:20SWR			13:32	13:32	13:36SWR			14:02
Guildford	arr	13:55C			13:45	13:45	14:11C			14:15
Guildford	dep	14:02SW			13:47	13:47	14:17SWR			14:17
Portchester	arr							13:17		
Portchester	dep							13:17		
Fareham	arr		12:46	12:46				13:22		
Fareham	dep		12:47	12:47				13:23		
Southampton Central	arr			13:08C						
Southampton Central	dep			13:30SW						
Botley	arr							13:30		
Botley	dep							13:30		
Hedge End	arr							13:34		
Hedge End	dep							13:35		
Eastleigh	arr		13:00C					13:41		
Eastleigh	dep		13:09SW					13:42		
Southampton Airport Parkway	arr			13:37						
Southampton Airport Parkway	dep			13:38						
Winchester	arr		13:17	13:47				13:53		
Winchester	dep		13:18	13:48				13:54		
Micheldever	arr							14:02		
Micheldever	dep							14:02		
Basingstoke	arr		13:34					14:15		
Basingstoke	dep		13:36					14:17		
Farnborough	arr							14:30		
Farnborough	dep							14:31		
Woking	arr	14:11		14:19	13:57	13:57	14:25	14:40		14:25
Woking	dep	14:12		14:21	13:59	13:59	14:26	14:41		14:26
Clapham Junction	arr	14:31	14:12					15:01	15:11C	
Clapham Junction	dep	14:32	14:13						15:21SW	
Vauxhall	arr								15:26	
Vauxhall	dep								15:26	
London Waterloo	arr	14:40	14:24	14:49	14:23	14:27	14:51	15:13	15:31	14:51

Chapter 5

Percentages

Why it is important to understand: Percentages

Engineers and scientists use percentages all the time in calculations; calculators are able to handle calculations with percentages. For example, percentage change is commonly used in engineering, statistics, physics, finance, chemistry and economics. When you feel able to do calculations with basic arithmetic, fractions, decimals and percentages, all with the aid of a calculator, then suddenly mathematics doesn't seem quite so difficult.

At the end of this chapter, you should be able to:

- understand the term 'percentage'
- convert decimals to percentages and vice versa
- calculate the percentage of a quantity
- express one quantity as a percentage of another quantity
- calculate percentage error and percentage change

5.1 Introduction

Percentages are used to give a common standard. The use of percentages is very common in many aspects of commercial life, as well as in engineering. Interest rates, sale reductions, pay rises, exams and VAT are all examples of situations in which percentages are used. For this chapter you will need to know about decimals and fractions and be able to use a calculator.

We are familiar with the symbol for percentage, i.e. %. Here are some examples.

- Interest rates indicate the cost at which we can borrow money. If you borrow £8000 at a **6.5% interest rate** for a year, it will cost you 6.5% of the amount borrowed to do so, which will need to be repaid along with the original money you borrowed. If you repay the loan in 1 year, how much interest will you have paid?

- A pair of trainers in a shop cost £60. They are advertised in a sale as **20% off**. How much will you pay?

- If you earn £20000 p.a. and you receive a **2.5% pay rise**, how much extra will you have to spend the following year?

- A book costing £18 can be purchased on the internet for **30% less**. What will be its cost?

When we have completed his chapter on percentages you will be able to understand how to perform the above calculations.

Percentages are fractions having 100 as their denominator. For example, the fraction $\frac{40}{100}$ is written as 40% and is read as 'forty per cent'.

The easiest way to understand percentages is to go through some worked examples.

5.2 Percentage calculations

5.2.1 To convert a decimal to a percentage

A decimal number is converted to a percentage by multiplying by 100.

> **Problem 1.** Express 0.015 as a percentage

To express a decimal number as a percentage, merely multiply by 100, i.e.

$$0.015 = 0.015 \times 100\%$$
$$= \mathbf{1.5\%}$$

Multiplying a decimal number by 100 means moving the decimal point 2 places **to the right**.

> **Problem 2.** Express 0.275 as a percentage

$$0.275 = 0.275 \times 100\%$$
$$= \mathbf{27.5\%}$$

5.2.2 To convert a percentage to a decimal

A percentage is converted to a decimal number by dividing by 100.

> **Problem 3.** Express 6.5% as a decimal number

$$6.5\% = \frac{6.5}{100} = \mathbf{0.065}$$

Dividing by 100 means moving the decimal point 2 places **to the left**.

> **Problem 4.** Express 17.5% as a decimal number

$$17.5\% = \frac{17.5}{100}$$
$$= \mathbf{0.175}$$

5.2.3 To convert a fraction to a percentage

A fraction is converted to a percentage by multiplying by 100.

> **Problem 5.** Express $\frac{5}{8}$ as a percentage

$$\frac{5}{8} = \frac{5}{8} \times 100\% = \frac{500}{8}\%$$
$$= \mathbf{62.5\%}$$

> **Problem 6.** Express $\frac{5}{19}$ as a percentage, correct to 2 decimal places

$$\frac{5}{19} = \frac{5}{19} \times 100\%$$
$$= \frac{500}{19}\%$$
$$= 26.3157889\ldots \text{ by calculator}$$
$$= \mathbf{26.32\%} \text{ correct to 2 decimal places}$$

> **Problem 7.** In two successive tests a student gains marks of 57/79 and 49/67. Is the second mark better or worse than the first?

$$57/79 = \frac{57}{79} = \frac{57}{79} \times 100\% = \frac{5700}{79}\%$$
$$= \mathbf{72.15\%} \text{ correct to 2 decimal places}$$
$$49/67 = \frac{49}{67} = \frac{49}{67} \times 100\% = \frac{4900}{67}\%$$
$$= \mathbf{73.13\%} \text{ correct to 2 decimal places}$$

Hence, **the second test is marginally better than the first test**. This question demonstrates how much easier it is to compare two fractions when they are expressed as percentages.

5.2.4 To convert a percentage to a fraction

A percentage is converted to a fraction by dividing by 100 and then, by cancelling, reducing it to its simplest form.

> **Problem 8.** Express 75% as a fraction

$$75\% = \frac{75}{100}$$
$$= \frac{3}{4}$$

The fraction $\frac{75}{100}$ is reduced to its simplest form by cancelling, i.e. dividing both numerator and denominator by 25.

Problem 9. Express 37.5% as a fraction

$$37.5\% = \frac{37.5}{100}$$

$$= \frac{375}{1000} \quad \text{by multiplying both numerator and denominator by 10}$$

$$= \frac{15}{40} \quad \text{by dividing both numerator and denominator by 25}$$

$$= \frac{3}{8} \quad \text{by dividing both numerator and denominator by 5}$$

Now try the following Practice Exercise

Practice Exercise 21   Percentages (answers on page 424)

In Problems 1 to 5, express the given numbers as percentages.

1. 0.0032

2. 1.734

3. 0.057

4. 0.374

5. 1.285

6. Express 20% as a decimal number.

7. Express 1.25% as a decimal number.

8. Express $\frac{11}{16}$ as a percentage.

9. Express $\frac{5}{13}$ as a percentage, correct to 3 decimal places.

10. Express as percentages, correct to 3 significant figures,

    (a) $\frac{7}{33}$ (b) $\frac{19}{24}$ (c) $1\frac{11}{16}$

11. Place the following in order of size, the smallest first, expressing each as a percentage correct to 1 decimal place.

    (a) $\frac{12}{21}$ (b) $\frac{9}{17}$ (c) $\frac{5}{9}$ (d) $\frac{6}{11}$

12. Express 65% as a fraction in its simplest form.

13. Express 31.25% as a fraction in its simplest form.

14. Express 56.25% as a fraction in its simplest form.

15. Evaluate A to J in the following table.

Decimal number	Fraction	Percentage
0.5	A	B
C	$\frac{1}{4}$	D
E	F	30
G	$\frac{3}{5}$	H
I	J	85

16. A resistor has a value of $820\,\Omega \pm 5\%$. Determine the range of resistance values expected.

5.3 Further percentage calculations

5.3.1 Finding a percentage of a quantity

To find a percentage of a quantity, convert the percentage to a fraction (by dividing by 100) and remember that 'of' means multiply.

Problem 10. Find 27% of £65

$$27\% \text{ of } £65 = \frac{27}{100} \times 65$$

$$= \textbf{£17.55} \text{ by calculator}$$

Problem 11. In a machine shop, it takes 32 minutes to machine a certain part. Using a new tool, the time can be reduced by 12.5%. Calculate the new time taken

$$12.5\% \text{ of } 32 \text{ minutes} = \frac{12.5}{100} \times 32$$

$$= 4 \text{ minutes}$$

Hence, **new time taken** $= 32 - 4 = \textbf{28 minutes}$.

Alternatively, if the time is reduced by 12.5%, it now takes $100\% - 12.5\% = 87.5\%$ of the original time, i.e.

$$87.5\% \text{ of } 32 \text{ minutes} = \frac{87.5}{100} \times 32$$

$$= \textbf{28 minutes}$$

Problem 12. A 160 GB iPod is advertised as costing £190 excluding VAT. If VAT is added at 20%, what will be the total cost of the iPod?

$$\text{VAT} = 20\% \text{ of } £190 = \frac{20}{100} \times 190 = £38$$

Total cost of iPod $= £190 + £38 = £228$

A quicker method to determine the total cost is: $1.20 \times £190 = £228$

5.3.2 Expressing one quantity as a percentage of another quantity

To express one quantity as a percentage of another quantity, divide the first quantity by the second then multiply by 100.

Problem 13. Express 23 cm as a percentage of 72 cm, correct to the nearest 1%

$$23 \text{ cm as a percentage of } 72 \text{ cm} = \frac{23}{72} \times 100\%$$
$$= 31.94444\ldots\%$$
$$= 32\% \text{ correct to the nearest 1\%}$$

Problem 14. Express 47 minutes as a percentage of 2 hours, correct to 1 decimal place

Note that it is essential that the two quantities are in the **same units**.

Working in minute units, 2 hours $= 2 \times 60$
$$= 120 \text{ minutes}$$
47 minutes as a percentage of 120 min $= \frac{47}{120} \times 100\%$
$$= 39.2\% \text{ correct to 1 decimal place}$$

5.3.3 Percentage change

Percentage change is given by
$$\frac{\text{new value} - \text{original value}}{\text{original value}} \times 100\%.$$

Problem 15. A box of resistors increases in price from £45 to £52. Calculate the percentage change in cost, correct to 3 significant figures

$$\% \text{ change} = \frac{\text{new value} - \text{original value}}{\text{original value}} \times 100\%$$
$$= \frac{52 - 45}{45} \times 100\% = \frac{7}{45} \times 100$$
$$= 15.6\% = \textbf{percentage change in cost}$$

Problem 16. A drilling speed should be set to 400 rev/min. The nearest speed available on the machine is 412 rev/min. Calculate the percentage overspeed

$$\% \text{ overspeed} = \frac{\text{available speed} - \text{correct speed}}{\text{correct speed}} \times 100\%$$
$$= \frac{412 - 400}{400} \times 100\% = \frac{12}{400} \times 100\%$$
$$= 3\%$$

Now try the following Practice Exercise

Practice Exercise 22   Further percentages (answers on page 424)

1. Calculate 43.6% of 50 kg.

2. Determine 36% of 27 m.

3. Calculate, correct to 4 significant figures,
   (a)   18% of 2758 tonnes
   (b)   47% of 18.42 grams
   (c)   147% of 14.1 seconds.

4. When 1600 bolts are manufactured, 36 are unsatisfactory. Determine the percentage that is unsatisfactory.

5. Express
   (a)   140 kg as a percentage of 1 t.
   (b)   47 s as a percentage of 5 min.
   (c)   13.4 cm as a percentage of 2.5 m.

6. A block of Monel alloy consists of 70% nickel and 30% copper. If it contains 88.2 g of nickel, determine the mass of copper in the block.

7. An athlete runs 5000 m in 15 minutes 20 seconds. With intense training, he is able to reduce this time by 2.5%. Calculate his new time.

8. A copper alloy comprises 89% copper, 1.5% iron and the remainder aluminium. Find the amount of aluminium, in grams, in a 0.8 kg mass of the alloy.

9. A computer is advertised on the internet at £520, exclusive of VAT. If VAT is payable at 17.5%, what is the total cost of the computer?

10. Express 325 mm as a percentage of 867 mm, correct to 2 decimal places.

11. A child sleeps on average 9 hours 25 minutes per day. Express this as a percentage of the whole day, correct to 1 decimal place.

12. Express 408 g as a percentage of 2.40 kg.

13. When signing a new contract, a Premiership footballer's pay increases from £15 500 to £21 500 per week. Calculate the percentage pay increase, correct to 3 significant figures.

14. A metal rod 1.80 m long is heated and its length expands by 48.6 mm. Calculate the percentage increase in length.

15. 12.5% of a length of wood is 70 cm. What is the full length?

16. A metal rod, 1.20 m long, is heated and its length expands by 42 mm. Calculate the percentage increase in length.

17. For each of the following resistors, determine the (i) minimum value, (ii) maximum value:

    (a) $680 \, \Omega \pm 20\%$ (b) $47 \, k\Omega \pm 5\%$

18. An engine speed is 2400 rev/min. The speed is increased by 8%. Calculate the new speed.

5.4   More percentage calculations

5.4.1   Percentage error

$$\text{Percentage error} = \frac{\text{error}}{\text{correct value}} \times 100\%$$

Problem 17.   The length of a component is measured incorrectly as 64.5 mm. The actual length is 63 mm. What is the percentage error in the measurement?

$$\% \, \text{error} = \frac{\text{error}}{\text{correct value}} \times 100\%$$
$$= \frac{64.5 - 63}{63} \times 100\%$$
$$= \frac{1.5}{63} \times 100\% = \frac{150}{63}\%$$
$$= \mathbf{2.38\%}$$

The percentage measurement error is **2.38% too high**, which is sometimes written as **+ 2.38% error**.

Problem 18.   The voltage across a component in an electrical circuit is calculated as 50 V using Ohm's law. When measured, the actual voltage is 50.4 V. Calculate, correct to 2 decimal places, the percentage error in the calculation

$$\% \, \text{error} = \frac{\text{error}}{\text{correct value}} \times 100\%$$
$$= \frac{50.4 - 50}{50.4} \times 100\%$$
$$= \frac{0.4}{50.4} \times 100\% = \frac{40}{50.4}\%$$
$$= \mathbf{0.79\%}$$

The percentage error in the calculation is **0.79% too low**, which is sometimes written as **−0.79% error**.

5.4.2   Original value

$$\text{Original value} = \frac{\text{new value}}{100 \pm \% \, \text{change}} \times 100\%$$

Problem 19.   A man pays £149.50 in a sale for a DVD player which is labelled '35% off'. What was the original price of the DVD player?

In this case, it is a 35% reduction in price, so we use $\dfrac{\text{new value}}{100 - \% \, \text{change}} \times 100$, i.e. a minus sign in the denominator.

$$\text{Original price} = \frac{\text{new value}}{100 - \% \, \text{change}} \times 100$$
$$= \frac{149.5}{100 - 35} \times 100$$
$$= \frac{149.5}{65} \times 100 = \frac{14\,950}{65}$$
$$= \mathbf{£230}$$

Problem 20. A couple buys a flat and makes an 18% profit by selling it 3 years later for £153 400. Calculate the original cost of the flat

In this case, it is an 18% increase in price, so we use $\dfrac{\text{new value}}{100 + \% \text{ change}} \times 100$, i.e. a plus sign in the denominator.

$$\text{Original cost} = \frac{\text{new value}}{100 + \% \text{ change}} \times 100$$

$$= \frac{153\,400}{100 + 18} \times 100$$

$$= \frac{153\,400}{118} \times 100 = \frac{15\,340\,000}{118}$$

$$= \textbf{£130 000}$$

Problem 21. An electrical store makes 40% profit on each widescreen television it sells. If the selling price of a 32 inch HD television is £630, what was the cost to the dealer?

In this case, it is a 40% mark-up in price, so we use $\dfrac{\text{new value}}{100 + \% \text{ change}} \times 100$, i.e. a plus sign in the denominator.

$$\text{Dealer cost} = \frac{\text{new value}}{100 + \% \text{ change}} \times 100$$

$$= \frac{630}{100 + 40} \times 100$$

$$= \frac{630}{140} \times 100 = \frac{63\,000}{140}$$

$$= \textbf{£450}$$

The dealer buys from the manufacturer for £450 and sells to his customers for £630.

5.4.3 Percentage increase/decrease and interest

$$\text{New value} = \frac{100 + \% \text{ increase}}{100} \times \text{original value}$$

Problem 22. £3600 is placed in an ISA account which pays 3.5% interest per annum. How much is the investment worth after 1 year?

$$\text{Value after 1 year} = \frac{100 + 3.25}{100} \times £3600$$

$$= \frac{103.25}{100} \times £3600$$

$$= 1.0325 \times £3600$$

$$= \textbf{£3717}$$

Problem 23. The price of a fully installed combination condensing boiler is increased by 6.5%. It originally cost £2400. What is the new price?

$$\text{New price} = \frac{100 + 6.5}{100} \times £2400$$

$$= \frac{106.5}{100} \times £2400 = 1.065 \times £2400$$

$$= \textbf{£2556}$$

Now try the following Practice Exercise

Practice Exercise 23   Further percentages (answers on page 424)

1. A machine part has a length of 36 mm. The length is incorrectly measured as 36.9 mm. Determine the percentage error in the measurement.

2. When a resistor is removed from an electrical circuit the current flowing increases from 450 μA to 531 μA. Determine the percentage increase in the current.

3. In a shoe shop sale, everything is advertised as '40% off'. If a lady pays £186 for a pair of Jimmy Choo shoes, what was their original price?

4. Over a 4 year period a family home increases in value by 22.5% to £214 375. What was the value of the house 4 years ago?

5. An electrical retailer makes a 35% profit on all its products. What price does the retailer pay for a dishwasher which is sold for £351?

6. The cost of a sports car is £24 000 inclusive of VAT at 20%. What is the cost of the car without the VAT added?

7. £8000 is invested in bonds at a building society which is offering a rate of 4.75% per annum. Calculate the value of the investment after 2 years.

8. An electrical contractor earning £36 000 per annum receives a pay rise of 2.5%. He pays 22% of his income as tax and 11% on National Insurance contributions. Calculate the increase he will actually receive per month.

9. Five mates enjoy a meal out. With drinks, the total bill comes to £176. They add a 12.5% tip and divide the amount equally between them. How much does each pay?

10. In December a shop raises the cost of a 40 inch LCD TV costing £920 by 5%. It does not sell and in its January sale it reduces the TV by 5%. What is the sale price of the TV?

11. A man buys a business and makes a 20% profit when he sells it three years later for £222 000. What did he pay originally for the business?

12. A drilling machine should be set to 250 rev/min. The nearest speed available on the machine is 268 rev/min. Calculate the percentage overspeed.

13. Two kilograms of a compound contain 30% of element A, 45% of element B and 25% of element C. Determine the masses of the three elements present.

14. A concrete mixture contains seven parts by volume of ballast, four parts by volume of sand and two parts by volume of cement. Determine the percentage of each of these three constituents correct to the nearest 1% and the mass of cement in a two tonne dry mix, correct to 1 significant figure.

15. In a sample of iron ore, 18% is iron. How much ore is needed to produce 3600 kg of iron?

16. A screw's dimension is $12.5 \pm 8\%$ mm. Calculate the maximum and minimum possible length of the screw.

17. The output power of an engine is 450 kW. If the efficiency of the engine is 75%, determine the power input.

For fully worked solutions to each of the problems in Practice Exercises 21 to 23 in this chapter, go to the website:

www.routledge.com/cw/bird

Revision Test 2: Decimals, calculations and percentages

This assignment covers the material contained in Chapters 3–5. *The marks available are shown in brackets at the end of each question.*

1. Convert 0.048 to a proper fraction. (2)

2. Convert 6.4375 to a mixed number. (3)

3. Express $\dfrac{9}{32}$ as a decimal fraction. (2)

4. Express 0.0784 correct to 2 decimal places. (2)

5. Express 0.0572953 correct to 4 significant figures. (2)

6. Evaluate
   (a) $46.7 + 2.085 + 6.4 + 0.07$
   (b) $68.51 - 136.34$ (4)

7. Determine 2.37×1.2 (3)

8. Evaluate $250.46 \div 1.1$ correct to 1 decimal place. (3)

9. Evaluate $5.\dot{2} \times 12$ (2)

10. Evaluate the following, correct to 4 significant figures: $3.3^2 - 2.7^3 + 1.8^4$ (3)

11. Evaluate $\sqrt{6.72} - \sqrt[3]{2.54}$ correct to 3 decimal places. (3)

12. Evaluate $\dfrac{1}{0.0071} - \dfrac{1}{0.065}$ correct to 4 significant figures. (2)

13. The potential difference, V volts, available at battery terminals is given by $V = E - Ir$. Evaluate V when $E = 7.23$, $I = 1.37$ and $r = 3.60$ (3)

14. Evaluate $\dfrac{4}{9} + \dfrac{1}{5} - \dfrac{3}{8}$ as a decimal, correct to 3 significant figures. (3)

15. Evaluate $\dfrac{16 \times 10^{-6} \times 5 \times 10^9}{2 \times 10^7}$ in engineering form. (2)

16. Evaluate resistance, R, given
   $\dfrac{1}{R} = \dfrac{1}{R_1} + \dfrac{1}{R_2} + \dfrac{1}{R_3}$ when $R_1 = 3.6\,\text{k}\Omega$, $R_2 = 7.2\,\text{k}\Omega$ and $R_3 = 13.6\,\text{k}\Omega$. (3)

17. Evaluate $6\dfrac{2}{7} - 4\dfrac{5}{9}$ as a mixed number and as a decimal, correct to 3 decimal places. (3)

18. Evaluate, correct to 3 decimal places:
   $\sqrt{\left[\dfrac{2e^{1.7} \times 3.67^3}{4.61 \times \sqrt{3\pi}}\right]}$ (3)

19. If $a = 0.270, b = 15.85, c = 0.038, d = 28.7$ and $e = 0.680$, evaluate v correct to 3 significant figures, given that $v = \sqrt{\left(\dfrac{ab}{c} - \dfrac{d}{e}\right)}$ (4)

20. Evaluate the following, each correct to 2 decimal places.
   (a) $\left(\dfrac{36.2^2 \times 0.561}{27.8 \times 12.83}\right)^3$
   (b) $\sqrt{\left(\dfrac{14.69^2}{\sqrt{17.42} \times 37.98}\right)}$ (4)

21. If $1.6\,\text{km} = 1\,\text{mile}$, determine the speed of 45 miles/hour in kilometres per hour. (2)

22. The area A of a circle is given by $A = \pi r^2$. Find the area of a circle of radius $r = 3.73\,\text{cm}$, correct to 2 decimal places. (3)

23. Evaluate B, correct to 3 significant figures, when $W = 7.20, v = 10.0$ and $g = 9.81$, given that $B = \dfrac{Wv^2}{2g}$ (3)

24. Express 56.25% as a fraction in its simplest form. (3)

25. 12.5% of a length of wood is $90\,\text{cm}$. What is the full length? (3)

26. A metal rod, $1.50\,\text{m}$ long, is heated and its length expands by $45\,\text{mm}$. Calculate the percentage increase in length. (2)

27. A man buys a house and makes a 20% profit when he sells it three years later for £312 000. What did he pay for it originally? (3)

For lecturers/instructors/teachers, fully worked solutions to each of the problems in Revision Test 2, together with a full marking scheme, are available at the website:

www.routledge.com/cw/bird

Ratio and proportion

At the end of this chapter, you should be able to:

- define ratio
- perform calculations with ratios
- define direct proportion
- perform calculations with direct proportion, including Hooke's law, Charles's law and Ohm's law
- define inverse proportion
- perform calculations with inverse proportion, including Boyle's law

6.1 Introduction

Ratio is a way of comparing amounts of something; it shows how much bigger one thing is than the other. Some practical examples include mixing paint, sand and cement, or screen wash. Gears, map scales, food recipes, scale drawings and metal alloy constituents all use ratios.

Two quantities are in **direct proportion** when they increase or decrease in the **same ratio**. There are several practical engineering laws which rely on direct proportion. Also, calculating currency exchange rates and converting imperial to metric units rely on direct proportion.

Sometimes, as one quantity increases at a particular rate, another quantity decreases at the same rate; this is called **inverse proportion**. For example, the time taken to do a job is inversely proportional to the number of people in a team: double the people, half the time.

When we have completed this chapter on ratio and proportion you will be able to understand, and confidently perform, calculations on the above topics.

For this chapter you will need to know about decimals and fractions and to be able to use a calculator.

6.2 Ratios

Ratios are generally shown as numbers separated by a colon (:) so the ratio of 2 and 7 is written as $2:7$ and we read it as a ratio of 'two to seven.'

Some practical examples which are familiar include:

- Mixing 1 measure of screen wash to 6 measures of water; i.e. the ratio of screen wash to water is $1:6$

- Mixing 1 shovel of cement to 4 shovels of sand; i.e. the ratio of cement to sand is $1:4$

- Mixing 3 parts of red paint to 1 part white, i.e. the ratio of red to white paint is $3:1$

Ratio is the number of parts to a mix. The paint mix is 4 parts total, with 3 parts red and 1 part white. 3 parts red paint to 1 part white paint means there is

$$\frac{3}{4} \text{ red paint to } \frac{1}{4} \text{ white paint}$$

Here are some worked examples to help us understand more about ratios.

Problem 1. In a class, the ratio of female to male students is $6:27$. Reduce the ratio to its simplest form

(i) Both 6 and 27 can be divided by 3

(ii) Thus, $6:27$ is the same as **$2:9$**

$6:27$ and $2:9$ are called **equivalent ratios**.

It is normal to express ratios in their lowest, or simplest, form. In this example, the simplest form is **$2:9$** which means for every 2 females in the class there are 9 male students.

Problem 2. A gear wheel having 128 teeth is in mesh with a 48-tooth gear. What is the gear ratio?

$$\text{Gear ratio} = 128:48$$

A ratio can be simplified by finding common factors.

(i) 128 and 48 can both be divided by 2, i.e. $128:48$ is the same as $64:24$

(ii) 64 and 24 can both be divided by 8, i.e. $64:24$ is the same as $8:3$

(iii) There is no number that divides completely into both 8 and 3 so $8:3$ is the simplest ratio, i.e. **the gear ratio is $8:3$**

Thus, $128:48$ is equivalent to $64:24$ which is equivalent to $8:3$ and **$8:3$ is the simplest form**.

Problem 3. A wooden pole is 2.08 m long. Divide it in the ratio of 7 to 19

(i) Since the ratio is $7:19$, the total number of parts is $7 + 19 = 26$ parts.

(ii) 26 parts corresponds to 2.08 m $= 208$ cm, hence, 1 part corresponds to $\frac{208}{26} = 8$

(iii) Thus, 7 parts corresponds to $7 \times 8 = $ **56 cm** and 19 parts corresponds to $19 \times 8 = $ **152 cm**.

Hence, **2.08 m divides in the ratio of $7:19$ as 56 cm to 152 cm**.

(Check: $56 + 152$ must add up to 208, otherwise an error would have been made.)

Problem 4. In a competition, prize money of £828 is to be shared among the first three in the ratio $5:3:1$

(i) Since the ratio is $5:3:1$ the total number of parts is $5 + 3 + 1 = 9$ parts.

(ii) 9 parts corresponds to £828

(iii) 1 part corresponds to $\frac{828}{9} = $ **£92**, 3 parts corresponds to $3 \times £92 = $ **£276** and 5 parts corresponds to $5 \times £92 = $ **£460**

Hence, **£828 divides in the ratio of $5:3:1$ as £460 to £276 to £92**. (Check: $460 + 276 + 92$ must add up to 828, otherwise an error would have been made.)

Problem 5. A map scale is 1:30000. On the map the distance between two schools is 6 cm. Determine the actual distance between the schools, giving the answer in kilometres

Actual distance between schools

$$= 6 \times 30000 \, \text{cm} = 180000 \, \text{cm}$$

$$= \frac{180,000}{100} \, \text{m} = 1800 \, \text{m}$$

$$= \frac{1800}{1000} \, \text{m} = \textbf{1.80 km}$$

(1 mile $\approx$ 1.6 km, hence the schools are just over 1 mile apart.)

Now try the following Practice Exercise

Practice Exercise 24   Ratios (answers on page 424)

1. In a box of 333 paper clips, 9 are defective. Express the number of non-defective paper clips as a ratio of the number of defective paper clips, in its simplest form.

2. A gear wheel having 84 teeth is in mesh with a 24-tooth gear. Determine the gear ratio in its simplest form.

3. In a box of 2000 nails, 120 are defective. Express the number of non-defective nails as a ratio of the number of defective ones, in its simplest form.

4. A metal pipe 3.36 m long is to be cut into two in the ratio 6 to 15. Calculate the length of each piece.

5. The instructions for cooking a turkey say that it needs to be cooked 45 minutes for every kilogram. How long will it take to cook a 7 kg turkey?

6. In a will, £6440 is to be divided among three beneficiaries in the ratio 4:2:1. Calculate the amount each receives.

7. A local map has a scale of 1:22500. The distance between two motorways is 2.7 km. How far are they apart on the map?

8. Prize money in a lottery totals £3801 and is shared among three winners in the ratio 4:2:1. How much does the first prize winner receive?

Here are some further worked examples on ratios.

Problem 6. Express 45 p as a ratio of £7.65 in its simplest form

(i) Changing both quantities to the same units, i.e. to pence, gives a ratio of 45:765

(ii) Dividing both quantities by 5 gives
    $45:765 \equiv 9:153$

(iii) Dividing both quantities by 3 gives
    $9:153 \equiv 3:51$

(iv) Dividing both quantities by 3 again gives
    $3:51 \equiv 1:17$

Thus, **45 p as a ratio of £7.65 is 1:17**
$45:765, 9:153, 3:51$ and $1:17$ are **equivalent ratios** and **1:17 is the simplest ratio**.

Problem 7. A glass contains 30 ml of whisky which is 40% alcohol. If 45 ml of water is added and the mixture stirred, what is now the alcohol content?

(i) The 30 ml of whisky contains 40% alcohol $= \frac{40}{100} \times 30 = 12$ ml.

(ii) After 45 ml of water is added we have $30 + 45 = 75$ ml of fluid, of which alcohol is 12 ml.

(iii) Fraction of alcohol present $= \frac{12}{75}$

(iv) Percentage of alcohol present $= \frac{12}{75} \times 100\%$

$$= \textbf{16\%}$$

Problem 8. 20 tonnes of a mixture of sand and gravel is 30% sand. How many tonnes of sand must be added to produce a mixture which is 40% gravel?

(i)   Amount of sand in 20 tonnes $= 30\%$ of 20 t
$$= \frac{30}{100} \times 20 = 6\,\text{t}$$

(ii) If the mixture has 6 t of sand then amount of gravel $= 20 - 6 = 14\,\text{t}$

(iii) We want this 14 t of gravel to be 40% of the new mixture. 1% would be $\frac{14}{40}\,\text{t}$ and 100% of the mixture would be $\frac{14}{40} \times 100\,\text{t} = 35\,\text{t}$

(iv) If there is 14 t of gravel then amount of sand $= 35 - 14 = 21\,\text{t}$

(v)   We already have 6 t of sand, so **amount of sand to be added to produce a mixture with 40% gravel** $= 21 - 6 = \mathbf{15\,t}$

(Note 1 tonne $= 1000\,\text{kg}$.)

Now try the following Practice Exercise

Practice Exercise 25   Further ratios (answers on page 424)

1. Express 130 g as a ratio of 1.95 kg.

2. In a laboratory, acid and water are mixed in the ratio 2:5. How much acid is needed to make 266 ml of the mixture?

3. A glass contains 30 ml of gin which is 40% alcohol. If 18 ml of water is added and the mixture stirred, determine the new percentage alcoholic content.

4. A wooden beam 4 m long weighs 84 kg. Determine the mass of a similar beam that is 60 cm long.

5. An alloy is made up of metals P and Q in the ratio 3.25:1 by mass. How much of P has to be added to 4.4 kg of Q to make the alloy?

6. 15 000 kg of a mixture of sand and gravel is 20% sand. Determine the amount of sand that must be added to produce a mixture with 30% gravel.

6.3   Direct proportion

Two quantities are in **direct proportion** when they increase or decrease in the **same ratio**. For example, if 12 cans of lager have a mass of 4 kg, then 24 cans of lager will have a mass of 8 kg; i.e. if the quantity of cans doubles then so does the mass. This is direct proportion.

In the previous section we had an example of mixing 1 shovel of cement to 4 shovels of sand; i.e. the ratio of cement to sand was 1:4. So, if we have a mix of 10 shovels of cement and 40 shovels of sand and we wanted to double the amount of the mix then we would need to double both the cement and sand, i.e. 20 shovels of cement and 80 shovels of sand. This is another example of direct proportion.

Here are three laws in engineering which involve direct proportion:

(a) **Hooke's law*** states that, within the elastic limit of a material, the strain ε produced is directly proportional to the stress σ producing it, i.e. $\varepsilon \propto \sigma$ (note than '$\propto$' means 'is proportional to').

ROBERT HOOKE
1635-1703

*Who was Hooke? – **Robert Hooke** FRS (28 July 1635–3 March 1703) was an English natural philosopher, architect and polymath who, amongst other things, discovered the law of elasticity. To find out more go to www.routledge.com/cw/bird

(b) **Charles's law*** states that, for a given mass of gas at constant pressure, the volume V is directly proportional to its thermodynamic temperature T, i.e. $V \propto T$.

(c) **Ohm's law**† states that the current I flowing through a fixed resistance is directly proportional to the applied voltage V, i.e. $I \propto V$.

Here are some worked examples to help us understand more about direct proportion.

Problem 9. Three energy saving light bulbs cost £7.80. Determine the cost of seven such light bulbs

(i) 3 light bulbs cost £7.80

(ii) Therefore, 1 light bulb costs $\dfrac{7.80}{3} = £2.60$

Hence, **7 light bulbs cost** $7 \times £2.60 = £18.20$

Problem 10. If 56 litres of petrol costs £72.80, calculate the cost of 32 litres

(i) 56 litres of petrol costs £72.80

(ii) Therefore, 1 litre of petrol costs $\dfrac{72.80}{56} = £1.30$

Hence, **32 litres cost** $32 \times 1.30 = £41.60$

*Who was Charles? – **Jacques Alexandre César Charles** (12 November 1746 – 7 April 1823) was a French inventor, scientist, mathematician and balloonist. Charles's law describes how gases expand when heated. To find out more go to www.routledge.com/cw/bird

† Who was Ohm? – To find out more go to www. routledge. com/ cw/bird

Problem 11. Hooke's law states that stress, σ, is directly proportional to strain, ε, within the elastic limit of a material. When, for mild steel, the stress is 63 MPa, the strain is 0.0003. Determine (a) the value of strain when the stress is 42 MPa, (b) the value of stress when the strain is 0.00072

(a) Stress is directly proportional to strain.
  (i) When the stress is 63 MPa, the strain is 0.0003

  (ii) Hence, a stress of 1 MPa corresponds to a strain of $\dfrac{0.0003}{63}$

  (iii) Thus, **the value of strain when the stress is 42 MPa** $= \dfrac{0.0003}{63} \times 42 = \mathbf{0.0002}$

(b) Strain is proportional to stress.
  (i) When the strain is 0.0003, the stress is 63 MPa.

  (ii) Hence, a strain of 0.0001 corresponds to $\dfrac{63}{3}$ MPa.

  (iii) Thus, **the value of stress when the strain is 0.00072** $= \dfrac{63}{3} \times 7.2 = \mathbf{151.2\,MPa}$.

Problem 12. Charles's law states that for a given mass of gas at constant pressure, the volume is directly proportional to its thermodynamic temperature. A gas occupies a volume of 2.4 litres at 600 K. Determine (a) the temperature when the volume is 3.2 litres, (b) the volume at 540 K

(a) Volume is directly proportional to temperature.
  (i) When the volume is 2.4 litres, the temperature is 600 K.

  (ii) Hence, a volume of 1 litre corresponds to a temperature of $\dfrac{600}{2.4}$ K.

  (iii) Thus, **the temperature when the volume is 3.2 litres** $= \dfrac{600}{2.4} \times 3.2 = \mathbf{800\,K}$

(b) Temperature is proportional to volume.
  (i) When the temperature is 600 K, the volume is 2.4 litres.

(ii) Hence, a temperature of 1 K corresponds to a volume of $\dfrac{2.4}{600}$ litres.

(iii) Thus, **the volume at a temperature of 540 K** $= \dfrac{2.4}{600} \times 540 = \textbf{2.16 litres}$.

Now try the following Practice Exercise

Practice Exercise 26   Direct proportion (answers on page 424)

1. Three engine parts cost £208.50. Calculate the cost of eight such parts.

2. If 9 litres of gloss white paint costs £24.75, calculate the cost of 24 litres of the same paint.

3. The total mass of 120 house bricks is 57.6 kg. Determine the mass of 550 such bricks.

4. A simple machine has an effort : load ratio of 3 : 37. Determine the effort, in newtons, to lift a load of 5.55 kN.

5. If 16 cans of lager weighs 8.32 kg, what will 28 cans weigh?

6. Hooke's law states that stress is directly proportional to strain within the elastic limit of a material. When, for copper, the stress is 60 MPa, the strain is 0.000625. Determine (a) the strain when the stress is 24 MPa and (b) the stress when the strain is 0.0005

7. Charles's law states that volume is directly proportional to thermodynamic temperature for a given mass of gas at constant pressure. A gas occupies a volume of 4.8 litres at 330 K. Determine (a) the temperature when the volume is 6.4 litres and (b) the volume when the temperature is 396 K.

8. A machine produces 320 bolts in a day. Calculate the number of bolts produced by 4 machines in 7 days.

Here are some further worked examples on direct proportion.

Problem 13. Some guttering on a house has to decline by 3 mm for every 70 cm to allow rainwater to drain. The gutter spans 8.4 m. How much lower should the low end be?

(i) The guttering has to decline in the ratio 3 : 700 or $\dfrac{3}{700}$

(ii) If d is the vertical drop in 8.4 m or 8400 mm, then the decline must be in the ratio d : 8400 or $\dfrac{d}{8400}$

(iii) Now $\dfrac{d}{8400} = \dfrac{3}{700}$

(iv) Cross-multiplying gives $700 \times d = 8400 \times 3$ from which, $d = \dfrac{8400 \times 3}{700}$

i.e. $d = \textbf{36 mm}$**, which is how much lower the end should be to allow rainwater to drain**.

Problem 14. Ohm's law states that the current flowing in a fixed resistance is directly proportional to the applied voltage. When 90 mV is applied across a resistor the current flowing is 3 A. Determine (a) the current when the voltage is 60 mV and (b) the voltage when the current is 4.2 A

(a) Current is directly proportional to the voltage.

   (i) When voltage is 90 mV, the current is 3 A.

   (ii) Hence, a voltage of 1 mV corresponds to a current of $\dfrac{3}{90}$ A.

   (iii) Thus, **when the voltage is 60 mV, the current** $= 60 \times \dfrac{3}{90} = \textbf{2A}$.

(b) Voltage is directly proportional to the current.

   (i) When current is 3 A, the voltage is 90 mV.

   (ii) Hence, a current of 1 A corresponds to a voltage of $\dfrac{90}{3}$ mV $= 30$ mV.

   (iii) Thus, **when the current is 4.2 A, the voltage** $= 30 \times 4.2 = \textbf{126 mV}$.

Problem 15. Some approximate imperial to metric conversions are shown in Table 6.1. Use the table to determine

(a) the number of millimetres in 12.5 inches

(b) a speed of 50 miles per hour in kilometres per hour

(c) the number of miles in 300 km

(d) the number of kilograms in 20 pounds weight

(e) the number of pounds and ounces in 56 kilograms (correct to the nearest ounce)

(f) the number of litres in 24 gallons

(g) the number of gallons in 60 litres

Table 6.1

length	1 inch = 2.54 cm
	1 mile = 1.6 km
weight	2.2 lb = 1 kg
	(1 lb = 16 oz)
capacity	1.76 pints = 1 litre
	(8 pints = 1 gallon)

(a) 12.5 inches = 12.5 × 2.54 cm = 31.75 cm
31.73 cm = 31.75 × 10 mm = **317.5 mm**

(b) 50 m.p.h. = 50 × 1.6 km/h = **80 km/h**

(c) $300\,km = \dfrac{300}{1.6}$ miles = **186.5 miles**

(d) $20\,lb = \dfrac{20}{2.2}$ kg = **9.09 kg**

(e) 56 kg = 56 × 2.2 lb = 123.2 lb

0.2 lb = 0.2 × 16 oz = 3.2 oz = 3 oz, correct to the nearest ounce.

Thus, 56 kg = **123 lb 3 oz**, correct to the nearest ounce.

(f) 24 gallons = 24 × 8 pints = 192 pints
192 pints $= \dfrac{192}{1.76}$ litres = **109.1 litres**

(g) 60 litres = 60 × 1.76 pints = 105.6 pints
105.6 pints $= \dfrac{105.6}{8}$ gallons = **13.2 gallons**

Problem 16. Currency exchange rates for five countries are shown in Table 6.2. Calculate

(a) how many euros £55 will buy

(b) the number of Japanese yen which can be bought for £23

(c) the number of pounds sterling which can be exchanged for 6405 krone

(d) the number of American dollars which can be purchased for £92.50

(e) the number of pounds sterling which can be exchanged for 2925 Swiss francs

Table 6.2

France	£1 = 1.25 euros
Japan	£1 = 185 yen
Norway	£1 = 10.50 krone
Switzerland	£1 = 1.95 francs
USA	£1 = 1.80 dollars

(a) £1 = 1.25 euros, hence £55 = 55 × 1.25 euros = **68.75 euros**.

(b) £1 = 185 yen, hence £23 = 23 × 185 yen = **4255 yen**.

(c) £1 = 10.50 krone, hence 6405 krone = $£\dfrac{6405}{10.50}$ = **£610**

(d) £1 = 1.80 dollars, hence £92.50 = 92.50 × 1.80 dollars = **$166.50**

(e) £1 = 1.95 Swiss francs, hence 2925 Swiss francs = $£\dfrac{2925}{1.95}$ = **£1500**

Now try the following Practice Exercise

Practice Exercise 27   Further direct proportion (answers on page 424)

1. Ohm's law states that current is proportional to p.d. in an electrical circuit. When a p.d. of 60 mV is applied across a circuit a current of 24 μA flows. Determine (a) the current flowing when the p.d. is 5 V and (b) the p.d. when the current is 10 mA.

2. The tourist rate for the Swiss franc is quoted in a newspaper as £1 = 1.40 fr. How many francs can be purchased for £310?

3. If 1 inch = 2.54 cm, find the number of millimetres in 27 inches.

4. If 2.2 lb = 1 kg and 1 lb = 16 oz, determine the number of pounds and ounces in 38 kg (correct to the nearest ounce).

5. If 1 litre = 1.76 pints and 8 pints = 1 gallon, determine (a) the number of litres in 35 gallons and (b) the number of gallons in 75 litres.

6. Hooke's law states that stress is directly proportional to strain within the elastic limit of a material. When for brass the stress is 21 MPa, the strain is 0.00025. Determine the stress when the strain is 0.00035

7. If 12 inches = 30.48 cm, find the number of millimetres in 23 inches.

8. The tourist rate for the Canadian dollar is quoted in a newspaper as £1 = $1.84. How many Canadian dollars can be purchased for £550?

6.4 Inverse proportion

Two variables, x and y, are in inverse proportion to one another if y is proportional to $\dfrac{1}{x}$, i.e. $y \alpha \dfrac{1}{x}$ or $y = \dfrac{k}{x}$ or $k = xy$ where k is a constant, called the **coefficient of proportionality**.

Inverse proportion means that, as the value of one variable increases, the value of another decreases, and that their product is always the same.

For example, the time for a journey is inversely proportional to the speed of travel. So, if at 30 m.p.h. a journey is completed in 20 minutes, then at 60 m.p.h. the journey would be completed in 10 minutes. Double the speed, half the journey time. (Note that $30 \times 20 = 60 \times 10$)

In another example, the time needed to dig a hole is inversely proportional to the number of people digging. So, if four men take 3 hours to dig a hole, then two men (working at the same rate) would take 6 hours. Half the men, twice the time. (Note that $4 \times 3 = 2 \times 6$)

Here are some worked examples on inverse proportion.

Problem 17. It is estimated that a team of four designers would take a year to develop an engineering process. How long would three designers take?

If 4 designers take 1 year, then 1 designer would take 4 years to develop the process. Hence, 3 designers would take $\dfrac{4}{3}$ years, i.e. **1 year 4 months**.

Problem 18. A team of five people can deliver leaflets to every house in a particular area in four hours. How long will it take a team of three people?

If 5 people take 4 hours to deliver the leaflets, then 1 person would take $5 \times 4 = 20$ hours. Hence, 3 people would take $\dfrac{20}{3}$ hours, i.e. $6\dfrac{2}{3}$ hours, i.e. **6 hours 40 minutes**.

Problem 19. The electrical resistance R of a piece of wire is inversely proportional to the cross-sectional area A. When $A = 5\,\text{mm}^2$, $R = 7.02$ ohms. Determine (a) the coefficient of proportionality and (b) the cross-sectional area when the resistance is 4 ohms

(a) $R \alpha \dfrac{1}{A}$, i.e. $R = \dfrac{k}{A}$ or $k = RA$. Hence, when $R = 7.2$ and $A = 5$, the

coefficient of proportionality, $k = (7.2)(5) = 36$

(b) Since $k = RA$ then $A = \dfrac{k}{R}$. Hence, when $R = 4$,

the cross sectional area, $A = \dfrac{36}{4} = 9\,\text{mm}^2$

Problem 20. Boyle's law* states that, at constant temperature, the volume V of a fixed mass of gas is inversely proportional to its absolute pressure p. If a gas occupies a volume of $0.08\,\text{m}^3$ at a pressure of 1.5×10^6 pascals, determine (a) the coefficient of proportionality and (b) the volume if the pressure is changed to 4×10^6 pascals

---

* Who was Boyle? – **Robert Boyle** (25 January 1627–31 December 1691) was a natural philosopher, chemist, physicist, and inventor. Regarded today as the first modern chemist, he is best known for Boyle's law, which describes the inversely proportional relationship between the absolute pressure and volume of a gas, providing the temperature is kept constant within a closed system. To find out more go to www.routledge.com/cw/bird

(a) $V \propto \dfrac{1}{p}$ i.e. $V = \dfrac{k}{p}$ or $k = pV$. Hence, the

coefficient of proportionality, k

$$= (1.5 \times 10^6)(0.08) = \mathbf{0.12 \times 10^6}$$

(b) **Volume,** $V = \dfrac{k}{p} = \dfrac{0.12 \times 10^6}{4 \times 10^6} = \mathbf{0.03\,m^3}$

Now try the following Practice Exercise

Practice Exercise 28   Inverse proportion
(answers on page 424)

1. A 10 kg bag of potatoes lasts for a week with a family of seven people. Assuming all eat the same amount, how long will the potatoes last if there are only two in the family?

2. If eight men take 5 days to build a wall, how long would it take two men?

3. If y is inversely proportional to x and $y = 15.3$ when $x = 0.6$, determine (a) the coefficient of proportionality, (b) the value of y when x is 1.5 and (c) the value of x when y is 27.2

4. A car travelling at 50 km/h makes a journey in 70 minutes. How long will the journey take at 70 km/h?

5. Boyle's law states that, for a gas at constant temperature, the volume of a fixed mass of gas is inversely proportional to its absolute pressure. If a gas occupies a volume of 1.5 m^3 at a pressure of 200×10^3 pascals, determine (a) the constant of proportionality, (b) the volume when the pressure is 800×10^3 pascals and (c) the pressure when the volume is 1.25 m^3.

6. The energy received by a surface from a source of heat is inversely proportional to the square of the distance between the heat source and the surface. A surface 1 m from the heat source receives 200 J of energy. Calculate (a) the energy received when the distance is changed to 2.5 m, (b) the distance required if the surface is to receive 800 J of energy.

For fully worked solutions to each of the problems in Practice Exercises 24 to 28 in this chapter, go to the website:
www.routledge.com/cw/bird

Chapter 7

Powers, roots and laws of indices

Why it is important to understand: Powers, roots and laws of indices

Powers and roots are used extensively in mathematics and engineering, so it is important to get a good grasp of what they are and how, and why, they are used. Being able to multiply powers together by adding their indices is particularly useful for disciplines like engineering and electronics, where quantities are often expressed as a value multiplied by some power of ten. In the field of electrical engineering, for example, the relationship between electric current, voltage and resistance in an electrical system is critically important, and yet the typical unit values for these properties can differ by several orders of magnitude. Studying, or working, in an engineering discipline, you very quickly become familiar with powers and roots and laws of indices. This chapter provides an important lead into Chapter 8 which deals with units, prefixes and engineering notation.

At the end of this chapter, you should be able to:

- understand the terms base, index and power
- understand square roots
- perform calculations with powers and roots
- state the laws of indices
- perform calculations using the laws of indices

7.1 Introduction

The manipulation of powers and roots is a crucial underlying skill needed in algebra. In this chapter, powers and roots of numbers are explained, together with the laws of indices.

Many worked examples are included to help understanding.

7.2 Powers and roots

7.2.1 Indices

The number 16 is the same as $2 \times 2 \times 2 \times 2$, and $2 \times 2 \times 2 \times 2$ can be abbreviated to 2^4. When written as 2^4, 2 is called the **base** and the 4 is called the **index** or **power**. 2^4 is read as '**two to the power of four**'.

Similarly, 3^5 is read as '**three to the power of 5**'

When the indices are 2 and 3 they are given special names; i.e. 2 is called 'squared' and 3 is called 'cubed'. Thus,

4^2 is called '**four squared**' rather than '4 to the power of 2' and

5^3 is called '**five cubed**' rather than '5 to the power of 3'

When no index is shown, the power is 1. For example, 2 means 2^1

Problem 1. Evaluate (a) 2^6 (b) 3^4

(a) 2^6 means $2 \times 2 \times 2 \times 2 \times 2 \times 2$ (i.e. 2 multiplied by itself 6 times), and $2 \times 2 \times 2 \times 2 \times 2 \times 2 = \mathbf{64}$
    i.e. $\mathbf{2^6 = 64}$

(b) 3^4 means $3 \times 3 \times 3 \times 3$ (i.e. 3 multiplied by itself 4 times), and $3 \times 3 \times 3 \times 3 = \mathbf{81}$
    i.e. $\mathbf{3^4 = 81}$

Problem 2. Change the following to index form: (a) 32 (b) 625

(a) (i) To express 32 in its lowest factors, 32 is initially divided by the lowest prime number, i.e. 2

   (ii) $32 \div 2 = 16$, hence $32 = 2 \times 16$

   (iii) 16 is also divisible by 2, i.e. $16 = 2 \times 8$. Thus, $32 = 2 \times 2 \times 8$

   (iv) 8 is also divisible by 2, i.e. $8 = 2 \times 4$. Thus, $32 = 2 \times 2 \times 2 \times 4$

   (v)   4 is also divisible by 2, i.e. $4 = 2 \times 2$. Thus, $32 = 2 \times 2 \times 2 \times 2 \times 2$

   (vi) Thus, $32 = \mathbf{2^5}$

(b) (i) 625 is not divisible by the lowest prime number, i.e. 2. The next prime number is 3 and 625 is not divisible by 3 either. The next prime number is 5

   (ii) $625 \div 5 = 125$, i.e. $625 = 5 \times 125$

   (iii) 125 is also divisible by 5, i.e. $125 = 5 \times 25$. Thus, $625 = 5 \times 5 \times 25$

   (iv) 25 is also divisible by 5, i.e. $25 = 5 \times 5$. Thus, $625 = 5 \times 5 \times 5 \times 5$

   (v)   Thus, $\mathbf{625 = 5^4}$

Problem 3. Evaluate $3^3 \times 2^2$

$$3^3 \times 2^2 = 3 \times 3 \times 3 \times 2 \times 2$$
$$= 27 \times 4$$
$$= \mathbf{108}$$

7.2.2 Square roots

When a number is multiplied by itself the product is called a square.

For example, the square of 3 is $3 \times 3 = 3^2 = 9$

A square root is the reverse process; i.e. the value of the base which when multiplied by itself gives the number; i.e. the square root of 9 is 3

The symbol $\sqrt{}$ is used to denote a square root. Thus, $\sqrt{9} = 3$. Similarly, $\sqrt{4} = 2$ and $\sqrt{25} = 5$

Because $-3 \times -3 = 9$, $\sqrt{9}$ also equals -3. Thus, $\sqrt{9} = +3$ or -3 which is usually written as $\sqrt{9} = \pm 3$. Similarly, $\sqrt{16} = \pm 4$ and $\sqrt{36} = \pm 6$

The square root of, say, 9 may also be written in index form as $9^{\frac{1}{2}}$.

$$9^{\frac{1}{2}} \equiv \sqrt{9} = \pm 3$$

Problem 4. Evaluate $\dfrac{3^2 \times 2^3 \times \sqrt{36}}{\sqrt{16} \times 4}$ taking only positive square roots

$$\frac{3^2 \times 2^3 \times \sqrt{36}}{\sqrt{16} \times 4} = \frac{3 \times 3 \times 2 \times 2 \times 2 \times 6}{4 \times 4}$$
$$= \frac{9 \times 8 \times 6}{16} = \frac{9 \times 1 \times 6}{2}$$
$$= \frac{9 \times 1 \times 3}{1} \qquad \text{by cancelling}$$
$$= \mathbf{27}$$

Problem 5. Evaluate $\dfrac{10^4 \times \sqrt{100}}{10^3}$ taking the positive square root only

$$\frac{10^4 \times \sqrt{100}}{10^3} = \frac{10 \times 10 \times 10 \times 10 \times 10}{10 \times 10 \times 10}$$
$$= \frac{1 \times 1 \times 1 \times 10 \times 10}{1 \times 1 \times 1} \qquad \text{by cancelling}$$
$$= \frac{100}{1}$$
$$= \mathbf{100}$$

Now try the following Practice Exercise

Practice Exercise 29   **Powers and roots**
(answers on page 424)

Evaluate the following without the aid of a calculator.

1. 3^3

2. 2^7

3. 10^5

4. $2^4 \times 3^2 \times 2 \div 3$

5. Change 16 to index form.

6. $25^{\frac{1}{2}}$

7. $64^{\frac{1}{2}}$

8. $\dfrac{10^5}{10^3}$

9. $\dfrac{10^2 \times 10^3}{10^5}$

10. $\dfrac{2^5 \times 64^{\frac{1}{2}} \times 3^2}{\sqrt{144} \times 3}$
taking positive square roots only.

7.3   Laws of indices

There are six laws of indices.

(1)   From earlier, $2^2 \times 2^3 = (2 \times 2) \times (2 \times 2 \times 2)$

$$= 32$$
$$= 2^5$$

Hence, $2^2 \times 2^3 = 2^5$

or $2^2 \times 2^3 = 2^{2+3}$

This is the first law of indices, which demonstrates that **when multiplying two or more numbers having the same base, the indices are added.**

(2) $\dfrac{2^5}{2^3} = \dfrac{2 \times 2 \times 2 \times 2 \times 2}{2 \times 2 \times 2} = \dfrac{1 \times 1 \times 1 \times 2 \times 2}{1 \times 1 \times 1}$

$$= \dfrac{2 \times 2}{1} = 4 = 2^2$$

Hence, $\dfrac{2^5}{2^3} = 2^2$ or $\dfrac{2^5}{2^3} = 2^{5-3}$

This is the second law of indices, which demonstrates that **when dividing two numbers having the same base, the index in the denominator is subtracted from the index in the numerator.**

(3) $(3^5)^2 = 3^{5 \times 2} = 3^{10}$ and $(2^2)^3 = 2^{2 \times 3} = 2^6$

This is the third law of indices, which demonstrates that **when a number which is raised to a power is raised to a further power, the indices are multiplied.**

(4) $3^0 = 1$ and $17^0 = 1$

This is the fourth law of indices, which states that **when a number has an index of 0, its value is 1.**

(5) $3^{-4} = \dfrac{1}{3^4}$ and $\dfrac{1}{2^{-3}} = 2^3$

This is the fifth law of indices, which demonstrates that **a number raised to a negative power is the reciprocal of that number raised to a positive power.**

(6) $8^{\frac{2}{3}} = \sqrt[3]{8^2} = (2)^2 = 4$ and

$25^{\frac{1}{2}} = \sqrt[2]{25^1} = \sqrt{25^1} = \pm 5$

(Note that $\sqrt{} \equiv \sqrt[2]{}$)

This is the sixth law of indices, which demonstrates that **when a number is raised to a fractional power the denominator of the fraction is the root of the number and the numerator is the power.**

Here are some worked examples using the laws of indices.

Problem 6.   Evaluate in index form $5^3 \times 5 \times 5^2$

$5^3 \times 5 \times 5^2 = 5^3 \times 5^1 \times 5^2$ (Note that 5 means 5^1)

$$= 5^{3+1+2}$$ from law (1)

$$= \mathbf{5^6}$$

Problem 7.   Evaluate $\dfrac{3^5}{3^4}$

$$\dfrac{3^5}{3^4} = 3^{5-4}$$ from law (2)

$$= 3^1$$

$$= \mathbf{3}$$

Problem 8.   Evaluate $\dfrac{2^4}{2^4}$

$$\dfrac{2^4}{2^4} = 2^{4-4}$$ from law (2)

$$= 2^0$$

But $\dfrac{2^4}{2^4} = \dfrac{2 \times 2 \times 2 \times 2}{2 \times 2 \times 2 \times 2} = \dfrac{16}{16} = 1$

Hence, $\mathbf{2^0 = 1}$ from law (4)

Any number raised to the power of zero equals 1. For example, $6^0 = 1$, $128^0 = 1$, $13742^0 = 1$, and so on.

Problem 9. Evaluate $\dfrac{3 \times 3^2}{3^4}$

$$\dfrac{3 \times 3^2}{3^4} = \dfrac{3^1 \times 3^2}{3^4} = \dfrac{3^{1+2}}{3^4} = \dfrac{3^3}{3^4} = 3^{3-4} = 3^{-1}$$

from laws (1) and (2)

But $\dfrac{3^3}{3^4} = \dfrac{3 \times 3 \times 3}{3 \times 3 \times 3 \times 3} = \dfrac{1 \times 1 \times 1}{1 \times 1 \times 1 \times 3}$

(by cancelling)

$$= \dfrac{1}{3}$$

Hence, $\qquad \dfrac{3 \times 3^2}{3^4} = 3^{-1} = \dfrac{1}{3} \qquad$ from law (5)

Similarly, $\quad 2^{-1} = \dfrac{1}{2}, 2^{-5} = \dfrac{1}{2^5}, \dfrac{1}{5^4} = 5^{-4}$, and so on.

Problem 10. Evaluate $\dfrac{10^3 \times 10^2}{10^8}$

$$\dfrac{10^3 \times 10^2}{10^8} = \dfrac{10^{3+2}}{10^8} = \dfrac{10^5}{10^8} \qquad \text{from law (1)}$$

$$= 10^{5-8} = 10^{-3} \qquad \text{from law (2)}$$

$$= \dfrac{1}{10^{+3}} = \dfrac{1}{1000} \qquad \text{from law (5)}$$

Hence, $\quad \dfrac{10^3 \times 10^2}{10^8} = 10^{-3} = \dfrac{1}{1000} = 0.001$

Understanding powers of ten is important, especially when dealing with prefixes in Chapter 8. For example,

$$10^2 = 100, 10^3 = 1000, 10^4 = 10\,000,$$

$$10^5 = 100\,000, 10^6 = 1\,000\,000$$

$$10^{-1} = \dfrac{1}{10} = 0.1, 10^{-2} = \dfrac{1}{10^2} = \dfrac{1}{100} = 0.01,$$

and so on.

Problem 11. Evaluate (a) $5^2 \times 5^3 \div 5^4$ (b) $(3 \times 3^5) \div (3^2 \times 3^3)$

From laws (1) and (2):

(a) $5^2 \times 5^3 \div 5^4 = \dfrac{5^2 \times 5^3}{5^4} = \dfrac{5^{(2+3)}}{5^4}$

$$= \dfrac{5^5}{5^4} = 5^{(5-4)} = 5^1 = 5$$

(b) $(3 \times 3^5) \div (3^2 \times 3^3) = \dfrac{3 \times 3^5}{3^2 \times 3^3} = \dfrac{3^{(1+5)}}{3^{(2+3)}}$

$$= \dfrac{3^6}{3^5} = 3^{6-5} = 3^1 = 3$$

Problem 12. Simplify (a) $(2^3)^4$ (b) $(3^2)^5$, expressing the answers in index form

From law (3):

(a) $(2^3)^4 = 2^{3 \times 4} = 2^{12}$

(b) $(3^2)^5 = 3^{2 \times 5} = 3^{10}$

Problem 13. Evaluate: $\dfrac{(10^2)^3}{10^4 \times 10^2}$

From laws (1) to (4):

$$\dfrac{(10^2)^3}{10^4 \times 10^2} = \dfrac{10^{(2 \times 3)}}{10^{(4+2)}} = \dfrac{10^6}{10^6} = 10^{6-6} = 10^0 = 1$$

Problem 14. Find the value of (a) $\dfrac{2^3 \times 2^4}{2^7 \times 2^5}$ (b) $\dfrac{(3^2)^3}{3 \times 3^9}$

From the laws of indices:

(a) $\dfrac{2^3 \times 2^4}{2^7 \times 2^5} = \dfrac{2^{(3+4)}}{2^{(7+5)}} = \dfrac{2^7}{2^{12}} = 2^{7-12}$

$$= 2^{-5} = \dfrac{1}{2^5} = \dfrac{1}{32}$$

(b) $\dfrac{(3^2)^3}{3 \times 3^9} = \dfrac{3^{2 \times 3}}{3^{1+9}} = \dfrac{3^6}{3^{10}} = 3^{6-10}$

$$= 3^{-4} = \dfrac{1}{3^4} = \dfrac{1}{81}$$

Problem 15. Evaluate (a) $4^{1/2}$ (b) $16^{3/4}$ (c) $27^{2/3}$ (d) $9^{-1/2}$

(a) $4^{1/2} = \sqrt{4} = \pm 2$

(b) $16^{3/4} = \sqrt[4]{16^3} = (2)^3 = 8$
(Note that it does not matter whether the 4th root of 16 is found first or whether 16 cubed is found first – the same answer will result.)

(c) $27^{2/3} = \sqrt[3]{27^2} = (3)^2 = 9$

(d) $9^{-1/2} = \dfrac{1}{9^{1/2}} = \dfrac{1}{\sqrt{9}} = \dfrac{1}{\pm 3} = \pm \dfrac{1}{3}$

Now try the following Practice Exercise

Practice Exercise 30 Laws of indices (answers on page 424)

Evaluate the following without the aid of a calculator.

1. $2^2 \times 2 \times 2^4$

2. $3^5 \times 3^3 \times 3$ in index form

3. $\dfrac{2^7}{2^3}$

4. $\dfrac{3^3}{3^5}$

5. 7^0

6. $\dfrac{2^3 \times 2 \times 2^6}{2^7}$

7. $\dfrac{10 \times 10^6}{10^5}$

8. $10^4 \div 10$

9. $\dfrac{10^3 \times 10^4}{10^9}$

10. $5^6 \times 5^2 \div 5^7$

11. $(7^2)^3$ in index form

12. $(3^3)^2$

13. $\dfrac{3^7 \times 3^4}{3^5}$ in index form

14. $\dfrac{(9 \times 3^2)^3}{(3 \times 27)^2}$ in index form

15. $\dfrac{(16 \times 4)^2}{(2 \times 8)^3}$

16. $\dfrac{5^{-2}}{5^{-4}}$

17. $\dfrac{3^2 \times 3^{-4}}{3^3}$

18. $\dfrac{7^2 \times 7^{-3}}{7 \times 7^{-4}}$

19. $\dfrac{2^3 \times 2^{-4} \times 2^5}{2 \times 2^{-2} \times 2^6}$

20. $\dfrac{5^{-7} \times 5^2}{5^{-8} \times 5^3}$

Here are some further worked examples using the laws of indices.

Problem 16. Evaluate $\dfrac{3^3 \times 5^7}{5^3 \times 3^4}$

The laws of indices only apply to terms **having the same base**. Grouping terms having the same base and then applying the laws of indices to each of the groups independently gives

$$\frac{3^3 \times 5^7}{5^3 \times 3^4} = \frac{3^3}{3^4} \times \frac{5^7}{5^3} = 3^{(3-4)} \times 5^{(7-3)}$$

$$= 3^{-1} \times 5^4 = \frac{5^4}{3^1} = \frac{625}{3} = \mathbf{208\frac{1}{3}}$$

Problem 17. Find the value of $\dfrac{2^3 \times 3^5 \times (7^2)^2}{7^4 \times 2^4 \times 3^3}$

$$\frac{2^3 \times 3^5 \times (7^2)^2}{7^4 \times 2^4 \times 3^3} = 2^{3-4} \times 3^{5-3} \times 7^{2 \times 2 - 4}$$

$$= 2^{-1} \times 3^2 \times 7^0$$

$$= \frac{1}{2} \times 3^2 \times 1 = \frac{9}{2} = \mathbf{4\frac{1}{2}}$$

Problem 18. Evaluate $\dfrac{4^{1.5} \times 8^{1/3}}{2^2 \times 32^{-2/5}}$

$$4^{1.5} = 4^{3/2} = \sqrt{4^3} = 2^3 = 8, \, 8^{1/3} = \sqrt[3]{8} = 2,$$

$$2^2 = 4, \, 32^{-2/5} = \frac{1}{32^{2/5}} = \frac{1}{\sqrt[5]{32^2}} = \frac{1}{2^2} = \frac{1}{4}$$

Hence, $\dfrac{4^{1.5} \times 8^{1/3}}{2^2 \times 32^{-2/5}} = \dfrac{8 \times 2}{4 \times \dfrac{1}{4}} = \dfrac{16}{1} = \mathbf{16}$

Alternatively,

$$\frac{4^{1.5} \times 8^{1/3}}{2^2 \times 32^{-2/5}} = \frac{[(2)^2]^{3/2} \times (2^3)^{1/3}}{2^2 \times (2^5)^{-2/5}}$$

$$= \frac{2^3 \times 2^1}{2^2 \times 2^{-2}} = 2^{3+1-2-(-2)} = 2^4 = \mathbf{16}$$

Problem 19. Evaluate $\dfrac{3^2 \times 5^5 + 3^3 \times 5^3}{3^4 \times 5^4}$

Dividing each term by the HCF (highest common factor) of the three terms, i.e. $3^2 \times 5^3$, gives

$$\frac{3^2 \times 5^5 + 3^3 \times 5^3}{3^4 \times 5^4} = \frac{\dfrac{3^2 \times 5^5}{3^2 \times 5^3} + \dfrac{3^3 \times 5^3}{3^2 \times 5^3}}{\dfrac{3^4 \times 5^4}{3^2 \times 5^3}}$$

$$= \frac{3^{(2-2)} \times 5^{(5-3)} + 3^{(3-2)} \times 5^0}{3^{(4-2)} \times 5^{(4-3)}}$$

$$= \frac{3^0 \times 5^2 + 3^1 \times 5^0}{3^2 \times 5^1}$$

$$= \frac{1 \times 25 + 3 \times 1}{9 \times 5} = \frac{\mathbf{28}}{\mathbf{45}}$$

Problem 20. Find the value of $\dfrac{3^2 \times 5^5}{3^4 \times 5^4 + 3^3 \times 5^3}$

To simplify the arithmetic, each term is divided by the HCF of all the terms, i.e. $3^2 \times 5^3$. Thus,

$$\frac{3^2 \times 5^5}{3^4 \times 5^4 + 3^3 \times 5^3} = \frac{\dfrac{3^2 \times 5^5}{3^2 \times 5^3}}{\dfrac{3^4 \times 5^4}{3^2 \times 5^3} + \dfrac{3^3 \times 5^3}{3^2 \times 5^3}}$$

$$= \frac{3^{(2-2)} \times 5^{(5-3)}}{3^{(4-2)} \times 5^{(4-3)} + 3^{(3-2)} \times 5^{(3-3)}}$$

$$= \frac{3^0 \times 5^2}{3^2 \times 5^1 + 3^1 \times 5^0}$$

$$= \frac{1 \times 5^2}{3^2 \times 5 + 3 \times 1} = \frac{25}{45 + 3} = \mathbf{\frac{25}{48}}$$

Problem 21. Simplify $\dfrac{7^{-3} \times 3^4}{3^{-2} \times 7^5 \times 5^{-2}}$ expressing the answer in index form with positive indices

Since $7^{-3} = \dfrac{1}{7^3}$, $\dfrac{1}{3^{-2}} = 3^2$ and $\dfrac{1}{5^{-2}} = 5^2$, then

$$\frac{7^{-3} \times 3^4}{3^{-2} \times 7^5 \times 5^{-2}} = \frac{3^4 \times 3^2 \times 5^2}{7^3 \times 7^5}$$

$$= \frac{3^{(4+2)} \times 5^2}{7^{(3+5)}} = \mathbf{\frac{3^6 \times 5^2}{7^8}}$$

Problem 22. Simplify $\dfrac{16^2 \times 9^{-2}}{4 \times 3^3 - 2^{-3} \times 8^2}$ expressing the answer in index form with positive indices

Expressing the numbers in terms of their lowest prime numbers gives

$$\frac{16^2 \times 9^{-2}}{4 \times 3^3 - 2^{-3} \times 8^2} = \frac{(2^4)^2 \times (3^2)^{-2}}{2^2 \times 3^3 - 2^{-3} \times (2^3)^2}$$

$$= \frac{2^8 \times 3^{-4}}{2^2 \times 3^3 - 2^{-3} \times 2^6}$$

$$= \frac{2^8 \times 3^{-4}}{2^2 \times 3^3 - 2^3}$$

Dividing each term by the HCF (i.e. 2^2) gives

$$\frac{2^8 \times 3^{-4}}{2^2 \times 3^3 - 2^3} = \frac{2^6 \times 3^{-4}}{3^3 - 2} = \mathbf{\frac{2^6}{3^4(3^3 - 2)}}$$

Problem 23. Simplify $\dfrac{\left(\frac{4}{3}\right)^3 \times \left(\frac{3}{5}\right)^{-2}}{\left(\frac{2}{5}\right)^{-3}}$ giving the answer with positive indices

Raising a fraction to a power means that both the numerator and the denominator of the fraction are raised to that power, i.e. $\left(\dfrac{4}{3}\right)^3 = \dfrac{4^3}{3^3}$

A fraction raised to a negative power has the same value as the inverse of the fraction raised to a positive power.

Thus, $\left(\dfrac{3}{5}\right)^{-2} = \dfrac{1}{\left(\frac{3}{5}\right)^2} = \dfrac{1}{\dfrac{3^2}{5^2}} = 1 \times \dfrac{5^2}{3^2} = \dfrac{5^2}{3^2}$

Similarly, $\left(\dfrac{2}{5}\right)^{-3} = \left(\dfrac{5}{2}\right)^3 = \dfrac{5^3}{2^3}$

Thus, $\dfrac{\left(\frac{4}{3}\right)^3 \times \left(\frac{3}{5}\right)^{-2}}{\left(\frac{2}{5}\right)^{-3}} = \dfrac{\dfrac{4^3}{3^3} \times \dfrac{5^2}{3^2}}{\dfrac{5^3}{2^3}}$

$$= \frac{4^3}{3^3} \times \frac{5^2}{3^2} \times \frac{2^3}{5^3} = \frac{(2^2)^3 \times 2^3}{3^{(3+2)} \times 5^{(3-2)}}$$

$$= \mathbf{\frac{2^9}{3^5 \times 5}}$$

Now try the following Practice Exercise

Practice Exercise 31   Further problems on indices (answers on page 424)

In Problems 1 to 4, simplify the expressions given, expressing the answers in index form and with positive indices.

1. $\dfrac{3^3 \times 5^2}{5^4 \times 3^4}$

2. $\dfrac{7^{-2} \times 3^{-2}}{3^5 \times 7^4 \times 7^{-3}}$

3. $\dfrac{4^2 \times 9^3}{8^3 \times 3^4}$

4. $\dfrac{8^{-2} \times 5^2 \times 3^{-4}}{25^2 \times 2^4 \times 9^{-2}}$

In Problems 5 to 15, evaluate the expressions given.

5. $\left(\dfrac{1}{3^2}\right)^{-1}$

6. $81^{0.25}$

7. $16^{-\frac{1}{4}}$

8. $\left(\dfrac{4}{9}\right)^{1/2}$

9. $\dfrac{9^2 \times 7^4}{3^4 \times 7^4 + 3^3 \times 7^2}$

10. $\dfrac{3^3 \times 5^2}{2^3 \times 3^2 - 8^2 \times 9}$

11. $\dfrac{3^3 \times 7^2 - 5^2 \times 7^3}{3^2 \times 5 \times 7^2}$

12. $\dfrac{(2^4)^2 - 3^{-2} \times 4^4}{2^3 \times 16^2}$

13. $\dfrac{\left(\dfrac{1}{2}\right)^3 - \left(\dfrac{2}{3}\right)^{-2}}{\left(\dfrac{3}{2}\right)^2}$

14. $\dfrac{\left(\dfrac{4}{3}\right)^4}{\left(\dfrac{2}{9}\right)^2}$

15. $\dfrac{(3^2)^{3/2} \times (8^{1/3})^2}{(3)^2 \times (4^3)^{1/2} \times (9)^{-1/2}}$

For fully worked solutions to each of the problems in Practice Exercises 29 to 31 in this chapter, go to the website:
www.routledge.com/cw/bird

Units, prefixes and engineering notation

Why it is important to understand: Units, prefixes and engineering notation

In engineering there are many different quantities to get used to, and hence many units to become familiar with. For example, force is measured in newtons, electric current is measured in amperes and pressure is measured in pascals. Sometimes the units of these quantities are either very large or very small and hence prefixes are used. For example, 1000 pascals may be written as 10^3 Pa which is written as 1 kPa in prefix form, the k being accepted as a symbol to represent 1000 or 10^3. Studying, or working, in an engineering discipline, you very quickly become familiar with the standard units of measurement, the prefixes used and engineering notation. An electronic calculator is extremely helpful with engineering notation.

At the end of this chapter, you should be able to:

- state the seven SI units
- understand derived units
- recognise common engineering units
- understand common prefixes used in engineering
- express decimal numbers in standard form
- use engineering notation and prefix form with engineering units

8.1 Introduction

Of considerable importance in engineering is a knowledge of units of engineering quantities, the prefixes used with units, and engineering notation.

We need to know, for example, that

$$80\,\text{kV} = 80 \times 10^3\,\text{V}, \text{ which means } 80\,000 \text{ volts}$$

and $25\,\text{mA} = 25 \times 10^{-3}\,\text{A}$,

which means 0.025 amperes

and $50\,\text{nF} = 50 \times 10^{-9}\,\text{F}$,

which means 0.000000050 farads

This is explained in this chapter.

8.2 SI units

The system of units used in engineering and science is the Système Internationale d'Unités (**International System of Units**), usually abbreviated to SI units, and is based on the metric system. This was introduced in 1960 and has now been adopted by the majority of countries as the official system of measurement.

The basic seven units used in the SI system are listed in Table 8.1 with their symbols.

There are, of course, many units other than these seven. These other units are called **derived units** and are

Table 8.1 Basic SI units

Quantity	Unit	Symbol
Length	metre	m (1 m = 100 cm = 1000 mm)
Mass	kilogram	kg (1 kg = 1000 g)
Time	second	s
Electric current	ampere	A
Thermodynamic temperature	kelvin	K (K = °C + 273)
Luminous intensity	candela	cd
Amount of substance	mole	mol

defined in terms of the standard units listed in the table. For example, speed is measured in metres per second, therefore using two of the standard units, i.e. length and time.

*Who was Newton? – **Sir Isaac Newton** PRS MP (25 December 1642–20 March 1727) was an English polymath. Newton showed that the motions of objects are governed by the same set of natural laws, by demonstrating the consistency between Kepler's laws of planetary motion and his theory of gravitation. To find out more go to www.routledge.com/cw/bird

Some derived units are given **special names**. For example, force = mass × acceleration has units of kilogram metre per second squared, which uses three of the base units, i.e. kilograms, metres and seconds. The unit of kg m/s^2 is given the special name of a **newton***. Table 8.2 contains a list of some quantities and their units that are common in engineering.

8.3 Common prefixes

SI units may be made larger or smaller by using prefixes which denote multiplication or division by a particular amount.

The most common multiples are listed in Table 8.3. A knowledge of indices is needed since all of the prefixes are powers of 10 with indices that are a multiple of 3.

Here are some examples of prefixes used with engineering units.

A **frequency of 15 GHz** means 15×10^9 Hz, which is 15 000 000 000 hertz*,

i.e. 15 gigahertz is written as 15 GHz and is equal to 15 thousand million hertz.

*Who was Hertz? – **Heinrich Rudolf Hertz** (22 February 1857–1 January 1894) was the first person to conclusively prove the existence of electromagnetic waves. The scientific unit of frequency – cycles per second – was named the '**hertz**' in his honour. To find out more go to www.routledge.com/cw/bird

Table 8.2 Some quantities and their units that are common in engineering

Quantity	Unit	Symbol
Length	metre	m
Area	square metre	m^2
Volume	cubic metre	m^3
Mass	kilogram	kg
Time	second	s
Electric current	ampere	A
Speed, velocity	metre per second	m/s
Acceleration	metre per second squared	m/s^2
Density	kilogram per cubic metre	kg/m^3
Temperature	kelvin or Celsius	K or °C
Angle	radian or degree	rad or °
Angular velocity	radian per second	rad/s
Frequency	hertz	Hz
Force	newton	N
Pressure	pascal	Pa
Energy, work	joule	J
Power	watt	W
Charge, quantity of electricity	coulomb	C
Electric potential	volt	V
Capacitance	farad	F
Electrical resistance	ohm	Ω
Inductance	henry	H
Moment of force	newton metre	N m

Table 8.3 Common SI multiples

Prefix	Name	Meaning	
G	giga	multiply by 10^9	i.e. $\times 1\,000\,000\,000$
M	mega	multiply by 10^6	i.e. $\times 1\,000\,000$
k	kilo	multiply by 10^3	i.e. $\times 1\,000$
m	milli	multiply by 10^{-3}	i.e. $\times \dfrac{1}{10^3} = \dfrac{1}{1000} = 0.001$
μ	micro	multiply by 10^{-6}	i.e. $\times \dfrac{1}{10^6} = \dfrac{1}{1\,000\,000} = 0.000001$
n	nano	multiply by 10^{-9}	i.e. $\times \dfrac{1}{10^9} = \dfrac{1}{1\,000\,000\,000} = 0.000\,000\,001$
p	pico	multiply by 10^{-12}	i.e. $\times \dfrac{1}{10^{12}} = \dfrac{1}{1\,000\,000\,000\,000} = 0.000\,000\,000\,001$

(Instead of writing $15\,000\,000\,000$ hertz, it is much neater, takes up less space and prevents errors caused by having so many zeros, to write the frequency as 15 GHz.)

A **voltage of 40 MV** means 40×10^6 V, which is $40\,000\,000$ volts,

i.e. 40 megavolts is written as 40 MV and is equal to 40 million volts.

An **inductance of 12 mH** means 12×10^{-3} H or $\dfrac{12}{10^3}$ H or $\dfrac{12}{1000}$ H, which is 0.012 H,

i.e. 12 millihenrys is written as 12 mH and is equal to 12 thousandths of a henry*.

A **time of 150 ns** means 150×10^{-9} s or $\dfrac{150}{10^9}$ s, which is $0.000\,000\,150$ s,

i.e. 150 nanoseconds is written as 150 ns and is equal to 150 thousand millionths of a second.

A **force of 20 kN** means 20×10^3 N, which is $20\,000$ newtons,

i.e. 20 kilonewtons is written as 20 kN and is equal to 20 thousand newtons.

A **charge of 30 μC** means 30×10^{-6} C or $\dfrac{30}{10^6}$ C, which is $0.000\,030$ C,

i.e. 30 microcoulombs is written as 30 μC and is equal to 30 millionths of a coulomb.

A **capacitance of 45 pF** means 45×10^{-12} F or $\dfrac{45}{10^{12}}$ F, which is $0.000\,000\,000\,045$ F,

*Who was Henry? – **Joseph Henry** (17 December 1797–13 May 1878) was an American scientist who discovered the electromagnetic phenomenon of self-inductance. He also discovered mutual inductance independently of Michael Faraday, though Faraday was the first to publish his results. Henry was the inventor of a precursor to the electric doorbell and electric relay. The SI unit of inductance, the **henry**, is named in his honour. To find out more go to www.routledge.com/cw/bird

i.e. 45 picofarads is written as 45 pF and is equal to 45 million millionths of farad (named after Michael Faraday*).

In engineering it is important to understand what such quantities as 15 GHz, 40 MV, 12 mH, 150 ns, 20 kN, 30 µC and 45 pF mean.

Now try the following Practice Exercise

Practice Exercise 32   SI units and common prefixes (answers on page 425)

1. State the SI unit of volume.

2. State the SI unit of capacitance.

3. State the SI unit of area.

4. State the SI unit of velocity.

5. State the SI unit of density.

6. State the SI unit of energy.

*Who was Faraday? – **Michael Faraday**, FRS (22 September 1791–25 August 1867) was an English scientist who contributed to the fields of electromagnetism and electrochemistry. His main discoveries include those of electromagnetic induction, diamagnetism and electrolysis. To find out more go to www.routledge.com/cw/bird

7. State the SI unit of charge.

8. State the SI unit of power.

9. State the SI unit of angle.

10. State the SI unit of electric potential.

11. State which quantity has the unit kg.

12. State which quantity has the unit symbol Ω.

13. State which quantity has the unit Hz.

14. State which quantity has the unit m/s^2.

15. State which quantity has the unit symbol A.

16. State which quantity has the unit symbol H.

17. State which quantity has the unit symbol m.

18. State which quantity has the unit symbol K.

19. State which quantity has the unit Pa.

20. State which quantity has the unit rad/s.

21. What does the prefix G mean?

22. What is the symbol and meaning of the prefix milli?

23. What does the prefix p mean?

24. What is the symbol and meaning of the prefix mega?

8.4   Standard form

A number written with one digit to the left of the decimal point and multiplied by 10 raised to some power is said to be written in **standard form**.

For example, $43\,645 = 4.3645 \times 10^4$

in standard form

and $0.0534 = 5.34 \times 10^{-2}$

in standard form

Problem 1.   Express in standard form (a) 38.71 (b) 3746 (c) 0.0124

For a number to be in standard form, it is expressed with only one digit to the left of the decimal point. Thus,

(a)   38.71 must be divided by 10 to achieve one digit to the left of the decimal point and it must also be multiplied by 10 to maintain the equality, i.e.

$$38.71 = \frac{38.71}{10} \times 10 = \mathbf{3.871 \times 10} \text{ in standard form}$$

(b) $3746 = \frac{3746}{1000} \times 1000 = \mathbf{3.746 \times 10^3}$ in standard form.

(c) $0.0124 = 0.0124 \times \frac{100}{100} = \frac{1.24}{100} = \mathbf{1.24 \times 10^{-2}}$ in standard form.

Problem 2. Express the following numbers, which are in standard form, as decimal numbers:

(a) 1.725×10^{-2} (b) 5.491×10^4 (c) 9.84×10^0

(a) $1.725 \times 10^{-2} = \frac{1.725}{100} = \mathbf{0.01725}$ (i.e. move the decimal point 2 places to the left).

(b) $5.491 \times 10^4 = 5.491 \times 10000 = \mathbf{54910}$ (i.e. move the decimal point 4 places to the right).

(c) $9.84 \times 10^0 = 9.84 \times 1 = \mathbf{9.84}$ (since $10^0 = 1$).

Problem 3. Express in standard form, correct to 3 significant figures, (a) $\frac{3}{8}$ (b) $19\frac{2}{3}$ (c) $741\frac{9}{16}$

(a) $\frac{3}{8} = 0.375$, and expressing it in standard form gives

$$0.375 = \mathbf{3.75 \times 10^{-1}}$$

(b) $19\frac{2}{3} = 19.\dot{6} = \mathbf{1.97 \times 10}$ in standard form, correct to 3 significant figures.

(c) $741\frac{9}{16} = 741.5625 = \mathbf{7.42 \times 10^2}$ in standard form, correct to 3 significant figures.

Problem 4. Express the following numbers, given in standard form, as fractions or mixed numbers, (a) 2.5×10^{-1} (b) 6.25×10^{-2} (c) 1.354×10^2

(a) $2.5 \times 10^{-1} = \frac{2.5}{10} = \frac{25}{100} = \mathbf{\frac{1}{4}}$

(b) $6.25 \times 10^{-2} = \frac{6.25}{100} = \frac{625}{10000} = \mathbf{\frac{1}{16}}$

(c) $1.354 \times 10^2 = 135.4 = 135\frac{4}{10} = \mathbf{135\frac{2}{5}}$

Problem 5. Evaluate (a) $(3.75 \times 10^3)(6 \times 10^4)$ (b) $\frac{3.5 \times 10^5}{7 \times 10^2}$, expressing the answers in standard form

(a) $(3.75 \times 10^3)(6 \times 10^4) = (3.75 \times 6)(10^{3+4})$
$$= 22.50 \times 10^7$$
$$= \mathbf{2.25 \times 10^8}$$

(b) $\frac{3.5 \times 10^5}{7 \times 10^2} = \frac{3.5}{7} \times 10^{5-2} = 0.5 \times 10^3 = \mathbf{5 \times 10^2}$

Now try the following Practice Exercise

Practice Exercise 33   Standard form (answers on page 425)

In Problems 1 to 5, express in standard form.

1. (a) 73.9 (b) 28.4 (c) 197.62

2. (a) 2748 (b) 33170 (c) 274218

3. (a) 0.2401 (b) 0.0174 (c) 0.00923

4. (a) 1702.3 (b) 10.04 (c) 0.0109

5. (a) $\frac{1}{2}$ (b) $11\frac{7}{8}$
   (c) $\frac{1}{32}$ (d) $130\frac{3}{5}$

In Problems 6 and 7, express the numbers given as integers or decimal fractions.

6. (a) 1.01×10^3 (b) 9.327×10^2
   (c) 5.41×10^4 (d) 7×10^0

7. (a) 3.89×10^{-2} (b) 6.741×10^{-1}
   (c) 8×10^{-3}

In Problems 8 and 9, evaluate the given expressions, stating the answers in standard form.

8. (a) $(4.5 \times 10^{-2})(3 \times 10^3)$
   (b) $2 \times (5.5 \times 10^4)$

9. (a) $\frac{6 \times 10^{-3}}{3 \times 10^{-5}}$
   (b) $\frac{(2.4 \times 10^3)(3 \times 10^{-2})}{(4.8 \times 10^4)}$

10. Write the following statements in standard form.
    (a) The density of aluminium is 2710 $kg\,m^{-3}$.

(b)   Poisson's ratio for gold is 0.44

(c)   The impedance of free space is 376.73 Ω.

(d)   The electron rest energy is 0.511 MeV.

(e)   Proton charge–mass ratio is 95789700 $C\,kg^{-1}$.

(f)   The normal volume of a perfect gas is 0.02241 $m^3\,mol^{-1}$.

8.5   Engineering notation

In engineering, standard form is not as important as engineering notation. **Engineering notation** is similar to standard form except that the power of 10 **is always a multiple of 3**.

For example, $43645 = 43.645 \times 10^3$

in engineering notation

and $0.0534 = 53.4 \times 10^{-3}$

in engineering notation

From the list of engineering prefixes on page 62 it is apparent that all prefixes involve powers of 10 that are multiples of 3.

For example, a force of 43645 N can rewritten as 43.645×10^3 N and from the list of prefixes can then be expressed as 43.645 kN.

Thus,   **43645 N ≡ 43.645 kN**

To help further, on your calculator is an 'ENG' button. Enter the number 43645 into your calculator and then press =. Now press the ENG button and the answer is 43.645×10^3. We then have to appreciate that 10^3 is the prefix 'kilo', giving **43645 N ≡ 43.645 kN**.

In another example, let a current be 0.0745 A. Enter 0.0745 into your calculator. Press =. Now press ENG and the answer is 74.5×10^{-3}. We then have to appreciate that 10^{-3} is the prefix 'milli', giving **0.0745 A ≡ 74.5 mA**.

Problem 6.   Express the following in engineering notation and in prefix form:

(a) 300 000 W   (b) 0.000068 H

(a)   Enter 300 000 into the calculator. Press =

Now press ENG and the answer is 300×10^3

From the table of prefixes on page 62, 10^3 corresponds to kilo.

Hence, 300 000 W = **300×10^3 W** in engineering notation

= **300 kW** in prefix form.

(b)   Enter 0.000068 into the calculator. Press =

Now press ENG and the answer is 68×10^{-6}

From the table of prefixes on page 62, 10^{-6} corresponds to micro.

Hence, 0.000068 H = **68×10^{-6} H** in engineering notation

= **68 μH** in prefix form.

Problem 7.   Express the following in engineering notation and in prefix form:

(a) $42 \times 10^5\,\Omega$ (b) 4.7×10^{-10} F

(a)   Enter 42×10^5 into the calculator. Press =

Now press ENG and the answer is 4.2×10^6

From the table of prefixes on page 62, 10^6 corresponds to mega.

Hence, $42 \times 10^5\,\Omega$ = **$4.2 \times 10^6\,\Omega$** in engineering notation

= **4.2 MΩ** in prefix form.

(b)   Enter $47 \div 10^{10} = \dfrac{47}{10\,000\,000\,000}$ into the calculator. Press =

Now press ENG and the answer is 4.7×10^{-9}

From the table of prefixes on page 62, 10^{-9} corresponds to nano.

Hence, $47 \div 10^{10}$ F = **4.7×10^{-9} F** in engineering notation

= **4.7 nF** in prefix form.

Problem 8.   Rewrite (a) 0.056 mA in μA (b) 16 700 kHz as MHz

(a)   Enter $0.056 \div 1000$ into the calculator (since milli means $\div 1000$). Press =

Now press ENG and the answer is 56×10^{-6}

From the table of prefixes on page 62, 10^{-6} corresponds to micro.

Hence, $0.056\,\text{mA} = \dfrac{0.056}{1000}\,\text{A} = 56 \times 10^{-6}\,\text{A}$

$\qquad\qquad\qquad = \mathbf{56\,\mu A}.$

(b) Enter $16\,700 \times 1000$ into the calculator (since kilo means $\times 1000$). Press $=$

Now press ENG and the answer is 16.7×10^6

From the table of prefixes on page 62, 10^6 corresponds to mega.

Hence, $16\,700\,\text{kHz} = 16\,700 \times 1000\,\text{Hz}$

$\qquad\qquad\qquad = 16.7 \times 10^6\,\text{Hz}$

$\qquad\qquad\qquad = \mathbf{16.7\,MHz}$

Problem 9. Rewrite (a) 63×10^4 V in kV (b) $3100\,\text{pF}$ in nF

(a) Enter 63×10^4 into the calculator. Press $=$

Now press ENG and the answer is 630×10^3

From the table of prefixes on page 62, 10^3 corresponds to kilo.

Hence, $63 \times 10^4\,\text{V} = 630 \times 10^3\,\text{V} = \mathbf{630\,kV}.$

(b) Enter 3100×10^{-12} into the calculator. Press $=$

Now press ENG and the answer is 3.1×10^{-9}

From the table of prefixes on page 62, 10^{-9} corresponds to nano.

Hence, $3100\,\text{pF} = 31 \times 10^{-12}\,\text{F} = 3.1 \times 10^{-9}\,\text{F}$

$\qquad\qquad\qquad = \mathbf{3.1\,nF}$

Problem 10. Rewrite (a) $14\,700\,\text{mm}$ in metres (b) $276\,\text{cm}$ in metres (c) $3.375\,\text{kg}$ in grams

(a) $1\,\text{m} = 1000\,\text{mm}$, hence
$1\,\text{mm} = \dfrac{1}{1000} = \dfrac{1}{10^3} = 10^{-3}\,\text{m}.$

Hence, $14\,700\,\text{mm} = 14\,700 \times 10^{-3}\,\text{m} = \mathbf{14.7\,m}.$

(b) $1\,\text{m} = 100\,\text{cm}$, hence $1\,\text{cm} = \dfrac{1}{100} = \dfrac{1}{10^2} = 10^{-2}\,\text{m}.$

Hence, $276\,\text{cm} = 276 \times 10^{-2}\,\text{m} = \mathbf{2.76\,m}.$

(c) $1\,\text{kg} = 1000\,\text{g} = 10^3\,\text{g}$

Hence, $3.375\,\text{kg} = 3.375 \times 10^3\,\text{g} = \mathbf{3375\,g}.$

Now try the following Practice Exercise

Practice Exercise 34　Engineering notation (answers on page 425)

In Problems 1 to 12, express in engineering notation in prefix form.

1. $60\,000\,\text{Pa}$
2. $0.00015\,\text{W}$
3. $5 \times 10^7\,\text{V}$
4. $5.5 \times 10^{-8}\,\text{F}$
5. $100\,000\,\text{W}$
6. $0.00054\,\text{A}$
7. $15 \times 10^5\,\Omega$
8. $225 \times 10^{-4}\,\text{V}$
9. $35\,000\,000\,000\,\text{Hz}$
10. $1.5 \times 10^{-11}\,\text{F}$
11. $0.000017\,\text{A}$
12. $46\,200\,\Omega$
13. Rewrite $0.003\,\text{mA}$ in μA
14. Rewrite $2025\,\text{kHz}$ as MHz
15. Rewrite $5 \times 10^4\,\text{N}$ in kN
16. Rewrite $300\,\text{pF}$ in nF
17. Rewrite $6250\,\text{cm}$ in metres
18. Rewrite $34.6\,\text{g}$ in kg

In Problems 19 and 20, use a calculator to evaluate in engineering notation.

19. $4.5 \times 10^{-7} \times 3 \times 10^4$
20. $\dfrac{\left(1.6 \times 10^{-5}\right)\left(25 \times 10^3\right)}{\left(100 \times 10^{-6}\right)}$
21. The distance from Earth to the moon is around $3.8 \times 10^8\,\text{m}$. State the distance in kilometres.
22. The radius of a hydrogen atom is $0.53 \times 10^{-10}\,\text{m}$. State the radius in nanometres.
23. The tensile stress acting on a rod is $5600000\,\text{Pa}$. Write this value in engineering notation.
24. The expansion of a rod is $0.0043\,\text{m}$. Write this value in engineering notation.

For fully worked solutions to each of the problems in Practice Exercises 32 to 34 in this chapter, go to the website:
www.routledge.com/cw/bird

This assignment covers the material contained in Chapters 6–8. *The marks available are shown in brackets at the end of each question.*

1. In a box of 1500 nails, 125 are defective. Express the non-defective nails as a ratio of the defective ones, in its simplest form. (3)

2. Prize money in a lottery totals £4500 and is shared among three winners in the ratio $5 : 3 : 1$. How much does the first prize winner receive? (3)

3. A simple machine has an effort : load ratio of $3 : 41$. Determine the effort, in newtons, to lift a load of $6.15\,\text{kN}$. (3)

4. If 15 cans of lager weigh $7.8\,\text{kg}$, what will 24 cans weigh? (3)

5. Hooke's law states that stress is directly proportional to strain within the elastic limit of a material. When for brass the stress is $21\,\text{MPa}$, the strain is 250×10^{-6}. Determine the stress when the strain is 350×10^{-6}. (3)

6. If $12\,\text{inches} = 30.48\,\text{cm}$, find the number of millimetres in 17 inches. (3)

7. If x is inversely proportional to y and $x = 12$ when $y = 0.4$, determine
   (a) the value of x when y is 3.
   (b) the value of y when $x = 2$. (3)

8. Evaluate
   (a) $3 \times 2^3 \times 2^2$
   (b) $49^{\frac{1}{2}}$ (4)

9. Evaluate $\dfrac{3^2 \times \sqrt{36} \times 2^2}{3 \times 81^{\frac{1}{2}}}$ taking positive square roots only. (3)

10. Evaluate $6^4 \times 6 \times 6^2$ in index form. (3)

11. Evaluate
    (a) $\dfrac{2^7}{2^2}$ (b) $\dfrac{10^4 \times 10 \times 10^5}{10^6 \times 10^2}$ (4)

12. Evaluate
    (a) $\dfrac{2^3 \times 2 \times 2^2}{2^4}$
    (b) $\dfrac{\left(2^3 \times 16\right)^2}{(8 \times 2)^3}$
    (c) $\left(\dfrac{1}{4^2}\right)^{-1}$ (7)

13. Evaluate
    (a) $(27)^{-\frac{1}{3}}$ (b) $\dfrac{\left(\frac{3}{2}\right)^{-2} - \frac{2}{9}}{\left(\frac{2}{3}\right)^{2}}$ (5)

14. State the SI unit of (a) capacitance (b) electrical potential (c) work (3)

15. State the quantity that has an SI unit of (a) kilograms (b) henrys (c) hertz (d) m^3 (4)

16. Express the following in engineering notation in prefix form.
    (a) $250\,000\,\text{J}$ (b) $0.05\,\text{H}$
    (c) $2 \times 10^8\,\text{W}$ (d) $750 \times 10^{-8}\,\text{F}$ (4)

17. Rewrite (a) $0.0067\,\text{mA}$ in μA (b) $40 \times 10^4\,\text{kV}$ as MV (2)

Chapter 9

Basic algebra

Why it is important to understand: Basic algebra

Algebra is one of the most fundamental tools for engineers because it allows them to determine the value of something (length, material constant, temperature, mass, and so on) given values that they do know (possibly other length, material properties, mass). Although the types of problems that mechanical, chemical, civil, environmental or electrical engineers deal with vary, all engineers use algebra to solve problems. An example where algebra is frequently used is in simple electrical circuits, where the resistance is proportional to voltage. Using Ohm's law, or *V = IR*, an engineer simply multiplies the current in a circuit by the resistance to determine the voltage across the circuit. Engineers and scientists use algebra in many ways, and so frequently that they don't even stop the think about it. Depending on what type of engineer you choose to be, you will use varying degrees of algebra, but in all instances algebra lays the foundation for the mathematics you will need to become an engineer.

At the end of this chapter, you should be able to:

- understand basic operations in algebra
- add, subtract, multiply and divide using letters instead of numbers
- state the laws of indices in letters instead of numbers
- simplify algebraic expressions using the laws of indices

9.1 Introduction

We are already familiar with evaluating formulae using a calculator from Chapter 4.

For example, if the length of a football pitch is L and its width is b, then the formula for the area A is given by

$$A = L \times b$$

This is an **algebraic equation**.
If $L = 120$ m and $b = 60$ m, then the area $A = 120 \times 60 = 7200$ m^2.

The total resistance, R_T, of resistors R_1, R_2 and R_3 connected in series is given by

$$R_T = R_1 + R_2 + R_3$$

This is an **algebraic equation**.
If $R_1 = 6.3$ kΩ, $R_2 = 2.4$ kΩ and $R_3 = 8.5$ kΩ, then

$$R_T = 6.3 + 2.4 + 8.5 = 17.2 \, \text{k}\Omega$$

The temperature in Fahrenheit, F, is given by

$$F = \frac{9}{5}C + 32$$

where C is the temperature in Celsius*. This is an **algebraic equation**.

If $C = 100°C$, then $F = \dfrac{9}{5} \times 100 + 32$
$$= 180 + 32 = 212°F.$$

If you can cope with evaluating formulae then you will be able to cope with algebra.

9.2   Basic operations

Algebra merely uses letters to represent numbers.
If, say, a, b, c and d represent any four numbers then in algebra:

(a) $a + a + a + a = 4a$. For example, if $a = 2$, then
$2 + 2 + 2 + 2 = 4 \times 2 = 8$

(b) $5b$ means $5 \times b$. For example, if $b = 4$, then
$5b = 5 \times 4 = 20$

(c) $2a + 3b + a - 2b = 2a + a + 3b - 2b = 3a + b$

Only similar terms can be combined in algebra. The $2a$ and the $+a$ can be combined to give $3a$

*Who was Celsius? – **Anders Celsius** (27 November 1701–25 April 1744) was the Swedish astronomer that proposed the Celsius temperature scale in 1742 which takes his name. To find out more go to www.routledge.com/cw/bird

and the $3b$ and $-2b$ can be combined to give $1b$, which is written as b.

In addition, with terms separated by $+$ and $-$ signs, the order in which they are written does not matter. In this example, $2a + 3b + a - 2b$ is the same as $2a + a + 3b - 2b$, which is the same as $3b + a + 2a - 2b$, and so on. (Note that the first term, i.e. $2a$, means $+2a$)

(d) $4abcd = 4 \times a \times b \times c \times d$

For example, if $a = 3, b = -2, c = 1$ and $d = -5$, then $4abcd = 4 \times 3 \times -2 \times 1 \times -5 = 120$. (Note that $- \times - = +$)

(e) $(a)(c)(d)$ means $a \times c \times d$

Brackets are often used instead of multiplication signs.
For example, $(2)(5)(3)$ means $2 \times 5 \times 3 = 30$

(f) $ab = ba$
If $a = 2$ and $b = 3$ then 2×3 is exactly the same as 3×2, i.e. 6

(g) $b^2 = b \times b$. For example, if $b = 3$, then $3^2 = 3 \times 3 = 9$

(h) $a^3 = a \times a \times a$ For example, if $a = 2$, then $2^3 = 2 \times 2 \times 2 = 8$

Here are some worked examples to help get a feel for basic operations in this introduction to algebra.

9.2.1   Addition and subtraction

Problem 1.   Find the sum of $4x, 3x, -2x$ and $-x$

$4x + 3x + -2x + -x = 4x + 3x - 2x - x$
$$\text{(Note that } + \times - = -)$$
$$= 4x$$

Problem 2.   Find the sum of $5x, 3y, z, -3x, -4y$ and $6z$

$5x + 3y + z + -3x + -4y + 6z$
$$= 5x + 3y + z - 3x - 4y + 6z$$
$$= 5x - 3x + 3y - 4y + z + 6z$$
$$= 2x - y + 7z$$

Note that the order can be changed when terms are separated by $+$ and $-$ signs. Only similar terms can be combined.

Problem 3. Simplify $4x^2 - x - 2y + 5x + 3y$

$$4x^2 - x - 2y + 5x + 3y = 4x^2 + 5x - x + 3y - 2y$$
$$= \mathbf{4x^2 + 4x + y}$$

Problem 4. Simplify $3xy - 7x + 4xy + 2x$

$$3xy - 7x + 4xy + 2x = 3xy + 4xy + 2x - 7x$$
$$= \mathbf{7xy - 5x}$$

Now try the following Practice Exercise

Practice Exercise 35   Addition and subtraction in algebra (answers on page 425)

1. Find the sum of $4a, -2a, 3a$ and $-8a$

2. Find the sum of $2a, 5b, -3c, -a, -3b$ and $7c$

3. Simplify $2x - 3x^2 - 7y + x + 4y - 2y^2$

4. Simplify $5ab - 4a + ab + a$

5. Simplify $2x - 3y + 5z - x - 2y + 3z + 5x$

6. Simplify $3 + x + 5x - 2 - 4x$

7. Add $x - 2y + 3$ to $3x + 4y - 1$

8. Subtract $a - 2b$ from $4a + 3b$

9. From $a + b - 2c$ take $3a + 2b - 4c$

10. From $x^2 + xy - y^2$ take $xy - 2x^2$

9.2.2 Multiplication and division

Problem 5. Simplify $bc \times abc$

$$bc \times abc = a \times b \times b \times c \times c$$
$$= a \times b^2 \times c^2$$
$$= \mathbf{ab^2c^2}$$

Problem 6. Simplify $-2p \times -3p$

$$- \times - = + \text{ hence}, -2p \times -3p = \mathbf{6p^2}$$

Problem 7. Simplify $ab \times b^2c \times a$

$$ab \times b^2c \times a = a \times a \times b \times b \times b \times c$$
$$= a^2 \times b^3 \times c$$
$$= \mathbf{a^2b^3c}$$

Problem 8. Evaluate $3ab + 4bc - abc$ when $a = 3, b = 2$ and $c = 5$

$$3ab + 4bc - abc = 3 \times a \times b + 4 \times b \times c - a \times b \times c$$
$$= 3 \times 3 \times 2 + 4 \times 2 \times 5 - 3 \times 2 \times 5$$
$$= 18 + 40 - 30$$
$$= \mathbf{28}$$

Problem 9. Determine the value of $5pq^2r^3$, given that $p = 2, q = \dfrac{2}{5}$ and $r = 2\dfrac{1}{2}$

$$5pq^2r^3 = 5 \times p \times q^2 \times r^3$$
$$= 5 \times 2 \times \left(\frac{2}{5}\right)^2 \times \left(2\frac{1}{2}\right)^3$$
$$= 5 \times 2 \times \left(\frac{2}{5}\right)^2 \times \left(\frac{5}{2}\right)^3 \qquad \text{since} \quad 2\frac{1}{2} = \frac{5}{2}$$
$$= \frac{5}{1} \times \frac{2}{1} \times \frac{2}{5} \times \frac{2}{5} \times \frac{5}{2} \times \frac{5}{2} \times \frac{5}{2}$$
$$= \frac{1}{1} \times \frac{1}{1} \times \frac{1}{1} \times \frac{1}{1} \times \frac{1}{1} \times \frac{5}{1} \times \frac{5}{1} \qquad \text{by cancelling}$$
$$= 5 \times 5$$
$$= \mathbf{25}$$

Problem 10. Multiply $2a + 3b$ by $a + b$

Each term in the first expression is multiplied by a, then each term in the first expression is multiplied by b and the two results are added. The usual layout is shown below.

$$2a + 3b$$
$$a + b$$

Multiplying by a gives $2a^2 + 3ab$
Multiplying by b gives $2ab + 3b^2$

Adding gives $2a^2 + 5ab + 3b^2$

Thus, $(2a + 3b)(a + b) = 2a^2 + 5ab + 3b^2$

Problem 11. Multiply $3x - 2y^2 + 4xy$ by $2x - 5y$

$$3x \quad - 2y^2 \quad + 4xy$$
$$2x \quad - 5y$$

Multiplying
by $2x \rightarrow$ $6x^2 - 4xy^2 \quad + 8x^2y$

Multiplying
by $-5y \rightarrow$ $-20xy^2 \qquad -15xy + 10y^3$

Adding gives $6x^2 - 24xy^2 + 8x^2y - 15xy + 10y^3$

Thus, $(3x - 2y^2 + 4xy)(2x - 5y)$
$$= 6x^2 - 24xy^2 + 8x^2y - 15xy + 10y^3$$

Problem 12. Simplify $2x \div 8xy$

$2x \div 8xy$ means $\dfrac{2x}{8xy}$

$$\frac{2x}{8xy} = \frac{2 \times x}{8 \times x \times y}$$

$$= \frac{1 \times 1}{4 \times 1 \times y} \qquad \text{by cancelling}$$

$$= \frac{1}{4y}$$

Problem 13. Simplify $\dfrac{9a^2bc}{3ac}$

$$\frac{9a^2bc}{3ac} = \frac{9 \times a \times a \times b \times c}{3 \times a \times c}$$

$$= 3 \times a \times b$$

$$= 3ab$$

Problem 14. Divide $2x^2 + x - 3$ by $x - 1$

(i) $2x^2 + x - 3$ is called the **dividend** and $x - 1$ the **divisor**. The usual layout is shown below with the

dividend and divisor both arranged in descending powers of the symbols.

$$\begin{array}{r} 2x + 3 \\ x-1{\overline{\smash{\big)}\,2x^2 + x - 3}} \\ \underline{2x^2 - 2x} \\ 3x - 3 \\ \underline{3x - 3} \\ \cdot\quad\cdot \end{array}$$

(ii) Dividing the first term of the dividend by the first term of the divisor, i.e. $\dfrac{2x^2}{x}$ gives $2x$, which is put above the first term of the dividend as shown.

(iii) The divisor is then multiplied by $2x$, i.e. $2x(x - 1) = 2x^2 - 2x$, which is placed under the dividend as shown. Subtracting gives $3x - 3$

(iv) The process is then repeated, i.e. the first term of the divisor, x, is divided into $3x$, giving $+3$, which is placed above the dividend as shown.

(v) Then $3(x - 1) = 3x - 3$, which is placed under the $3x - 3$. The remainder, on subtraction, is zero, which completes the process.

Thus, $(2x^2 + x - 3) \div (x - 1) = (2x + 3)$

(A check can be made on this answer by multiplying $(2x + 3)$ by $(x - 1)$, which equals $2x^2 + x - 3$)

Problem 15. Simplify $\dfrac{x^3 + y^3}{x + y}$

$$\begin{array}{r} \text{(i) (iv) (vii)} \\ x^2 - xy + y^2 \\ x+y{\overline{\smash{\big)}\,x^3 + 0 \quad +0 \quad +y^3}} \\ \underline{x^3 + x^2y} \\ -x^2y \qquad + y^3 \\ \underline{-x^2y - xy^2} \\ xy^2 + y^3 \\ \underline{xy^2 + y^3} \\ \cdot\quad\cdot \end{array}$$

(i) x into x^3 goes x^2. Put x^2 above x^3.

(ii) $x^2(x + y) = x^3 + x^2y$

(iii) Subtract.

(iv) x into $-x^2y$ goes $-xy$. Put $-xy$ above the dividend.

(v) $-xy(x+y) = -x^2y - xy^2$

(vi) Subtract.

(vii) x into xy^2 goes y^2. Put y^2 above the dividend.

(viii) $y^2(x+y) = xy^2 + y^3$

(ix) Subtract.

Thus, $\dfrac{x^3 + y^3}{x+y} = x^2 - xy + y^2$

The zeros shown in the dividend are not normally shown, but are included to clarify the subtraction process and to keep similar terms in their respective columns.

Problem 16. Divide $4a^3 - 6a^2b + 5b^3$ by $2a - b$

$$
\begin{array}{r}
2a^2 - 2ab - b^2 \\
2a - b\overline{\smash{)}\,4a^3 - 6a^2b \qquad\quad + 5b^3} \\
\underline{4a^3 - 2a^2b} \\
-4a^2b \qquad\quad + 5b^3 \\
\underline{-4a^2b + 2ab^2} \\
-2ab^2 + 5b^3 \\
\underline{-2ab^2 + b^3} \\
4b^3
\end{array}
$$

Thus, $\dfrac{4a^3 - 6a^2b + 5b^3}{2a - b} = 2a^2 - 2ab - b^2$, **remainder $4b^3$.**

Alternatively, the answer may be expressed as

$$\dfrac{4a^3 - 6a^2b + 5b^3}{2a - b} = 2a^2 - 2ab - b^2 + \dfrac{4b^3}{2a - b}$$

Now try the following Practice Exercise

Practice Exercise 36   Basic operations in algebra (answers on page 425)

1. Simplify $pq \times pq^2r$

2. Simplify $-4a \times -2a$

3. Simplify $3 \times -2q \times -q$

4. Evaluate $3pq - 5qr - pqr$ when $p = 3$, $q = -2$ and $r = 4$

5. Determine the value of $3x^2yz^3$, given that $x = 2, y = 1\frac{1}{2}$ and $z = \frac{2}{3}$

6. If $x = 5$ and $y = 6$, evaluate $\dfrac{23(x - y)}{y + xy + 2x}$

7. If $a = 4, b = 3, c = 5$ and $d = 6$, evaluate $\dfrac{3a + 2b}{3c - 2d}$

8. Simplify $2x \div 14xy$

9. Simplify $\dfrac{25x^2yz^3}{5xyz}$

10. Multiply $3a - b$ by $a + b$

11. Multiply $2a - 5b + c$ by $3a + b$

12. Simplify $3a \div 9ab$

13. Simplify $4a^2b \div 2a$

14. Divide $6x^2y$ by $2xy$

15. Divide $2x^2 + xy - y^2$ by $x + y$

16. Divide $3p^2 - pq - 2q^2$ by $p - q$

17. Simplify $(a + b)^2 + (a - b)^2$

9.3 Laws of indices

The laws of indices with numbers were covered in Chapter 7; the laws of indices in algebraic terms are as follows:

(1) $a^m \times a^n = a^{m+n}$

For example, $a^3 \times a^4 = a^{3+4} = a^7$

(2) $\dfrac{a^m}{a^n} = a^{m-n}$ For example, $\dfrac{c^5}{c^2} = c^{5-2} = c^3$

(3) $(a^m)^n = a^{mn}$ For example, $(d^2)^3 = d^{2\times3} = d^6$

(4) $a^{\frac{m}{n}} = \sqrt[n]{a^m}$ For example, $x^{\frac{4}{3}} = \sqrt[3]{x^4}$

(5) $a^{-n} = \dfrac{1}{a^n}$ For example, $3^{-2} = \dfrac{1}{3^2} = \dfrac{1}{9}$

(6) $a^0 = 1$ For example, $17^0 = 1$

Here are some worked examples to demonstrate these laws of indices.

Problem 17. Simplify $a^2b^3c \times ab^2c^5$

$$a^2b^3c \times ab^2c^5 = a^2 \times b^3 \times c \times a \times b^2 \times c^5$$
$$= a^2 \times b^3 \times c^1 \times a^1 \times b^2 \times c^5$$

Grouping together like terms gives

$$a^2 \times a^1 \times b^3 \times b^2 \times c^1 \times c^5$$

Using law (1) of indices gives

$$a^{2+1} \times b^{3+2} \times c^{1+5} = a^3 \times b^5 \times c^6$$

i.e. $$\mathbf{a^2b^3c \times ab^2c^5 = a^3\,b^5\,c^6}$$

Problem 18. Simplify $a^{\frac{1}{3}} b^{\frac{3}{2}} c^{-2} \times a^{\frac{1}{6}} b^{\frac{1}{2}} c$

Using law (1) of indices,

$$a^{\frac{1}{3}} b^{\frac{3}{2}} c^{-2} \times a^{\frac{1}{6}} b^{\frac{1}{2}} c = a^{\frac{1}{3}+\frac{1}{6}} \times b^{\frac{3}{2}+\frac{1}{2}} \times c^{-2+1}$$
$$= \boldsymbol{a^{\frac{1}{2}} b^2 c^{-1}}$$

Problem 19. Simplify $\dfrac{x^5y^2z}{x^2yz^3}$

$$\frac{x^5y^2z}{x^2yz^3} = \frac{x^5 \times y^2 \times z}{x^2 \times y \times z^3}$$
$$= \frac{x^5}{x^2} \times \frac{y^2}{y^1} \times \frac{z}{z^3}$$
$$= x^{5-2} \times y^{2-1} \times z^{1-3} \quad \text{by law (2) of indices}$$
$$= x^3 \times y^1 \times z^{-2}$$
$$= \boldsymbol{x^3\,y\,z^{-2}} \text{ or } \frac{\boldsymbol{x^3 y}}{\boldsymbol{z^2}}$$

Problem 20. Simplify $\dfrac{a^3b^2c^4}{abc^{-2}}$ and evaluate when $a=3, b=\dfrac{1}{4}$ and $c=2$

Using law (2) of indices,

$$\frac{a^3}{a} = a^{3-1} = a^2, \frac{b^2}{b} = b^{2-1} = b \text{ and}$$
$$\frac{c^4}{c^{-2}} = c^{4--2} = c^6$$

Thus, $\dfrac{a^3b^2c^4}{abc^{-2}} = \boldsymbol{a^2bc^6}$

When $a=3, b=\dfrac{1}{4}$ and $c=2$,

$$a^2bc^6 = (3)^2\left(\frac{1}{4}\right)(2)^6 = (9)\left(\frac{1}{4}\right)(64) = \mathbf{144}$$

Problem 21. Simplify $(p^3)^2(q^2)^4$

Using law (3) of indices gives

$$(p^3)^2(q^2)^4 = p^{3\times2} \times q^{2\times4}$$
$$= \boldsymbol{p^6q^8}$$

Problem 22. Simplify $\dfrac{(mn^2)^3}{(m^{1/2}n^{1/4})^4}$

The brackets indicate that each letter in the bracket must be raised to the power outside. Using law (3) of indices gives

$$\frac{(mn^2)^3}{(m^{1/2}n^{1/4})^4} = \frac{m^{1\times3}n^{2\times3}}{m^{(1/2)\times4}n^{(1/4)\times4}} = \frac{m^3n^6}{m^2n^1}$$

Using law (2) of indices gives

$$\frac{m^3n^6}{m^2n^1} = m^{3-2}n^{6-1} = \boldsymbol{mn^5}$$

Problem 23. Simplify $(a^3b)(a^{-4}b^{-2})$, expressing the answer with positive indices only

Using law (1) of indices gives $a^{3+-4}b^{1+-2} = a^{-1}b^{-1}$

Using law (5) of indices gives $a^{-1}b^{-1} = \dfrac{1}{a^{+1}b^{+1}} = \dfrac{\mathbf{1}}{\mathbf{ab}}$

Problem 24. Simplify $\dfrac{d^2e^2f^{1/2}}{(d^{3/2}ef^{5/2})^2}$ expressing the answer with positive indices only

Using law (3) of indices gives

$$\frac{d^2e^2f^{1/2}}{(d^{3/2}e\,f^{5/2})^2} = \frac{d^2e^2f^{1/2}}{d^3e^2f^5}$$

Using law (2) of indices gives

$$d^{2-3}e^{2-2}f^{\frac{1}{2}-5} = d^{-1}e^0f^{-\frac{9}{2}}$$

$$= d^{-1}f^{-\frac{9}{2}} \quad \text{since } e^0 = 1 \text{ from law}$$
$$\text{(6) of indices}$$

$$= \frac{1}{df^{9/2}} \quad \text{from law (5) of indices}$$

Now try the following Practice Exercise

Practice Exercise 37   Laws of indices
(answers on page 425)

In Problems 1 to 18, simplify the following, giving each answer as a power.

1. $z^2 \times z^6$

2. $a \times a^2 \times a^5$

3. $n^8 \times n^{-5}$

4. $b^4 \times b^7$

5. $b^2 \div b^5$

6. $c^5 \times c^3 \div c^4$

7. $\dfrac{m^5 \times m^6}{m^4 \times m^3}$

8. $\dfrac{(x^2)(x)}{x^6}$

9. $(x^3)^4$

10. $(y^2)^{-3}$

11. $(t \times t^3)^2$

12. $(c^{-7})^{-2}$

13. $\left(\dfrac{a^2}{a^5}\right)^3$

14. $\left(\dfrac{1}{b^3}\right)^4$

15. $\left(\dfrac{b^2}{b^7}\right)^{-2}$

16. $\dfrac{1}{(s^3)^3}$

17. $p^3qr^2 \times p^2q^5r \times pqr^2$

18. $\dfrac{x^3y^2z}{x^5yz^3}$

19. Simplify $(x^2y^3z)(x^3yz^2)$ and evaluate when $x = \dfrac{1}{2}, y = 2$ and $z = 3$

20. Simplify $\dfrac{a^5bc^3}{a^2b^3c^2}$ and evaluate when $a = \dfrac{3}{2}, b = \dfrac{1}{2}$ and $c = \dfrac{2}{3}$

Here are some further worked examples on the laws of indices.

Problem 25. Simplify $\dfrac{p^{1/2}q^2r^{2/3}}{p^{1/4}q^{1/2}r^{1/6}}$ and evaluate when $p = 16, q = 9$ and $r = 4$, taking positive roots only

Using law (2) of indices gives $p^{\frac{1}{2}-\frac{1}{4}}q^{2-\frac{1}{2}}r^{\frac{2}{3}-\frac{1}{6}}$

$$p^{\frac{1}{2}-\frac{1}{4}}q^{2-\frac{1}{2}}r^{\frac{2}{3}-\frac{1}{6}} = p^{\frac{1}{4}}q^{\frac{3}{2}}r^{\frac{1}{2}}$$

When $p = 16, q = 9$ and $r = 4$,

$$p^{\frac{1}{4}}q^{\frac{3}{2}}r^{\frac{1}{2}} = 16^{\frac{1}{4}}9^{\frac{3}{2}}4^{\frac{1}{2}}$$

$$= (\sqrt[4]{16})(\sqrt{9^3})(\sqrt{4}) \text{ from law (4) of indices}$$

$$= (2)(3^3)(2) = \mathbf{108}$$

Problem 26. Simplify $\dfrac{x^2y^3 + xy^2}{xy}$

Algebraic expressions of the form $\dfrac{a+b}{c}$ can be split into $\dfrac{a}{c} + \dfrac{b}{c}$. Thus,

$$\frac{x^2y^3 + xy^2}{xy} = \frac{x^2y^3}{xy} + \frac{xy^2}{xy} = x^{2-1}y^{3-1} + x^{1-1}y^{2-1}$$

$$= \boldsymbol{xy^2 + y}$$

$$(\text{since } x^0 = 1, \text{ from law (6) of indices}).$$

Problem 27. Simplify $\dfrac{x^2y}{xy^2 - xy}$

The highest common factor (HCF) of each of the three terms comprising the numerator and denominator is xy. Dividing each term by xy gives

$$\frac{x^2y}{xy^2 - xy} = \frac{\dfrac{x^2y}{xy}}{\dfrac{xy^2}{xy} - \dfrac{xy}{xy}} = \boldsymbol{\frac{x}{y-1}}$$

Problem 28. Simplify $\dfrac{a^2b}{ab^2 - a^{1/2}b^3}$

The HCF of each of the three terms is $a^{1/2}b$. Dividing each term by $a^{1/2}b$ gives

$$\frac{a^2b}{ab^2 - a^{1/2}b^3} = \frac{\dfrac{a^2b}{a^{1/2}b}}{\dfrac{ab^2}{a^{1/2}b} - \dfrac{a^{1/2}b^3}{a^{1/2}b}} = \boldsymbol{\frac{a^{3/2}}{a^{1/2}b - b^2}}$$

Problem 29. Simplify $(a^3\sqrt{b}\sqrt{c^5})(\sqrt{a}\sqrt[3]{b^2}c^3)$ and evaluate when $a = \dfrac{1}{4}, b = 6$ and $c = 1$

Using law (4) of indices, the expression can be written as

$$(a^3\sqrt{b}\sqrt{c^5})(\sqrt{a}\sqrt[3]{b^2}c^3) = \left(a^3 b^{\frac{1}{2}} c^{\frac{5}{2}}\right)\left(a^{\frac{1}{2}} b^{\frac{2}{3}} c^3\right)$$

Using law (1) of indices gives

$$\left(a^3 b^{\frac{1}{2}} c^{\frac{5}{2}}\right)\left(a^{\frac{1}{2}} b^{\frac{2}{3}} c^3\right) = a^{3+\frac{1}{2}} b^{\frac{1}{2}+\frac{2}{3}} c^{\frac{5}{2}+3}$$

$$= a^{\frac{7}{2}} b^{\frac{7}{6}} c^{\frac{11}{2}}$$

It is usual to express the answer in the same form as the question. Hence,

$$a^{\frac{7}{2}} b^{\frac{7}{6}} c^{\frac{11}{2}} = \sqrt{a^7}\sqrt[6]{b^7}\sqrt{c^{11}}$$

When $a = \dfrac{1}{4}, b = 64$ and $c = 1$,

$$\sqrt{a^7}\sqrt[6]{b^7}\sqrt{c^{11}} = \sqrt{\left(\frac{1}{4}\right)^7}\left(\sqrt[6]{64^7}\right)\left(\sqrt{1^{11}}\right)$$

$$= \left(\frac{1}{2}\right)^7 (2)^7(1) = \mathbf{1}$$

Problem 30. Simplify $\dfrac{(x^2 y^{1/2})(\sqrt{x}\sqrt[3]{y^2})}{(x^5 y^3)^{1/2}}$

Using laws (3) and (4) of indices gives

$$\frac{(x^2 y^{1/2})\left(\sqrt{x}\sqrt[3]{y^2}\right)}{(x^5 y^3)^{1/2}} = \frac{(x^2 y^{1/2})(x^{1/2} y^{2/3})}{x^{5/2} y^{3/2}}$$

Using laws (1) and (2) of indices gives

$$x^{2+\frac{1}{2}-\frac{5}{2}} y^{\frac{1}{2}+\frac{2}{3}-\frac{3}{2}} = x^0 y^{-\frac{1}{3}} = \mathbf{y^{-\frac{1}{3}}} \text{ or } \frac{\mathbf{1}}{\mathbf{y^{1/3}}} \text{ or } \frac{\mathbf{1}}{\sqrt[3]{\mathbf{y}}}$$

from laws (5) and (6) of indices.

Now try the following Practice Exercise

Practice Exercise 38   Laws of indices (answers on page 425)

1. Simplify $(a^{3/2}bc^{-3})(a^{1/2}b^{-1/2}c)$ and evaluate when $a = 3, b = 4$ and $c = 2$.

In Problems 2 to 5, simplify the given expressions.

2. $\dfrac{a^2 b + a^3 b}{a^2 b^2}$

3. $(a^2)^{1/2}(b^2)^3\left(c^{1/2}\right)^3$

4. $\dfrac{(abc)^2}{(a^2 b^{-1} c^{-3})^3}$

5. $\dfrac{p^3 q^2}{pq^2 - p^2 q}$

6. $(\sqrt{x}\sqrt{y^3}\sqrt[3]{z^2})(\sqrt{x}\sqrt{y^3}\sqrt{z^3})$

7. $(e^2 f^3)(e^{-3} f^{-5})$, expressing the answer with positive indices only.

8. $\dfrac{(a^3 b^{1/2} c^{-1/2})(ab)^{1/3}}{(\sqrt{a^3}\sqrt{b}\, c)}$

For fully worked solutions to each of the problems in Practice Exercises 35 to 38 in this chapter, go to the website:
www.routledge.com/cw/bird

Chapter 10

Further algebra

Why it is important to understand: **Further algebra**

Algebra is a form of mathematics that allows you to work with unknowns. If you do not know what a number is, arithmetic does not allow you to use it in calculations. Algebra has variables. Variables are labels for numbers and measurements you do not yet know. Algebra lets you use these variables in equations and formulae. A basic form of mathematics, algebra is nevertheless among the most commonly used forms of mathematics in the workforce. Although relatively simple, algebra possesses a powerful problem-solving tool used in many fields of engineering. For example, in designing a rocket to go to the moon, an engineer must use algebra to solve for flight trajectory, how long to burn each thruster and at what intensity, and at what angle to lift off. An engineer uses mathematics all the time – and in particular, algebra. Becoming familiar with algebra will make all engineering mathematics studies so much easier.

At the end of this chapter, you should be able to:

- use brackets with basic operations in algebra
- understand factorisation
- factorise simple algebraic expressions
- use the laws of precedence to simplify algebraic expressions

10.1 Introduction

In this chapter, the use of brackets and factorisation with algebra is explained, together with further practice with the laws of precedence. Understanding of these topics is often necessary when solving and transposing equations.

10.2 Brackets

With algebra

(a) $2(a+b) = 2a + 2b$

(b) $(a+b)(c+d) = a(c+d) + b(c+d)$
$$= ac + ad + bc + bd$$

Here are some worked examples to help understanding of brackets with algebra.

Problem 1. Determine $2b(a - 5b)$

$$2b(a - 5b) = 2b \times a + 2b \times -5b$$
$$= 2ba - 10b^2$$
$$= \mathbf{2ab - 10b^2} \quad \text{(Note that } 2ba \text{ is the same as } 2ab)$$

Problem 2. Determine $(3x+4y)(x-y)$

$$
\begin{aligned}
(3x+4y)(x-y) &= 3x(x-y)+4y(x-y) \\
&= 3x^2 - 3xy + 4yx - 4y^2 \\
&= 3x^2 - 3xy + 4xy - 4y^2
\end{aligned}
$$

(Note that $4yx$ is the same as $4xy$)

$$= 3x^2 + xy - 4y^2$$

Problem 3. Simplify $3(2x-3y)-(3x-y)$

$$3(2x-3y)-(3x-y) = 3 \times 2x - 3 \times 3y - 3x - -y$$

(Note that $-(3x-y) = -1(3x-y)$ and the -1 multiplies **both** terms in the bracket)

$$
\begin{aligned}
&= 6x - 9y - 3x + y \\
&\text{(Note: } - \times - = +) \\
&= 6x - 3x + y - 9y \\
&= 3x - 8y
\end{aligned}
$$

Problem 4. Remove the brackets and simplify the expression $(a-2b)+5(b-c)-3(c+2d)$

$$
\begin{aligned}
&(a-2b)+5(b-c)-3(c+2d) \\
&= a - 2b + 5 \times b + 5 \times -c - 3 \times c - 3 \times 2d \\
&= a - 2b + 5b - 5c - 3c - 6d \\
&= a + 3b - 8c - 6d
\end{aligned}
$$

Problem 5. Simplify $(p+q)(p-q)$

$$
\begin{aligned}
(p+q)(p-q) &= p(p-q)+q(p-q) \\
&= p^2 - pq + qp - q^2 \\
&= p^2 - q^2
\end{aligned}
$$

Problem 6. Simplify $(2x-3y)^2$

$$
\begin{aligned}
(2x-3y)^2 &= (2x-3y)(2x-3y) \\
&= 2x(2x-3y)-3y(2x-3y) \\
&= 2x \times 2x + 2x \times -3y - 3y \times 2x \\
&\qquad\qquad\qquad\qquad\qquad -3y \times -3y \\
&= 4x^2 - 6xy - 6xy + 9y^2 \\
&\text{(Note: } + \times - = - \text{ and } - \times - = +) \\
&= 4x^2 - 12xy + 9y^2
\end{aligned}
$$

Problem 7. Remove the brackets from the expression and simplify $2[x^2 - 3x(y+x)+4xy]$

$$2[x^2 - 3x(y+x)+4xy] = 2[x^2 - 3xy - 3x^2 + 4xy]$$

(Whenever more than one type of brackets is involved, always **start with the inner brackets**)

$$
\begin{aligned}
&= 2[-2x^2 + xy] \\
&= -4x^2 + 2xy \\
&= 2xy - 4x^2
\end{aligned}
$$

Problem 8. Remove the brackets and simplify the expression $2a - [3\{2(4a-b) - 5(a+2b)\} + 4a]$

(i) Removing the innermost brackets gives

$$2a - [3\{8a - 2b - 5a - 10b\} + 4a]$$

(ii) Collecting together similar terms gives

$$2a - [3\{3a - 12b\} + 4a]$$

(iii) Removing the 'curly' brackets gives

$$2a - [9a - 36b + 4a]$$

(iv) Collecting together similar terms gives

$$2a - [13a - 36b]$$

(v) Removing the outer brackets gives

$$2a - 13a + 36b$$

(vi) i.e. $-11a + 36b$ or $36b - 11a$

Now try the following Practice Exercise

Practice Exercise 39   Brackets (answers on page 426)

Expand the brackets in Problems 1 to 28.

1. $(x+2)(x+3)$
2. $(x+4)(2x+1)$
3. $(2x+3)^2$
4. $(2j-4)(j+3)$
5. $(2x+6)(2x+5)$
6. $(pq+r)(r+pq)$
7. $(a+b)(a+b)$
8. $(x+6)^2$
9. $(a-c)^2$
10. $(5x+3)^2$
11. $(2x-6)^2$
12. $(2x-3)(2x+3)$
13. $(8x+4)^2$
14. $(rs+t)^2$
15. $3a(b-2a)$
16. $2x(x-y)$
17. $(2a-5b)(a+b)$

18. $3(3p - 2q) - (q - 4p)$

19. $(3x - 4y) + 3(y - z) - (z - 4x)$

20. $(2a + 5b)(2a - 5b)$

21. $(x - 2y)^2$ 22. $(3a - b)^2$

23. $2x + [y - (2x + y)]$

24. $3a + 2[a - (3a - 2)]$

25. $4[a^2 - 3a(2b + a) + 7ab]$

26. $3[x^2 - 2x(y + 3x) + 3xy(1 + x)]$

27. $2 - 5[a(a - 2b) - (a - b)^2]$

28. $24p - [2\{3(5p - q) - 2(p + 2q)\} + 3q]$

10.3 Factorisation

The **factors** of 8 are 1, 2, 4 and 8 because 8 divides by 1, 2, 4 and 8

The factors of 24 are 1, 2, 3, 4, 6, 8, 12 and 24 because 24 divides by 1, 2, 3, 4, 6, 8, 12 and 24

The **common factors** of 8 and 24 are 1, 2, 4 and 8 since 1, 2, 4 and 8 are factors of both 8 and 24

The **highest common factor (HCF)** is the largest number that divides into two or more terms.

Hence, the HCF of 8 and 24 is 8, as explained in Chapter 1.

When two or more terms in an algebraic expression contain a common factor, then this factor can be shown outside of a bracket. For example,

$$df + dg = d(f + g)$$

which is just the reverse of

$$d(f + g) = df + dg$$

This process is called **factorisation**.

Here are some worked examples to help understanding of factorising in algebra.

Problem 9. Factorise $ab - 5ac$

a is common to both terms ab and $-5ac$. a is therefore taken outside of the bracket. What goes inside the bracket?

(i)   What multiplies a to make ab? Answer: b

(ii) What multiplies a to make $-5ac$? Answer: $-5c$

Hence, $b - 5c$ appears in the bracket. Thus,

$$ab - 5ac = a(b - 5c)$$

Problem 10. Factorise $2x^2 + 14xy^3$

For the numbers 2 and 14, the highest common factor (HCF) is 2 (i.e. 2 is the largest number that divides into both 2 and 14).

For the x terms, x^2 and x, the HCF is x

Thus, the HCF of $2x^2$ and $14xy^3$ is $2x$

$2x$ is therefore taken outside of the bracket. What goes inside the bracket?

(i)   What multiplies $2x$ to make $2x^2$? Answer: x

(ii) What multiplies $2x$ to make $14xy^3$? Answer: $7y^3$

Hence $x + 7y^3$ appears inside the bracket. Thus,

$$2x^2 + 14xy^3 = 2x(x + 7y^3)$$

Problem 11. Factorise $3x^3y - 12xy^2 + 15xy$

For the numbers 3, 12 and 15, the highest common factor is 3 (i.e. 3 is the largest number that divides into 3, 12 and 15).

For the x terms, x^3, x and x, the HCF is x

For the y terms, y, y^2 and y, the HCF is y

Thus, the HCF of $3x^3y$ and $12xy^2$ and $15xy$ is $3xy$

$3xy$ is therefore taken outside of the bracket. What goes inside the bracket?

(i)   What multiplies $3xy$ to make $3x^3y$? Answer: x^2

(ii) What multiplies $3xy$ to make $-12xy^2$? Answer: $-4y$

(iii) What multiplies $3xy$ to make $15xy$? Answer: 5

Hence, $x^2 - 4y + 5$ appears inside the bracket. Thus,

$$3x^3y - 12xy^2 + 15xy = 3xy(x^2 - 4y + 5)$$

Problem 12. Factorise $25a^2b^5 - 5a^3b^2$

For the numbers 25 and 5, the highest common factor is 5 (i.e. 5 is the largest number that divides into 25 and 5).

For the a terms, a^2 and a^3, the HCF is a^2

For the b terms, b^5 and b^2, the HCF is b^2

Thus, the HCF of $25a^2b^5$ and $5a^3b^2$ is $5a^2b^2$

$5a^2b^2$ is therefore taken outside of the bracket. What goes inside the bracket?

(i) What multiplies $5a^2b^2$ to make $25a^2b^5$? Answer: $5b^3$

(ii) What multiplies $5a^2b^2$ to make $-5a^3b^2$? Answer: $-a$

Hence, $5b^3 - a$ appears in the bracket. Thus,

$$25a^2b^5 - 5a^3b^2 = 5a^2b^2(5b^3 - a)$$

Problem 13. Factorise $ax - ay + bx - by$

The first two terms have a common factor of a and the last two terms a common factor of b. Thus,

$$ax - ay + bx - by = a(x - y) + b(x - y)$$

The two newly formed terms have a common factor of $(x - y)$. Thus,

$$a(x - y) + b(x - y) = (x - y)(a + b)$$

Problem 14. Factorise $2ax - 3ay + 2bx - 3by$

a is a common factor of the first two terms and b a common factor of the last two terms. Thus,

$$2ax - 3ay + 2bx - 3by = a(2x - 3y) + b(2x - 3y)$$

$(2x - 3y)$ is now a common factor. Thus,

$$a(2x - 3y) + b(2x - 3y) = (2x - 3y)(a + b)$$

Alternatively, $2x$ is a common factor of the original first and third terms and $-3y$ is a common factor of the second and fourth terms. Thus,

$$2ax - 3ay + 2bx - 3by = 2x(a + b) - 3y(a + b)$$

$(a + b)$ is now a common factor. Thus,

$$2x(a + b) - 3y(a + b) = (a + b)(2x - 3y)$$

as before

Problem 15. Factorise $x^3 + 3x^2 - x - 3$

x^2 is a common factor of the first two terms. Thus,

$$x^3 + 3x^2 - x - 3 = x^2(x + 3) - x - 3$$

-1 is a common factor of the last two terms. Thus,

$$x^2(x + 3) - x - 3 = x^2(x + 3) - 1(x + 3)$$

$(x + 3)$ is now a common factor. Thus,

$$x^2(x + 3) - 1(x - 3) = (x + 3)(x^2 - 1)$$

Now try the following Practice Exercise

Practice Exercise 40 Factorisation (answers on page 426)

Factorise and simplify the following.

1. $2x + 4$ 2. $2xy - 8xz$

3. $pb + 2pc$ 4. $2x + 4xy$

5. $4d^2 - 12df^5$ 6. $4x + 8x^2$

7. $2q^2 + 8qn$ 8. $rs + rp + rt$

9. $x + 3x^2 + 5x^3$ 10. $abc + b^3c$

11. $3x^2y^4 - 15xy^2 + 18xy$

12. $4p^3q^2 - 10pq^3$ 13. $21a^2b^2 - 28ab$

14. $2xy^2 + 6x^2y + 8x^3y$

15. $2x^2y - 4xy^3 + 8x^3y^4$

16. $28y + 7y^2 + 14xy$

17. $\dfrac{3x^2 + 6x - 3xy}{xy + 2y - y^2}$ 18. $\dfrac{abc + 2ab}{2c + 4} - \dfrac{abc}{2c}$

19. $\dfrac{5rs + 15r^3t + 20r}{6r^2t^2 + 8t + 2ts} - \dfrac{r}{2t}$

20. $ay + by + a + b$ 21. $px + qx + py + qy$

22. $ax - ay + bx - by$

23. $2ax + 3ay - 4bx - 6by$

24. $\dfrac{A^3}{p^2g^3} - \dfrac{A^2}{pg^2} + \dfrac{A^5}{pg}$

10.4 Laws of precedence

Sometimes addition, subtraction, multiplication, division, powers and brackets can all be involved in an algebraic expression. With mathematics there is a definite order of precedence (first met in Chapter 1) which we need to adhere to.

With the **laws of precedence** the order is

Brackets

Order (or p**O**wer)

Division

Multiplication

Addition

Subtraction

The first letter of each word spells **BODMAS**.
Here are some examples to help understanding of BODMAS with algebra.

Problem 16. Simplify $2x + 3x \times 4x - x$

$$
\begin{aligned}
2x + 3x \times 4x - x &= 2x + 12x^2 - x && \text{(M)} \\
&= 2x - x + 12x^2 \\
&= \mathbf{x + 12x^2} && \text{(S)} \\
&\text{or } \mathbf{x(1 + 12x)} && \text{by factorising}
\end{aligned}
$$

Problem 17. Simplify $(y + 4y) \times 3y - 5y$

$$
\begin{aligned}
(y + 4y) \times 3y - 5y &= 5y \times 3y - 5y && \text{(B)} \\
&= \mathbf{15y^2 - 5y} && \text{(M)} \\
&\text{or } \mathbf{5y(3y - 1)} && \text{by factorising}
\end{aligned}
$$

Problem 18. Simplify $p + 2p \times (4p - 7p)$

$$
\begin{aligned}
p + 2p \times (4p - 7p) &= p + 2p \times -3p && \text{(B)} \\
&= \mathbf{p - 6p^2} && \text{(M)} \\
&\text{or } \mathbf{p(1 - 6p)} && \text{by factorising}
\end{aligned}
$$

Problem 19. Simplify $t \div 2t + 3t - 5t$

$$
\begin{aligned}
t \div 2t + 3t - 5t &= \frac{t}{2t} + 3t - 5t && \text{(D)} \\
&= \frac{1}{2} + 3t - 5t && \text{by cancelling} \\
&= \mathbf{\frac{1}{2} - 2t} && \text{(S)}
\end{aligned}
$$

Problem 20. Simplify $x \div (4x + x) - 3x$

$$
\begin{aligned}
x \div (4x + x) - 3x &= x \div 5x - 3x && \text{(B)} \\
&= \frac{x}{5x} - 3x && \text{(D)} \\
&= \mathbf{\frac{1}{5} - 3x} && \text{by cancelling}
\end{aligned}
$$

Problem 21. Simplify $2y \div (6y + 3y - 5y)$

$$
\begin{aligned}
2y \div (6y + 3y - 5y) &= 2y \div 4y && \text{(B)} \\
&= \frac{2y}{4y} && \text{(D)} \\
&= \mathbf{\frac{1}{2}} && \text{by cancelling}
\end{aligned}
$$

Problem 22. Simplify
$$5a + 3a \times 2a + a \div 2a - 7a$$

$$
\begin{aligned}
5a + 3a \times 2a &+ a \div 2a - 7a \\
&= 5a + 3a \times 2a + \frac{a}{2a} - 7a && \text{(D)} \\
&= 5a + 3a \times 2a + \frac{1}{2} - 7a && \text{by cancelling} \\
&= 5a + 6a^2 + \frac{1}{2} - 7a && \text{(M)} \\
&= -2a + 6a^2 + \frac{1}{2} && \text{(S)} \\
&= \mathbf{6a^2 - 2a + \frac{1}{2}}
\end{aligned}
$$

Problem 23. Simplify
$$(4y + 3y)2y + y \div 4y - 6y$$

$$
\begin{aligned}
(4y + 3y)2y &+ y \div 4y - 6y \\
&= 7y \times 2y + y \div 4y - 6y && \text{(B)} \\
&= 7y \times 2y + \frac{y}{4y} - 6y && \text{(D)} \\
&= 7y \times 2y + \frac{1}{4} - 6y && \text{by cancelling} \\
&= \mathbf{14y^2 + \frac{1}{4} - 6y} && \text{(M)}
\end{aligned}
$$

Problem 24. Simplify
$$5b + 2b \times 3b + b \div (4b - 7b)$$

$$
\begin{aligned}
5b + 2b \times 3b &+ b \div (4b - 7b) \\
&= 5b + 2b \times 3b + b \div -3b && \text{(B)} \\
&= 5b + 2b \times 3b + \frac{b}{-3b} && \text{(D)} \\
&= 5b + 2b \times 3b + \frac{1}{-3} && \text{by cancelling} \\
&= 5b + 2b \times 3b - \frac{1}{3} \\
&= \mathbf{5b + 6b^2 - \frac{1}{3}} && \text{(M)}
\end{aligned}
$$

Problem 25. Simplify
$$(5p + p)(2p + 3p) \div (4p - 5p)$$

$(5p + p)(2p + 3p) \div (4p - 5p)$

$\quad = (6p)(5p) \div (-p)$ \hfill (B)

$\quad = 6p \times 5p \div -p$

$\quad = 6p \times \dfrac{5p}{-p}$ \hfill (D)

$\quad = 6p \times \dfrac{5}{-1}$ \hfill by cancelling

$\quad = 6p \times -5$

$\quad = \mathbf{-30p}$

Now try the following Practice Exercise

Practice Exercise 41   Laws of precedence
(answers on page 426)

Simplify the following.

1. $3x + 2x \times 4x - x$

2. $(2y + y) \times 4y - 3y$

3. $4b + 3b \times (b - 6b)$

4. $8a \div 2a + 6a - 3a$

5. $6x \div (3x + x) - 4x$

6. $4t \div (5t - 3t + 2t)$

7. $3y + 2y \times 5y + 2y \div 8y - 6y$

8. $(x + 2x)3x + 2x \div 6x - 4x$

9. $5a + 2a \times 3a + a \div (2a - 9a)$

11. $(3t + 2t)(5t + t) \div (t - 3t)$

12. $x \div 5x - x + (2x - 3x)x$

12. $3a + 2a \times 5a + 4a \div 2a - 6a$

For fully worked solutions to each of the problems in Practice Exercises 39 to 41 in this chapter,
go to the website:
www.routledge.com/cw/bird

Chapter 11

Solving simple equations

Why it is important to understand: Solving simple equations

In mathematics, engineering and science, formulae are used to relate physical quantities to each other. They provide rules so that if we know the values of certain quantities, we can calculate the values of others. Equations occur in all branches of engineering. Simple equations always involve one unknown quantity which we try to find when we solve the equation. In reality, we all solve simple equations in our heads all the time without even noticing it. If, for example, you have bought two CDs, for the same price, and a DVD, and know that you spent £25 in total and that the DVD was £11, then you actually solve the linear equation $2x + 11 = 25$ to find out that the price of each CD was £7. It is probably true to say that there is no branch of engineering, physics, economics, chemistry and computer science which does not require the solution of simple equations. The ability to solve simple equations is another stepping stone on the way to having confidence to handle engineering mathematics.

At the end of this chapter, you should be able to:

- distinguish between an algebraic expression and an algebraic equation
- maintain the equality of a given equation whilst applying arithmetic operations
- solve linear equations in one unknown including those involving brackets and fractions
- form and solve linear equations involved with practical situations
- evaluate an unknown quantity in a formula by substitution of data

11.1 Introduction

$3x - 4$ is an example of an **algebraic expression**.
$3x - 4 = 2$ is an example of an **algebraic equation** (i.e. it contains an '=' sign).

An equation is simply a statement that two expressions are equal.

Hence, $A = \pi r^2$ (where A is the area of a circle of radius r)

$$F = \frac{9}{5}C + 32 \text{ (which relates Fahrenheit and Celsius temperatures)}$$

and $y = 3x + 2$ (which is the equation of a straight line graph)

are all examples of equations.

11.2 Solving equations

To '**solve an equation**' means '**to find the value of the unknown**'. For example, solving $3x - 4 = 2$ means that the value of x is required.
In this example, $x = 2$. How did we arrive at $x = 2$? This is the purpose of this chapter – to show how to solve such equations.

Many equations occur in engineering and it is essential that we can solve them when needed.

Here are some examples to demonstrate how simple equations are solved.

Problem 1.   Solve the equation $4x = 20$

Dividing each side of the equation by 4 gives

$$\frac{4x}{4} = \frac{20}{4}$$

i.e. $x = 5$ by cancelling, which is the solution to the equation $4x = 20$

The same operation **must** be applied to both sides of an equation so that the equality is maintained.

We can do anything we like to an equation, **as long as we do the same to both sides.** This is, in fact, the only rule to remember when solving simple equations (and also when transposing formulae, which we do in Chapter 12).

Problem 2.   Solve the equation $\dfrac{2x}{5} = 6$

Multiplying both sides by 5 gives $5\left(\dfrac{2x}{5}\right) = 5(6)$

Cancelling and removing brackets gives $2x = 30$

Dividing both sides of the equation by 2 gives

$$\frac{2x}{2} = \frac{30}{2}$$

Cancelling gives $x = 15$

which is the solution of the equation $\dfrac{2x}{5} = 6$

Problem 3.   Solve the equation $a - 5 = 8$

Adding 5 to both sides of the equation gives

$$a - 5 + 5 = 8 + 5$$

i.e. $a = 8 + 5$

i.e. $a = 13$

which is the solution of the equation $a - 5 = 8$

Note that adding 5 to both sides of the above equation results in the -5 moving from the LHS to the RHS, but the sign is changed to $+$

Problem 4.   Solve the equation $x + 3 = 7$

Subtracting 3 from both sides gives $x + 3 - 3 = 7 - 3$

i.e. $x = 7 - 3$

i.e. $x = 4$

which is the solution of the equation $x + 3 = 7$

Note that subtracting 3 from both sides of the above equation results in the $+3$ moving from the LHS to the RHS, but the sign is changed to $-$. So, we can move straight from $x + 3 = 7$ to $x = 7 - 3$

Thus, a term can be moved from one side of an equation to the other **as long as a change in sign is made.**

Problem 5.   Solve the equation $6x + 1 = 2x + 9$

In such equations the terms containing x are grouped on one side of the equation and the remaining terms grouped on the other side of the equation. As in Problems 3 and 4, changing from one side of an equation to the other must be accompanied by a change of sign.

Since $6x + 1 = 2x + 9$

then $6x - 2x = 9 - 1$

i.e. $4x = 8$

Dividing both sides by 4 gives $\dfrac{4x}{4} = \dfrac{8}{4}$

Cancelling gives $x = 2$

which is the solution of the equation $6x + 1 = 2x + 9$.

In the above examples, the solutions can be checked. Thus, in Problem 5, where $6x + 1 = 2x + 9$, if $x = 2$, then

$$\text{LHS of equation} = 6(2) + 1 = 13$$

$$\text{RHS of equation} = 2(2) + 9 = 13$$

Since the left hand side (LHS) equals the right hand side (RHS) then $x = 2$ must be the correct solution of the equation.

When solving simple equations, always check your answers by substituting your solution back into the original equation.

Problem 6.   Solve the equation $4 - 3p = 2p - 11$

In order to keep the p term positive the terms in p are moved to the RHS and the constant terms to the LHS. Similar to Problem 5, if $4 - 3p = 2p - 11$

then $4 + 11 = 2p + 3p$

i.e. $15 = 5p$

Dividing both sides by 5 gives $\dfrac{15}{5} = \dfrac{5p}{5}$

Cancelling gives $3 = p$ or $p = 3$

which is the solution of the equation $4 - 3p = 2p - 11$.

By substituting $p = 3$ into the original equation, the solution may be checked.

$$\text{LHS} = 4 - 3(3) = 4 - 9 = -5$$

$$\text{RHS} = 2(3) - 11 = 6 - 11 = -5$$

Since LHS = RHS, the solution $p = 3$ must be correct. If, in this example, the unknown quantities had been grouped initially on the LHS instead of the RHS, then $-3p - 2p = -11 - 4$

i.e. $-5p = -15$

from which, $\dfrac{-5p}{-5} = \dfrac{-15}{-5}$

and $\boldsymbol{p = 3}$

as before.

It is often easier, however, to work with positive values where possible.

Problem 7. Solve the equation $3(x - 2) = 9$

Removing the bracket gives $3x - 6 = 9$

Rearranging gives $3x = 9 + 6$

i.e. $3x = 15$

Dividing both sides by 3 gives $\boldsymbol{x = 5}$

which is the solution of the equation $3(x - 2) = 9$.
The equation may be checked by substituting $x = 5$ back into the original equation.

Problem 8. Solve the equation
$4(2r - 3) - 2(r - 4) = 3(r - 3) - 1$

Removing brackets gives

$$8r - 12 - 2r + 8 = 3r - 9 - 1$$

Rearranging gives $8r - 2r - 3r = -9 - 1 + 12 - 8$

i.e. $3r = -6$

Dividing both sides by 3 gives $r = \dfrac{-6}{3} = \boldsymbol{-2}$

which is the solution of the equation
$4(2r - 3) - 2(r - 4) = 3(r - 3) - 1$

The solution may be checked by substituting $r = -2$ back into the original equation.

$$\text{LHS} = 4(-4 - 3) - 2(-2 - 4) = -28 + 12 = -16$$

$$\text{RHS} = 3(-2 - 3) - 1 = -15 - 1 = -16$$

Since LHS = RHS then $r = -2$ is the correct solution.

___

Now try the following Practice Exercise

Practice Exercise 42   Solving simple equations (answers on page 426)

Solve the following equations.

1. $2x + 5 = 7$

2. $8 - 3t = 2$

3. $\dfrac{2}{3}c - 1 = 3$

4. $2x - 1 = 5x + 11$

5. $7 - 4p = 2p - 5$

6. $2.6x - 1.3 = 0.9x + 0.4$

7. $2a + 6 - 5a = 0$

8. $3x - 2 - 5x = 2x - 4$

9. $20d - 3 + 3d = 11d + 5 - 8$

10. $2(x - 1) = 4$

11. $16 = 4(t + 2)$

12. $5(f - 2) - 3(2f + 5) + 15 = 0$

13. $2x = 4(x - 3)$

14. $6(2 - 3y) - 42 = -2(y - 1)$

15. $2(3g - 5) - 5 = 0$

16. $4(3x + 1) = 7(x + 4) - 2(x + 5)$

17. $11 + 3(r - 7) = 16 - (r + 2)$

18. $8 + 4(x - 1) - 5(x - 3) = 2(5 - 2x)$

Here are some further worked examples on solving simple equations.

Problem 9. Solve the equation $\dfrac{4}{x} = \dfrac{2}{5}$

The lowest common multiple (LCM) of the denominators, i.e. the lowest algebraic expression that both x and 5 will divide into, is $5x$

Multiplying both sides by $5x$ gives

$$5x\left(\dfrac{4}{x}\right) = 5x\left(\dfrac{2}{5}\right)$$

Cancelling gives $\quad 5(4) = x(2)$

i.e. $\qquad\qquad 20 = 2x$ $\qquad$ (1)

Dividing both sides by 2 gives $\quad \dfrac{20}{2} = \dfrac{2x}{2}$

Cancelling gives $\qquad\qquad \mathbf{10 = x}$ or $\mathbf{x = 10}$

which is the solution of the equation $\dfrac{4}{x} = \dfrac{2}{5}$

When there is just one fraction on each side of the equation as in this example, there is a quick way to arrive at equation (1) without needing to find the LCM of the denominators.

We can move from $\dfrac{4}{x} = \dfrac{2}{5}$ to $4 \times 5 = 2 \times x$ by what is called '**cross-multiplication**'.

In general, if $\dfrac{a}{b} = \dfrac{c}{d}$ then $ad = bc$

We can use cross-multiplication when there is one fraction only on each side of the equation.

Problem 10. Solve the equation $\dfrac{3}{t-2} = \dfrac{4}{3t+4}$

Cross-multiplication gives $\; 3(3t+4) = 4(t-2)$

Removing brackets gives $\quad 9t + 12 = 4t - 8$

Rearranging gives $\qquad 9t - 4t = -8 - 12$

i.e. $\qquad\qquad\qquad 5t = -20$

Dividing both sides by 5 gives $\quad t = \dfrac{-20}{5} = \mathbf{-4}$

which is the solution of the equation $\dfrac{3}{t-2} = \dfrac{4}{3t+4}$

Problem 11. Solve the equation

$$\dfrac{2y}{5} + \dfrac{3}{4} + 5 = \dfrac{1}{20} - \dfrac{3y}{2}$$

The lowest common multiple (LCM) of the denominators is 20; i.e. the lowest number that 4, 5, 20 and 2 will divide into.
Multiplying each term by 20 gives

$$20\left(\dfrac{2y}{5}\right) + 20\left(\dfrac{3}{4}\right) + 20(5) = 20\left(\dfrac{1}{20}\right) - 20\left(\dfrac{3y}{2}\right)$$

Cancelling gives $\quad 4(2y) + 5(3) + 100 = 1 - 10(3y)$

i.e. $\qquad\qquad 8y + 15 + 100 = 1 - 30y$

Rearranging gives $\qquad 8y + 30y = 1 - 15 - 100$

i.e. $\qquad\qquad\qquad 38y = -114$

Dividing both sides by 38 gives $\quad \dfrac{38y}{38} = \dfrac{-114}{38}$

Cancelling gives $\qquad\qquad \mathbf{y = -3}$

which is the solution of the equation

$$\dfrac{2y}{5} + \dfrac{3}{4} + 5 = \dfrac{1}{20} - \dfrac{3y}{2}$$

Problem 12. Solve the equation $\sqrt{x} = 2$

Whenever square root signs are involved in an equation, both sides of the equation must be squared.

Squaring both sides gives $\quad \left(\sqrt{x}\right)^2 = (2)^2$

i.e. $\qquad\qquad\qquad \mathbf{x = 4}$

which is the solution of the equation $\sqrt{x} = 2$

Problem 13. Solve the equation $2\sqrt{d} = 8$

Whenever square roots are involved in an equation, the square root term needs to be isolated on its own before squaring both sides.

Cross-multiplying gives $\qquad \sqrt{d} = \dfrac{8}{2}$

Cancelling gives $\qquad\qquad \sqrt{d} = 4$

Squaring both sides gives $\quad \left(\sqrt{d}\right)^2 = (4)^2$

i.e. $\qquad\qquad\qquad \mathbf{d = 16}$

which is the solution of the equation $2\sqrt{d} = 8$

Problem 14. Solve the equation $\left(\dfrac{\sqrt{b}+3}{\sqrt{b}}\right) = 2$

Cross-multiplying gives $\; \sqrt{b} + 3 = 2\sqrt{b}$

Rearranging gives $\qquad 3 = 2\sqrt{b} - \sqrt{b}$

i.e. $\qquad\qquad\qquad 3 = \sqrt{b}$

Squaring both sides gives $\quad \mathbf{9 = b}$

which is the solution of the equation $\left(\dfrac{\sqrt{b}+3}{\sqrt{b}}\right) = 2$

Problem 15. Solve the equation $x^2 = 25$

Whenever a square term is involved, the square root of both sides of the equation must be taken.

Taking the square root of both sides gives $\sqrt{x^2} = \sqrt{25}$

i.e. $$x = \pm 5$$

which is the solution of the equation $x^2 = 25$

Problem 16. Solve the equation $\dfrac{15}{4t^2} = \dfrac{2}{3}$

We need to rearrange the equation to get the t^2 term on its own.

Cross-multiplying gives $\quad 15(3) = 2(4t^2)$

i.e. $$45 = 8t^2$$

Dividing both sides by 8 gives $\dfrac{45}{8} = \dfrac{8t^2}{8}$

By cancelling $\quad\quad\quad 5.625 = t^2$

or $$t^2 = 5.625$$

Taking the square root of both sides gives

$$\sqrt{t^2} = \sqrt{5.625}$$

i.e. $$t = \pm 2.372$$

correct to 4 significant figures, which is the solution of the equation $\dfrac{15}{4t^2} = \dfrac{2}{3}$

Now try the following Practice Exercise

Practice Exercise 43   Solving simple equations (answers on page 426)

Solve the following equations.

1. $\dfrac{1}{5}d + 3 = 4$

2. $2 + \dfrac{3}{4}y = 1 + \dfrac{2}{3}y + \dfrac{5}{6}$

3. $\dfrac{1}{4}(2x - 1) + 3 = \dfrac{1}{2}$

4. $\dfrac{1}{5}(2f - 3) + \dfrac{1}{6}(f - 4) + \dfrac{2}{15} = 0$

5. $\dfrac{1}{3}(3m - 6) - \dfrac{1}{4}(5m + 4) + \dfrac{1}{5}(2m - 9) = -3$

6. $\dfrac{x}{3} - \dfrac{x}{5} = 2$

7. $1 - \dfrac{y}{3} = 3 + \dfrac{y}{3} - \dfrac{y}{6}$

8. $\dfrac{2}{a} = \dfrac{3}{8}$

9. $\dfrac{1}{3n} + \dfrac{1}{4n} = \dfrac{7}{24}$

10. $\dfrac{x + 3}{4} = \dfrac{x - 3}{5} + 2$

11. $\dfrac{3t}{20} = \dfrac{6 - t}{12} + \dfrac{2t}{15} - \dfrac{3}{2}$

12. $\dfrac{y}{5} + \dfrac{7}{20} = \dfrac{5 - y}{4}$

13. $\dfrac{v - 2}{2v - 3} = \dfrac{1}{3}$

14. $\dfrac{2}{a - 3} = \dfrac{3}{2a + 1}$

15. $\dfrac{x}{4} - \dfrac{x + 6}{5} = \dfrac{x + 3}{2}$

16. $3\sqrt{t} = 9$

17. $2\sqrt{y} = 5$

18. $4 = \sqrt{\left(\dfrac{3}{a}\right)} + 3$

19. $\dfrac{3\sqrt{x}}{1 - \sqrt{x}} = -6$

20. $10 = 5\sqrt{\left(\dfrac{x}{2} - 1\right)}$

21. $16 = \dfrac{t^2}{9}$

22. $\sqrt{\left(\dfrac{y + 2}{y - 2}\right)} = \dfrac{1}{2}$

23. $\dfrac{6}{a} = \dfrac{2a}{3}$

24. $\dfrac{11}{2} = 5 + \dfrac{8}{x^2}$

11.3 Practical problems involving simple equations

There are many practical situations in engineering in which solving equations is needed. Here are some worked examples to demonstrate typical practical situations.

Problem 17. Applying the principle of moments to a beam results in the equation

$$F \times 3 = (7.5 - F) \times 2$$

where F is the force in newtons. Determine the value of F

Removing brackets gives $\qquad 3F = 15 - 2F$

Rearranging gives $\qquad 3F + 2F = 15$

i.e. $\qquad\qquad\qquad 5F = 15$

Dividing both sides by 5 gives $\qquad \dfrac{5F}{5} = \dfrac{15}{5}$

from which, force, $\boldsymbol{F = 3N}$

Problem 18. A copper wire has a length L of 1.5 km, a resistance R of 5 Ω and a resistivity ρ of $17.2 \times 10^{-6}\,\Omega\,\text{mm}$. Find the cross-sectional area, a, of the wire, given that $R = \dfrac{\rho L}{a}$

Since $R = \dfrac{\rho L}{a}$ then

$$5\Omega = \frac{(17.2 \times 10^{-6}\,\Omega\,\text{mm})(1500 \times 10^3\,\text{mm})}{a}.$$

From the units given, a is measured in mm^2.

Thus, $5a = 17.2 \times 10^{-6} \times 1500 \times 10^3$

and $\qquad a = \dfrac{17.2 \times 10^{-6} \times 1500 \times 10^3}{5}$

$$= \frac{17.2 \times 1500 \times 10^3}{10^6 \times 5} = \frac{17.2 \times 15}{10 \times 5} = 5.16$$

Hence, the cross-sectional area of the wire is $\boldsymbol{5.16\,mm^2}$.

Problem 19. $PV = mRT$ is the characteristic gas equation. Find the value of gas constant R when pressure $P = 3 \times 10^6$ Pa, volume $V = 0.90\,\text{m}^3$, mass $m = 2.81$ kg and temperature $T = 231$ K

Dividing both sides of $PV = mRT$ by mT gives

$$\frac{PV}{mT} = \frac{mRT}{mT}$$

Cancelling gives $\qquad \dfrac{PV}{mT} = R$

Substituting values gives $\qquad R = \dfrac{(3 \times 10^6)(0.90)}{(2.81)(231)}$

Using a calculator, **gas constant, $R = 4160\,J/(kg\ K)$**, correct to 4 significant figures.

Problem 20. A rectangular box with square ends has its length 15 cm greater than its breadth and the total length of its edges is 2.04 m. Find the width of the box and its volume

Let x cm = width = height of box. Then the length of the box is $(x + 15)$ cm, as shown in Fig. 11.1.

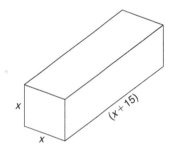

x $\qquad$ $(x+15)$

x

Figure 11.1

The length of the edges of the box is $2(4x) + 4(x + 15)$ cm, which equals 2.04 m or 204 cm.

Hence, $\qquad\qquad 204 = 2(4x) + 4(x + 15)$

$$204 = 8x + 4x + 60$$

$$204 - 60 = 12x$$

i.e. $\qquad\qquad\qquad 144 = 12x$

and $\qquad\qquad\qquad x = 12\,\text{cm}$

Hence, **the width of the box is 12 cm**.

Volume of box = length $\times$ width $\times$ height

$$= (x + 15)(x)(x) = (12 + 15)(12)(12)$$

$$= (27)(12)(12)$$

$$= \boldsymbol{3888\,cm^3}$$

Problem 21. The temperature coefficient of resistance α may be calculated from the formula $R_t = R_0(1 + \alpha t)$. Find α, given $R_t = 0.928$, $R_0 = 0.80$ and $t = 40$

Since $R_t = R_0(1 + \alpha t)$, then

$$0.928 = 0.80[1 + \alpha(40)]$$

$$0.928 = 0.80 + (0.8)(\alpha)(40)$$

$$0.928 - 0.80 = 32\alpha$$

$$0.128 = 32\alpha$$

Hence, $$\alpha = \frac{0.128}{32} = 0.004$$

Problem 22. The distance s metres travelled in time t seconds is given by the formula $s = ut + \frac{1}{2}at^2$, where u is the initial velocity in m/s and a is the acceleration in m/s^2. Find the acceleration of the body if it travels 168 m in 6 s, with an initial velocity of 10 m/s

$s = ut + \frac{1}{2}at^2$, and $s = 168, u = 10$ and $t = 6$

Hence, $$168 = (10)(6) + \frac{1}{2}a(6)^2$$

$$168 = 60 + 18a$$

$$168 - 60 = 18a$$

$$108 = 18a$$

$$a = \frac{108}{18} = 6$$

Hence, **the acceleration of the body is 6 m/s^2**

Problem 23. When three resistors in an electrical circuit are connected in parallel the total resistance R_T is given by $\frac{1}{R_T} = \frac{1}{R_1} + \frac{1}{R_2} + \frac{1}{R_3}$. Find the total resistance when $R_1 = 5\,\Omega, R_2 = 10\,\Omega$ and $R_3 = 30\,\Omega$

$$\frac{1}{R_T} = \frac{1}{5} + \frac{1}{10} + \frac{1}{30} = \frac{6+3+1}{30} = \frac{10}{30} = \frac{1}{3}$$

Taking the reciprocal of both sides gives $R_T = 3\,\Omega$

Alternatively, if $\frac{1}{R_T} = \frac{1}{5} + \frac{1}{10} + \frac{1}{30}$, the LCM of the denominators is $30R_T$

Hence, $$30R_T\left(\frac{1}{R_T}\right) = 30R_T\left(\frac{1}{5}\right)$$
$$+ 30R_T\left(\frac{1}{10}\right) + 30R_T\left(\frac{1}{30}\right)$$

Cancelling gives $30 = 6R_T + 3R_T + R_T$

i.e. $30 = 10R_T$

and $$R_T = \frac{30}{10} = 3\,\Omega, \text{ as above.}$$

Now try the following Practice Exercise

Practice Exercise 44   Practical problems involving simple equations (answers on page 426)

1. A formula used for calculating resistance of a cable is $R = \frac{\rho L}{a}$. Given $R = 1.25, L = 2500$ and $a = 2 \times 10^{-4}$, find the value of ρ.

2. Force F newtons is given by $F = ma$, where m is the mass in kilograms and a is the acceleration in metres per second squared. Find the acceleration when a force of 4 kN is applied to a mass of 500 kg.

3. $PV = mRT$ is the characteristic gas equation. Find the value of m when $P = 100 \times 10^3, V = 3.00, R = 288$ and $T = 300$

4. When three resistors R_1, R_2 and R_3 are connected in parallel, the total resistance R_T is determined from $\frac{1}{R_T} = \frac{1}{R_1} + \frac{1}{R_2} + \frac{1}{R_3}$
   (a) Find the total resistance when $R_1 = 3\,\Omega, R_2 = 6\,\Omega$ and $R_3 = 18\,\Omega$.
   (b) Find the value of R_3 given that $R_T = 3\,\Omega, R_1 = 5\,\Omega$ and $R_2 = 10\,\Omega$.

5. Six digital camera batteries and three camcorder batteries cost £96. If a camcorder battery costs £5 more than a digital camera battery, find the cost of each.

6. Ohm's law may be represented by $I = V/R$, where I is the current in amperes, V is the voltage in volts and R is the resistance in ohms. A soldering iron takes a current of 0.30 A from a 240 V supply. Find the resistance of the element.

7. The distance, s, travelled in time t seconds is given by the formula $s = ut + \frac{1}{2}at^2$ where u is the initial velocity in m/s and a is the acceleration in m/s^2. Calculate the acceleration of the body if it travels 165 m in 3 s, with an initial velocity of 10 m/s.

8. The stress, σ pascals, acting on the reinforcing rod in a concrete column is given

in the following equation:
$$500 \times 10^{-6}\sigma + 2.67 \times 10^5 = 3.55 \times 10^5$$
Find the value of the stress in MPa.

Here are some further worked examples on solving simple equations in practical situations.

Problem 24. The extension x m of an aluminium tie bar of length l m and cross-sectional area A m^2 when carrying a load of F newtons is given by the modulus of elasticity $E = Fl/Ax$. Find the extension of the tie bar (in mm) if $E = 70 \times 10^9$ N/m^2, $F = 20 \times 10^6$ N, $A = 0.1$ m^2 and $l = 1.4$ m

$E = Fl/Ax$, hence
$$70 \times 10^9 \frac{\text{N}}{\text{m}^2} = \frac{(20 \times 10^6 \,\text{N})(1.4 \,\text{m})}{(0.1 \,\text{m}^2)(x)}$$

(the unit of x is thus metres)
$$70 \times 10^9 \times 0.1 \times x = 20 \times 10^6 \times 1.4$$
$$x = \frac{20 \times 10^6 \times 1.4}{70 \times 10^9 \times 0.1}$$

Cancelling gives
$$x = \frac{2 \times 1.4}{7 \times 100} \,\text{m}$$
$$= \frac{2 \times 1.4}{7 \times 100} \times 1000 \,\text{mm}$$
$$= 4 \,\text{mm}$$

Hence, **the extension of the tie bar, $x = 4$ mm**.

Problem 25. Power in a d.c. circuit is given by $P = \dfrac{V^2}{R}$ where V is the supply voltage and R is the circuit resistance. Find the supply voltage if the circuit resistance is 1.25 Ω and the power measured is 320 W

Since $P = \dfrac{V^2}{R}$, then $320 = \dfrac{V^2}{1.25}$
$$(320)(1.25) = V^2$$
i.e. $V^2 = 400$

Supply voltage, $\qquad$ **$V = \sqrt{400} = \pm 20$ V**

Problem 26. A painter is paid £6.30 per hour for a basic 36 hour week and overtime is paid at one and a third times this rate. Determine how many hours the painter has to work in a week to earn £319.20

Basic rate per hour $= £6.30$ and overtime rate per hour $= 1\frac{1}{3} \times £6.30 = £8.40$ Let the number of overtime hours worked $= x$

Then, $\qquad (36)(6.30) + (x)(8.40) = 319.20$
$$226.80 + 8.40x = 319.20$$
$$8.40x = 319.20 - 226.80 = 92.40$$
$$x = \frac{92.40}{8.40} = 11$$

Thus, 11 hours overtime would have to be worked to earn £319.20 per week. Hence, **the total number of hours worked** is $36 + 11$, i.e. **47 hours**.

Problem 27. A formula relating initial and final states of pressures, P_1 and P_2, volumes, V_1 and V_2, and absolute temperatures, T_1 and T_2, of an ideal gas is $\dfrac{P_1 V_1}{T_1} = \dfrac{P_2 V_2}{T_2}$. Find the value of P_2 given $P_1 = 100 \times 10^3$, $V_1 = 1.0$, $V_2 = 0.266$, $T_1 = 423$ and $T_2 = 293$

Since $\dfrac{P_1 V_1}{T_1} = \dfrac{P_2 V_2}{T_2}$

then $\dfrac{(100 \times 10^3)(1.0)}{423} = \dfrac{P_2(0.266)}{293}$

Cross-multiplying gives
$$(100 \times 10^3)(1.0)(293) = P_2(0.266)(423)$$
$$P_2 = \frac{(100 \times 10^3)(1.0)(293)}{(0.266)(423)}$$

Hence, $\qquad$ **$P_2 = 260 \times 10^3$ or 2.6×10^5**

Problem 28. The stress, f, in a material of a thick cylinder can be obtained from $\dfrac{D}{d} = \sqrt{\left(\dfrac{f+p}{f-p}\right)}$. Calculate the stress, given that $D = 21.5$, $d = 10.75$ and $p = 1800$

Since $\dfrac{D}{d} = \sqrt{\left(\dfrac{f+p}{f-p}\right)}$ then $\dfrac{21.5}{10.75} = \sqrt{\left(\dfrac{f+1800}{f-1800}\right)}$

i.e. $\qquad\qquad 2 = \sqrt{\left(\dfrac{f+1800}{f-1800}\right)}$

Squaring both sides gives

$$4 = \frac{f + 1800}{f - 1800}$$

Cross-multiplying gives

$$4(f - 1800) = f + 1800$$

$$4f - 7200 = f + 1800$$

$$4f - f = 1800 + 7200$$

$$3f = 9000$$

$$f = \frac{9000}{3} = 3000$$

Hence,   **stress, $f = 3000$**

Problem 29.   12 workmen employed on a building site earn between them a total of £4035 per week. Labourers are paid £275 per week and craftsmen are paid £380 per week. How many craftsmen and how many labourers are employed?

Let the number of craftsmen be c. The number of labourers is therefore $(12 - c)$.

The wage bill equation is

$$380c + 275(12 - c) = 4035$$

$$380c + 3300 - 275c = 4035$$

$$380c - 275c = 4035 - 3300$$

$$105c = 735$$

$$c = \frac{735}{105} = 7$$

Hence, there are **7 craftsmen** and $(12 - 7)$, i.e. **5 labourers** on the site.

Now try the following Practice Exercise

Practice Exercise 45   Practical problems involving simple equations (answers on page 426)

1.   A rectangle has a length of 20 cm and a width b cm. When its width is reduced by 4 cm its area becomes 160 cm^2. Find the original width and area of the rectangle.

2.   Given $R_2 = R_1(1 + \alpha t)$, find α given $R_1 = 5.0, R_2 = 6.03$ and $t = 51.5$

3.   If $v^2 = u^2 + 2as$, find u given $v = 24$, $a = -40$ and $s = 4.05$

4.   The relationship between the temperature on a Fahrenheit scale and that on a Celsius scale is given by $F = \frac{9}{5}C + 32$. Express 113°F in degrees Celsius.

5.   If $t = 2\pi\sqrt{\dfrac{w}{Sg}}$, find the value of S given $w = 1.219, g = 9.81$ and $t = 0.3132$

6.   Two joiners and five mates earn £1824 between them for a particular job. If a joiner earns £72 more than a mate, calculate the earnings for a joiner and for a mate.

7.   An alloy contains 60% by weight of copper, the remainder being zinc. How much copper must be mixed with 50 kg of this alloy to give an alloy containing 75% copper?

8.   A rectangular laboratory has a length equal to one and a half times its width and a perimeter of 40 m. Find its length and width.

9.   Applying the principle of moments to a beam results in the following equation:

$$F \times 3 = (5 - F) \times 7$$

where F is the force in newtons. Determine the value of F.

For fully worked solutions to each of the problems in Practice Exercises 42 to 45 in this chapter, go to the website:

Revision Test 4: Algebra and simple equations

This assignment covers the material contained in Chapters 9–11. *The marks available are shown in brackets at the end of each question.*

1. Evaluate $3pqr^3 - 2p^2qr + pqr$ when $p = \dfrac{1}{2}$, $q = -2$ and $r = 1$ (3)

   In Problems 2 to 7, simplify the expressions.

2. $\dfrac{9p^2qr^3}{3pq^2r}$ (3)

3. $2(3x - 2y) - (4y - 3x)$ (3)

4. $(x - 2y)(2x + y)$ (3)

5. $p^2q^{-3}r^4 \times pq^2r^{-3}$ (3)

6. $(3a - 2b)^2$ (3)

7. $\dfrac{a^4b^2c}{ab^3c^2}$ (3)

8. Factorise
   (a) $2x^2y^3 - 10xy^2$
   (b) $21ab^2c^3 - 7a^2bc^2 + 28a^3bc^4$ (5)

9. Factorise and simplify
   $\dfrac{2x^2y + 6xy^2}{x + 3y} - \dfrac{x^3y^2}{x^2y}$ (5)

10. Remove the brackets and simplify
    $10a - [3(2a - b) - 4(b - a) + 5b]$ (4)

11. Simplify $x \div 5x - x + (2x - 3x)x$ (4)

12. Simplify $3a + 2a \times 5a + 4a \div 2a - 6a$ (4)

13. Solve the equations
    (a) $3a = 39$
    (b) $2x - 4 = 9$ (3)

14. Solve the equations
    (a) $\dfrac{4}{9}y = 8$
    (b) $6x - 1 = 4x + 5$ (4)

15. Solve the equation
    $5(t - 2) - 3(4 - t) = 2(t + 3) - 40$ (4)

16. Solve the equations
    (a) $\dfrac{3}{2x + 1} = \dfrac{1}{4x - 3}$
    (b) $2x^2 = 162$ (7)

17. Kinetic energy is given by the formula, $E_k = \dfrac{1}{2}mv^2$ joules, where m is the mass in kilograms and v is the velocity in metres per second. Evaluate the velocity when $E_k = 576 \times 10^{-3}$ J and the mass is 5 kg.   (4)

18. An approximate relationship between the number of teeth T on a milling cutter, the diameter of the cutter D and the depth of cut d is given by $T = \dfrac{12.5D}{D + 4d}$. Evaluate d when $T = 10$ and $D = 32$ (5)

19. The modulus of elasticity E is given by the formula $E = \dfrac{FL}{xA}$ where F is force in newtons, L is the length in metres, x is the extension in metres and A the cross-sectional area in square metres. Evaluate A, in square centimetres, when $E = 80 \times 10^9$ N/m^2, $x = 2$ mm, $F = 100 \times 10^3$ N and $L = 2.0$ m.   (5)

Multiple choice questions Test 1
Number and algebra
This test covers the material in Chapters 1 to 11

All questions have only one correct answer (Answers on page 440).

1. $1\frac{1}{3} + 1\frac{2}{3} \div 2\frac{2}{3} - \frac{1}{3}$ is equal to:

   (a) $1\frac{5}{8}$ (b) $\frac{19}{24}$ (c) $2\frac{1}{21}$ (d) $1\frac{2}{7}$

2. The value of $\frac{2^{-3}}{2^{-4}} - 1$ is equal to:

   (a) $-\frac{1}{2}$ (b) 2 (c) 1 (d) $\frac{1}{2}$

3. Four engineers can complete a task in 5 hours. Assuming the rate of work remains constant, six engineers will complete the task in:

   (a) 126 h (b) 4 h 48 min
   (c) 3 h 20 min (d) 7 h 30 min

4. In an engineering equation $\frac{3^4}{3^r} = \frac{1}{9}$. The value of r is:

   (a) -6 (b) 2 (c) 6 (d) -2

5. When $p = 3$, $q = -\frac{1}{2}$ and $r = -2$, the engineering expression $2p^2 q^3 r^4$ is equal to:

   (a) 1296 (b) -36 (c) 36 (d) 18

6. $\frac{3}{4} \div 1\frac{3}{4}$ is equal to:

   (a) $\frac{3}{7}$ (b) $1\frac{9}{16}$ (c) $1\frac{5}{16}$ (d) $2\frac{1}{2}$

7. $(2e - 3f)(e + f)$ is equal to:

   (a) $2e^2 - 3f^2$ (b) $2e^2 - 5ef - 3f^2$
   (c) $2e^2 + 3f^2$ (d) $2e^2 - ef - 3f^2$

8. $16^{-\frac{3}{4}}$ is equal to:

   (a) 8 (b) $-\frac{1}{2^3}$ (c) 4 (d) $\frac{1}{8}$

9. If $x = \dfrac{57.06 \times 0.0711}{\sqrt{0.0635}}$ cm, which of the following statements is correct?

   (a) $x = 16.09$ cm, correct to 4 significant figures
   (b) $x = 16$ cm, correct to 2 significant figures

   (c) $x = 1.61 \times 10^1$ cm, correct to 3 decimal places
   (d) $x = 16.099$ cm, correct to 3 decimal places

10. $(5.5 \times 10^2)(2 \times 10^3)$ cm in standard form is equal to:

    (a) 11×10^6 cm (b) 1.1×10^6 cm
    (c) 11×10^5 cm (d) 1.1×10^5 cm

11. The engineering expression $\dfrac{(16 \times 4)^2}{(8 \times 2)^4}$ is equal to:

    (a) 4 (b) 2^{-4} (c) $\frac{1}{2^2}$ (d) 1

12. $\left(16^{-\frac{1}{4}} - 27^{-\frac{2}{3}}\right)$ is equal to:

    (a) -7 (b) $\frac{7}{18}$ (c) $1\frac{8}{9}$ (d) $-8\frac{1}{2}$

13. The value of $\frac{2}{5}$ of $\left(4\frac{1}{2} - 3\frac{1}{4}\right) + 5 \div \frac{5}{16} - \frac{1}{4}$ is:

    (a) $17\frac{7}{20}$ (b) $80\frac{1}{2}$ (c) $16\frac{1}{4}$ (d) 88

14. $(2x - y)^2$ is equal to:

    (a) $4x^2 + y^2$ (b) $2x^2 - 2xy + y^2$
    (c) $4x^2 - y^2$ (d) $4x^2 - 4xy + y^2$

15. $\left(\sqrt{x}\right)\left(y^{3/2}\right)\left(x^2 y\right)$ is equal to:

    (a) $\sqrt{(xy)^5}$ (b) $x^{\sqrt{2}} y^{5/2}$
    (c) $xy^{5/2}$ (d) $x\sqrt{y^3}$

16. $p \times \left(\dfrac{1}{p^{-2}}\right)^{-3}$ is equivalent to:

    (a) $\dfrac{1}{p^{\frac{3}{2}}}$ (b) p^{-5} (c) p^5 (d) $\dfrac{1}{p^{-5}}$

17. $45 + 30 \div (21 - 6) - 2 \times 5 + 1$ is equal to:

    (a) -4 (b) 35 (c) -7 (d) 38

18. Factorising $2xy^2 + 6x^3y - 8x^3y^2$ gives:

    (a) $2x\left(y^2 + 3x^2 - 4x^2y\right)$

    (b) $2xy\left(y + 3x^2y - 4x^2y^2\right)$

    (c) $96x^7y^5$

    (d) $2xy\left(y + 3x^2 - 4x^2y\right)$

19. $\dfrac{62.91}{0.12} + \sqrt{\left(\dfrac{6.36\pi}{2e^{-3} \times \sqrt{73.81}}\right)}$ is equal to:

    (a) 529.08 correct to 5 significant figures

    (b) 529.082 correct to 3 decimal places

    (c) 5.29×10^2

    (d) 529.0 correct to 1 decimal place

20. Expanding $(m - n)^2$ gives:

    (a) $m^2 + n^2$ (b) $m^2 - 2mn - n^2$

    (c) $m^2 - 2mn + n^2$ (d) $m^2 - n^2$

21. $2x^2 - (x - xy) - x(2y - x)$ simplifies to:

    (a) $x(3x - 1 - y)$ (b) $x^2 - 3xy - xy$

    (c) $x(xy - y - 1)$ (d) $3x^2 - x + xy$

22. 11 mm expressed as a percentage of 41 mm is:

    (a) 2.68, correct to 3 significant figures

    (b) 2.6, correct to 2 significant figures

    (c) 26.83, correct to 2 decimal places

    (d) 0.2682, correct to 4 decimal places

23. The current in a component in an electrical circuit is calculated as 25 mA using Ohm's law. When measured, the actual current is 25.2 mA. Correct to 2 decimal places, the percentage error in the calculation is:

    (a) 0.80% (b) 1.25% (c) 0.79% (d) 1.26%

24. $\left(\dfrac{a^2}{a^{-3} \times a}\right)^{-2}$ is equivalent to:

    (a) a^{-8} (b) $\dfrac{1}{a^{-10}}$ (c) $\dfrac{1}{a^{-8}}$ (d) a^{-10}

25. Resistance, R, is given by the formula $R = \dfrac{\rho\ell}{A}$. When resistivity $\rho = 0.017\,\mu\Omega\text{m}$, length $\ell = 5$ km and cross-sectional area $A = 25$ mm^2, the resistance is:

    (a) 3.4 mΩ (b) 0.034 Ω (c) 3.4 kΩ (d) 3.4 Ω

The companion website for this book contains the above multiple-choice test. If you prefer to attempt the test online then visit:

www.routledge.com/cw/bird

For a copy of this multiple choice test, go to:

www.routledge.com/cw/bird

COMPANION @ WEBSITE

Transposing formulae

At the end of this chapter, you should be able to:

- define 'subject of the formula'
- transpose equations whose terms are connected by plus and/or minus signs
- transpose equations that involve fractions
- transpose equations that contain a root or power
- transpose equations in which the subject appears in more than one term

12.1 Introduction

In the formula $I = \dfrac{V}{R}$, I is called the **subject of the formula**.

Similarly, in the formula $y = mx + c$, y is the subject of the formula.

When a symbol other than the subject is required to be the subject, the formula needs to be rearranged to make a new subject. This rearranging process is called **transposing the formula** or **transposition**.

For example, in the above formulae,

$$\text{if } I = \frac{V}{R} \text{ then } V = IR$$

$$\text{and if } y = mx + c \text{ then } x = \frac{y - c}{m}$$

How did we arrive at these transpositions? This is the purpose of this chapter – to show how to transpose formulae. A great many equations occur in engineering and it is essential that we can transpose them when needed.

12.2 Transposing formulae

There are no new rules for transposing formulae. The same rules as were used for simple equations in Chapter 11 are used; i.e. **the balance of an equation must be maintained**: whatever is done to one side of an equation must be done to the other.

It is best that you cover simple equations before trying this chapter.

Here are some worked examples to help understanding of transposing formulae.

Problem 1. Transpose $p = q + r + s$ to make r the subject

The object is to obtain r on its own on the LHS of the equation. Changing the equation around so that r is on the LHS gives

$$q + r + s = p \qquad (1)$$

From Chapter 11 on simple equations, a term can be moved from one side of an equation to the other side as long as the sign is changed.
Rearranging gives $r = p - q - s$
Mathematically, we have subtracted $q + s$ from both sides of equation (1).

Problem 2. If $a + b = w - x + y$, express x as the subject

As stated in Problem 1, a term can be moved from one side of an equation to the other side but with a change of sign.
Hence, rearranging gives $x = w + y - a - b$

Problem 3. Transpose $v = f\lambda$ to make λ the subject

$v = f\lambda$ relates velocity v, frequency f and wavelength λ

Rearranging gives $\qquad f\lambda = v$

Dividing both sides by f gives $\quad \dfrac{f\lambda}{f} = \dfrac{v}{f}$

Cancelling gives $\qquad \lambda = \dfrac{v}{f}$

Problem 4. When a body falls freely through a height h, the velocity v is given by $v^2 = 2gh$. Express this formula with h as the subject

Rearranging gives $\qquad 2gh = v^2$

Dividing both sides by $2g$ gives $\quad \dfrac{2gh}{2g} = \dfrac{v^2}{2g}$

Cancelling gives $\qquad h = \dfrac{v^2}{2g}$

Problem 5. If $I = \dfrac{V}{R}$, rearrange to make V the subject

$I = \dfrac{V}{R}$ is Ohm's law, where I is the current, V is the voltage and R is the resistance.

Rearranging gives $\qquad \dfrac{V}{R} = I$

Multiplying both sides by R gives $\quad R\left(\dfrac{V}{R}\right) = R(I)$

Cancelling gives $\qquad \boldsymbol{V = IR}$

Problem 6. Transpose $a = \dfrac{F}{m}$ for m

$a = \dfrac{F}{m}$ relates acceleration a, force F and mass m.

Rearranging gives $\qquad \dfrac{F}{m} = a$

Multiplying both sides by m gives $\quad m\left(\dfrac{F}{m}\right) = m(a)$

Cancelling gives $\qquad F = ma$

Rearranging gives $\qquad ma = F$

Dividing both sides by a gives $\quad \dfrac{ma}{a} = \dfrac{F}{a}$

i.e. $\qquad \boldsymbol{m = \dfrac{F}{a}}$

Problem 7. Rearrange the formula $R = \dfrac{\rho L}{A}$ to make (a) A the subject and (b) L the subject

$R = \dfrac{\rho L}{A}$ relates resistance R of a conductor, resistivity ρ, conductor length L and conductor cross-sectional area A.

(a) Rearranging gives $\qquad \dfrac{\rho L}{A} = R$

Multiplying both sides by A gives

$$A\left(\dfrac{\rho L}{A}\right) = A(R)$$

Cancelling gives $\qquad \rho L = AR$

Rearranging gives $\qquad AR = \rho l$

Dividing both sides by R gives $\quad \dfrac{AR}{R} = \dfrac{\rho L}{R}$

Cancelling gives $\qquad \boldsymbol{A = \dfrac{\rho L}{R}}$

(b) Multiplying both sides of $\dfrac{\rho L}{A} = R$ by A gives

$$\rho L = AR$$

Dividing both sides by ρ gives $\dfrac{\rho L}{\rho} = \dfrac{AR}{\rho}$

Cancelling gives $L = \dfrac{AR}{\rho}$

Problem 8. Transpose $y = mx + c$ to make m the subject

$y = mx + c$ is the equation of a straight line graph, where y is the vertical axis variable, x is the horizontal axis variable, m is the gradient of the graph and c is the y-axis intercept.

Subtracting c from both sides gives $\qquad y - c = mx$

or $\qquad mx = y - c$

Dividing both sides by x gives $\qquad m = \dfrac{y - c}{x}$

Now try the following Practice Exercise

Practice Exercise 46 Transposing formulae
(answers on page 426)

Make the symbol indicated the subject of each of the formulae shown and express each in its simplest form.

1. $a + b = c - d - e$ $\qquad (d)$
2. $y = 7x$ $\qquad (x)$
3. $pv = c$ $\qquad (v)$
4. $v = u + at$ $\qquad (a)$
5. $V = IR$ $\qquad (R)$
6. $x + 3y = t$ $\qquad (y)$
7. $c = 2\pi r$ $\qquad (r)$
8. $y = mx + c$ $\qquad (x)$
9. $I = PRT$ $\qquad (T)$
10. $X_L = 2\pi f L$ $\qquad (L)$
11. $I = \dfrac{E}{R}$ $\qquad (R)$
12. $y = \dfrac{x}{a} + 3$ $\qquad (x)$
13. $F = \dfrac{9}{5}C + 32$ $\qquad (C)$
14. $X_C = \dfrac{1}{2\pi f C}$ $\qquad (f)$

12.3 Further transposing of formulae

Here are some more transposition examples to help us further understand how more difficult formulae are transposed.

Problem 9. Transpose the formula $v = u + \dfrac{Ft}{m}$ to make F the subject

$v = u + \dfrac{Ft}{m}$ relates final velocity v, initial velocity u, force F, mass m and time t. $\left(\dfrac{F}{m} \text{ is acceleration } a.\right)$

Rearranging gives $\qquad u + \dfrac{Ft}{m} = v$

and $\qquad \dfrac{Ft}{m} = v - u$

Multiplying each side by m gives

$$m\left(\dfrac{Ft}{m}\right) = m(v - u)$$

Cancelling gives $\qquad Ft = m(v - u)$

Dividing both sides by t gives $\qquad \dfrac{Ft}{t} = \dfrac{m(v - u)}{t}$

Cancelling gives $F = \dfrac{m(v - u)}{t}$ or $F = \dfrac{m}{t}(v - u)$

This shows two ways of expressing the answer. There is often more than one way of expressing a transposed answer. In this case, these equations for F are equivalent; neither one is more correct than the other.

Problem 10. The final length L_2 of a piece of wire heated through $\theta°C$ is given by the formula $L_2 = L_1(1 + \alpha\theta)$ where L_1 is the original length. Make the coefficient of expansion α the subject

Rearranging gives $\qquad L_1(1 + \alpha\theta) = L_2$

Removing the bracket gives $\qquad L_1 + L_1\alpha\theta = L_2$

Rearranging gives $\qquad L_1\alpha\theta = L_2 - L_1$

Dividing both sides by $L_1\theta$ gives $\quad \dfrac{L_1\alpha\theta}{L_1\theta} = \dfrac{L_2 - L_1}{L_1\theta}$

Cancelling gives $\qquad\qquad \boldsymbol{\alpha = \dfrac{L_2 - L_1}{L_1\theta}}$

An alternative method of transposing $L_2 = L_1(1 + \alpha\theta)$ for α is:

Dividing both sides by L_1 gives $\quad \dfrac{L_2}{L_1} = 1 + \alpha\theta$

Subtracting 1 from both sides gives $\dfrac{L_2}{L_1} - 1 = \alpha\theta$

or $\qquad\qquad\qquad\qquad \alpha\theta = \dfrac{L_2}{L_1} - 1$

Dividing both sides by θ gives $\quad \boldsymbol{\alpha = \dfrac{\dfrac{L_2}{L_1} - 1}{\theta}}$

The two answers $\alpha = \dfrac{L_2 - L_1}{L_1\theta}$ and $\alpha = \dfrac{\dfrac{L_2}{L_1} - 1}{\theta}$ look quite different. They are, however, equivalent. The first answer looks tidier but is no more correct than the second answer.

Problem 11. A formula for the distance s moved by a body is given by $s = \dfrac{1}{2}(v + u)t$. Rearrange the formula to make u the subject

Rearranging gives $\qquad \dfrac{1}{2}(v + u)t = s$

Multiplying both sides by 2 gives $\quad (v + u)t = 2s$

Dividing both sides by t gives $\quad \dfrac{(v + u)t}{t} = \dfrac{2s}{t}$

Cancelling gives $\qquad\qquad v + u = \dfrac{2s}{t}$

Rearranging gives $\quad \boldsymbol{u = \dfrac{2s}{t} - v}$ or $\boldsymbol{u = \dfrac{2s - vt}{t}}$

Problem 12. A formula for kinetic energy is $k = \dfrac{1}{2}mv^2$. Transpose the formula to make v the subject

Rearranging gives $\quad \dfrac{1}{2}mv^2 = k$

Whenever the prospective new subject is a squared term, that term is isolated on the LHS and then the square root of both sides of the equation is taken.

Multiplying both sides by 2 gives $\quad mv^2 = 2k$

Dividing both sides by m gives $\quad \dfrac{mv^2}{m} = \dfrac{2k}{m}$

Cancelling gives $\qquad\qquad v^2 = \dfrac{2k}{m}$

Taking the square root of both sides gives

$$\sqrt{v^2} = \sqrt{\left(\dfrac{2k}{m}\right)}$$

i.e. $\qquad\qquad v = \sqrt{\left(\dfrac{2k}{m}\right)}$

Problem 13. In a right-angled triangle having sides x, y and hypotenuse z, Pythagoras' theorem* states $z^2 = x^2 + y^2$. Transpose the formula to find x

*Who was Pythagoras? – **Pythagoras of Samos** (Born about 570 BC and died about 495 BC) was an Ionian Greek philosopher and mathematician. He is best known for the Pythagorean theorem, which states that in a right-angled triangle $a^2 + b^2 = c^2$. To find out more go to www.routledge.com/cw/bird

Rearranging gives $\qquad x^2 + y^2 = z^2$

and $\qquad\qquad\qquad x^2 = z^2 - y^2$

Taking the square root of both sides gives

$$x = \sqrt{z^2 - y^2}$$

Problem 14. Transpose $y = \dfrac{ML^2}{8EI}$ to make L the subject

Multiplying both sides by $8EI$ gives $\quad 8EIy = ML^2$

Dividing both sides by M gives $\qquad \dfrac{8EIy}{M} = L^2$

or $\qquad\qquad\qquad\qquad L^2 = \dfrac{8EIy}{M}$

Taking the square root of both sides gives

$$\sqrt{L^2} = \sqrt{\dfrac{8EIy}{M}}$$

i.e. $\qquad\qquad L = \sqrt{\dfrac{8EIy}{M}}$

Problem 15. Given $t = 2\pi\sqrt{\dfrac{l}{g}}$, find g in terms of t, l and π

Whenever the prospective new subject is within a square root sign, it is best to isolate that term on the LHS and then to square both sides of the equation.

Rearranging gives $\qquad\qquad 2\pi\sqrt{\dfrac{l}{g}} = t$

Dividing both sides by 2π gives $\qquad \sqrt{\dfrac{l}{g}} = \dfrac{t}{2\pi}$

Squaring both sides gives $\qquad \dfrac{l}{g} = \left(\dfrac{t}{2\pi}\right)^2 = \dfrac{t^2}{4\pi^2}$

Cross-multiplying, (i.e. multiplying

each term by $4\pi^2 g$), gives $\qquad 4\pi^2 l = gt^2$

or $\qquad\qquad\qquad\qquad gt^2 = 4\pi^2 l$

Dividing both sides by t^2 gives $\qquad \dfrac{gt^2}{t^2} = \dfrac{4\pi^2 l}{t^2}$

Cancelling gives $\qquad\qquad g = \dfrac{4\pi^2 l}{t^2}$

Problem 16. The impedance Z of an a.c. circuit is given by $Z = \sqrt{R^2 + X^2}$ where R is the resistance. Make the reactance, X, the subject

Rearranging gives $\qquad \sqrt{R^2 + X^2} = Z$

Squaring both sides gives $\qquad R^2 + X^2 = Z^2$

Rearranging gives $\qquad\qquad X^2 = Z^2 - R^2$

Taking the square root of both sides gives

$$X = \sqrt{Z^2 - R^2}$$

Problem 17. The volume V of a hemisphere of radius r is given by $V = \dfrac{2}{3}\pi r^3$. (a) Find r in terms of V. (b) Evaluate the radius when $V = 32\,\text{cm}^3$

(a) Rearranging gives $\qquad\qquad \dfrac{2}{3}\pi r^3 = V$

Multiplying both sides by 3 gives $\quad 2\pi r^3 = 3V$

Dividing both sides by 2π gives $\quad \dfrac{2\pi r^3}{2\pi} = \dfrac{3V}{2\pi}$

Cancelling gives $\qquad\qquad r^3 = \dfrac{3V}{2\pi}$

Taking the cube root of both sides gives

$$\sqrt[3]{r^3} = \sqrt[3]{\left(\dfrac{3V}{2\pi}\right)}$$

i.e. $\qquad\qquad r = \sqrt[3]{\left(\dfrac{3V}{2\pi}\right)}$

(b) When $V = 32\text{cm}^3$,

$$\text{radius } r = \sqrt[3]{\left(\dfrac{3V}{2\pi}\right)} = \sqrt[3]{\left(\dfrac{3 \times 32}{2\pi}\right)} = \textbf{2.48 cm}.$$

Now try the following Practice Exercise

Practice Exercise 47   Further transposing formulae (answers on page 427)

Make the symbol indicated the subject of each of the formulae shown in Problems 1 to 13 and express each in its simplest form.

1. $S = \dfrac{a}{1-r}$ (r)

2. $y = \dfrac{\lambda(x-d)}{d}$ (x)

3. $A = \dfrac{3(F-f)}{L}$ (f)

4. $y = \dfrac{AB^2}{5CD}$ (D)

5. $R = R_0(1+\alpha t)$ (t)

6. $\dfrac{1}{R} = \dfrac{1}{R_1} + \dfrac{1}{R_2}$ (R_2)

7. $I = \dfrac{E-e}{R+r}$ (R)

8. $y = 4ab^2c^2$ (b)

9. $\dfrac{a^2}{x^2} + \dfrac{b^2}{y^2} = 1$ (x)

10. $t = 2\pi\sqrt{\dfrac{L}{g}}$ (L)

11. $v^2 = u^2 + 2as$ (u)

12. $A = \dfrac{\pi R^2 \theta}{360}$ (R)

13. $N = \sqrt{\left(\dfrac{a+x}{y}\right)}$ (a)

14. Transpose $Z = \sqrt{R^2 + (2\pi f L)^2}$ for L and evaluate L when $Z = 27.82$, $R = 11.76$ and $f = 50$.

15. The lift force, L, on an aircraft is given by: $L = \dfrac{1}{2}\rho v^2 a c$ where ρ is the density, v is the velocity, a is the area and c is the lift coefficient. Transpose the equation to make the velocity the subject.

16. The angular deflection θ of a beam of electrons due to a magnetic field is given by:

$\theta = k\left(\dfrac{HL}{V^{\frac{1}{2}}}\right)$ Transpose the equation for V.

12.4 More difficult transposing of formulae

Here are some more transposition examples to help us further understand how more difficult formulae are transposed.

Problem 18. (a) Transpose $S = \sqrt{\dfrac{3d(L-d)}{8}}$ to make L the subject. (b) Evaluate L when $d = 1.65$ and $S = 0.82$

The formula $S = \sqrt{\dfrac{3d(L-d)}{8}}$ represents the sag S at the centre of a wire.

(a) Squaring both sides gives $S^2 = \dfrac{3d(L-d)}{8}$

Multiplying both sides by 8 gives

$$8S^2 = 3d(L-d)$$

Dividing both sides by $3d$ gives $\dfrac{8S^2}{3d} = L - d$

Rearranging gives $\mathbf{L = d + \dfrac{8S^2}{3d}}$

(b) When $d = 1.65$ and $S = 0.82$,

$$\mathbf{L} = d + \dfrac{8S^2}{3d} = 1.65 + \dfrac{8 \times 0.82^2}{3 \times 1.65} = \mathbf{2.737}$$

Problem 19. Transpose the formula $p = \dfrac{a^2x^2 + a^2y}{r}$ to make a the subject

Rearranging gives $\dfrac{a^2x^2 + a^2y}{r} = p$

Multiplying both sides by r gives

$$a^2x + a^2y = rp$$

Factorising the LHS gives $a^2(x+y) = rp$

Dividing both sides by $(x+y)$ gives

$$\dfrac{a^2(x+y)}{(x+y)} = \dfrac{rp}{(x+y)}$$

Cancelling gives $a^2 = \dfrac{rp}{(x+y)}$

Taking the square root of both sides gives

$$\mathbf{a} = \sqrt{\left(\dfrac{\mathbf{rp}}{\mathbf{x+y}}\right)}$$

Whenever the letter required as the new subject occurs more than once in the original formula, after rearranging, **factorising** will always be needed.

Problem 20. Make b the subject of the formula
$$a = \frac{x-y}{\sqrt{bd+be}}$$

Rearranging gives $\qquad \dfrac{x-y}{\sqrt{bd+be}} = a$

Multiplying both sides by $\sqrt{bd+be}$ gives

$$x-y = a\sqrt{bd+be}$$

or $\qquad\qquad\qquad a\sqrt{bd+be} = x-y$

Dividing both sides by a gives $\sqrt{bd+be} = \dfrac{x-y}{a}$

Squaring both sides gives $\qquad bd+be = \left(\dfrac{x-y}{a}\right)^2$

Factorising the LHS gives $\qquad b(d+e) = \left(\dfrac{x-y}{a}\right)^2$

Dividing both sides by $(d+e)$ gives

$$b = \frac{\left(\dfrac{x-y}{a}\right)^2}{(d+e)} \quad \text{or} \quad b = \frac{(x-y)^2}{a^2(d+e)}$$

Problem 21. If $a = \dfrac{b}{1+b}$, make b the subject of the formula

Rearranging gives $\qquad \dfrac{b}{1+b} = a$

Multiplying both sides by $(1+b)$ gives

$$b = a(1+b)$$

Removing the bracket gives $\qquad b = a + ab$

Rearranging to obtain terms in b on the LHS gives

$$b - ab = a$$

Factorising the LHS gives $\qquad b(1-a) = a$

Dividing both sides by $(1-a)$ gives $b = \dfrac{a}{1-a}$

Problem 22. Transpose the formula $V = \dfrac{Er}{R+r}$ to make r the subject

Rearranging gives $\qquad \dfrac{Er}{R+r} = V$

Multiplying both sides by $(R+r)$ gives

$$Er = V(R+r)$$

Removing the bracket gives $\qquad Er = VR + Vr$

Rearranging to obtain terms in r on the LHS gives

$$Er - Vr = VR$$

Factorising gives $\qquad r(E-V) = VR$

Dividing both sides by $(E-V)$ gives $r = \dfrac{VR}{E-V}$

Problem 23. Transpose the formula
$$y = \frac{pq^2}{r+q^2} - t \text{ to make } q \text{ the subject}$$

Rearranging gives $\qquad \dfrac{pq^2}{r+q^2} - t = y$

and $\qquad\qquad\qquad \dfrac{pq^2}{r+q^2} = y + t$

Multiplying both sides by $(r+q^2)$ gives

$$pq^2 = (r+q^2)(y+t)$$

Removing brackets gives $\quad pq^2 = ry + rt + q^2y + q^2t$

Rearranging to obtain terms in q on the LHS gives

$$pq^2 - q^2y - q^2t = ry + rt$$

Factorising gives $q^2(p-y-t) = r(y+t)$

Dividing both sides by $(p-y-t)$ gives

$$q^2 = \frac{r(y+t)}{(p-y-t)}$$

Taking the square root of both sides gives

$$q = \sqrt{\left(\frac{r(y+t)}{p-y-t}\right)}$$

Problem 24. Given that $\dfrac{D}{d} = \sqrt{\left(\dfrac{f+p}{f-p}\right)}$ express p in terms of D, d and f

Rearranging gives $\qquad \sqrt{\left(\dfrac{f+p}{f-p}\right)} = \dfrac{D}{d}$

Squaring both sides gives $\qquad \left(\dfrac{f+p}{f-p}\right) = \dfrac{D^2}{d^2}$

Cross-multiplying, i.e. multiplying each term by $d^2(f-p)$, gives

$$d^2(f+p) = D^2(f-p)$$

Removing brackets gives $d^2f + d^2p = D^2f - D^2p$

Rearranging, to obtain terms in p on the LHS gives

$$d^2p + D^2p = D^2f - d^2f$$

Factorising gives $p(d^2 + D^2) = f(D^2 - d^2)$

Dividing both sides by $(d^2 + D^2)$ gives

$$p = \frac{f(D^2 - d^2)}{(d^2 + D^2)}$$

Now try the following Practice Exercise

Practice Exercise 48 Further transposing formulae (answers on page 427)

Make the symbol indicated the subject of each of the formulae shown in Problems 1 to 7 and express each in its simplest form.

1. $y = \dfrac{a^2m - a^2n}{x}$ (a)

2. $M = \pi(R^4 - r^4)$ (R)

3. $x + y = \dfrac{r}{3+r}$ (r)

4. $m = \dfrac{\mu L}{L + rCR}$ (L)

5. $a^2 = \dfrac{b^2 - c^2}{b^2}$ (b)

6. $\dfrac{x}{y} = \dfrac{1 + r^2}{1 - r^2}$ (r)

7. $\dfrac{p}{q} = \sqrt{\left(\dfrac{a + 2b}{a - 2b}\right)}$ (b)

8. A formula for the focal length, f, of a convex lens is $\dfrac{1}{f} = \dfrac{1}{u} + \dfrac{1}{v}$. Transpose the formula to make v the subject and evaluate v when $f = 5$ and $u = 6$

9. The quantity of heat, Q, is given by the formula $Q = mc(t_2 - t_1)$. Make t_2 the subject of the formula and evaluate t_2 when $m = 10, t_1 = 15$, $c = 4$ and $Q = 1600$

10. The velocity, v, of water in a pipe appears in the formula $h = \dfrac{0.03Lv^2}{2dg}$. Express v

as the subject of the formula and evaluate v when $h = 0.712, L = 150, d = 0.30$ and $g = 9.81$

11. The sag, S, at the centre of a wire is given by the formula $S = \sqrt{\left(\dfrac{3d(L - d)}{8}\right)}$. Make L the subject of the formula and evaluate L when $d = 1.75$ and $S = 0.80$

12. In an electrical alternating current circuit the impedance Z is given by $Z = \sqrt{\left\{R^2 + \left(\omega L - \dfrac{1}{\omega C}\right)^2\right\}}$. Transpose the formula to make C the subject and hence evaluate C when $Z = 130, R = 120, \omega = 314$ and $L = 0.32$

13. An approximate relationship between the number of teeth, T, on a milling cutter, the diameter of cutter, D, and the depth of cut, d, is given by $T = \dfrac{12.5\,D}{D + 4d}$. Determine the value of D when $T = 10$ and $d = 4\,\text{mm}$.

14. Make λ, the wavelength of X-rays, the subject of the following formula: $\dfrac{\mu}{\rho} = \dfrac{CZ^4\sqrt{\lambda^5}\,n}{a}$

15. A simply supported beam of length L has a centrally applied load F and a uniformly distributed load of w per metre length of beam. The reaction at the beam support is given by:

$$R = \frac{1}{2}(F + wL)$$

Rearrange the equation to make w the subject. Hence determine the value of w when $L = 4$ m, $F = 8$ kN and $R = 10$ kN

16. The rate of heat conduction through a slab of material, Q, is given by the formula $Q = \dfrac{kA(t_1 - t_2)}{d}$ where t_1 and t_2 are the temperatures of each side of the material, A is the area of the slab, d is the thickness of the slab, and k is the thermal conductivity of the material. Rearrange the formula to obtain an expression for t_2

17. The slip, s, of a vehicle is given by: $s = \left(1 - \dfrac{r\omega}{v}\right) \times 100\%$ where r is the tyre

radius, ω is the angular velocity and v the velocity. Transpose to make r the subject of the formula.

18. The critical load, F newtons, of a steel column may be determined from the formula $L\sqrt{\dfrac{F}{EI}} = n\pi$ where L is the length, EI is the flexural rigidity, and n is a positive integer. Transpose for F and hence determine the value of F when $n = 1$, $E = 0.25 \times 10^{12}\,\text{N/m}^2$, $I = 6.92 \times 10^{-6}\,\text{m}^4$ and $L = 1.12\,\text{m}$

19. The flow of slurry along a pipe in a coal processing plant is given by: $V = \dfrac{\pi p r^4}{8\eta\ell}$ Transpose the equation for r

20. The deflection head H of a metal structure is given by: $H = \sqrt{\dfrac{I\rho^4 D^2 \ell^{\frac{3}{2}}}{20g}}$ Transpose the formula for length ℓ

For fully worked solutions to each of the problems in Practice Exercises 46 to 48 in this chapter, go to the website:
www.routledge.com/cw/bird

Chapter 13

Solving simultaneous equations

Why it is important to understand: Solving simultaneous equations

Simultaneous equations arise a great deal in engineering and science, some applications including theory of structures, data analysis, electrical circuit analysis and air traffic control. Systems that consist of a small number of equations can be solved analytically using standard methods from algebra (as explained in this chapter). Systems of large numbers of equations require the use of numerical methods and computers. Solving simultaneous equations is an important skill required in all aspects of engineering.

At the end of this chapter, you should be able to:

- solve simultaneous equations in two unknowns by substitution
- solve simultaneous equations in two unknowns by elimination
- solve simultaneous equations involving practical situations
- solve simultaneous equations in three unknowns

13.1 Introduction

Only one equation is necessary when finding the value of a **single unknown quantity** (as with simple equations in Chapter 11). However, when an equation contains **two unknown quantities** it has an infinite number of solutions. When two equations are available connecting the same two unknown values then a unique solution is possible. Similarly, for three unknown quantities it is necessary to have three equations in order to solve for a particular value of each of the unknown quantities, and so on.

Equations which have to be solved together to find the unique values of the unknown quantities, which are true for each of the equations, are called **simultaneous equations**.

Two methods of solving simultaneous equations analytically are:

(a) by **substitution**, and

(b) by **elimination**.

(A graphical solution of simultaneous equations is shown in Chapter 19.)

13.2 Solving simultaneous equations in two unknowns

The method of solving simultaneous equations is demonstrated in the following worked problems.

Problem 1. Solve the following equations for x and y, (a) by substitution and (b) by elimination

$$x + 2y = -1 \qquad (1)$$
$$4x - 3y = 18 \qquad (2)$$

(a) By substitution

From equation (1): $x = -1 - 2y$
Substituting this expression for x into equation (2) gives

$$4(-1 - 2y) - 3y = 18$$

This is now a simple equation in y.
Removing the bracket gives

$$-4 - 8y - 3y = 18$$
$$-11y = 18 + 4 = 22$$
$$y = \frac{22}{-11} = -2$$

Substituting $y = -2$ into equation (1) gives

$$x + 2(-2) = -1$$
$$x - 4 = -1$$
$$x = -1 + 4 = 3$$

Thus, $x = 3$ and $y = -2$ is the solution to the simultaneous equations.
Check: in equation (2), since $x = 3$ and $y = -2$,

$$\text{LHS} = 4(3) - 3(-2) = 12 + 6 = 18 = \text{RHS}$$

(b) By elimination

$$x + 2y = -1 \qquad (1)$$
$$4x - 3y = 18 \qquad (2)$$

If equation (1) is multiplied throughout by 4, the coefficient of x will be the same as in equation (2), giving

$$4x + 8y = -4 \qquad (3)$$

Subtracting equation (3) from equation (2) gives

$$4x - 3y = 18 \qquad (2)$$
$$\underline{4x + 8y = -4 \qquad (3)}$$
$$0 - 11y = 22$$

Hence, $y = \dfrac{22}{-11} = \mathbf{-2}$

(Note: in the above subtraction,

$$18 - -4 = 18 + 4 = 22)$$

Substituting $y = -2$ into either equation (1) or equation (2) will give $x = 3$ as in method (a). The solution $x = 3$, $y = -2$ is the only pair of values that satisfies both of the original equations.

Problem 2. Solve, by a substitution method, the simultaneous equations

$$3x - 2y = 12 \qquad (1)$$
$$x + 3y = -7 \qquad (2)$$

From equation (2), $x = -7 - 3y$
Substituting for x in equation (1) gives

$$3(-7 - 3y) - 2y = 12$$

i.e. $\qquad -21 - 9y - 2y = 12$
$$-11y = 12 + 21 = 33$$

Hence, $y = \dfrac{33}{-11} = -3$
Substituting $y = -3$ in equation (2) gives

$$x + 3(-3) = -7$$

i.e. $\qquad x - 9 = -7$

Hence $\qquad x = -7 + 9 = 2$

Thus, $x = 2$, $y = -3$ is the solution of the simultaneous equations. (Such solutions should always be checked by substituting values into each of the original two equations.)

Problem 3. Use an elimination method to solve the following simultaneous equations

$$3x + 4y = 5 \qquad (1)$$
$$2x - 5y = -12 \qquad (2)$$

If equation (1) is multiplied throughout by 2 and equation (2) by 3, the coefficient of x will be the same in the newly formed equations. Thus,

$2 \times$ equation (1) gives $\quad 6x + 8y = 10 \qquad (3)$

$3 \times$ equation (2) gives $\quad 6x - 15y = -36 \qquad (4)$

Equation (3) – equation (4) gives

$$0 + 23y = 46$$

i.e. $\qquad y = \dfrac{46}{23} = \mathbf{2}$

(Note $+8y - -15y = 8y + 15y = 23y$ and $10 - -36 = 10 + 36 = 46$)

Substituting $y = 2$ in equation (1) gives

$$3x + 4(2) = 5$$

from which $\qquad 3x = 5 - 8 = -3$

and $\qquad\qquad x = -1$

Checking, by substituting $x = -1$ and $y = 2$ in equation (2), gives

$$\text{LHS} = 2(-1) - 5(2) = -2 - 10 = -12 = \text{RHS}$$

Hence, $x = -1$ and $y = 2$ is the solution of the simultaneous equations.

The elimination method is the most common method of solving simultaneous equations.

Problem 4. Solve

$$7x - 2y = 26 \qquad (1)$$
$$6x + 5y = 29 \qquad (2)$$

When equation (1) is multiplied by 5 and equation (2) by 2, the coefficients of y in each equation are numerically the same, i.e. 10, but are of opposite sign.

$5 \times$ equation (1) gives $\qquad 35x - 10y = 130 \qquad (3)$

$2 \times$ equation (2) gives $\qquad 12x + 10y = 58 \qquad (4)$

Adding equations (3)
and (4) gives $\qquad\qquad 47x + 0 = 188$

Hence, $\qquad\qquad x = \dfrac{188}{47} = 4$

Note that when the signs of common coefficients are **different** the two equations are **added** and when the signs of common coefficients are the **same** the two equations are **subtracted** (as in Problems 1 and 3).

Substituting $x = 4$ in equation (1) gives

$$7(4) - 2y = 26$$
$$28 - 2y = 26$$
$$28 - 26 = 2y$$
$$2 = 2y$$

Hence, $\qquad\qquad y = 1$

Checking, by substituting $x = 4$ and $y = 1$ in equation (2), gives

$$\text{LHS} = 6(4) + 5(1) = 24 + 5 = 29 = \text{RHS}$$

Thus, the solution is $x = 4, y = 1$

Now try the following Practice Exercise

Solve the following simultaneous equations and verify the results.

1. $2x - y = 6$
    $x + y = 6$

2. $2x - y = 2$
    $x - 3y = -9$

3. $x - 4y = -4$
    $5x - 2y = 7$

4. $3x - 2y = 10$
    $5x + y = 21$

5. $5p + 4q = 6$
    $2p - 3q = 7$

6. $7x + 2y = 11$
    $3x - 5y = -7$

7. $2x - 7y = -8$
    $3x + 4y = 17$

8. $a + 2b = 8$
    $b - 3a = -3$

9. $a + b = 7$
    $a - b = 3$

10. $2x + 5y = 7$
    $x + 3y = 4$

11. $3s + 2t = 12$
    $4s - t = 5$

12. $3x - 2y = 13$
    $2x + 5y = -4$

13. $5m - 3n = 11$
    $3m + n = 8$

14. $8a - 3b = 51$
    $3a + 4b = 14$

15. $5x = 2y$
    $3x + 7y = 41$

16. $5c = 1 - 3d$
    $2d + c + 4 = 0$

13.3 Further solving of simultaneous equations

Here are some further worked problems on solving simultaneous equations.

Problem 5. Solve

$$3p = 2q \qquad (1)$$
$$4p + q + 11 = 0 \qquad (2)$$

Rearranging gives

$$3p - 2q = 0 \qquad (3)$$
$$4p + q = -11 \qquad (4)$$

Multiplying equation (4) by 2 gives

$$8p + 2q = -22 \qquad (5)$$

Adding equations (3) and (5) gives

$$11p + 0 = -22$$

$$p = \frac{-22}{11} = -2$$

Substituting $p = -2$ into equation (1) gives

$$3(-2) = 2q$$

$$-6 = 2q$$

$$q = \frac{-6}{2} = -3$$

Checking, by substituting $p = -2$ and $q = -3$ into equation (2), gives

$$\text{LHS} = 4(-2) + (-3) + 11 = -8 - 3 + 11 = 0 = \text{RHS}$$

Hence, the solution is $p = -2, q = -3$

Problem 6.   Solve

$$\frac{x}{8} + \frac{5}{2} = y \qquad (1)$$

$$13 - \frac{y}{3} = 3x \qquad (2)$$

Whenever fractions are involved in simultaneous equations it is often easier to firstly remove them. Thus, multiplying equation (1) by 8 gives

$$8\left(\frac{x}{8}\right) + 8\left(\frac{5}{2}\right) = 8y$$

i.e. $\qquad\qquad x + 20 = 8y \qquad (3)$

Multiplying equation (2) by 3 gives

$$39 - y = 9x \qquad (4)$$

Rearranging equations (3) and (4) gives

$$x - 8y = -20 \qquad (5)$$

$$9x + y = 39 \qquad (6)$$

Multiplying equation (6) by 8 gives

$$72x + 8y = 312 \qquad (7)$$

Adding equations (5) and (7) gives

$$73x + 0 = 292$$

$$x = \frac{292}{73} = 4$$

Substituting $x = 4$ into equation (5) gives

$$4 - 8y = -20$$

$$4 + 20 = 8y$$

$$24 = 8y$$

$$y = \frac{24}{8} = 3$$

Checking, substituting $x = 4$ and $y = 3$ in the original equations, gives

(1): $\text{LHS} = \dfrac{4}{8} + \dfrac{5}{2} = \dfrac{1}{2} + 2\dfrac{1}{2} = 3 = y = \text{RHS}$

(2): $\text{LHS} = 13 - \dfrac{3}{3} = 13 - 1 = 12$

$\qquad\quad \text{RHS} = 3x = 3(4) = 12$

Hence, the solution is $x = 4, y = 3$

Problem 7.   Solve

$$2.5x + 0.75 - 3y = 0$$

$$1.6x = 1.08 - 1.2y$$

It is often easier to remove decimal fractions. Thus, multiplying equations (1) and (2) by 100 gives

$$250x + 75 - 300y = 0 \qquad (1)$$

$$160x = 108 - 120y \qquad (2)$$

Rearranging gives

$$250x - 300y = -75 \qquad (3)$$

$$160x + 120y = 108 \qquad (4)$$

Multiplying equation (3) by 2 gives

$$500x - 600y = -150 \qquad (5)$$

Multiplying equation (4) by 5 gives

$$800x + 600y = 540 \qquad (6)$$

Adding equations (5) and (6) gives

$$1300x + 0 = 390$$

$$x = \frac{390}{1300} = \frac{39}{130} = \frac{3}{10} = 0.3$$

Substituting $x = 0.3$ into equation (1) gives

$$250(0.3) + 75 - 300y = 0$$

$$75 + 75 = 300y$$

$$150 = 300y$$

$$y = \frac{150}{300} = 0.5$$

Checking, by substituting $x = 0.3$ and $y = 0.5$ in equation (2), gives

$$\text{LHS} = 160(0.3) = 48$$

$$\text{RHS} = 108 - 120(0.5) = 108 - 60 = 48$$

Hence, the solution is $x = 0.3$, $y = 0.5$

Now try the following Practice Exercise

Practice Exercise 50   Solving simultaneous equations (answers on page 427)

Solve the following simultaneous equations and verify the results.

1. $7p + 11 + 2q = 0$
    $-1 = 3q - 5p$

2. $\dfrac{x}{2} + \dfrac{y}{3} = 4$
    $\dfrac{x}{6} - \dfrac{y}{9} = 0$

3. $\dfrac{a}{2} - 7 = -2b$
    $12 = 5a + \dfrac{2}{3}b$

4. $\dfrac{3}{2}s - 2t = 8$
    $\dfrac{s}{4} + 3t = -2$

5. $\dfrac{x}{5} + \dfrac{2y}{3} = \dfrac{49}{15}$
    $\dfrac{3x}{7} - \dfrac{y}{2} + \dfrac{5}{7} = 0$

6. $v - 1 = \dfrac{u}{12}$
    $u + \dfrac{v}{4} - \dfrac{25}{2} = 0$

7. $1.5x - 2.2y = -18$
    $2.4x + 0.6y = 33$

8. $3b - 2.5a = 0.45$
    $1.6a + 0.8b = 0.8$

13.4 Solving more difficult simultaneous equations

Here are some further worked problems on solving more difficult simultaneous equations.

Problem 8. Solve

$$\frac{2}{x} + \frac{3}{y} = 7 \tag{1}$$

$$\frac{1}{x} - \frac{4}{y} = -2 \tag{2}$$

In this type of equation the solution is easier if a substitution is initially made. Let $\dfrac{1}{x} = a$ and $\dfrac{1}{y} = b$

Thus equation (1) becomes $2a + 3b = 7$ $\quad$ (3)

and equation (2) becomes $a - 4b = -2$ $\quad$ (4)

Multiplying equation (4) by 2 gives

$$2a - 8b = -4 \tag{5}$$

Subtracting equation (5) from equation (3) gives

$$0 + 11b = 11$$

i.e. $\qquad\qquad\qquad\quad$ $b = 1$

Substituting $b = 1$ in equation (3) gives

$$2a + 3 = 7$$

$$2a = 7 - 3 = 4$$

i.e. $\qquad\qquad\qquad\quad$ $a = 2$

Checking, substituting $a = 2$ and $b = 1$ in equation (4), gives

$$\text{LHS} = 2 - 4(1) = 2 - 4 = -2 = \text{RHS}$$

Hence, $a = 2$ and $b = 1$

However, since $\quad \dfrac{1}{x} = a, \quad x = \dfrac{1}{a} = \dfrac{1}{2}$ or 0.5

and since $\qquad \dfrac{1}{y} = b, \quad y = \dfrac{1}{b} = \dfrac{1}{1} = 1$

Hence, the solution is $x = 0.5$, $y = 1$

Problem 9. Solve

$$\frac{1}{2a} + \frac{3}{5b} = 4 \tag{1}$$

$$\frac{4}{a} + \frac{1}{2b} = 10.5 \tag{2}$$

Let $\dfrac{1}{a} = x$ $\quad$ and $\quad \dfrac{1}{b} = y$

then

$$\frac{x}{2} + \frac{3}{5}y = 4 \tag{3}$$

$$4x + \frac{1}{2}y = 10.5 \tag{4}$$

To remove fractions, equation (3) is multiplied by 10, giving

$$10\left(\frac{x}{2}\right) + 10\left(\frac{3}{5}y\right) = 10(4)$$

i.e. $5x + 6y = 40$ (5)

Multiplying equation (4) by 2 gives

$$8x + y = 21 \quad (6)$$

Multiplying equation (6) by 6 gives

$$48x + 6y = 126 \quad (7)$$

Subtracting equation (5) from equation (7) gives

$$43x + 0 = 86$$

$$x = \frac{86}{43} = 2$$

Substituting $x = 2$ into equation (3) gives

$$\frac{2}{2} + \frac{3}{5}y = 4$$

$$\frac{3}{5}y = 4 - 1 = 3$$

$$y = \frac{5}{3}(3) = 5$$

Since $\frac{1}{a} = x$, $a = \frac{1}{x} = \frac{1}{2}$ or 0.5

and since $\frac{1}{b} = y$, $b = \frac{1}{y} = \frac{1}{5}$ or 0.2

Hence, the solution is $a = 0.5$, $b = 0.2$, which may be checked in the original equations.

Problem 10. Solve

$$\frac{1}{x+y} = \frac{4}{27} \quad (1)$$

$$\frac{1}{2x-y} = \frac{4}{33} \quad (2)$$

To eliminate fractions, both sides of equation (1) are multiplied by $27(x + y)$, giving

$$27(x+y)\left(\frac{1}{x+y}\right) = 27(x+y)\left(\frac{4}{27}\right)$$

i.e. $27(1) = 4(x + y)$

$$27 = 4x + 4y \quad (3)$$

Similarly, in equation (2) $33 = 4(2x - y)$

i.e. $33 = 8x - 4y$ (4)

Equation (3) + equation (4) gives

$$60 = 12x \text{ and } x = \frac{60}{12} = 5$$

Substituting $x = 5$ in equation (3) gives

$$27 = 4(5) + 4y$$

from which $4y = 27 - 20 = 7$

and $y = \frac{7}{4} = 1\frac{3}{4}$ or **1.75**

Hence, $x = 5$, $y = 1.75$ is the required solution, which may be checked in the original equations.

Problem 11. Solve

$$\frac{x-1}{3} + \frac{y+2}{5} = \frac{2}{15} \quad (1)$$

$$\frac{1-x}{6} + \frac{5+y}{2} = \frac{5}{6} \quad (2)$$

Before equations (1) and (2) can be simultaneously solved, the fractions need to be removed and the equations rearranged.

Multiplying equation (1) by 15 gives

$$15\left(\frac{x-1}{3}\right) + 15\left(\frac{y+2}{5}\right) = 15\left(\frac{2}{15}\right)$$

i.e. $5(x - 1) + 3(y + 2) = 2$

$$5x - 5 + 3y + 6 = 2$$

$$5x + 3y = 2 + 5 - 6$$

Hence, $5x + 3y = 1$ (3)

Multiplying equation (2) by 6 gives

$$6\left(\frac{1-x}{6}\right) + 6\left(\frac{5+y}{2}\right) = 6\left(\frac{5}{6}\right)$$

i.e. $(1 - x) + 3(5 + y) = 5$

$$1 - x + 15 + 3y = 5$$

$$-x + 3y = 5 - 1 - 15$$

Hence, $-x + 3y = -11$ (4)

Thus the initial problem containing fractions can be expressed as

$$5x + 3y = 1 \qquad (3)$$

$$-x + 3y = -11 \qquad (4)$$

Subtracting equation (4) from equation (3) gives

$$6x + 0 = 12$$

$$x = \frac{12}{6} = 2$$

Substituting $x = 2$ into equation (3) gives

$$5(2) + 3y = 1$$

$$10 + 3y = 1$$

$$3y = 1 - 10 = -9$$

$$y = \frac{-9}{3} = -3$$

Checking, substituting $x = 2$, $y = -3$ in equation (4) gives

$$\text{LHS} = -2 + 3(-3) = -2 - 9 = -11 = \text{RHS}$$

Hence, the solution is $x = 2, y = -3$

Now try the following Practice Exercise

In Problems 1 to 7, solve the simultaneous equations and verify the results

1. $\dfrac{3}{x} + \dfrac{2}{y} = 14$ 2. $\dfrac{4}{a} - \dfrac{3}{b} = 18$

   $\dfrac{5}{x} - \dfrac{3}{y} = -2$ $\dfrac{2}{a} + \dfrac{5}{b} = -4$

3. $\dfrac{1}{2p} + \dfrac{3}{5q} = 5$ 4. $\dfrac{5}{x} + \dfrac{3}{y} = 1.1$

   $\dfrac{5}{p} - \dfrac{1}{2q} = \dfrac{35}{2}$ $\dfrac{3}{x} - \dfrac{7}{y} = -1.1$

5. $\dfrac{c+1}{4} - \dfrac{d+2}{3} + 1 = 0$

   $\dfrac{1-c}{5} + \dfrac{3-d}{4} + \dfrac{13}{20} = 0$

6. $\dfrac{3r+2}{5} - \dfrac{2s-1}{4} = \dfrac{11}{5}$ 7. $\dfrac{5}{x+y} = \dfrac{20}{27}$

   $\dfrac{3+2r}{4} + \dfrac{5-s}{3} = \dfrac{15}{4}$ $\dfrac{4}{2x-y} = \dfrac{16}{33}$

8. If $5x - \dfrac{3}{y} = 1$ and $x + \dfrac{4}{y} = \dfrac{5}{2}$, find the value of $\dfrac{xy+1}{y}$

13.5 Practical problems involving simultaneous equations

There are a number of situations in engineering and science in which the solution of simultaneous equations is required. Some are demonstrated in the following worked problems.

Problem 12. The law connecting friction F and load L for an experiment is of the form $F = aL + b$ where a and b are constants. When $F = 5.6\,\text{N}$, $L = 8.0\,\text{N}$ and when $F = 4.4\,\text{N}$, $L = 2.0\,\text{N}$. Find the values of a and b and the value of F when $L = 6.5\,\text{N}$

Substituting $F = 5.6$ and $L = 8.0$ into $F = aL + b$ gives

$$5.6 = 8.0a + b \qquad (1)$$

Substituting $F = 4.4$ and $L = 2.0$ into $F = aL + b$ gives

$$4.4 = 2.0a + b \qquad (2)$$

Subtracting equation (2) from equation (1) gives

$$1.2 = 6.0a$$

$$a = \frac{1.2}{6.0} = \frac{1}{5} \text{ or } 0.2$$

Substituting $a = \dfrac{1}{5}$ into equation (1) gives

$$5.6 = 8.0\left(\frac{1}{5}\right) + b$$

$$5.6 = 1.6 + b$$

$$5.6 - 1.6 = b$$

i.e. $\qquad\qquad b = 4$

Checking, substituting $a = \dfrac{1}{5}$ and $b = 4$ in equation (2), gives

$$\text{RHS} = 2.0\left(\frac{1}{5}\right) + 4 = 0.4 + 4 = 4.4 = \text{LHS}$$

Hence, $a = \dfrac{1}{5}$ and $b = 4$

When $L = 6.5$, $F = aL + b = \dfrac{1}{5}(6.5) + 4 = 1.3 + 4$, i.e. $F = 5.30\,\text{N}$.

Problem 13. The equation of a straight line, of gradient m and intercept on the y-axis c, is $y = mx + c$. If a straight line passes through the point where $x = 1$ and $y = -2$, and also through the point where $x = 3.5$ and $y = 10.5$, find the values of the gradient and the y-axis intercept

Substituting $x = 1$ and $y = -2$ into $y = mx + c$ gives

$$-2 = m + c \qquad\qquad (1)$$

Substituting $x = 3.5$ and $y = 10.5$ into $y = mx + c$ gives

$$10.5 = 3.5m + c \qquad\qquad (2)$$

Subtracting equation (1) from equation (2) gives

$$12.5 = 2.5m, \quad \text{from which,} \quad m = \frac{12.5}{2.5} = 5$$

Substituting $m = 5$ into equation (1) gives

$$-2 = 5 + c$$

$$c = -2 - 5 = -7$$

Checking, substituting $m = 5$ and $c = -7$ in equation (2), gives

$$\text{RHS} = (3.5)(5) + (-7) = 17.5 - 7 = 10.5 = \text{LHS}$$

Hence, the **gradient** $m = 5$ and the **y-axis intercept** $c = -7$

Problem 14. When Kirchhoff's laws* are applied to the electrical circuit shown in Fig. 13.1, the currents I_1 and I_2 are connected by the equations

$$27 = 1.5I_1 + 8(I_1 - I_2) \qquad\qquad (1)$$
$$-26 = 2I_2 - 8(I_1 - I_2) \qquad\qquad (2)$$

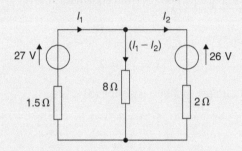

Figure 13.1

Solve the equations to find the values of currents I_1 and I_2

*Who was Kirchhoff? – **Gustav Robert Kirchhoff** (12 March 1824–17 October 1887) was a German physicist. Concepts in circuit theory and thermal emission are named 'Kirchhoff's laws' after him, as well as a law of thermochemistry. To find out more go to www.routledge.com/cw/bird

Removing the brackets from equation (1) gives

$$27 = 1.5I_1 + 8I_1 - 8I_2$$

Rearranging gives

$$9.5I_1 - 8I_2 = 27 \qquad (3)$$

Removing the brackets from equation (2) gives

$$-26 = 2I_2 - 8I_1 + 8I_2$$

Rearranging gives

$$-8I_1 + 10I_2 = -26 \qquad (4)$$

Multiplying equation (3) by 5 gives

$$47.5I_1 - 40I_2 = 135 \qquad (5)$$

Multiplying equation (4) by 4 gives

$$-32I_1 + 40I_2 = -104 \qquad (6)$$

Adding equations (5) and (6) gives

$$15.5I_1 + 0 = 31$$

$$I_1 = \frac{31}{15.5} = \mathbf{2}$$

Substituting $I_1 = 2$ into equation (3) gives

$$9.5(2) - 8I_1 = 27$$

$$19 - 8I_2 = 27$$

$$19 - 27 = 8I_2$$

$$-8 = 8I_2$$

and $\qquad\qquad\qquad I_2 = -1$

Hence, the solution is $I_1 = 2$ and $I_2 = -1$ (which may be checked in the original equations).

Problem 15. The distance s metres from a fixed point of a vehicle travelling in a straight line with constant acceleration, $a\,\text{m/s}^2$, is given by $s = ut + \frac{1}{2}at^2$, where u is the initial velocity in m/s and t the time in seconds. Determine the initial velocity and the acceleration given that $s = 42\,\text{m}$ when $t = 2\,\text{s}$, and $s = 144\,\text{m}$ when $t = 4\,\text{s}$. Also find the distance travelled after 3 s

Substituting $s = 42$ and $t = 2$ into $s = ut + \frac{1}{2}at^2$ gives

$$42 = 2u + \frac{1}{2}a(2)^2$$

i.e. $\qquad\qquad 42 = 2u + 2a \qquad (1)$

Substituting $s = 144$ and $t = 4$ into $s = ut + \frac{1}{2}at^2$ gives

$$144 = 4u + \frac{1}{2}a(4)^2$$

i.e. $\qquad\qquad 144 = 4u + 8a \qquad (2)$

Multiplying equation (1) by 2 gives

$$84 = 4u + 4a \qquad (3)$$

Subtracting equation (3) from equation (2) gives

$$60 = 0 + 4a$$

and $\qquad\qquad a = \frac{60}{4} = \mathbf{15}$

Substituting $a = 15$ into equation (1) gives

$$42 = 2u + 2(15)$$

$$42 - 30 = 2u$$

$$u = \frac{12}{2} = \mathbf{6}$$

Substituting $a = 15$ and $u = 6$ in equation (2) gives

$$\text{RHS} = 4(6) + 8(15) = 24 + 120 = 144 = \text{LHS}$$

Hence, **the initial velocity $u = 6\,\text{m/s}$ and the acceleration $a = 15\,\text{m/s}^2$**.

Distance travelled after 3 s is given by $s = ut + \frac{1}{2}at^2$ where $t = 3, u = 6$ and $a = 15$

Hence, $s = (6)(3) + \frac{1}{2}(15)(3)^2 = 18 + 67.5$

i.e. **distance travelled after 3 s = 85.5 m**.

Problem 16. The resistance $R\,\Omega$ of a length of wire at $t^\circ\text{C}$ is given by $R = R_0(1 + \alpha t)$, where R_0 is the resistance at 0°C and α is the temperature coefficient of resistance in $/^\circ\text{C}$. Find the values of α and R_0 if $R = 30\,\Omega$ at 50°C and $R = 35\,\Omega$ at 100°C

Substituting $R = 30$ and $t = 50$ into $R = R_0(1 + \alpha t)$ gives

$$30 = R_0(1 + 50\alpha) \qquad (1)$$

Substituting $R = 35$ and $t = 100$ into $R = R_0(1 + \alpha t)$ gives

$$35 = R_0(1 + 100\alpha) \qquad (2)$$

Although these equations may be solved by the conventional substitution method, an easier way is to eliminate

R_0 by division. Thus, dividing equation (1) by equation (2) gives

$$\frac{30}{35} = \frac{R_0(1 + 50\alpha)}{R_0(1 + 100\alpha)} = \frac{1 + 50\alpha}{1 + 100\alpha}$$

Cross-multiplying gives

$$30(1 + 100\alpha) = 35(1 + 50\alpha)$$

$$30 + 3000\alpha = 35 + 1750\alpha$$

$$3000\alpha - 1750\alpha = 35 - 30$$

$$1250\alpha = 5$$

i.e. $$\alpha = \frac{5}{1250} = \frac{1}{250} \quad \text{or} \quad \mathbf{0.004}$$

Substituting $\alpha = \frac{1}{250}$ into equation (1) gives

$$30 = R_0 \left\{ 1 + (50) \left(\frac{1}{250} \right) \right\}$$

$$30 = R_0(1.2)$$

$$R_0 = \frac{30}{1.2} = 25$$

Checking, substituting $\alpha = \frac{1}{250}$ and $R_0 = 25$ in equation (2), gives

$$\text{RHS} = 25 \left\{ 1 + (100) \left(\frac{1}{250} \right) \right\}$$

$$= 25(1.4) = 35 = \text{LHS}$$

Thus, the solution is $\boldsymbol{\alpha = 0.004/°C}$ and $\boldsymbol{R_0 = 25\,\Omega}$.

Problem 17. The molar heat capacity of a solid compound is given by the equation $c = a + bT$, where a and b are constants. When $c = 52, T = 100$ and when $c = 172, T = 400$. Determine the values of a and b

When $c = 52, \quad T = 100$, hence

$$52 = a + 100b \tag{1}$$

When $c = 172, \quad T = 400$, hence

$$172 = a + 400b \tag{2}$$

Equation (2) − equation (1) gives

$$120 = 300b$$

from which, $$b = \frac{120}{300} = \mathbf{0.4}$$

Substituting $b = 0.4$ in equation (1) gives

$$52 = a + 100(0.4)$$

$$a = 52 - 40 = \mathbf{12}$$

Hence, $\boldsymbol{a = 12}$ and $\boldsymbol{b = 0.4}$

Now try the following Practice Exercise

Practice Exercise 52   Practical problems involving simultaneous equations (answers on page 428)

1. In a system of pulleys, the effort P required to raise a load W is given by $P = aW + b$, where a and b are constants. If $W = 40$ when $P = 12$ and $W = 90$ when $P = 22$, find the values of a and b.

2. Applying Kirchhoff's laws to an electrical circuit produces the following equations:

$$5 = 0.2I_1 + 2(I_1 - I_2)$$

$$12 = 3I_2 + 0.4I_2 - 2(I_1 - I_2)$$

Determine the values of currents I_1 and I_2

3. Velocity v is given by the formula $v = u + at$. If $v = 20$ when $t = 2$ and $v = 40$ when $t = 7$, find the values of u and a. Then, find the velocity when $t = 3.5$

4. Three new cars and four new vans supplied to a dealer together cost £97 700 and five new cars and two new vans of the same models cost £103 100. Find the respective costs of a car and a van.

5. $y = mx + c$ is the equation of a straight line of slope m and y-axis intercept c. If the line passes through the point where $x = 2$ and $y = 2$, and also through the point where $x = 5$ and $y = 0.5$, find the slope and y-axis intercept of the straight line.

6. The resistance R ohms of copper wire at $t°C$ is given by $R = R_0(1 + \alpha t)$, where R_0 is the resistance at $0°C$ and α is the temperature coefficient of resistance. If $R = 25.44\,\Omega$ at $30°C$ and $R = 32.17\,\Omega$ at $100°C$, find α and R_0

7. The molar heat capacity of a solid compound is given by the equation $c = a + bT$. When

$c = 52, T = 100$ and when $c = 172, T = 400$. Find the values of a and b.

8. In an engineering process, two variables p and q are related by $q = ap + b/p$, where a and b are constants. Evaluate a and b if $q = 13$ when $p = 2$ and $q = 22$ when $p = 5$

9. In a system of forces, the relationship between two forces F_1 and F_2 is given by

$$5F_1 + 3F_2 + 6 = 0$$
$$3F_1 + 5F_2 + 18 = 0$$

Solve for F_1 and F_2

10. For a balanced beam, the equilibrium of forces is given by: $R_1 + R_2 = 12.0\,\text{kN}$
As a result of taking moments:
$0.2R_1 + 7 \times 0.3 + 3 \times 0.6 = 0.8R_2$
Determine the values of the reaction forces R_1 and R_1

13.6 Solving simultaneous equations in three unknowns

Equations containing three unknowns may be solved using exactly the same procedures as those used with two equations and two unknowns, providing that there are three equations to work with. The method is demonstrated in the following worked problem.

Problem 18. Solve the simultaneous equations.

$$x + y + z = 4 \qquad (1)$$
$$2x - 3y + 4z = 33 \qquad (2)$$
$$3x - 2y - 2z = 2 \qquad (3)$$

There are a number of ways of solving these equations. One method is shown below.
The initial object is to produce two equations with two unknowns. For example, multiplying equation (1) by 4 and then subtracting this new equation from equation (2) will produce an equation with only x and y involved.
Multiplying equation (1) by 4 gives

$$4x + 4y + 4z = 16 \qquad (4)$$

Equation (2) – equation (4) gives

$$-2x - 7y = 17 \qquad (5)$$

Similarly, multiplying equation (3) by 2 and then adding this new equation to equation (2) will produce another equation with only x and y involved.

Multiplying equation (3) by 2 gives

$$6x - 4y - 4z = 4 \qquad (6)$$

Equation (2) + equation (6) gives

$$8x - 7y = 37 \qquad (7)$$

Rewriting equation (5) gives

$$-2x - 7y = 17 \qquad (5)$$

Now we can use the previous method for solving simultaneous equations in two unknowns.

Equation (7) – equation (5) gives $\quad 10x = 20$

from which, $\qquad\qquad\qquad\qquad x = 2$

(Note that $8x - -2x = 8x + 2x = 10x$)

Substituting $x = 2$ into equation (5) gives

$$-4 - 7y = 17$$

from which, $\qquad -7y = 17 + 4 = 21$

and $\qquad\qquad\qquad y = -3$

Substituting $x = 2$ and $y = -3$ into equation (1) gives

$$2 - 3 + z = 4$$

from which, $\qquad\qquad z = 5$

Hence, the solution of the simultaneous equations is $x = 2, y = -3$ and $z = 5$

Now try the following Practice Exercise

Practice Exercise 53 Simultaneous equations in three unknowns (answers on page 428)

In Problems 1 to 9, solve the simultaneous equations in 3 unknowns.

1. $x + 2y + 4z = 16$
   $2x - y + 5z = 18$
   $3x + 2y + 2z = 14$

2. $2x + y - z = 0$
   $3x + 2y + z = 4$
   $5x + 3y + 2z = 8$

3. $3x + 5y + 2z = 6$
   $x - y + 3z = 0$
   $2x + 7y + 3z = -3$

4. $2x + 4y + 5z = 23$
   $3x - y - 2z = 6$
   $4x + 2y + 5z = 31$

5. $2x + 3y + 4z = 36$
   $3x + 2y + 3z = 29$
   $x + y + z = 11$

6. $4x + y + 3z = 31$
   $2x - y + 2z = 10$
   $3x + 3y - 2z = 7$

7. $5x + 5y - 4z = 37$
   $2x - 2y + 9z = 20$
   $-4x + y + z = -14$

8. $6x + 7y + 8z = 13$
   $3x + y - z = -11$
   $2x - 2y - 2z = -18$

9. $3x + 2y + z = 14$
   $7x + 3y + z = 22.5$
   $4x - 4y - z = -8.5$

10. Kirchhoff's laws are used to determine the current equations in an electrical network and result in the following:

$$i_1 + 8i_2 + 3i_3 = -31$$

$$3i_1 - 2i_2 + i_3 = -5$$

$$2i_1 - 3i_2 + 2i_3 = 6$$

Determine the values of i_1, i_2 and i_3

11. The forces in three members of a framework are F_1, F_2 and F_3. They are related by the following simultaneous equations.

$$1.4F_1 + 2.8F_2 + 2.8F_3 = 5.6$$

$$4.2F_1 - 1.4F_2 + 5.6F_3 = 35.0$$

$$4.2F_1 + 2.8F_2 - 1.4F_3 = -5.6$$

Find the values of F_1, F_2 and F_3

For fully worked solutions to each of the problems in Practice Exercises 49 to 53 in this chapter, go to the website:
www.routledge.com/cw/bird

This assignment covers the material contained in Chapters 12 and 13. *The marks available are shown in brackets at the end of each question.*

1. Transpose $p - q + r = a - b$ for b. (2)

2. Make π the subject of the formula $r = \dfrac{c}{2\pi}$ (2)

3. Transpose $V = \dfrac{1}{3}\pi r^2 h$ for h. (2)

4. Transpose $I = \dfrac{E - e}{R + r}$ for E. (3)

5. Transpose $k = \dfrac{b}{ad - 1}$ for d. (4)

6. Make g the subject of the formula $t = 2\pi\sqrt{\dfrac{L}{g}}$ (3)

7. Transpose $A = \dfrac{\pi R^2 \theta}{360}$ for R. (2)

8. Make r the subject of the formula $x + y = \dfrac{r}{3 + r}$ (5)

9. Make L the subject of the formula $m = \dfrac{\mu L}{L + rCR}$ (5)

10. The surface area A of a rectangular prism is given by the formula $A = 2(bh + hl + lb)$. Evaluate b when $A = 11\,750\,\text{mm}^2, h = 25\,\text{mm}$ and $l = 75\,\text{mm}$. (4)

11. The velocity v of water in a pipe appears in the formula $h = \dfrac{0.03\,Lv^2}{2dg}$. Evaluate v when $h = 0.384, d = 0.20, L = 80$ and $g = 10$ (5)

12. A formula for the focal length f of a convex lens is $\dfrac{1}{f} = \dfrac{1}{u} + \dfrac{1}{v}$. Evaluate v when $f = 4$ and $u = 20$ (4)

13. Impedance in an a.c. circuit is given by: $Z = \sqrt{\left[R^2 + \left(2\pi nL - \dfrac{1}{2\pi nC}\right)^2\right]}$ where R is the resistance in ohms, L is the inductance in henrys, C is the capacitance in farads and n is the frequency of oscillations per second. Given $n = 50$, $R = 20, L = 0.40$ and $Z = 25$, determine the value of capacitance. (9)

In Problems 14 and 15, solve the simultaneous equations.

14. (a) $2x + y = 6$ (b) $4x - 3y = 11$
     $5x - y = 22$ $3x + 5y = 30$ (9)

15. (a) $3a - 8 + \dfrac{b}{8} = 0$

     $b + \dfrac{a}{2} = \dfrac{21}{4}$

(b) $\dfrac{2p + 1}{5} - \dfrac{1 - 4q}{2} = \dfrac{5}{2}$

     $\dfrac{1 - 3p}{7} + \dfrac{2q - 3}{5} + \dfrac{32}{35} = 0$ (18)

16. In an engineering process two variables x and y are related by the equation $y = ax + \dfrac{b}{x}$, where a and b are constants. Evaluate a and b if $y = 15$ when $x = 1$ and $y = 13$ when $x = 3$ (5)

17. Kirchhoff's laws are used to determine the current equations in an electrical network and result in the following:

$$i_1 + 8i_2 + 3i_3 = -31$$
$$3i_1 - 2i_2 + i_3 = -5$$
$$2i_1 - 3i_2 + 2i_3 = 6$$

Determine the values of i_1, i_2 and i_3 (10)

18. The forces acting on a beam are given by:
$R_1 + R_2 = 3.3\,\text{kN}$
and $22 \times 2.7 + 61 \times 0.4 - 12R_1 = 46R_2$
Calculate the reaction forces R_1 and R_2 (8)

For lecturers/instructors/teachers, fully worked solutions to each of the problems in Revision Test 5, together with a full marking scheme, are available at the website:
www.routledge.com/cw/bird

Chapter 14

Solving quadratic equations

Why it is important to understand: Solving quadratic equations

Quadratic equations have many applications in engineering and science; they are used in describing the trajectory of a ball, determining the height of a throw, and in the concept of acceleration, velocity, ballistics and stopping power. In addition, the quadratic equation has been found to be widely evident in a number of natural processes; some of these include the processes by which light is reflected off a lens, water flows down a rocky stream, or even the manner in which fur, spots, or stripes develop on wild animals. When traffic policemen arrive at the scene of a road accident, they measure the length of the skid marks and assess the road conditions. They can then use a quadratic equation to calculate the speed of the vehicles and hence reconstruct exactly what happened. The U-shape of a parabola can describe the trajectories of water jets in a fountain and a bouncing ball, or be incorporated into structures like the parabolic reflectors that form the base of satellite dishes and car headlights. Quadratic functions can help plot the course of moving objects and assist in determining minimum and maximum values. Most of the objects we use every day, from cars to clocks, would not exist if someone somewhere hadn't applied quadratic functions to their design. Solving quadratic equations is an important skill required in all aspects of engineering.

At the end of this chapter, you should be able to:

- define a quadratic equation
- solve quadratic equations by factorisation
- solve quadratic equations by 'completing the square'
- solve quadratic equations involving practical situations
- solve linear and quadratic equations simultaneously

14.1 Introduction

As stated in Chapter 11, an **equation** is a statement that two quantities are equal and to '**solve an equation**' means 'to find the value of the unknown'. The value of the unknown is called the **root** of the equation.

A **quadratic equation** is one in which the highest power of the unknown quantity is 2. For example, $x^2 - 3x + 1 = 0$ is a quadratic equation.

There are four methods of **solving quadratic equations**. These are:

(a) by factorisation (where possible),

(b) by 'completing the square',

(c) by using the 'quadratic formula', or

(d) graphically (see Chapter 19).

14.2 Solution of quadratic equations by factorisation

Multiplying out $(x+1)(x-3)$ gives $x^2 - 3x + x - 3$ i.e. $x^2 - 2x - 3$. The reverse process of moving from $x^2 - 2x - 3$ to $(x+1)(x-3)$ is called **factorising**.

If the quadratic expression can be factorised this provides the simplest method of solving a quadratic equation.

For example, if $x^2 - 2x - 3 = 0$, then, by factorising
$$(x+1)(x-3) = 0$$

Hence, either $\quad (x+1) = 0$, i.e. $x = -1$

or $\qquad\qquad (x-3) = 0$, i.e. $x = 3$

Hence, $x = -1$ and $x = 3$ are the roots of the quadratic equation $x^2 - 2x - 3 = 0$

The technique of factorising is often one of trial and error.

Problem 1. Solve the equation $x^2 + x - 6 = 0$ by factorisation

---

The factors of x^2 are x and x. These are placed in brackets: $\quad (x \quad)(x \quad)$

The factors of -6 are $+6$ and -1, or -6 and $+1$, or $+3$ and -2, or -3 and $+2$

The only combination to give a middle term of $+x$ is $+3$ and -2,

i.e. $\qquad x^2 + x - 6 = (x+3)(x-2)$

The quadratic equation $x^2 + x - 6 = 0$ thus becomes
$$(x+3)(x-2) = 0$$

Since the only way that this can be true is for either the first or the second or both factors to be zero, then

either $\qquad (x+3) = 0$, i.e. $x = -3$

or $\qquad (x-2) = 0$, i.e. $x = 2$

Hence, **the roots of $x^2 + x - 6 = 0$ are $x = -3$ and $x = 2$**

Problem 2. Solve the equation $x^2 + 2x - 8 = 0$ by factorisation

---

The factors of x^2 are x and x. These are placed in brackets: $\quad (x \quad)(x \quad)$

The factors of -8 are $+8$ and -1, or -8 and $+1$, or $+4$ and -2, or -4 and $+2$

The only combination to give a middle term of $+2x$ is $+4$ and -2,

i.e. $\qquad x^2 + 2x - 8 = (x+4)(x-2)$

(Note that the product of the two inner terms $(4x)$ added to the product of the two outer terms $(-2x)$ must equal the middle term, $+2x$ in this case.)

The quadratic equation $x^2 + 2x - 8 = 0$ thus becomes
$$(x+4)(x-2) = 0$$

Since the only way that this can be true is for either the first or the second or both factors to be zero,

either $\qquad (x+4) = 0$, i.e. $x = -4$

or $\qquad (x-2) = 0$, i.e. $x = 2$

Hence, **the roots of $x^2 + 2x - 8 = 0$ are $x = -4$ and $x = 2$**

Problem 3. Determine the roots of $x^2 - 6x + 9 = 0$ by factorisation

---

$$x^2 - 6x + 9 = (x-3)(x-3), \quad \text{i.e. } (x-3)^2 = 0$$

The LHS is known as a **perfect square**.

Hence, $x = 3$ is the only root of the equation $x^2 - 6x + 9 = 0$

Problem 4. Solve the equation $x^2 - 4x = 0$

Factorising gives $\quad x(x-4) = 0$

If $\qquad\qquad x(x-4) = 0$,

either $\qquad\qquad x = 0$ or $x - 4 = 0$

i.e. $\qquad\qquad\quad x = 0$ or $x = 4$

These are the two roots of the given equation. Answers can always be checked by substitution into the original equation.

Problem 5. Solve the equation $x^2 + 3x - 4 = 0$

Factorising gives $\qquad (x-1)(x+4) = 0$

Hence, either $\qquad x - 1 = 0$ or $x + 4 = 0$

i.e. $\qquad\qquad\qquad x = 1$ or $x = -4$

Problem 6. Determine the roots of $4x^2 - 25 = 0$ by factorisation

---

The LHS of $4x^2 - 25 = 0$ is **the difference of two squares**, $(2x)^2$ and $(5)^2$

By factorising, $4x^2 - 25 = (2x + 5)(2x - 5)$, i.e. $(2x + 5)(2x - 5) = 0$

Hence, either $\quad (2x + 5) = 0$, i.e. $\quad x = -\dfrac{5}{2} = -2.5$

or $\qquad\qquad (2x - 5) = 0$, i.e. $\quad x = \dfrac{5}{2} = 2.5$

Problem 7. Solve the equation $x^2 - 5x + 6 = 0$

Factorising gives $\qquad (x - 3)(x - 2) = 0$

Hence, either $\qquad x - 3 = 0 \quad$ or $\quad x - 2 = 0$

i.e. $\qquad\qquad\qquad \mathbf{x = 3} \quad$ or $\quad \mathbf{x = 2}$

Problem 8. Solve the equation $x^2 = 15 - 2x$

Rearranging gives $\qquad x^2 + 2x - 15 = 0$

Factorising gives $\qquad (x + 5)(x - 3) = 0$

Hence, either $\qquad x + 5 = 0 \quad$ or $\quad x - 3 = 0$

i.e. $\qquad\qquad\qquad \mathbf{x = -5} \quad$ or $\quad \mathbf{x = 3}$

Problem 9. Solve the equation $3x^2 - 11x - 4 = 0$ by factorisation

The factors of $3x^2$ are $3x$ and x. These are placed in brackets: $\quad (3x \quad)(x \quad)$

The factors of -4 are -4 and $+1$, or $+4$ and -1, or -2 and 2.

Remembering that the product of the two inner terms added to the product of the two outer terms must equal $-11x$, the only combination to give this is $+1$ and -4,

i.e. $\qquad 3x^2 - 11x - 4 = (3x + 1)(x - 4)$

The quadratic equation $3x^2 - 11x - 4 = 0$ thus becomes $\qquad\qquad\qquad (3x + 1)(x - 4) = 0$

Hence, either $(3x + 1) = 0$, i.e. $\quad x = -\dfrac{1}{3}$

or $\qquad\qquad (x - 4) = 0$, i.e. $\quad x = \ \ 4$

and both solutions may be checked in the original equation.

Problem 10. Solve the quadratic equation $4x^2 + 8x + 3 = 0$ by factorising

The factors of $4x^2$ are $4x$ and x, or $2x$ and $2x$

The factors of 3 are 3 and 1, or -3 and -1

Remembering that the product of the inner terms added to the product of the two outer terms must equal $+8x$, the only combination that is true (by trial and error) is

$$(4x^2 + 8x + 3) = (2x + 3)(2x + 1)$$

Hence, $(2x + 3)(2x + 1) = 0$, from which either $(2x + 3) = 0$ or $(2x + 1) = 0$

Thus, $\quad 2x = -3$, from which $\quad x = -\dfrac{3}{2} \quad$ or $\quad \mathbf{-1.5}$

or $\qquad 2x = -1$, from which $\quad x = -\dfrac{1}{2} \quad$ or $\quad \mathbf{-0.5}$

which may be checked in the original equation.

Problem 11. Solve the quadratic equation $15x^2 + 2x - 8 = 0$ by factorising

The factors of $15x^2$ are $15x$ and x or $5x$ and $3x$

The factors of -8 are -4 are $+2$, or 4 and -2, or -8 and $+1$, or 8 and -1

By trial and error the only combination that works is

$$15x^2 + 2x - 8 = (5x + 4)(3x - 2)$$

Hence, $(5x + 4)(3x - 2) = 0$, from which either $5x + 4 = 0$ or $3x - 2 = 0$

Hence, $x = -\dfrac{4}{5} \quad$ or $\quad x = \dfrac{2}{3}$

which may be checked in the original equation.

Problem 12. The roots of a quadratic equation are $\dfrac{1}{3}$ and -2. Determine the equation in x

If the roots of a quadratic equation are, say, α and β, then $(x - \alpha)(x - \beta) = 0$

Hence, if $\alpha = \dfrac{1}{3}$ and $\beta = -2$,

$$\left(x - \dfrac{1}{3}\right)(x - (-2)) = 0$$

$$\left(x - \dfrac{1}{3}\right)(x + 2) = 0$$

$$x^2 - \dfrac{1}{3}x + 2x - \dfrac{2}{3} = 0$$

$$x^2 + \dfrac{5}{3}x - \dfrac{2}{3} = 0$$

or $\qquad\qquad 3x^2 + 5x - 2 = 0$

Problem 13. Find the equation in x whose roots are 5 and -5

If 5 and -5 are the roots of a quadratic equation then

$$(x-5)(x+5) = 0$$

i.e. $\qquad x^2 - 5x + 5x - 25 = 0$

i.e. $\qquad \mathbf{x^2 - 25 = 0}$

Problem 14. Find the equation in x whose roots are 1.2 and -0.4

If 1.2 and -0.4 are the roots of a quadratic equation then

$$(x-1.2)(x+0.4) = 0$$

i.e. $\qquad x^2 - 1.2x + 0.4x - 0.48 = 0$

i.e. $\qquad \mathbf{x^2 - 0.8x - 0.48 = 0}$

Now try the following Practice Exercise

Practice Exercise 54   Solving quadratic equations by factorisation (answers on page 428)

In Problems 1 to 30, solve the given equations by factorisation.

1. $x^2 - 16 = 0$ 2. $x^2 + 4x - 32 = 0$

3. $(x+2)^2 = 16$ 4. $4x^2 - 9 = 0$

5. $3x^2 + 4x = 0$ 6. $8x^2 - 32 = 0$

7. $x^2 - 8x + 16 = 0$ 8. $x^2 + 10x + 25 = 0$

9. $x^2 - 2x + 1 = 0$ 10. $x^2 + 5x + 6 = 0$

11. $x^2 + 10x + 21 = 0$ 12. $x^2 - x - 2 = 0$

13. $y^2 - y - 12 = 0$ 14. $y^2 - 9y + 14 = 0$

15. $x^2 + 8x + 16 = 0$ 16. $x^2 - 4x + 4 = 0$

17. $x^2 + 6x + 9 = 0$ 18. $x^2 - 9 = 0$

19. $3x^2 + 8x + 4 = 0$ 20. $4x^2 + 12x + 9 = 0$

21. $4z^2 - \dfrac{1}{16} = 0$ 22. $x^2 + 3x - 28 = 0$

23. $2x^2 - x - 3 = 0$ 24. $6x^2 - 5x + 1 = 0$

25. $10x^2 + 3x - 4 = 0$ 26. $21x^2 - 25x = 4$

27. $8x^2 + 13x - 6 = 0$ 28. $5x^2 + 13x - 6 = 0$

29. $6x^2 - 5x - 4 = 0$ 30. $8x^2 + 2x - 15 = 0$

In Problems 31 to 36, determine the quadratic equations in x whose roots are

31. 3 and 1   32. 2 and -5

33. -1 and -4 34. 2.5 and -0.5

35. 6 and -6 36. 2.4 and -0.7

14.3 Solution of quadratic equations by 'completing the square'

An expression such as x^2 or $(x+2)^2$ or $(x-3)^2$ is called a **perfect square**.

If $x^2 = 3$ then $x = \pm\sqrt{3}$

If $(x+2)^2 = 5$ then $x+2 = \pm\sqrt{5}$ and $x = -2 \pm\sqrt{5}$

If $(x-3)^2 = 8$ then $x-3 = \pm\sqrt{8}$ and $x = 3 \pm\sqrt{8}$

Hence, if a quadratic equation can be rearranged so that one side of the equation is a perfect square and the other side of the equation is a number, then the solution of the equation is readily obtained by taking the square roots of each side as in the above examples. The process of rearranging one side of a quadratic equation into a perfect square before solving is called '**completing the square**'.

$$(x+a)^2 = x^2 + 2ax + a^2$$

Thus, in order to make the quadratic expression $x^2 + 2ax$ into a perfect square, it is necessary to add (half the coefficient of x)^2, i.e. $\left(\dfrac{2a}{2}\right)^2$ or a^2

For example, $x^2 + 3x$ becomes a perfect square by adding $\left(\dfrac{3}{2}\right)^2$, i.e.

$$x^2 + 3x + \left(\frac{3}{2}\right)^2 = \left(x + \frac{3}{2}\right)^2$$

The method of completing the square is demonstrated in the following worked problems.

Problem 15. Solve $2x^2 + 5x = 3$ by completing the square

The procedure is as follows.

(i) Rearrange the equation so that all terms are on the same side of the equals sign (and the coefficient of the x^2 term is positive). Hence,

$$2x^2 + 5x - 3 = 0$$

(ii)   Make the coefficient of the x^2 term unity. In this case this is achieved by dividing throughout by 2. Hence,

$$\frac{2x^2}{2} + \frac{5x}{2} - \frac{3}{2} = 0$$

i.e. $$x^2 + \frac{5}{2}x - \frac{3}{2} = 0$$

(iii)   Rearrange the equations so that the x^2 and x terms are on one side of the equals sign and the constant is on the other side. Hence,

$$x^2 + \frac{5}{2}x = \frac{3}{2}$$

(iv)   Add to both sides of the equation (half the coefficient of x)². In this case the coefficient of x is $\frac{5}{2}$

Half the coefficient squared is therefore $\left(\frac{5}{4}\right)^2$

Thus,

$$x^2 + \frac{5}{2}x + \left(\frac{5}{4}\right)^2 = \frac{3}{2} + \left(\frac{5}{4}\right)^2$$

The LHS is now a perfect square, i.e.

$$\left(x + \frac{5}{4}\right)^2 = \frac{3}{2} + \left(\frac{5}{4}\right)^2$$

(v)   Evaluate the RHS. Thus,

$$\left(x + \frac{5}{4}\right)^2 = \frac{3}{2} + \frac{25}{16} = \frac{24 + 25}{16} = \frac{49}{16}$$

(vi)   Take the square root of both sides of the equation (remembering that the square root of a number gives a ± answer). Thus,

$$\sqrt{\left(x + \frac{5}{4}\right)^2} = \sqrt{\left(\frac{49}{16}\right)}$$

i.e. $$x + \frac{5}{4} = \pm\frac{7}{4}$$

(vii)   Solve the simple equation. Thus,

$$x = -\frac{5}{4} \pm \frac{7}{4}$$

i.e. $$x = -\frac{5}{4} + \frac{7}{4} = \frac{2}{4} = \frac{1}{2} \text{ or } 0.5$$

or $$x = -\frac{5}{4} - \frac{7}{4} = -\frac{12}{4} = -3$$

Hence, $x = 0.5$ or $x = -3$; i.e. **the roots of the equation** $2x^2 + 5x = 3$ **are 0.5 and −3**

Problem 16.   Solve $2x^2 + 9x + 8 = 0$, correct to 3 significant figures, by completing the square

Making the coefficient of x^2 unity gives

$$x^2 + \frac{9}{2}x + 4 = 0$$

Rearranging gives $$x^2 + \frac{9}{2}x = -4$$

Adding to both sides (half the coefficient of x)² gives

$$x^2 + \frac{9}{2}x + \left(\frac{9}{4}\right)^2 = \left(\frac{9}{4}\right)^2 - 4$$

The LHS is now a perfect square. Thus,

$$\left(x + \frac{9}{4}\right)^2 = \frac{81}{16} - 4 = \frac{81}{16} - \frac{64}{16} = \frac{17}{16}$$

Taking the square root of both sides gives

$$x + \frac{9}{4} = \sqrt{\left(\frac{17}{16}\right)} = \pm 1.031$$

Hence, $$x = -\frac{9}{4} \pm 1.031$$

i.e. $x = -1.22$ or -3.28, correct to 3 significant figures.

Problem 17.   By completing the square, solve the quadratic equation $4.6y^2 + 3.5y - 1.75 = 0$, correct to 3 decimal places

$$4.6y^2 + 3.5y - 1.75 = 0$$

Making the coefficient of y^2 unity gives

$$y^2 + \frac{3.5}{4.6}y - \frac{1.75}{4.6} = 0$$

and rearranging gives $$y^2 + \frac{3.5}{4.6}y = \frac{1.75}{4.6}$$

Adding to both sides (half the coefficient of y)² gives

$$y^2 + \frac{3.5}{4.6}y + \left(\frac{3.5}{9.2}\right)^2 = \frac{1.75}{4.6} + \left(\frac{3.5}{9.2}\right)^2$$

The LHS is now a perfect square. Thus,

$$\left(y + \frac{3.5}{9.2}\right)^2 = 0.5251654$$

Taking the square root of both sides gives

$$y + \frac{3.5}{9.2} = \sqrt{0.5251654} = \pm 0.7246830$$

Hence,

$$y = -\frac{3.5}{9.2} \pm 0.7246830$$

i.e. $y = 0.344$ or -1.105

Now try the following Practice Exercise

Practice Exercise 55   Solving quadratic equations by completing the square (answers on page 428)

Solve the following equations correct to 3 decimal places by completing the square.

1. $x^2 + 4x + 1 = 0$ 2. $2x^2 + 5x - 4 = 0$
3. $3x^2 - x - 5 = 0$ 4. $5x^2 - 8x + 2 = 0$
5. $4x^2 - 11x + 3 = 0$ 6. $2x^2 + 5x = 2$

14.4   Solution of quadratic equations by formula

Let the general form of a quadratic equation be given by $ax^2 + bx + c = 0$, where a, b and c are constants. Dividing $ax^2 + bx + c = 0$ by a gives

$$x^2 + \frac{b}{a}x + \frac{c}{a} = 0$$

Rearranging gives $x^2 + \frac{b}{a}x = -\frac{c}{a}$

Adding to each side of the equation the square of half the coefficient of the term in x to make the LHS a perfect square gives

$$x^2 + \frac{b}{a}x + \left(\frac{b}{2a}\right)^2 = \left(\frac{b}{2a}\right)^2 - \frac{c}{a}$$

Rearranging gives $\left(x + \frac{b}{a}\right)^2 = \frac{b^2}{4a^2} - \frac{c}{a} = \frac{b^2 - 4ac}{4a^2}$

Taking the square root of both sides gives

$$x + \frac{b}{2a} = \sqrt{\left(\frac{b^2 - 4ac}{4a^2}\right)} = \frac{\pm\sqrt{b^2 - 4ac}}{2a}$$

Hence, $x = -\frac{b}{2a} \pm \frac{\sqrt{b^2 - 4ac}}{2a}$

i.e. the quadratic formula is $x = \dfrac{-b \pm \sqrt{b^2 - 4ac}}{2a}$

(This method of obtaining the formula is completing the square — as shown in the previous section.)
In summary,

if $ax^2 + bx + c = 0$ then $x = \dfrac{-b \pm \sqrt{b^2 - 4ac}}{2a}$

This is known as the **quadratic formula**.

Problem 18. Solve $x^2 + 2x - 8 = 0$ by using the quadratic formula

Comparing $x^2 + 2x - 8 = 0$ with $ax^2 + bx + c = 0$ gives $a = 1, b = 2$ and $c = -8$
Substituting these values into the quadratic formula

$$x = \frac{-b \pm \sqrt{b^2 - 4ac}}{2a}$$

gives

$$x = \frac{-2 \pm \sqrt{2^2 - 4(1)(-8)}}{2(1)} = \frac{-2 \pm \sqrt{4 + 32}}{2}$$

$$= \frac{-2 \pm \sqrt{36}}{2} = \frac{-2 \pm 6}{2}$$

$$= \frac{-2 + 6}{2} \text{ or } \frac{-2 - 6}{2}$$

Hence, $x = \dfrac{4}{2}$ or $\dfrac{-8}{2}$, i.e. $x = 2$ or $x = -4$

Problem 19. Solve $3x^2 - 11x - 4 = 0$ by using the quadratic formula

Comparing $3x^2 - 11x - 4 = 0$ with $ax^2 + bx + c = 0$ gives $a = 3, b = -11$ and $c = -4$. Hence,

$$x = \frac{-(-11) \pm \sqrt{(-11)^2 - 4(3)(-4)}}{2(3)}$$

$$= \frac{+11 \pm \sqrt{121 + 48}}{6} = \frac{11 \pm \sqrt{169}}{6}$$

$$= \frac{11 \pm 13}{6} = \frac{11 + 13}{6} \text{ or } \frac{11 - 13}{6}$$

Hence, $x = \dfrac{24}{6}$ or $\dfrac{-2}{6}$, i.e. $x = 4$ or $x = -\dfrac{1}{3}$

Problem 20. Solve $4x^2 + 7x + 2 = 0$ giving the roots correct to 2 decimal places

Comparing $4x^2 + 7x + 2 = 0$ with $ax^2 + bx + c$ gives $a = 4, b = 7$ and $c = 2$. Hence,

$$x = \frac{-7 \pm \sqrt{7^2 - 4(4)(2)}}{2(4)}$$

$$= \frac{-7 \pm \sqrt{17}}{8} = \frac{-7 \pm 4.123}{8}$$

$$= \frac{-7 + 4.123}{8} \text{ or } \frac{-7 - 4.123}{8}$$

Hence, $x = -0.36$ or -1.39, correct to 2 decimal places.

Problem 21. Use the quadratic formula to solve $\dfrac{x+2}{4} + \dfrac{3}{x-1} = 7$ correct to 4 significant figures

Multiplying throughout by $4(x-1)$ gives

$$4(x-1)\frac{(x+2)}{4} + 4(x-1)\frac{3}{(x-1)} = 4(x-1)(7)$$

Cancelling gives $(x-1)(x+2) + (4)(3) = 28(x-1)$

$$x^2 + x - 2 + 12 = 28x - 28$$

Hence, $x^2 - 27x + 38 = 0$

Using the quadratic formula,

$$x = \frac{-(-27) \pm \sqrt{(-27)^2 - 4(1)(38)}}{2}$$

$$= \frac{27 \pm \sqrt{577}}{2} = \frac{27 \pm 24.0208}{2}$$

Hence, $x = \dfrac{27 + 24.0208}{2} = 25.5104$

or $x = \dfrac{27 - 24.0208}{2} = 1.4896$

Hence, $x = 25.51$ or 1.490, correct to 4 significant figures.

Now try the following Practice Exercise

Practice Exercise 56 Solving quadratic equations by formula (answers on page 428)

Solve the following equations by using the quadratic formula, correct to 3 decimal places.

1. $2x^2 + 5x - 4 = 0$
2. $5.76x^2 + 2.86x - 1.35 = 0$
3. $2x^2 - 7x + 4 = 0$
4. $4x + 5 = \dfrac{3}{x}$
5. $(2x + 1) = \dfrac{5}{x-3}$
6. $3x^2 - 5x + 1 = 0$

7. $4x^2 + 6x - 8 = 0$
8. $5.6x^2 - 11.2x - 1 = 0$
9. $3x(x + 2) + 2x(x - 4) = 8$
10. $4x^2 - x(2x + 5) = 14$
11. $\dfrac{5}{x-3} + \dfrac{2}{x-2} = 6$
12. $\dfrac{3}{x-7} + 2x = 7 + 4x$
13. $\dfrac{x+1}{x-1} = x - 3$

14.5 Practical problems involving quadratic equations

There are many **practical problems** in which a quadratic equation has first to be obtained, from given information, before it is solved.

Problem 22. The area of a rectangle is $23.6\,\text{cm}^2$ and its width is $3.10\,\text{cm}$ shorter than its length. Determine the dimensions of the rectangle, correct to 3 significant figures

Let the length of the rectangle be x cm. Then the width is $(x - 3.10)$ cm.

$$\text{Area} = \text{length} \times \text{width} = x(x - 3.10) = 23.6$$

i.e. $x^2 - 3.10x - 23.6 = 0$

Using the quadratic formula,

$$x = \frac{-(-3.10) \pm \sqrt{(-3.10)^2 - 4(1)(-23.6)}}{2(1)}$$

$$= \frac{3.10 \pm \sqrt{9.61 + 94.4}}{2} = \frac{3.10 \pm 10.20}{2}$$

$$= \frac{13.30}{2} \text{ or } \frac{-7.10}{2}$$

Hence, $x = 6.65\,\text{cm}$ or $-3.55\,\text{cm}$. The latter solution is neglected since length cannot be negative.

Thus, length $x = 6.65\,\text{cm}$ and width $= x - 3.10 = 6.65 - 3.10 = 3.55\,\text{cm}$, i.e. **the dimensions of the rectangle are 6.65 cm by 3.55 cm.**

(Check: Area $= 6.65 \times 3.55 = 23.6\,\text{cm}^2$, correct to 3 significant figures.)

Problem 23.   Calculate the diameter of a solid cylinder which has a height of 82.0 cm and a total surface area of 2.0 m²

Total surface area of a cylinder

$= $ curved surface area $+ 2$ circular ends

$= 2\pi r h + 2\pi r^2$ (where $r =$ radius and $h =$ height)

Since the total surface area $= 2.0\,\text{m}^2$ and the height $h = 82\,\text{cm}$ or $0.82\,\text{m}$,

$$2.0 = 2\pi r(0.82) + 2\pi r^2$$

i.e. $$2\pi r^2 + 2\pi r(0.82) - 2.0 = 0$$

Dividing throughout by 2π gives $r^2 + 0.82r - \dfrac{1}{\pi} = 0$

Using the quadratic formula,

$$r = \frac{-0.82 \pm \sqrt{(0.82)^2 - 4(1)\left(-\frac{1}{\pi}\right)}}{2(1)}$$

$$= \frac{-0.82 \pm \sqrt{1.94564}}{2} = \frac{-0.82 \pm 1.39486}{2}$$

$$= 0.2874 \text{ or } -1.1074$$

Thus, the radius r of the cylinder is $0.2874\,\text{m}$ (the negative solution being neglected).

Hence, the diameter of the cylinder

$$= 2 \times 0.2874$$

$$= \mathbf{0.5748\,m} \text{ or } \mathbf{57.5\,cm}$$

correct to 3 significant figures.

Problem 24.   The height s metres of a mass projected vertically upwards at time t seconds is $s = ut - \dfrac{1}{2}gt^2$. Determine how long the mass will take after being projected to reach a height of 16 m (a) on the ascent and (b) on the descent, when $u = 30\,\text{m/s}$ and $g = 9.81\,\text{m/s}^2$

When height $s = 16\,\text{m}$, $16 = 30t - \dfrac{1}{2}(9.81)t^2$

i.e. $$4.905t^2 - 30t + 16 = 0$$

Using the quadratic formula,

$$t = \frac{-(-30) \pm \sqrt{(-30)^2 - 4(4.905)(16)}}{2(4.905)}$$

$$= \frac{30 \pm \sqrt{586.1}}{9.81} = \frac{30 \pm 24.21}{9.81} = 5.53 \text{ or } 0.59$$

Hence, the mass will reach a height of 16 m after 0.59 s on the ascent and after 5.53 s on the descent.

Problem 25.   A shed is 4.0 m long and 2.0 m wide. A concrete path of constant width is laid all the way around the shed. If the area of the path is 9.50 m², calculate its width to the nearest centimetre

Fig. 14.1 shows a plan view of the shed with its surrounding path of width t metres.

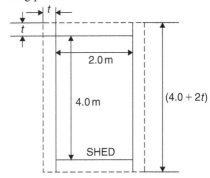

Figure 14.1

$$\text{Area of path} = 2(2.0 \times t) + 2t(4.0 + 2t)$$

i.e. $$9.50 = 4.0t + 8.0t + 4t^2$$

or $$4t^2 + 12.0t - 9.50 = 0$$

Hence,

$$t = \frac{-(12.0) \pm \sqrt{(12.0)^2 - 4(4)(-9.50)}}{2(4)}$$

$$= \frac{-12.0 \pm \sqrt{296.0}}{8} = \frac{-12.0 \pm 17.20465}{8}$$

i.e. $t = 0.6506\,\text{m}$ or $-3.65058\,\text{m}$.

Neglecting the negative result, which is meaningless, the width of the path, $t = \mathbf{0.651\,m}$ or $\mathbf{65\,cm}$ correct to the nearest centimetre.

Problem 26.   If the total surface area of a solid cone is 486.2 cm² and its slant height is 15.3 cm, determine its base diameter.

From Chapter 27, page 278, the total surface area A of a solid cone is given by $A = \pi r l + \pi r^2$, where l is the slant height and r the base radius.

If $A = 482.2$ and $l = 15.3$, then

$$482.2 = \pi r(15.3) + \pi r^2$$

i.e. $\qquad \pi r^2 + 15.3\pi r - 482.2 = 0$

or $\qquad r^2 + 15.3r - \dfrac{482.2}{\pi} = 0$

Using the quadratic formula,

$$r = \frac{-15.3 \pm \sqrt{\left[(15.3)^2 - 4\left(\dfrac{-482.2}{\pi}\right)\right]}}{2}$$

$$= \frac{-15.3 \pm \sqrt{848.0461}}{2} = \frac{-15.3 \pm 29.12123}{2}$$

Hence, radius $r = 6.9106$ cm (or -22.21 cm, which is meaningless and is thus ignored).
Thus, the **diameter of the base** $= 2r = 2(6.9106)$
$$= \mathbf{13.82\,cm}.$$

Now try the following Practice Exercise

Practice Exercise 57   Practical problems involving quadratic equations (answers on page 428)

1. The angle a rotating shaft turns through in t seconds is given by $\theta = \omega t + \dfrac{1}{2}at^2$. Determine the time taken to complete 4 radians if ω is 3.0 rad/s and α is 0.60 rad/s^2.

2. The power P developed in an electrical circuit is given by $P = 10I - 8I^2$, where I is the current in amperes. Determine the current necessary to produce a power of 2.5 watts in the circuit.

3. The area of a triangle is 47.6 cm^2 and its perpendicular height is 4.3 cm more than its base length. Determine the length of the base correct to 3 significant figures.

4. The sag, l, in metres in a cable stretched between two supports, distance x m apart, is given by $l = \dfrac{12}{x} + x$. Determine the distance between the supports when the sag is 20 m.

5. The acid dissociation constant K_a of ethanoic acid is 1.8×10^{-5} mol dm^{-3} for a particular solution. Using the Ostwald dilution law,

$K_a = \dfrac{x^2}{v(1-x)}$, determine x, the degree of ionisation, given that $v = 10$ dm^3.

6. A rectangular building is 15 m long by 11 m wide. A concrete path of constant width is laid all the way around the building. If the area of the path is 60.0 m^2, calculate its width correct to the nearest millimetre.

7. The total surface area of a closed cylindrical container is 20.0 m^3. Calculate the radius of the cylinder if its height is 2.80 m.

8. The bending moment M at a point in a beam is given by $M = \dfrac{3x(20-x)}{2}$, where x metres is the distance from the point of support. Determine the value of x when the bending moment is 50 N m.

9. A tennis court measures 24 m by 11 m. In the layout of a number of courts an area of ground must be allowed for at the ends and at the sides of each court. If a border of constant width is allowed around each court and the total area of the court and its border is 950 m^2, find the width of the borders.

10. Two resistors, when connected in series, have a total resistance of 40 ohms. When connected in parallel their total resistance is 8.4 ohms. If one of the resistors has a resistance of R_x, ohms,

    (a) show that $R_x^2 - 40R_x + 336 = 0$ and

    (b) calculate the resistance of each.

11. When a ball is thrown vertically upwards its height h varies with time t according to the equation $h = 25t - 4t^2$. Determine the times, correct to 3 significant figures, when the height is 12 m.

12. In an RLC electrical circuit, reactance X is given by: $\qquad X = \omega L - \dfrac{1}{\omega C}$
    $X = 220\ \Omega$, inductance $L = 800$ mH and capacitance $C = 25\ \mu$F. The angular velocity ω is measured in radians per second. Calculate the value of ω.

14.6 Solution of linear and quadratic equations simultaneously

Sometimes a linear equation and a quadratic equation need to be solved simultaneously. An algebraic method of solution is shown in Problem 27; a graphical solution is shown in Chapter 19, page 182.

Problem 27. Determine the values of x and y which simultaneously satisfy the equations $y = 5x - 4 - 2x^2$ and $y = 6x - 7$

For a simultaneous solution the values of y must be equal, hence the RHS of each equation is equated.

Thus, $5x - 4 - 2x^2 = 6x - 7$

Rearranging gives $5x - 4 - 2x^2 - 6x + 7 = 0$

i.e. $-x + 3 - 2x^2 = 0$

or $2x^2 + x - 3 = 0$

Factorising gives $(2x + 3)(x - 1) = 0$

i.e. $x = -\dfrac{3}{2}$ or $x = 1$

In the equation $y = 6x - 7$,

when $x = -\dfrac{3}{2}$, $y = 6\left(-\dfrac{3}{2}\right) - 7 = -16$

and when $x = 1$, $y = 6 - 7 = -1$

(Checking the result in $y = 5x - 4 - 2x^2$:

when $x = -\dfrac{3}{2}$, $y = 5\left(-\dfrac{3}{2}\right) - 4 - 2\left(-\dfrac{3}{2}\right)^2$

$$= -\dfrac{15}{2} - 4 - \dfrac{9}{2} = -16, \text{ as above,}$$

and when $x = 1$, $y = 5 - 4 - 2 = -1$, as above.)

Hence, the simultaneous solutions occur when

$$x = -\dfrac{3}{2}, y = -16 \text{ and when } x = 1, y = -1$$

Now try the following Practice Exercise

Practice Exercise 58   Solving linear and quadratic equations simultaneously (answers on page 428)

Determine the solutions of the following simultaneous equations.

1. $y = x^2 + x + 1$ 2. $y = 15x^2 + 21x - 11$

    $y = 4 - x$ $y = 2x - 1$

3. $2x^2 + y = 4 + 5x$

    $x + y = 4$

For fully worked solutions to each of the problems in Practice Exercises 54 to 58 in this chapter, go to the website:

www.routledge.com/cw/bird

Chapter 15

Logarithms

Why it is important to understand: Logarithms

All types of engineers use natural and common logarithms. Chemical engineers use them to measure radioactive decay and pH solutions, both of which are measured on a logarithmic scale. The Richter scale which measures earthquake intensity is a logarithmic scale. Biomedical engineers use logarithms to measure cell decay and growth, and also to measure light intensity for bone mineral density measurements. In electrical engineering, a dB (decibel) scale is very useful for expressing attenuations in radio propagation and circuit gains, and logarithms are used for implementing arithmetic operations in digital circuits. Logarithms are especially useful when dealing with the graphical analysis of non-linear relationships and logarithmic scales are used to linearise data to make data analysis simpler. Understanding and using logarithms is clearly important in all branches of engineering.

At the end of this chapter, you should be able to:

- define base, power, exponent and index
- define a logarithm
- distinguish between common and Napierian (i.e. hyperbolic or natural) logarithms
- evaluate logarithms to any base
- state the laws of logarithms
- simplify logarithmic expressions
- solve equations involving logarithms
- solve indicial equations
- sketch graphs of $\log_{10} x$ and $\log_e x$

15.1 Introduction to logarithms

With the use of calculators firmly established, logarithmic tables are now rarely used for calculation. However, the theory of logarithms is important, for there are several scientific and engineering laws that involve the rules of logarithms.

From Chapter 7, we know that $\qquad 16 = 2^4$

The number 4 is called the **power** or the **exponent** or the **index**. In the expression 2^4, the number 2 is called the **base**.

In another example, we know that $\qquad 64 = 8^2$

In this example, 2 is the power, or exponent, or index. The number 8 is the base.

15.1.1 What is a logarithm?

Consider the expression $16 = 2^4$
An alternative, yet equivalent, way of writing this expression is $\log_2 16 = 4$

This is stated as 'log to the base 2 of 16 equals 4'
We see that the logarithm is the same as the power or index in the original expression. It is the base in the original expression that becomes the base of the logarithm.

The two statements $16 = 2^4$ and

$\log_2 16 = 4$ are equivalent

If we write either of them, we are automatically implying the other.
In general, if a number y can be written in the form a^x, then the index x is called the 'logarithm of y to the base of a', i.e.

if $y = a^x$ then $x = \log_a y$

In another example, if we write down that $64 = 8^2$ then the equivalent statement using logarithms is $\log_8 64 = 2$. In another example, if we write down that $\log_3 27 = 3$ then the equivalent statement using powers is $3^3 = 27$. So the two sets of statements, one involving powers and one involving logarithms, are equivalent.

15.1.2 Common logarithms

From the above, if we write down that $1000 = 10^3$, then $3 = \log_{10} 1000$. This may be checked using the 'log' button on your calculator.
Logarithms having a base of 10 are called **common logarithms** and $\log_{10}$ is usually abbreviated to lg. The following values may be checked using a calculator.

$$\lg 27.5 = 1.4393\ldots$$
$$\lg 378.1 = 2.5776\ldots$$
$$\lg 0.0204 = -1.6903\ldots$$

15.1.3 Napierian logarithms

Logarithms having a base of e (where e is a mathematical constant approximately equal to 2.7183) are called **hyperbolic, Napierian** or **natural logarithms**, and $\log_e$ is usually abbreviated to ln. The following values may be checked using a calculator.

$$\ln 3.65 = 1.2947\ldots$$
$$\ln 417.3 = 6.0338\ldots$$
$$\ln 0.182 = -1.7037\ldots$$

Napierian logarithms are explained further in Chapter 16, following.

Here are some worked problems to help understanding of logarithms.

Problem 1. Evaluate $\log_3 9$

Let $x = \log_3 9$ then $3^x = 9$ from the definition of a logarithm,
i.e. $3^x = 3^2$, from which $x = 2$
Hence, **$\log_3 9 = 2$**

Problem 2. Evaluate $\log_{10} 10$

Let $x = \log_{10} 10$ then $10^x = 10$ from the definition of a logarithm,
i.e. $10^x = 10^1$, from which $x = 1$
Hence, **$\log_{10} 10 = 1$** (which may be checked using a calculator).

Problem 3. Evaluate $\log_{16} 8$

Let $x = \log_{16} 8$ then $16^x = 8$ from the definition of a logarithm,
i.e. $(2^4)^x = 2^3$ i.e. $2^{4x} = 2^3$ from the laws of indices,
from which, $4x = 3$ and $x = \dfrac{3}{4}$
Hence, **$\log_{16} 8 = \dfrac{3}{4}$**

Problem 4. Evaluate $\lg 0.001$

Let $x = \lg 0.001 = \log_{10} 0.001$ then $10^x = 0.001$
i.e. $10^x = 10^{-3}$
from which, $x = -3$
Hence, **$\lg 0.001 = -3$** (which may be checked using a calculator)

Problem 5. Evaluate $\ln e$

Let $x = \ln e = \log_e e$ then $e^x = e$
i.e. $e^x = e^1$, from which $x = 1$
Hence, **$\ln e = 1$** (which may be checked by a calculator)

Problem 6. Evaluate $\log_3 \dfrac{1}{81}$

Let $x = \log_3 \dfrac{1}{81}$ then $3^x = \dfrac{1}{81} = \dfrac{1}{3^4} = 3^{-4}$

from which $x = -4$

Hence, $\log_3 \dfrac{1}{81} = -4$

Problem 7. Solve the equation $\lg x = 3$

If $\lg x = 3$ then $\log_{10} x = 3$

and $x = 10^3$ i.e. $x = 1000$

Problem 8. Solve the equation $\log_2 x = 5$

If $\log_2 x = 5$ then $x = 2^5 = 32$

Problem 9. Solve the equation $\log_5 x = -2$

If $\log_5 x = -2$ then $x = 5^{-2} = \dfrac{1}{5^2} = \dfrac{1}{25}$

Now try the following Practice Exercise

Practice Exercise 59 Laws of logarithms
(answers on page 428)

In Problems 1 to 11, evaluate the given expressions.

1. $\log_{10} 10000$ 2. $\log_2 16$ 3. $\log_5 125$

4. $\log_2 \dfrac{1}{8}$ 5. $\log_8 2$ 6. $\log_7 343$

7. $\lg 100$ 8. $\lg 0.01$ 9. $\log_4 8$

10. $\log_{27} 3$ 11. $\ln e^2$

In Problems 12 to 18, solve the equations.

12. $\log_{10} x = 4$ 13. $\lg x = 5$

14. $\log_3 x = 2$ 15. $\log_4 x = -2\dfrac{1}{2}$

16. $\lg x = -2$ 17. $\log_8 x = -\dfrac{4}{3}$

18. $\ln x = 3$

15.2 Laws of logarithms

There are three laws of logarithms, which apply to any base:

(1) To multiply two numbers:

$$\log (A \times B) = \log A + \log B$$

The following may be checked by using a calculator.

$\lg 10 = 1$

Also, $\lg 5 + \lg 2 = 0.69897\ldots + 0.301029\ldots = 1$

Hence, $\lg (5 \times 2) = \lg 10 = \lg 5 + \lg 2$

(2) To divide two numbers:

$$\log \left(\dfrac{A}{B}\right) = \log A - \log B$$

The following may be checked using a calculator.

$\ln \left(\dfrac{5}{2}\right) = \ln 2.5 = 0.91629\ldots$

Also, $\ln 5 - \ln 2 = 1.60943\ldots - 0.69314\ldots$

$= 0.91629\ldots$

Hence, $\ln \left(\dfrac{5}{2}\right) = \ln 5 - \ln 2$

(3) To raise a number to a power:

$$\log A^n = n \log A$$

The following may be checked using a calculator.

$\lg 5^2 = \lg 25 = 1.39794\ldots$

Also, $2\lg 5 = 2 \times 0.69897\ldots = 1.39794\ldots$

Hence, $\lg 5^2 = 2\lg 5$

Here are some worked problems to help understanding of the laws of logarithms.

Problem 10. Write $\log 4 + \log 7$ as the logarithm of a single number

$\log 4 + \log 7 = \log(7 \times 4)$ by the first law of logarithms

$= \log 28$

Problem 11. Write $\log 16 - \log 2$ as the logarithm of a single number

$\log 16 - \log 2 = \log\left(\dfrac{16}{2}\right)$ by the second law of logarithms

$\qquad\qquad\quad = \mathbf{\log 8}$

Problem 12. Write $2\log 3$ as the logarithm of a single number

$2\log 3 = \log 3^2$ by the third law of logarithms

$\qquad\quad = \mathbf{\log 9}$

Problem 13. Write $\dfrac{1}{2}\log 25$ as the logarithm of a single number

$\dfrac{1}{2}\log 25 = \log 25^{\frac{1}{2}}$ by the third law of logarithms

$\qquad\quad = \log\sqrt{25} = \mathbf{\log 5}$

Problem 14. Simplify $\log 64 - \log 128 + \log 32$

$64 = 2^6, 128 = 2^7$ and $32 = 2^5$

Hence, $\log 64 - \log 128 + \log 32$

$\qquad\qquad = \log 2^6 - \log 2^7 + \log 2^5$

$\qquad\qquad = 6\log 2 - 7\log 2 + 5\log 2$

$\qquad\qquad\qquad$ by the third law of logarithms

$\qquad\qquad = \mathbf{4\log 2}$

Problem 15. Write $\dfrac{1}{2}\log 16 + \dfrac{1}{3}\log 27 - 2\log 5$ as the logarithm of a single number

$\dfrac{1}{2}\log 16 + \dfrac{1}{3}\log 27 - 2\log 5$

$\qquad\qquad = \log 16^{\frac{1}{2}} + \log 27^{\frac{1}{3}} - \log 5^2$

$\qquad\qquad\qquad$ by the third law of logarithms

$\qquad\qquad = \log\sqrt{16} + \log\sqrt[3]{27} - \log 25$

$\qquad\qquad\qquad$ by the laws of indices

$\qquad\qquad = \log 4 + \log 3 - \log 25$

$\qquad\qquad = \log\left(\dfrac{4 \times 3}{25}\right)$ by the first and second laws of logarithms

$\qquad\qquad = \log\left(\dfrac{12}{25}\right) = \mathbf{\log 0.48}$

Problem 16. Write (a) log 30 (b) log 450 in terms of log 2, log 3 and log 5 to any base

(a) $\log 30 = \log(2 \times 15) = \log(2 \times 3 \times 5)$

$\qquad\qquad = \mathbf{\log 2 + \log 3 + \log 5}$

$\qquad\qquad\qquad$ by the first law of logarithms

(b) $\log 450 = \log(2 \times 225) = \log(2 \times 3 \times 75)$

$\qquad\qquad = \log(2 \times 3 \times 3 \times 25)$

$\qquad\qquad = \log(2 \times 3^2 \times 5^2)$

$\qquad\qquad = \log 2 + \log 3^2 + \log 5^2$

$\qquad\qquad\qquad$ by the first law of logarithms

i.e. $\log 450 = \mathbf{\log 2 + 2\log 3 + 2\log 5}$

$\qquad\qquad\qquad$ by the third law of logarithms

Problem 17. Write $\log\left(\dfrac{8 \times \sqrt[4]{5}}{81}\right)$ in terms of log 2, log 3 and log 5 to any base

$\log\left(\dfrac{8 \times \sqrt[4]{5}}{81}\right) = \log 8 + \log\sqrt[4]{5} - \log 81$

$\qquad\qquad\qquad$ by the first and second laws of logarithms

$\qquad\qquad = \log 2^3 + \log 5^{\frac{1}{4}} - \log 3^4$

$\qquad\qquad\qquad$ by the laws of indices

i.e. $\log\left(\dfrac{8 \times \sqrt[4]{5}}{81}\right) = \mathbf{3\log 2 + \dfrac{1}{4}\log 5 - 4\log 3}$

$\qquad\qquad\qquad$ by the third law of logarithms

Problem 18. Evaluate

$$\dfrac{\log 25 - \log 125 + \dfrac{1}{2}\log 625}{3\log 5}$$

$$\frac{\log 25 - \log 125 + \frac{1}{2}\log 625}{3\log 5}$$

$$= \frac{\log 5^2 - \log 5^3 + \frac{1}{2}\log 5^4}{3\log 5}$$

$$= \frac{2\log 5 - 3\log 5 + \frac{4}{2}\log 5}{3\log 5}$$

$$= \frac{1\log 5}{3\log 5} = \frac{1}{3}$$

Problem 19. Solve the equation
$$\log(x-1) + \log(x+8) = 2\log(x+2)$$

LHS $= \log(x-1) + \log(x+8) = \log(x-1)(x+8)$
from the first
law of logarithms

$$= \log(x^2 + 7x - 8)$$

RHS $= 2\log(x+2) = \log(x+2)^2$
from the first
law of logarithms

$$= \log(x^2 + 4x + 4)$$

Hence, $\qquad \log(x^2 + 7x - 8) = \log(x^2 + 4x + 4)$

from which, $\qquad x^2 + 7x - 8 = x^2 + 4x + 4$

i.e. $\qquad 7x - 8 = 4x + 4$

i.e. $\qquad 3x = 12$

and $\qquad \boldsymbol{x = 4}$

Problem 20. Solve the equation $\frac{1}{2}\log 4 = \log x$

$$\frac{1}{2}\log 4 = \log 4^{\frac{1}{2}} \quad \text{from the third law of}$$
$$\text{logarithms}$$
$$= \log \sqrt{4} \text{ from the laws of indices}$$

Hence, $\qquad \frac{1}{2}\log 4 = \log x$

becomes $\qquad \log \sqrt{4} = \log x$

i.e. $\qquad \log 2 = \log x$

from which, $\qquad 2 = x$

i.e. the solution of the equation is $\boldsymbol{x = 2}$

Problem 21. Solve the equation
$$\log(x^2 - 3) - \log x = \log 2$$

$$\log(x^2 - 3) - \log x = \log\left(\frac{x^2 - 3}{x}\right) \text{ from the second}$$
$$\text{law of logarithms}$$

Hence, $\qquad \log\left(\frac{x^2 - 3}{x}\right) = \log 2$

from which, $\qquad \frac{x^2 - 3}{x} = 2$

Rearranging gives $\qquad x^2 - 3 = 2x$

and $\qquad x^2 - 2x - 3 = 0$

Factorising gives $\quad (x-3)(x+1) = 0$

from which, $\qquad x = 3 \text{ or } x = -1$

$x = -1$ is not a valid solution since the logarithm of a negative number has no real root.

Hence, the solution of the equation is $\boldsymbol{x = 3}$

Now try the following Practice Exercise

In Problems 1 to 11, write as the logarithm of a single number.

1. $\log 2 + \log 3$
2. $\log 3 + \log 5$

3. $\log 3 + \log 4 - \log 6$

4. $\log 7 + \log 21 - \log 49$

5. $2\log 2 + \log 3$
6. $2\log 2 + 3\log 5$

7. $2\log 5 - \frac{1}{2}\log 81 + \log 36$

8. $\frac{1}{3}\log 8 - \frac{1}{2}\log 81 + \log 27$

9. $\frac{1}{2}\log 4 - 2\log 3 + \log 45$

10. $\frac{1}{4}\log 16 + 2\log 3 - \log 18$

11. $2\log 2 + \log 5 - \log 10$

Simplify the expressions given in Problems 12 to 14.

12. $\log 27 - \log 9 + \log 81$

13. $\log 64 + \log 32 - \log 128$

14. $\log 8 - \log 4 + \log 32$

Evaluate the expressions given in Problems 15 and 16.

15. $\dfrac{\frac{1}{2}\log 16 - \frac{1}{3}\log 8}{\log 4}$

16. $\dfrac{\log 9 - \log 3 + \frac{1}{2}\log 81}{2\log 3}$

Solve the equations given in Problems 17 to 22.

17. $\log x^4 - \log x^3 = \log 5x - \log 2x$

18. $\log 2t^3 - \log t = \log 16 + \log t$

19. $2\log b^2 - 3\log b = \log 8b - \log 4b$

20. $\log(x+1) + \log(x-1) = \log 3$

21. $\dfrac{1}{3}\log 27 = \log(0.5a)$

22. $\log(x^2 - 5) - \log x = \log 4$

15.3 Indicial equations

The laws of logarithms may be used to solve certain equations involving powers, called **indicial equations**.

For example, to solve, say, $3^x = 27$, logarithms to a base of 10 are taken of both sides,

i.e. $\log_{10} 3^x = \log_{10} 27$

and $x\log_{10} 3 = \log_{10} 27$

by the third law of logarithms

Rearranging gives $x = \dfrac{\log_{10} 27}{\log_{10} 3} = \dfrac{1.43136\ldots}{0.47712\ldots}$

$= 3$ which may be readily checked.

$\left(\text{Note, } \dfrac{\log 27}{\log 3} \text{ is \textbf{not} equal to } \log\dfrac{27}{3}\right)$

Problem 22. Solve the equation $2^x = 5$, correct to 4 significant figures

Taking logarithms to base 10 of both sides of $2^x = 5$ gives

$$\log_{10} 2^x = \log_{10} 5$$

i.e. $x\log_{10} 2 = \log_{10} 5$

by the third law of logarithms

Rearranging gives $x = \dfrac{\log_{10} 5}{\log_{10} 2} = \dfrac{0.6989700\ldots}{0.3010299\ldots}$

$= \mathbf{2.322}$, correct to 4 significant figures.

Problem 23. Solve the equation $2^{x+1} = 3^{2x-5}$ correct to 2 decimal places

Taking logarithms to base 10 of both sides gives

$$\log_{10} 2^{x+1} = \log_{10} 3^{2x-5}$$

i.e. $(x+1)\log_{10} 2 = (2x-5)\log_{10} 3$

$x\log_{10} 2 + \log_{10} 2 = 2x\log_{10} 3 - 5\log_{10} 3$

$x(0.3010) + (0.3010) = 2x(0.4771) - 5(0.4771)$

i.e. $0.3010x + 0.3010 = 0.9542x - 2.3855$

Hence, $2.3855 + 0.3010 = 0.9542x - 0.3010x$

$2.6865 = 0.6532x$

from which $x = \dfrac{2.6865}{0.6532} = \mathbf{4.11}$,

correct to 2 decimal places.

Problem 24. Solve the equation $x^{2.7} = 34.68$, correct to 4 significant figures

Taking logarithms to base 10 of both sides gives

$$\log_{10} x^{2.7} = \log_{10} 34.68$$

$2.7\log_{10} x = \log_{10} 34.68$

Hence, $\log_{10} x = \dfrac{\log_{10} 34.68}{2.7} = 0.57040$

Thus, $x = \text{antilog } 0.57040 = 10^{0.57040}$

$= \textbf{3.719},$

correct to 4 significant figures.

Now try the following Practice Exercise

Practice Exercise 61   Indicial equations
(answers on page 429)

In Problems 1 to 8, solve the indicial equations for x, each correct to 4 significant figures.

1. $3^x = 6.4$ 2. $2^x = 9$

3. $2^{x-1} = 3^{2x-1}$ 4. $x^{1.5} = 14.91$

5. $25.28 = 4.2^x$ 6. $4^{2x-1} = 5^{x+2}$

7. $x^{-0.25} = 0.792$ 8. $0.027^x = 3.26$

9. The decibel gain n of an amplifier is given by $n = 10 \log_{10}\left(\dfrac{P_2}{P_1}\right)$, where P_1 is the power input and P_2 is the power output. Find the power gain $\dfrac{P_2}{P_1}$ when $n = 25$ decibels.

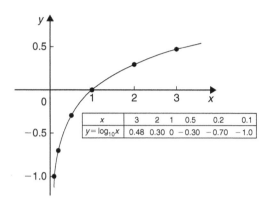

Figure 15.1

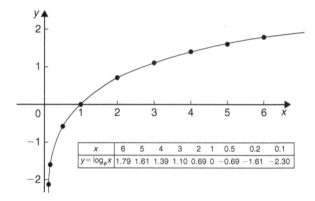

Figure 15.2

15.4   Graphs of logarithmic functions

A graph of $y = \log_{10} x$ is shown in Fig. 15.1 and a graph of $y = \log_e x$ is shown in Fig. 15.2. Both can be seen to be of similar shape; in fact, the same general shape occurs for a logarithm to any base.

In general, with a logarithm to any base, a, it is noted that

(a) **$\log_a 1 = 0$**

Let $\log_a = x$ then $a^x = 1$ from the definition of the logarithm.

If $a^x = 1$ then $x = 0$ from the laws of logarithms.

Hence, **$\log_a 1 = 0$**. In the above graphs it is seen that $\log_{10} 1 = 0$ and $\log_e 1 = 0$

(b) **$\log_a a = 1$**

Let $\log_a a = x$ then $a^x = a$ from the definition of a logarithm.

If $a^x = a$ then $x = 1$

Hence, **$\log_a a = 1$**. (Check with a calculator that $\log_{10} 10 = 1$ and $\log_e e = 1$)

(c) **$\log_a 0 \rightarrow -\infty$**

Let $\log_a 0 = x$ then $a^x = 0$ from the definition of a logarithm.

If $a^x = 0$, and a is a positive real number, then x must approach minus infinity. (For example, check with a calculator, $2^{-2} = 0.25, 2^{-20} = 9.54 \times 10^{-7}, 2^{-200} = 6.22 \times 10^{-61}$, and so on.)

Hence, **$\log_a 0 \rightarrow -\infty$**

For fully worked solutions to each of the problems in Practice Exercises 59 to 61 in this chapter, go to the website:
www.routledge.com/cw/bird

Chapter 16

Exponential functions

Why it is important to understand: Exponential functions

Exponential functions are used in engineering, physics, biology and economics. There are many quantities that grow exponentially; some examples are population, compound interest and charge in a capacitor. With exponential growth, the rate of growth increases as time increases. We also have exponential decay; some examples are radioactive decay, atmospheric pressure, Newton's law of cooling and linear expansion. Understanding and using exponential functions is important in many branches of engineering.

At the end of this chapter, you should be able to:

- evaluate exponential functions using a calculator
- state the exponential series for e^x
- plot graphs of exponential functions
- evaluate Napierian logarithms using a calculator
- solve equations involving Napierian logarithms
- appreciate the many examples of laws of growth and decay in engineering and science
- perform calculations involving the laws of growth and decay

16.1 Introduction to exponential functions

An exponential function is one which contains e^x, e being a constant called the exponent and having an approximate value of 2.7183. The exponent arises from the natural laws of growth and decay and is used as a base for natural or Napierian logarithms.

The most common method of evaluating an exponential function is by using a scientific notation **calculator**. Use your calculator to check the following values.

$$e^1 = 2.7182818, \text{ correct to 8 significant figures,}$$
$$e^{-1.618} = 0.1982949, \text{ correct to 7 significant figures,}$$
$$e^{0.12} = 1.1275, \text{ correct to 5 significant figures,}$$
$$e^{-1.47} = 0.22993, \text{ correct to 5 decimal places,}$$
$$e^{-0.431} = 0.6499, \text{ correct to 4 decimal places,}$$
$$e^{9.32} = 11159, \text{ correct to 5 significant figures,}$$
$$e^{-2.785} = 0.0617291, \text{ correct to 7 decimal places.}$$

Problem 1. Evaluate the following correct to 4 decimal places, using a calculator:

$$0.0256(e^{5.21} - e^{2.49})$$

$0.0256(e^{5.21} - e^{2.49})$
$$= 0.0256(183.094058\ldots - 12.0612761\ldots)$$
$$= \mathbf{4.3784}, \text{ correct to 4 decimal places.}$$

Problem 2. Evaluate the following correct to 4 decimal places, using a calculator:

$$5\left(\frac{e^{0.25}-e^{-0.25}}{e^{0.25}+e^{-0.25}}\right)$$

$$5\left(\frac{e^{0.25}-e^{-0.25}}{e^{0.25}+e^{-0.25}}\right)$$

$$=5\left(\frac{1.28402541\ldots-0.77880078\ldots}{1.28402541\ldots+0.77880078\ldots}\right)$$

$$=5\left(\frac{0.5052246\ldots}{2.0628262\ldots}\right)$$

$$=\mathbf{1.2246},\text{ correct to 4 decimal places.}$$

Problem 3. The instantaneous voltage v in a capacitive circuit is related to time t by the equation $v = Ve^{-t/CR}$ where V, C and R are constants. Determine v, correct to 4 significant figures, when $t = 50$ ms, $C = 10\,\mu$F, $R = 47\,$kΩ and $V = 300$ volts

$$v = Ve^{-t/CR} = 300e^{(-50\times10^{-3})/(10\times10^{-6}\times47\times10^{3})}$$

Using a calculator, $\quad v = 300e^{-0.1063829\ldots}$

$$= 300(0.89908025\ldots)$$

$$= \mathbf{269.7\,volts}.$$

Now try the following Practice Exercise

Practice Exercise 62   Evaluating exponential functions (answers on page 429)

1. Evaluate the following, correct to 4 significant figures.
   (a) $e^{-1.8}$ (b) $e^{-0.78}$ (c) e^{10}

2. Evaluate the following, correct to 5 significant figures.
   (a) $e^{1.629}$ (b) $e^{-2.7483}$ (c) $0.62e^{4.178}$
   In Problems 3 and 4, evaluate correct to 5 decimal places.

3. (a) $\dfrac{1}{7}e^{3.4629}$ (b) $8.52e^{-1.2651}$
   (c) $\dfrac{5e^{2.6921}}{3e^{1.1171}}$

4. (a) $\dfrac{5.6823}{e^{-2.1347}}$ (b) $\dfrac{e^{2.1127}-e^{-2.1127}}{2}$
   (c) $\dfrac{4\left(e^{-1.7295}-1\right)}{e^{3.6817}}$

5. The length of a bar, l, at a temperature, θ, is given by $l = l_0 e^{\alpha\theta}$, where l_0 and α are constants. Evaluate l, correct to 4 significant figures, where $l_0 = 2.587$, $\theta = 321.7$ and $\alpha = 1.771 \times 10^{-4}$

6. When a chain of length $2L$ is suspended from two points, $2D$ metres apart on the same horizontal level, $D = k\left\{\ln\left(\dfrac{L+\sqrt{L^2+k^2}}{k}\right)\right\}$. Evaluate D when $k = 75$ m and $L = 180$ m.

16.2   The power series for e^x

The value of e^x can be calculated to any required degree of accuracy since it is defined in terms of the following **power series**:

$$e^x = 1 + x + \frac{x^2}{2!} + \frac{x^3}{3!} + \frac{x^4}{4!} + \cdots \qquad (1)$$

(where $3! = 3 \times 2 \times 1$ and is called '**factorial 3**').
The series is valid for all values of x.
The series is said to **converge**; i.e. if all the terms are added, an actual value for e^x (where x is a real number) is obtained. The more terms that are taken, the closer will be the value of e^x to its actual value. The value of the exponent e, correct to say 4 decimal places, may be determined by substituting $x = 1$ in the power series of equation (1). Thus,

$$e^1 = 1 + 1 + \frac{(1)^2}{2!} + \frac{(1)^3}{3!} + \frac{(1)^4}{4!} + \frac{(1)^5}{5!} + \frac{(1)^6}{6!}$$

$$+ \frac{(1)^7}{7!} + \frac{(1)^8}{8!} + \cdots$$

$$= 1 + 1 + 0.5 + 0.16667 + 0.04167 + 0.00833$$

$$+ 0.00139 + 0.00020 + 0.00002 + \cdots$$

$$= 2.71828$$

i.e. $e = \mathbf{2.7183}$, correct to 4 decimal places.

The value of $e^{0.05}$, correct to say 8 significant figures, is found by substituting $x = 0.05$ in the power series for e^x. Thus,

$$e^{0.05} = 1 + 0.05 + \frac{(0.05)^2}{2!} + \frac{(0.05)^3}{3!} + \frac{(0.05)^4}{4!}$$
$$+ \frac{(0.05)^5}{5!} + \cdots$$
$$= 1 + 0.05 + 0.00125 + 0.000020833$$
$$+ 0.000000260 + 0.0000000026$$

i.e. $e^{0.05} = 1.0512711$, correct to 8 significant figures.

In this example, successive terms in the series grow smaller very rapidly and it is relatively easy to determine the value of $e^{0.05}$ to a high degree of accuracy. However, when x is nearer to unity or larger than unity, a very large number of terms are required for an accurate result.

If, in the series of equation (1), x is replaced by $-x$, then

$$e^{-x} = 1 + (-x) + \frac{(-x)^2}{2!} + \frac{(-x)^3}{3!} + \cdots$$

i.e. $e^{-x} = 1 - x + \frac{x^2}{2!} - \frac{x^3}{3!} + \cdots$

In a similar manner the power series for e^x may be used to evaluate any exponential function of the form ae^{kx}, where a and k are constants.

In the series of equation (1), let x be replaced by kx. Then

$$ae^{kx} = a \left\{ 1 + (kx) + \frac{(kx)^2}{2!} + \frac{(kx)^3}{3!} + \cdots \right\}$$

Thus, $5e^{2x} = 5 \left\{ 1 + (2x) + \frac{(2x)^2}{2!} + \frac{(2x)^3}{3!} + \cdots \right\}$

$$= 5 \left\{ 1 + 2x + \frac{4x^2}{2} + \frac{8x^3}{6} + \cdots \right\}$$

i.e. $5e^{2x} = 5 \left\{ 1 + 2x + 2x^2 + \frac{4}{3}x^3 + \cdots \right\}$

Problem 4. Determine the value of $5e^{0.5}$, correct to 5 significant figures, by using the power series for e^x

From equation (1),

$$e^x = 1 + x + \frac{x^2}{2!} + \frac{x^3}{3!} + \frac{x^4}{4!} + \cdots$$

Hence, $e^{0.5} = 1 + 0.5 + \frac{(0.5)^2}{(2)(1)} + \frac{(0.5)^3}{(3)(2)(1)}$

$$+ \frac{(0.5)^4}{(4)(3)(2)(1)} + \frac{(0.5)^5}{(5)(4)(3)(2)(1)}$$
$$+ \frac{(0.5)^6}{(6)(5)(4)(3)(2)(1)}$$
$$= 1 + 0.5 + 0.125 + 0.020833$$
$$+ 0.0026042 + 0.0002604$$
$$+ 0.0000217$$

i.e. $e^{0.5} = 1.64872$, correct to 6 significant figures

Hence, $5e^{0.5} = 5(1.64872) = 8.2436$, correct to 5 significant figures.

Problem 5. Determine the value of $3e^{-1}$, correct to 4 decimal places, using the power series for e^x

Substituting $x = -1$ in the power series

$$e^x = 1 + x + \frac{x^2}{2!} + \frac{x^3}{3!} + \frac{x^4}{4!} + \cdots$$

gives $e^{-1} = 1 + (-1) + \frac{(-1)^2}{2!} + \frac{(-1)^3}{3!}$

$$+ \frac{(-1)^4}{4!} + \cdots$$
$$= 1 - 1 + 0.5 - 0.166667 + 0.041667$$
$$- 0.008333 + 0.001389$$
$$- 0.000198 + \cdots$$
$$= 0.367858 \text{ correct to 6 decimal places}$$

Hence, $3e^{-1} = (3)(0.367858) = 1.1036$, correct to 4 decimal places.

Problem 6. Expand $e^x(x^2 - 1)$ as far as the term in x^5

The power series for e^x is

$$e^x = 1 + x + \frac{x^2}{2!} + \frac{x^3}{3!} + \frac{x^4}{4!} + \frac{x^5}{5!} + \cdots$$

Hence,

$e^x(x^2 - 1)$

$$= \left(1 + x + \frac{x^2}{2!} + \frac{x^3}{3!} + \frac{x^4}{4!} + \frac{x^5}{5!} + \cdots\right)(x^2 - 1)$$

$$= \left(x^2 + x^3 + \frac{x^4}{2!} + \frac{x^5}{3!} + \cdots\right)$$

$$\quad - \left(1 + x + \frac{x^2}{2!} + \frac{x^3}{3!} + \frac{x^4}{4!} + \frac{x^5}{5!} + \cdots\right)$$

Grouping like terms gives

$e^x(x^2 - 1)$

$$= -1 - x + \left(x^2 - \frac{x^2}{2!}\right) + \left(x^3 - \frac{x^3}{3!}\right)$$

$$\quad + \left(\frac{x^4}{2!} - \frac{x^4}{4!}\right) + \left(\frac{x^5}{3!} - \frac{x^5}{5!}\right) + \cdots$$

$$= \mathbf{-1 - x + \frac{1}{2}x^2 + \frac{5}{6}x^3 + \frac{11}{24}x^4 + \frac{19}{120}x^5}$$

when expanded as far as the term in x^5.

Now try the following Practice Exercise

Practice Exercise 63   Power series for e^x
(answers on page 429)

1. Evaluate $5.6e^{-1}$, correct to 4 decimal places, using the power series for e^x.

2. Use the power series for e^x to determine, correct to 4 significant figures, (a) e^2 (b) $e^{-0.3}$ and check your results using a calculator.

3. Expand $(1 - 2x)e^{2x}$ as far as the term in x^4.

4. Expand $(2e^{x^2})(x^{1/2})$ to six terms.

16.3   Graphs of exponential functions

Values of e^x and e^{-x} obtained from a calculator, correct to 2 decimal places, over a range $x = -3$ to $x = 3$, are shown in Table 16.1.

Fig. 16.1 shows graphs of $y = e^x$ and $y = e^{-x}$

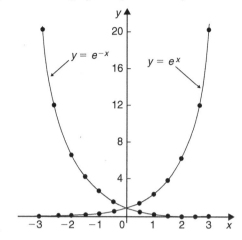

Figure 16.1

Problem 7.   Plot a graph of $y = 2e^{0.3x}$ over a range of $x = -3$ to $x = 3$. Then determine the value of y when $x = 2.2$ and the value of x when $y = 1.6$

A table of values is drawn up as shown below.

x	-3	-2	-1	0	1	2	3
$2e^{0.3x}$	0.81	1.10	1.48	2.00	2.70	3.64	4.92

A graph of $y = 2e^{0.3x}$ is shown plotted in Fig. 16.2. From the graph, **when $x = 2.2$, $y = 3.87$** and **when $y = 1.6$, $x = -0.74$**

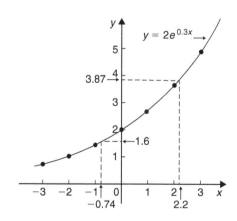

Figure 16.2

Table 16.1

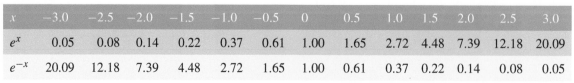

x	-3.0	-2.5	-2.0	-1.5	-1.0	-0.5	0	0.5	1.0	1.5	2.0	2.5	3.0
e^x	0.05	0.08	0.14	0.22	0.37	0.61	1.00	1.65	2.72	4.48	7.39	12.18	20.09
e^{-x}	20.09	12.18	7.39	4.48	2.72	1.65	1.00	0.61	0.37	0.22	0.14	0.08	0.05

Problem 8. Plot a graph of $y = \frac{1}{3}e^{-2x}$ over the range $x = -1.5$ to $x = 1.5$. Determine from the graph the value of y when $x = -1.2$ and the value of x when $y = 1.4$

A table of values is drawn up as shown below.

x	−1.5	−1.0	−0.5	0	0.5	1.0	1.5
$\frac{1}{3}e^{-2x}$	6.70	2.46	0.91	0.33	0.12	0.05	0.02

A graph of $\frac{1}{3}e^{-2x}$ is shown in Fig. 16.3.
From the graph, **when $x = -1.2$, $y = 3.67$ and when $y = 1.4$, $x = -0.72$**

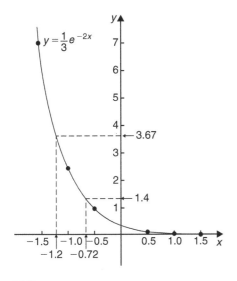

Figure 16.3

Problem 9. The decay of voltage, v volts, across a capacitor at time t seconds is given by $v = 250e^{-t/3}$. Draw a graph showing the natural decay curve over the first 6 seconds. From the graph, find (a) the voltage after 3.4 s and (b) the time when the voltage is 150 V

A table of values is drawn up as shown below.

t	0	1	2	3
$e^{-t/3}$	1.00	0.7165	0.5134	0.3679
$v = 250e^{-t/3}$	250.0	179.1	128.4	91.97

t	4	5	6
$e^{-t/3}$	0.2636	0.1889	0.1353
$v = 250e^{-t/3}$	65.90	47.22	33.83

The natural decay curve of $v = 250e^{-t/3}$ is shown in Fig. 16.4.

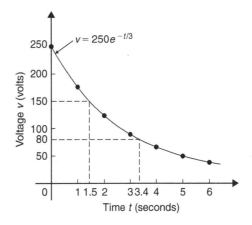

Figure 16.4

From the graph,

(a) when time $t = 3.4$ s, voltage $v = 80$ V, and

(b) when **voltage $v = 150$ V, time $t = 1.5$ s**.

Now try the following Practice Exercise

Practice Exercise 64   Exponential graphs (answers on page 429)

1. Plot a graph of $y = 3e^{0.2x}$ over the range $x = -3$ to $x = 3$. Hence determine the value of y when $x = 1.4$ and the value of x when $y = 4.5$

2. Plot a graph of $y = \frac{1}{2}e^{-1.5x}$ over a range $x = -1.5$ to $x = 1.5$ and then determine the value of y when $x = -0.8$ and the value of x when $y = 3.5$

3. In a chemical reaction the amount of starting material C cm^3 left after t minutes is given by $C = 40e^{-0.006t}$. Plot a graph of C against t and determine

   (a) the concentration C after 1 hour.

   (b) the time taken for the concentration to decrease by half.

4. The rate at which a body cools is given by $\theta = 250e^{-0.05t}$ where the excess of temperature of a body above its surroundings at time t minutes is $\theta\,°C$. Plot a graph showing the natural decay curve for the first hour of cooling. Then determine

   (a) the temperature after 25 minutes.

   (b) the time when the temperature is $195\,°C$.

16.4 Napierian logarithms

Logarithms having a base of e are called **hyperbolic**, **Napierian** or **natural logarithms** and the Napierian logarithm of x is written as $\log_e x$ or, more commonly, as $\ln x$. Logarithms were invented by **John Napier***, a Scotsman.

The most common method of evaluating a Napierian logarithm is by a scientific notation **calculator**. Use your calculator to check the following values:

$\ln 4.328 = 1.46510554\ldots = 1.4651$, correct to 4 decimal places

*Who was Napier? – **John Napier** of Merchiston (1550–4 April 1617) is best known as the discoverer of logarithms. The inventor of the so-called 'Napier's bones', Napier also made common the use of the decimal point in arithmetic and mathematics. To find out more go to www.routledge.com/cw/bird

$\ln 1.812 = 0.59443$, correct to 5 significant figures

$\ln 1 = 0$

$\ln 527 = 6.2672$, correct to 5 significant figures

$\ln 0.17 = -1.772$, correct to 4 significant figures

$\ln 0.00042 = -7.77526$, correct to 6 significant figures

$\ln e^3 = 3$

$\ln e^1 = 1$

From the last two examples we can conclude that

$$\log_e e^x = x$$

This is useful when **solving equations involving exponential functions**. For example, to solve $e^{3x} = 7$, take Napierian logarithms of both sides, which gives

$$\ln e^{3x} = \ln 7$$

i.e. $\qquad 3x = \ln 7$

from which $\qquad x = \dfrac{1}{3}\ln 7 = \mathbf{0.6486}$,

correct to 4 decimal places.

Problem 10. Evaluate the following, each correct to 5 significant figures: (a) $\dfrac{1}{2}\ln 4.7291$ (b) $\dfrac{\ln 7.8693}{7.8693}$ (c) $\dfrac{3.17\ln 24.07}{e^{-0.1762}}$

(a) $\dfrac{1}{2}\ln 4.7291 = \dfrac{1}{2}(1.5537349\ldots) = \mathbf{0.77687}$, correct to 5 significant figures.

(b) $\dfrac{\ln 7.8693}{7.8693} = \dfrac{2.06296911\ldots}{7.8693} = \mathbf{0.26215}$, correct to 5 significant figures.

(c) $\dfrac{3.17\ln 24.07}{e^{-0.1762}} = \dfrac{3.17(3.18096625\ldots)}{0.83845027\ldots} = \mathbf{12.027}$, correct to 5 significant figures.

Problem 11. Evaluate the following: (a) $\dfrac{\ln e^{2.5}}{\lg 10^{0.5}}$ (b) $\dfrac{5e^{2.23}\lg 2.23}{\ln 2.23}$ (correct to 3 decimal places)

(a) $\dfrac{\ln e^{2.5}}{\lg 10^{0.5}} = \dfrac{2.5}{0.5} = \mathbf{5}$

(b) $\dfrac{5e^{2.23}\lg 2.23}{\ln 2.23}$

$= \dfrac{5(9.29986607\ldots)(0.34830486\ldots)}{(0.80200158\ldots)}$

$= \mathbf{20.194}$, correct to 3 decimal places.

Problem 12. Solve the equation $9 = 4e^{-3x}$ to find x, correct to 4 significant figures

Rearranging $9 = 4e^{-3x}$ gives $\dfrac{9}{4} = e^{-3x}$

Taking the reciprocal of both sides gives

$$\dfrac{4}{9} = \dfrac{1}{e^{-3x}} = e^{3x}$$

Taking Napierian logarithms of both sides gives

$$\ln\left(\dfrac{4}{9}\right) = \ln(e^{3x})$$

Since $\log_e e^\alpha = \alpha$, then $\ln\left(\dfrac{4}{9}\right) = 3x$

Hence, $x = \dfrac{1}{3}\ln\left(\dfrac{4}{9}\right) = \dfrac{1}{3}(-0.81093) = \mathbf{-0.2703}$, correct to 4 significant figures.

Problem 13. Given $32 = 70\left(1 - e^{-\frac{t}{2}}\right)$, determine the value of t, correct to 3 significant figures

Rearranging $32 = 70\left(1 - e^{-\frac{t}{2}}\right)$ gives

$$\dfrac{32}{70} = 1 - e^{-\frac{t}{2}}$$

and

$$e^{-\frac{t}{2}} = 1 - \dfrac{32}{70} = \dfrac{38}{70}$$

Taking the reciprocal of both sides gives

$$e^{\frac{t}{2}} = \dfrac{70}{38}$$

Taking Napierian logarithms of both sides gives

$$\ln e^{\frac{t}{2}} = \ln\left(\dfrac{70}{38}\right)$$

i.e. $\dfrac{t}{2} = \ln\left(\dfrac{70}{38}\right)$

from which, $t = 2\ln\left(\dfrac{70}{38}\right) = \mathbf{1.22}$, correct to 3 significant figures.

Problem 14. Solve the equation

$$2.68 = \ln\left(\dfrac{4.87}{x}\right) \text{ to find } x$$

From the definition of a logarithm, since $2.68 = \ln\left(\dfrac{4.87}{x}\right)$ then $e^{2.68} = \dfrac{4.87}{x}$

Rearranging gives $x = \dfrac{4.87}{e^{2.68}} = 4.87e^{-2.68}$

i.e. $x = \mathbf{0.3339}$,

correct to 4 significant figures.

Problem 15. Solve $\dfrac{7}{4} = e^{3x}$ correct to 4 significant figures

Taking natural logs of both sides gives

$$\ln\dfrac{7}{4} = \ln e^{3x}$$

$$\ln\dfrac{7}{4} = 3x \ln e$$

Since $\ln e = 1$, $\ln\dfrac{7}{4} = 3x$

i.e. $0.55962 = 3x$

i.e. $x = \mathbf{0.1865}$,

correct to 4 significant figures.

Problem 16. Solve $e^{x-1} = 2e^{3x-4}$ correct to 4 significant figures

Taking natural logarithms of both sides gives

$$\ln\left(e^{x-1}\right) = \ln\left(2e^{3x-4}\right)$$

and by the first law of logarithms,

$$\ln\left(e^{x-1}\right) = \ln 2 + \ln\left(e^{3x-4}\right)$$

i.e. $x - 1 = \ln 2 + 3x - 4$

Rearranging gives

$$4 - 1 - \ln 2 = 3x - x$$

i.e. $3 - \ln 2 = 2x$

from which, $x = \dfrac{3 - \ln 2}{2} = \mathbf{1.153}$

Problem 17. Solve, correct to 4 significant figures, $\ln(x-2)^2 = \ln(x-2) - \ln(x+3) + 1.6$

Rearranging gives

$$\ln(x-2)^2 - \ln(x-2) + \ln(x+3) = 1.6$$

and by the laws of logarithms,

$$\ln\left\{\frac{(x-2)^2(x+3)}{(x-2)}\right\} = 1.6$$

Cancelling gives

$$\ln\{(x-2)(x+3)\} = 1.6$$

and

$$(x-2)(x+3) = e^{1.6}$$

i.e.

$$x^2 + x - 6 = e^{1.6}$$

or

$$x^2 + x - 6 - e^{1.6} = 0$$

i.e.

$$x^2 + x - 10.953 = 0$$

Using the quadratic formula,

$$x = \frac{-1 \pm \sqrt{1^2 - 4(1)(-10.953)}}{2}$$

$$= \frac{-1 \pm \sqrt{44.812}}{2}$$

$$= \frac{-1 \pm 6.6942}{2}$$

i.e. $x = 2.847$ or -3.8471

$x = -3.8471$ is not valid since the logarithm of a negative number has no real root.

Hence, **the solution of the equation is $x = 2.847$**

Now try the following Practice Exercise

Practice Exercise 65   Evaluating Napierian logarithms (answers on page 429)

In Problems 1 and 2, evaluate correct to 5 significant figures.

1. (a) $\dfrac{1}{3}\ln 5.2932$ (b) $\dfrac{\ln 82.473}{4.829}$

    (c) $\dfrac{5.62 \ln 321.62}{e^{1.2942}}$

2. (a) $\dfrac{1.786 \ln e^{1.76}}{\lg 10^{1.41}}$ (b) $\dfrac{5e^{-0.1629}}{2\ln 0.00165}$

    (c) $\dfrac{\ln 4.8629 - \ln 2.4711}{5.173}$

In Problems 3 to 16, solve the given equations, each correct to 4 significant figures.

3. $1.5 = 4e^{2t}$

4. $7.83 = 2.91e^{-1.7x}$

5. $16 = 24\left(1 - e^{-\frac{t}{2}}\right)$

6. $5.17 = \ln\left(\dfrac{x}{4.64}\right)$

7. $3.72 \ln\left(\dfrac{1.59}{x}\right) = 2.43$

8. $\ln x = 2.40$

9. $24 + e^{2x} = 45$

10. $5 = e^{x+1} - 7$

11. $5 = 8\left(1 - e^{\frac{-x}{2}}\right)$

12. $\ln(x+3) - \ln x = \ln(x-1)$

13. $\ln(x-1)^2 - \ln 3 = \ln(x-1)$

14. $\ln(x+3) + 2 = 12 - \ln(x-2)$

15. $e^{(x+1)} = 3e^{(2x-5)}$

16. $\ln(x+1)^2 = 1.5 - \ln(x-2) + \ln(x+1)$

17. Transpose $b = \ln t - a \ln D$ to make t the subject.

18. If $\dfrac{P}{Q} = 10 \log_{10}\left(\dfrac{R_1}{R_2}\right)$, find the value of R_1 when $P = 160$, $Q = 8$ and $R_2 = 5$

19. If $U_2 = U_1 e^{\left(\frac{W}{PV}\right)}$, make W the subject of the formula.

20. The velocity v_2 of a rocket is given by:
     $v_2 = v_1 + C \ln\left(\dfrac{m_1}{m_2}\right)$ where v_1 is the initial rocket velocity, C is the velocity of the jet exhaust gases, m_1 is the mass of the rocket before the jet engine is fired, and m_2 is the mass of the rocket after the jet engine is switched off. Calculate the velocity of the rocket given $v_1 = 600$ m/s, $C = 3500$ m/s, $m_1 = 8.50 \times 10^4$ kg and $m_2 = 7.60 \times 10^4$ kg.

21. The work done in an isothermal expansion of a gas from pressure p_1 to p_2 is given by:

$$w = w_0 \ln\left(\frac{p_1}{p_2}\right)$$

If the initial pressure $p_1 = 7.0$ kPa, calculate the final pressure p_2 if $w = 3\,w_0$

16.5 Laws of growth and decay

Laws of exponential growth and decay are of the form $y = Ae^{-kx}$ and $y = A(1 - e^{-kx})$, where A and k are constants. When plotted, the form of these equations is as shown in Fig. 16.5.

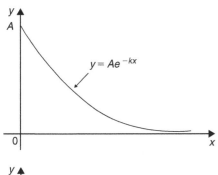

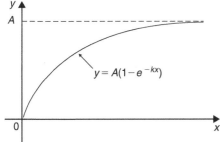

Figure 16.5

The laws occur frequently in engineering and science and examples of quantities related by a natural law include:

(a) Linear expansion $\qquad l = l_0 e^{\alpha\theta}$

(b) Change in electrical resistance with temperature $R_\theta = R_0 e^{\alpha\theta}$

(c) Tension in belts $\qquad T_1 = T_0 e^{\mu\theta}$

(d) Newton's law of cooling $\qquad \theta = \theta_0 e^{-kt}$

(e) Biological growth $\qquad y = y_0 e^{kt}$

(f) Discharge of a capacitor $\qquad q = Q e^{-t/CR}$

(g) Atmospheric pressure $\qquad p = p_0 e^{-h/c}$

(h) Radioactive decay $\qquad N = N_0 e^{-\lambda t}$

(i) Decay of current in an inductive circuit
$$i = I e^{-Rt/L}$$

(j) Growth of current in a capacitive circuit
$$i = I(1 - e^{-t/CR})$$

Here are some worked problems to demonstrate the laws of growth and decay.

Problem 18. The resistance R of an electrical conductor at temperature $\theta°C$ is given by $R = R_0 e^{\alpha\theta}$, where α is a constant and $R_0 = 5\,k\Omega$. Determine the value of α correct to 4 significant figures when $R = 6\,k\Omega$ and $\theta = 1500°C$. Also, find the temperature, correct to the nearest degree, when the resistance R is $5.4\,k\Omega$

Transposing $R = R_0 e^{\alpha\theta}$ gives $\dfrac{R}{R_0} = e^{\alpha\theta}$

Taking Napierian logarithms of both sides gives

$$\ln \frac{R}{R_0} = \ln e^{\alpha\theta} = \alpha\theta$$

Hence, $\alpha = \dfrac{1}{\theta} \ln \dfrac{R}{R_0} = \dfrac{1}{1500} \ln\left(\dfrac{6 \times 10^3}{5 \times 10^3}\right)$

$$= \frac{1}{1500}(0.1823215\ldots)$$

$$= 1.215477\ldots \times 10^{-4}$$

Hence, $\alpha = \mathbf{1.215 \times 10^{-4}}$ correct to 4 significant figures.

From above, $\ln \dfrac{R}{R_0} = \alpha\theta$ hence $\theta = \dfrac{1}{\alpha} \ln \dfrac{R}{R_0}$

When $R = 5.4 \times 10^3$, $\alpha = 1.215477\ldots \times 10^{-4}$ and $R_0 = 5 \times 10^3$

$$\theta = \frac{1}{1.215477\ldots \times 10^{-4}} \ln\left(\frac{5.4 \times 10^3}{5 \times 10^3}\right)$$

$$= \frac{10^4}{1.215477\ldots}(7.696104\ldots \times 10^{-2})$$

$$= \mathbf{633°C} \text{ correct to the nearest degree.}$$

Problem 19. In an experiment involving Newton's law of cooling*, the temperature $\theta(°C)$ is given by $\theta = \theta_0 e^{-kt}$. Find the value of constant k when $\theta_0 = 56.6°C$, $\theta = 16.5°C$ and $t = 79.0$ seconds

---

*Who was Newton? To find out more go to www.routledge.com/cw/bird

Transposing $\theta = \theta_0 e^{-kt}$ gives $\dfrac{\theta}{\theta_0} = e^{-kt}$, from which

$$\frac{\theta_0}{\theta} = \frac{1}{e^{-kt}} = e^{kt}$$

Taking Napierian logarithms of both sides gives

$$\ln \frac{\theta_0}{\theta} = kt$$

from which,

$$k = \frac{1}{t} \ln \frac{\theta_0}{\theta} = \frac{1}{79.0} \ln \left(\frac{56.6}{16.5} \right)$$

$$= \frac{1}{79.0} (1.2326486\ldots)$$

Hence, $k = 0.01560$ or 15.60×10^{-3}

Problem 20. The current i amperes flowing in a capacitor at time t seconds is given by $i = 8.0(1 - e^{-\frac{t}{CR}})$, where the circuit resistance R is $25\,k\Omega$ and capacitance C is $16\,\mu F$. Determine (a) the current i after 0.5 seconds and (b) the time, to the nearest millisecond, for the current to reach $6.0\,A$. Sketch the graph of current against time

---

(a) Current $i = 8.0\left(1 - e^{-\frac{t}{CR}}\right)$

$$= 8.0[1 - e^{-0.5/(16 \times 10^{-6})(25 \times 10^3)}]$$

$$= 8.0(1 - e^{-1.25})$$

$$= 8.0(1 - 0.2865047\ldots)$$

$$= 8.0(0.7134952\ldots)$$

$$= \mathbf{5.71\,amperes}$$

(b) Transposing $i = 8.0\left(1 - e^{-\frac{t}{CR}}\right)$ gives

$$\frac{i}{8.0} = 1 - e^{-\frac{t}{CR}}$$

from which, $e^{-\frac{t}{CR}} = 1 - \dfrac{i}{8.0} = \dfrac{8.0 - i}{8.0}$

Taking the reciprocal of both sides gives

$$e^{\frac{t}{CR}} = \frac{8.0}{8.0 - i}$$

Taking Napierian logarithms of both sides gives

$$\frac{t}{CR} = \ln \left(\frac{8.0}{8.0 - i} \right)$$

Hence,

$$t = CR \ln \left(\frac{8.0}{8.0 - i} \right)$$

When $i = 6.0\,A$,

$$t = (16 \times 10^{-6})(25 \times 10^3) \ln \left(\frac{8.0}{8.0 - 6.0} \right)$$

i.e. $t = \dfrac{400}{10^3} \ln \left(\dfrac{8.0}{2.0} \right) = 0.4 \ln 4.0$

$$= 0.4(1.3862943\ldots)$$

$$= 0.5545\,s$$

$$= \mathbf{555\,ms} \text{ correct to the nearest ms.}$$

A graph of current against time is shown in Fig. 16.6.

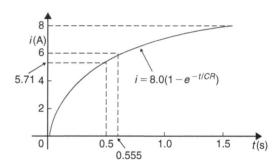

Figure 16.6

Problem 21. The temperature θ_2 of a winding which is being heated electrically at time t is given by $\theta_2 = \theta_1(1 - e^{-\frac{t}{\tau}})$, where θ_1 is the temperature (in degrees Celsius*) at time $t = 0$ and τ is a constant. Calculate

(a) θ_1, correct to the nearest degree, when θ_2 is $50°C$, t is $30\,s$ and τ is $60\,s$ and

(b) the time t, correct to 1 decimal place, for θ_2 to be half the value of θ_1

(a) Transposing the formula to make θ_1 the subject gives

$$\theta_1 = \frac{\theta_2}{(1 - e^{-t/\tau})} = \frac{50}{1 - e^{-30/60}}$$

$$= \frac{50}{1 - e^{-0.5}} = \frac{50}{0.393469\ldots}$$

i.e. $\boldsymbol{\theta_1 = 127°C}$ correct to the nearest degree.

---

*Who was Celsius? To find out more go to
www.routledge.com/cw/bird

(b) Transposing to make t the subject of the formula gives

$$\frac{\theta_2}{\theta_1} = 1 - e^{-\frac{t}{\tau}}$$

from which, $\quad e^{-\frac{t}{\tau}} = 1 - \frac{\theta_2}{\theta_1}$

Hence, $\quad -\frac{t}{\tau} = \ln\left(1 - \frac{\theta_2}{\theta_1}\right)$

i.e. $\quad t = -\tau \ln\left(1 - \frac{\theta_2}{\theta_1}\right)$

Since $\quad \theta_2 = \frac{1}{2}\theta_1$

$$t = -60\ln\left(1 - \frac{1}{2}\right) = -60\ln 0.5$$

$$= 41.59\,\text{s}$$

Hence, **the time for the temperature θ_2 to be one half of the value of θ_1 is 41.6 s, correct to 1 decimal place**.

Now try the following Practice Exercise

Practice Exercise 66   Laws of growth and decay (answers on page 429)

1. The temperature, $T\,°C$, of a cooling object varies with time, t minutes, according to the equation $T = 150e^{-0.04t}$. Determine the temperature when (a) $t = 0$, (b) $t = 10$ minutes.

2. The pressure p pascals at height h metres above ground level is given by $p = p_0 e^{-h/C}$, where p_0 is the pressure at ground level and C is a constant. Find pressure p when $p_0 = 1.012 \times 10^5\,\text{Pa}$, height $h = 1420\,\text{m}$ and $C = 71500$

3. The voltage drop, v volts, across an inductor L henrys at time t seconds is given by $v = 200e^{-\frac{Rt}{L}}$, where $R = 150\,\Omega$ and $L = 12.5 \times 10^{-3}\,\text{H}$. Determine (a) the voltage when $t = 160 \times 10^{-6}\,\text{s}$ and (b) the time for the voltage to reach 85 V.

4. The length l metres of a metal bar at temperature $t\,°C$ is given by $l = l_0 e^{\alpha t}$, where l_0 and α are constants. Determine (a) the value of l when $l_0 = 1.894$, $\alpha = 2.038 \times 10^{-4}$ and $t = 250°C$ and (b) the value of l_0 when $l = 2.416, t = 310°C$ and $\alpha = 1.682 \times 10^{-4}$

5. The temperature $\theta_2°C$ of an electrical conductor at time t seconds is given by $\theta_2 = \theta_1(1 - e^{-t/T})$, where θ_1 is the initial temperature and T seconds is a constant. Determine (a) θ_2 when $\theta_1 = 159.9°C$, $t = 30\,\text{s}$ and $T = 80\,\text{s}$ and (b) the time t for θ_2 to fall to half the value of θ_1 if T remains at 80 s.

6. A belt is in contact with a pulley for a sector of $\theta = 1.12$ radians and the coefficient of friction between these two surfaces is $\mu = 0.26$. Determine the tension on the taut side of the belt, T newtons, when tension on the slack side is given by $T_0 = 22.7\,\text{newtons}$, given that these quantities are related by the law $T = T_0 e^{\mu\theta}$

7. The instantaneous current i at time t is given by $i = 10e^{-t/CR}$ when a capacitor is being charged. The capacitance C is 7×10^{-6} farads and the resistance R is 0.3×10^6 ohms. Determine (a) the instantaneous current when t is 2.5 seconds and (b) the time for the instantaneous current to fall to 5 amperes. Sketch a curve of current against time from $t = 0$ to $t = 6$ seconds.

8. The amount of product x (in mol/cm^3) found in a chemical reaction starting with 2.5 mol/cm^3 of reactant is given by $x = 2.5(1 - e^{-4t})$ where t is the time, in minutes, to form product x. Plot a graph at 30 second intervals up to 2.5 minutes and determine x after 1 minute.

9. The current i flowing in a capacitor at time t is given by $i = 12.5(1 - e^{-t/CR})$, where resistance R is $30\,\text{k}\Omega$ and the capacitance C is $20\,\mu\text{F}$. Determine (a) the current flowing after 0.5 seconds and (b) the time for the current to reach 10 amperes.

10. The amount A after n years of a sum invested P is given by the compound interest law $A = Pe^{rn/100}$, when the per unit interest rate r is added continuously. Determine, correct to the nearest pound, the amount after 8 years for a sum of £1500 invested if the interest rate is 4% per annum.

11. The percentage concentration C of the starting material in a chemical reaction varies with time t according to the equation $C = 100\,e^{-0.004t}$. Determine the concentration when (a) $t = 0$, (b) $t = 100$ s, (c) $t = 1000$ s.

12. The current i flowing through a diode at room temperature is given by the equation $i = i_s\left(e^{40V} - 1\right)$ amperes. Calculate the current flowing in a silicon diode when the reverse saturation current $i_s = 50$ nA and the forward voltage $V = 0.27$ V.

13. A formula for chemical decomposition is given by: $C = A\left(1 - e^{-\frac{t}{10}}\right)$ where t is the time in seconds. Calculate the time, in milliseconds, for a compound to decompose to a value of $C = 0.12$ given $A = 8.5$

14. The mass, m, of pollutant in a water reservoir decreases according to the law $m = m_0\,e^{-0.1t}$ where t is the time in days and m_0 is the initial mass. Calculate the percentage decrease in the mass after 60 days, correct to 3 decimal places.

15. A metal bar is cooled with water. Its temperature, in °C, is given by the equation $\theta = 15 + 1300e^{-0.2t}$ where t is the time in minutes. Calculate how long it will take for the temperature, θ, to decrease to 36°C, correct to the nearest second.

For fully worked solutions to each of the problems in Practice Exercises 62 to 66 in this chapter, go to the website:
www.routledge.com/cw/bird

This assignment covers the material contained in Chapters 14–16. *The marks available are shown in brackets at the end of each question.*

1. Solve the following equations by factorisation.
   (a) $x^2 - 9 = 0$ (b) $x^2 + 12x + 36 = 0$
   (c) $x^2 + 3x - 4 = 0$ (d) $3z^2 - z - 4 = 0$
   (9)

2. Solve the following equations, correct to 3 decimal places.
   (a) $5x^2 + 7x - 3 = 0$ (b) $3a^2 + 4a - 5 = 0$
   (8)

3. Solve the equation $3x^2 - x - 4 = 0$ by completing the square.
   (6)

4. Determine the quadratic equation in x whose roots are 1 and -3.
   (3)

5. The bending moment M at a point in a beam is given by $M = \dfrac{3x(20-x)}{2}$ where x metres is the distance from the point of support. Determine the value of x when the bending moment is 50 Nm.
   (5)

6. The current i flowing through an electronic device is given by $i = (0.005v^2 + 0.014v)$ amperes where v is the voltage. Calculate the values of v when $i = 3 \times 10^{-3}$
   (6)

7. Evaluate the following, correct to 4 significant figures.
   (a) $3.2 \ln 4.92 - 5 \lg 17.9$ (b) $\dfrac{5(1 - e^{-2.65})}{e^{1.73}}$ (4)

8. Solve the following equations.
   (a) $\lg x = 4$ (b) $\ln x = 2$
   (c) $\log_2 x = 6$ (d) $5^x = 2$
   (e) $3^{2t-1} = 7^{t+2}$ (f) $3e^{2x} = 4.2$ (18)

9. Evaluate $\log_{16}\left(\dfrac{1}{8}\right)$
   (4)

10. Write the following as the logarithm of a single number.
    (a) $3 \log 2 + 2 \log 5 - \dfrac{1}{2} \log 16$

    (b) $3 \log 3 + \dfrac{1}{4} \log 16 - \dfrac{1}{3} \log 27$ (8)

11. Solve the equation
    $\log(x^2 + 8) - \log(2x) = \log 3$ (5)

12. Evaluate the following, each correct to 3 decimal places.
    (a) $\ln 462.9$

    (b) $\ln 0.0753$

    (c) $\dfrac{\ln 3.68 - \ln 2.91}{4.63}$ (3)

13. Expand xe^{3x} to six terms.   (5)

14. Evaluate v given that $v = E\left(1 - e^{-\frac{t}{CR}}\right)$ volts when $E = 100$ V, $C = 15 \,\mu\text{F}$, $R = 50 \,\text{k}\Omega$ and $t = 1.5$ s. Also, determine the time when the voltage is 60 V.
    (8)

15. Plot a graph of $y = \dfrac{1}{2} e^{-1.2x}$ over the range $x = -2$ to $x = +1$ and hence determine, correct to 1 decimal place,
    (a) the value of y when $x = -0.75$, and
    (b) the value of x when $y = 4.0$ (8)

Multiple choice questions Test 2
Equations, transposition, logarithms and exponentials
This test covers the material in Chapters 12 to 16

All questions have only one correct answer (Answers on page 440).

1. The relationship between the temperature in degrees Fahrenheit (F) and the temperature in degrees Celsius (C) is given by: $F = \dfrac{9}{5}C + 32$. 135°F is equivalent to:

   (a) 43°C (b) 57.2°C (c) 185.4°C (d) 184°C

2. Transposing $I = \dfrac{V}{R}$ for resistance R gives:

   (a) $I - V$ (b) $\dfrac{V}{I}$ (c) $\dfrac{I}{V}$ (d) VI

3. $\log_{16} 8$ is equal to:

   (a) $\dfrac{1}{2}$ (b) 144 (c) $\dfrac{3}{4}$ (d) 2

4. When two resistors R_1 and R_2 are connected in parallel the formula $\dfrac{1}{R_T} = \dfrac{1}{R_1} + \dfrac{1}{R_2}$ is used to determine the total resistance R_T. If $R_1 = 470\,\Omega$ and $R_2 = 2.7\,\text{k}\Omega$, R_T (correct to 3 significant figures) is equal to:

   (a) 2.68 Ω (b) 400 Ω (c) 473 Ω (d) 3170 Ω

5. Transposing $v = f\lambda$ to make wavelength λ the subject gives:

   (a) $\dfrac{v}{f}$ (b) $v + f$ (c) $f - v$ (d) $\dfrac{f}{v}$

6. The value of $\dfrac{\ln 2}{e^2 \lg 2}$, correct to 3 significant figures, is:

   (a) 0.0588 (b) 0.312 (c) 17.0 (d) 3.209

7. Transposing the formula $R = R_0(1 + \alpha t)$ for t gives:

   (a) $\dfrac{R - R_0}{(1 + \alpha)}$ (b) $\dfrac{R - R_0 - 1}{\alpha}$

   (c) $\dfrac{R - R_0}{\alpha R_0}$ (d) $\dfrac{R}{R_0 \alpha}$

8. The current I in an a.c. circuit is given by:

   $I = \dfrac{V}{\sqrt{R^2 + X^2}}$

   When $R = 4.8$, $X = 10.5$ and $I = 15$, the value of voltage V is:

   (a) 173.18 (b) 1.30 (c) 0.98 (d) 229.50

9. If $\log_2 x = 3$ then:

   (a) $x = 8$ (b) $x = \dfrac{3}{2}$ (c) $x = 9$ (d) $x = \dfrac{2}{3}$

10. The height s of a mass projected vertically upwards at time t is given by: $s = ut - \dfrac{1}{2}gt^2$. When $g = 10$, $t = 1.5$ and $s = 3.75$, the value of u is:

    (a) 10 (b) -5 (c) $+5$ (d) -10

11. The quantity of heat Q is given by the formula $Q = mc(t_2 - t_1)$. When $m = 5$, $t_1 = 20$, $c = 8$ and $Q = 1200$, the value of t_2 is:

    (a) 10 (b) 1.5 (c) 21.5 (d) 50

12. The pressure p pascals at height h metres above ground level is given by $p = p_0 e^{-h/k}$, where p_0 is the pressure at ground level and k is a constant. When p_0 is 1.01×10^5 Pa and the pressure at a height of 1500 m is 9.90×10^4 Pa, the value of k, correct to 3 significant figures is:

    (a) 1.33×10^{-5} (b) 75000 (c) 173000 (d) 197

13. Electrical resistance $R = \dfrac{\rho \ell}{a}$; transposing this equation for ℓ gives:

    (a) $\dfrac{\rho a}{R}$ (b) $\dfrac{R}{a\rho}$ (c) $\dfrac{a}{R\rho}$ (d) $\dfrac{Ra}{\rho}$

14. The solution of the simultaneous equations $3x - 2y = 13$ and $2x + 5y = -4$ is:

    (a) $x = -2$, $y = 3$ (b) $x = 1$, $y = -5$

    (c) $x = 3$, $y = -2$ (d) $x = -7$, $y = 2$

15. The value of $\dfrac{3.67 \ln 21.28}{e^{-0.189}}$, correct to 4 significant figures, is:

    (a) 9.289 (b) 13.56

    (c) 13.5566 (d) -3.844×10^9

16. A formula for the focal length f of a convex lens is $\dfrac{1}{f} = \dfrac{1}{u} + \dfrac{1}{v}$. When $f = 4$ and $u = 6$, v is:

    (a) -2 (b) 12 (c) $\dfrac{1}{12}$ (d) $-\dfrac{1}{2}$

17. Volume $= \dfrac{\text{mass}}{\text{density}}$. The density (in kg/m³) when the mass is 2.532 kg and the volume is 162 cm³ is:

    (a) 0.01563 kg/m³ (b) 410.2 kg/m³

    (c) 15 630 kg/m³ (d) 64.0 kg/m³

18. In the equation $5.0 = 3.0 \ln\left(\dfrac{2.9}{x}\right)$, x has a value correct to 3 significant figures of:

    (a) 1.59   (b) 0.392   (c) 0.0625   (d) 0.548

19. $PV = mRT$ is the characteristic gas equation. When $P = 100 \times 10^3$, $V = 4.0$, $R = 288$ and $T = 300$, the value of m is:

    (a) 4.630   (b) 313 600   (c) 0.216   (d) 100 592

20. The quadratic equation in x whose roots are -2 and $+5$ is:

    (a) $x^2 - 3x - 10 = 0$ (b) $x^2 + 7x + 10 = 0$

    (c) $x^2 + 3x - 10 = 0$ (d) $x^2 - 7x - 10 = 0$

21. The current i amperes flowing in a capacitor at time t seconds is given by $i = 10(1 - e^{-t/CR})$, where resistance R is 25×10^3 ohms and capacitance C is 16×10^{-6} farads. When current i reaches 7 amperes, the time t is:

    (a) $-0.48\,s$ (b) $0.14\,s$ (c) $0.21\,s$ (d) $0.48\,s$

22. In a system of pulleys, the effort P required to raise a load W is given by $P = aW + b$, where a and b are constants. If $W = 40$ when $P = 12$ and $W = 90$ when $P = 22$, the values of a and b are:

    (a) $a = 5$, $b = \dfrac{1}{4}$

    (b) $a = 1$, $b = -28$

    (c) $a = \dfrac{1}{3}$, $b = -8$

    (d) $a = \dfrac{1}{5}$, $b = 4$

23. Resistance R ohms varies with temperature t according to the formula $R = R_0(1 + \alpha t)$. Given $R = 21\ \Omega$, $\alpha = 0.004$ and $t = 100$, R_0 has a value of:

    (a) $21.4\ \Omega$ (b) $29.4\ \Omega$ (c) $15\ \Omega$ (d) $0.067\ \Omega$

24. $8x^2 + 13x - 6 = (x + p)(qx - 3)$. The values of p and q are:

    (a) $p = -2$, $q = 4$

    (b) $p = 3$, $q = 2$

    (c) $p = 2$, $q = 8$

    (d) $p = 1$, $q = 8$

25. $(1/4)\log 16 - 2\log 5 + (1/3)\log 27$ is equivalent to:

    (a) $-\log 20$ (b) $\log 0.24$ (c) $-\log 5$ (d) $\log 3$

26. The height S metres of a mass thrown vertically upwards at time t seconds is given by $S = 80t - 16t^2$. To reach a height of 50 metres on the descent will take the mass:

    (a) $0.73\,s$ (b) $5.56\,s$ (c) $4.27\,s$ (d) $81.77\,s$

27. The roots of the quadratic equation $8x^2 + 10x - 3 = 0$ are:

    (a) $-\dfrac{1}{4}$ and $\dfrac{3}{2}$ (b) 4 and $\dfrac{2}{3}$

    (c) $-\dfrac{3}{2}$ and $\dfrac{1}{4}$ (d) $\dfrac{2}{3}$ and -4

28. The volume V_2 of a material when the temperature is increased is given by $V_2 = V_1\left[1 + \gamma(t_2 - t_1)\right]$. The value of t_2 when $V_2 = 61.5\,\mathrm{cm}^3$, $V_1 = 60\,\mathrm{cm}^3$, $\gamma = 54 \times 10^{-6}$ and $t_1 = 250$ is:

    (a) 713   (b) 463   (c) 213   (d) 28 028

29. The roots of the quadratic equation $2x^2 - 5x + 1 = 0$, correct to 2 decimal places, are:

    (a) -0.22 and -2.28

    (b) 2.69 and -0.19

    (c) 0.19 and -2.69

    (d) 2.28 and 0.22

30. Transposing $t = 2\pi\sqrt{\dfrac{l}{g}}$ for g gives:

    (a) $\dfrac{(t - 2\pi)^2}{l}$ (b) $\left(\dfrac{2\pi}{t}\right)l^2$

    (c) $\dfrac{\sqrt{\dfrac{t}{2\pi}}}{l}$ (d) $\dfrac{4\pi^2 l}{t^2}$

The companion website for this book contains the above multiple-choice test. If you prefer to attempt the test online then visit:

www.routledge.com/cw/bird

For a copy of this multiple choice test, go to:

www.routledge.com/cw/bird

Chapter 17

Straight line graphs

Why it is important to understand: **Straight line graphs**

Graphs have a wide range of applications in engineering and in physical sciences because of their inherent simplicity. A graph can be used to represent almost any physical situation involving discrete objects and the relationship among them. If two quantities are directly proportional and one is plotted against the other, a straight line is produced. Examples include an applied force on the end of a spring plotted against spring extension, the speed of a flywheel plotted against time, and strain in a wire plotted against stress (Hooke's law). In engineering, the straight line graph is the most basic graph to draw and evaluate.

At the end of this chapter, you should be able to:

- understand rectangular axes, scales and co-ordinates
- plot given co-ordinates and draw the best straight line graph
- determine the gradient of a straight line graph
- estimate the vertical axis intercept
- state the equation of a straight line graph
- plot straight line graphs involving practical engineering examples

17.1 Introduction to graphs

A graph is a visual representation of information, showing how one quantity varies with another related quantity.

We often see graphs in newspapers or in business reports, in travel brochures and government publications. For example, a graph of the share price (in pence) over a 6 month period for a drinks company, Fizzy Pops, is shown in Fig. 17.1.

Generally, we see that the share price increases to a high of 400 p in June, but dips down to around 280 p in August before recovering slightly in September.

A graph should convey information more quickly to the reader than if the same information was explained in words.

When this chapter is completed you should be able to draw up a table of values, plot co-ordinates, determine the gradient and state the equation of a straight line graph. Some typical practical examples are included in which straight lines are used.

17.2 Axes, scales and co-ordinates

We are probably all familiar with reading a map to locate a town, or a local map to locate a particular street. For example, a street map of central Portsmouth is shown in Fig. 17.2. Notice the squares drawn horizontally and vertically on the map; this is called a **grid** and enables us to locate a place of interest or a particular road. Most maps contain such a grid.

We locate places of interest on the map by stating a letter and a number – this is called the **grid reference**.

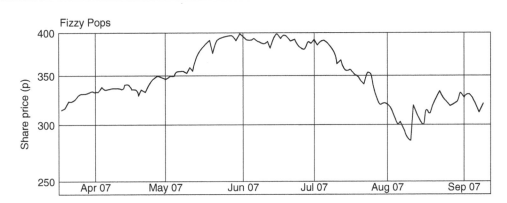

Figure 17.1

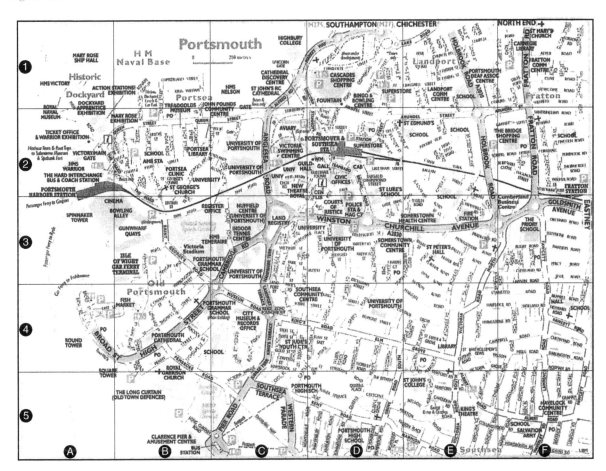

Figure 17.2 Reprinted with permission from AA Media Ltd.

For example, on the map, the Portsmouth & Southsea station is in square D2, King's Theatre is in square E5, HMS Warrior is in square A2, Gunwharf Quays is in square B3 and High Street is in square B4.

Portsmouth & Southsea station is located by moving horizontally along the bottom of the map until the square labelled D is reached and then moving vertically upwards until square 2 is met.

The letter/number, D2, is referred to as **co-ordinates**; i.e. co-ordinates are used to locate the position of a point on a map. If you are familiar with using a map in this way then you should have no difficulties with graphs, because similar co-ordinates are used with graphs.

As stated earlier, a **graph** is a visual representation of information, showing how one quantity varies with another related quantity. The most common method of showing a relationship between two sets of data is to use a pair of reference axes – these are two lines drawn at right angles to each other (often called **Cartesian**, named after Descartes*, or **rectangular axes**), as shown in Fig. 17.3.

The horizontal axis is labelled the x-axis and the vertical axis is labelled the y-axis. The point where x is 0 and y is 0 is called the **origin**.

x values have **scales** that are positive to the right of the origin and negative to the left. y values have scales that are positive up from the origin and negative down from the origin.

Co-ordinates are written with brackets and a comma in between two numbers. For example, point A is shown with co-ordinates $(3, 2)$ and is located by starting at the

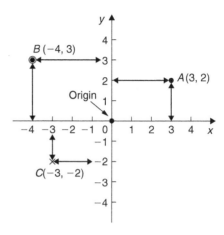

Figure 17.3

origin and moving 3 units in the positive x direction (i.e. to the right) and then 2 units in the positive y direction (i.e. up).

When co-ordinates are stated the first number is always the x value and the second number is always the y value. In Fig. 17.3, point B has co-ordinates $(-4, 3)$ and point C has co-ordinates $(-3, -2)$.

17.3 Straight line graphs

The distances travelled by a car in certain periods of time are shown in the following table of values.

Time (s)	10	20	30	40	50	60
Distance travelled (m)	50	100	150	200	250	300

We will plot time on the horizontal (or x) axis with a scale of 1 cm = 10 s.

We will plot distance on the vertical (or y) axis with a scale of 1 cm = 50 m.

(When choosing scales it is better to choose ones such as 1 cm = 1 unit, 1 cm = 2 units or 1 cm = 10 units because doing so makes reading values between these values easier.)

With the above data, the (x, y) co-ordinates become (time, distance) co-ordinates; i.e. the co-ordinates are $(10, 50)$, $(20, 100)$, $(30, 150)$, and so on.

The co-ordinates are shown plotted in Fig. 17.4 using crosses. (Alternatively, a dot or a dot and circle may be used, as shown in Fig. 17.3.)

A straight line is drawn through the plotted co-ordinates as shown in Fig. 17.4.

*Who was Descartes? – **René Descartes** (31 March 1596–11 February 1650) was a French philosopher, mathematician, and writer. He wrote many influential texts including *Meditations on First Philosophy*. Descartes is best known for the philosophical statement '*Cogito ergo sum*' (I think, therefore I am), found in part IV of *Discourse on the Method*. To find out more go to www.routledge.com/cw/bird

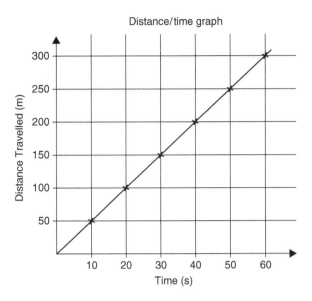

Distance/time graph

Figure 17.4

Student task

The following table gives the force F newtons which, when applied to a lifting machine, overcomes a corresponding load of L newtons.

F (newtons)	19	35	50	93	125	147
L (newtons)	40	120	230	410	540	680

1. Plot L horizontally and F vertically.

2. Scales are normally chosen such that the graph occupies as much space as possible on the graph paper. So in this case, the following scales are chosen.

   > Horizontal axis (i.e. L): 1 cm = 50 N
   >
   > Vertical axis (i.e. F): 1 cm = 10 N

3. Draw the axes and label them L (newtons) for the horizontal axis and F (newtons) for the vertical axis.

4. Label the origin as 0

5. Write on the horizontal scaling at 100, 200, 300, and so on, every 2 cm.

6. Write on the vertical scaling at 10, 20, 30, and so on, every 1 cm.

7. Plot on the graph the co-ordinates (40, 19), (120, 35), (230, 50), (410, 93), (540, 125) and (680, 147), marking each with a cross or a dot.

8. Using a ruler, draw the best straight line through the points. You will notice that not all of the points lie exactly on a straight line. This is quite normal with experimental values. In a practical situation it would be surprising if all of the points lay exactly on a straight line.

9. Extend the straight line at each end.

10. From the graph, determine the force applied when the load is 325 N. It should be close to 75 N. This process of finding an equivalent value within the given data is called **interpolation**. Similarly, determine the load that a force of 45 N will overcome. It should be close to 170 N.

11. From the graph, determine the force needed to overcome a 750 N load. It should be close to 161 N. This process of finding an equivalent value outside the given data is called **extrapolation**. To extrapolate we need to have extended the straight line drawn. Similarly, determine the force applied when the load is zero. It should be close to 11 N. The point where the straight line crosses the vertical axis is called the **vertical axis intercept**. So, in this case, the vertical-axis intercept = 11 N at co-ordinates (0, 11).

The graph you have drawn should look something like Fig. 17.5 shown below.

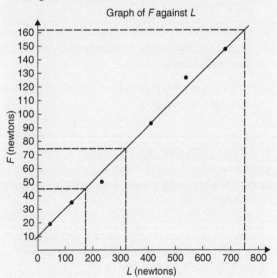

Graph of F against L

Figure 17.5

In another example, let the relationship between two variables x and y be $y = 3x + 2$

When $x = 0$, $y = 0 + 2 = 2$
When $x = 1$, $y = 3 + 2 = 5$
When $x = 2$, $y = 6 + 2 = 8$, and so on.

The co-ordinates $(0, 2)$, $(1, 5)$ and $(2, 8)$ have been produced and are plotted, as shown in Fig. 17.6.
When the points are joined together **a straight line graph results**, i.e. $y = 3x + 2$ is a straight line graph.

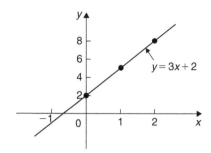

Figure 17.6 Graph of y/x

17.3.1 Summary of general rules to be applied when drawing graphs

(a) Give the graph a title clearly explaining what is being illustrated.

(b) Choose scales such that the graph occupies as much space as possible on the graph paper being used.

(c) Choose scales so that interpolation is made as easy as possible. Usually scales such as $1\,\text{cm} = 1$ unit, $1\,\text{cm} = 2$ units or $1\,\text{cm} = 10$ units are used. Awkward scales such as $1\,\text{cm} = 3$ units or $1\,\text{cm} = 7$ units should not be used.

(d) The scales need not start at zero, particularly when starting at zero produces an accumulation of points within a small area of the graph paper.

(e) The co-ordinates, or points, should be clearly marked. This is achieved by a cross, or a dot and circle, or just by a dot (see Fig. 17.3).

(f) A statement should be made next to each axis explaining the numbers represented with their appropriate units.

(g) Sufficient numbers should be written next to each axis without cramping.

> Problem 1. Plot the graph $y = 4x + 3$ in the range $x = -3$ to $x = +4$. From the graph, find (a) the value of y when $x = 2.2$ and (b) the value of x when $y = -3$

Whenever an equation is given and a graph is required, a table giving corresponding values of the variable is necessary. The table is achieved as follows:

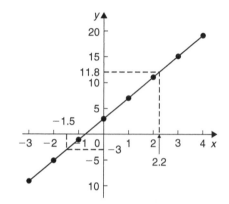

Figure 17.7

When $x = -3$, $y = 4x + 3 = 4(-3) + 3$
$$= -12 + 3 = -9$$

When $x = -2$, $y = 4(-2) + 3$
$$= -8 + 3 = -5, \text{and so on.}$$

Such a table is shown below.

x	-3	-2	-1	0	1	2	3	4
y	-9	-5	-1	3	7	11	15	19

The co-ordinates $(-3, -9)$, $(-2, -5)$, $(-1, -1)$, and so on, are plotted and joined together to produce the straight line shown in Fig. 17.7. (Note that the scales used on the x and y axes do not have to be the same.) From the graph:

(a) when $x = 2.2$, $y = 11.8$, and

(b) when $y = -3$, $x = -1.5$

Now try the following Practice Exercise

Practice Exercise 67   Straight line graphs
(answers on page 429)

1. Assuming graph paper measuring 20 cm by 20 cm is available, suggest suitable scales for the following ranges of values.
    (a) Horizontal axis: 3 V to 55 V; vertical axis: 10 Ω to 180 Ω.

    (b) Horizontal axis: 7 m to 86 m; vertical axis: 0.3 V to 1.69 V.

    (c) Horizontal axis: 5 N to 150 N; vertical axis: 0.6 mm to 3.4 mm.

2. Corresponding values obtained experimentally for two quantities are

x	−5	−3	−1	0	2	4
y	−13	−9	−5	−3	1	5

    Plot a graph of y (vertically) against x (horizontally) to scales of 2 cm = 1 for the horizontal x-axis and 1 cm = 1 for the vertical y-axis. (This graph will need the whole of the graph paper with the origin somewhere in the centre of the paper).
    From the graph, find
    (a) the value of y when $x = 1$

    (b) the value of y when $x = -2.5$

    (c) the value of x when $y = -6$

    (d) the value of x when $y = 5$

3. Corresponding values obtained experimentally for two quantities are

x	−2.0	−0.5	0	1.0	2.5	3.0	5.0
y	−13.0	−5.5	−3.0	2.0	9.5	12.0	22.0

    Use a horizontal scale for x of 1 cm = $\frac{1}{2}$ unit and a vertical scale for y of 1 cm = 2 units and draw a graph of x against y. Label the graph and each of its axes. By interpolation, find from the graph the value of y when x is 3.5

4. Draw a graph of $y - 3x + 5 = 0$ over a range of $x = -3$ to $x = 4$. Hence determine
    (a) the value of y when $x = 1.3$

    (b) the value of x when $y = -9.2$

5. The speed n rev/min of a motor changes when the voltage V across the armature is varied. The results are shown in the following table.

n (rev/min)	560	720	900	1010	1240	1410
V (volts)	80	100	120	140	160	180

    It is suspected that one of the readings taken of the speed is inaccurate. Plot a graph of speed (horizontally) against voltage (vertically) and find this value. Find also
    (a) the speed at a voltage of 132 V.

    (b) the voltage at a speed of 1300 rev/min.

17.4 Gradients, intercepts and equations of graphs

17.4.1 Gradients

The **gradient or slope** of a straight line is the ratio of the change in the value of y to the change in the value of x between any two points on the line. If, as x increases, ($\rightarrow$), y also increases, ($\uparrow$), then the gradient is positive. In Fig. 17.8(a), a straight line graph $y = 2x + 1$ is shown. To find the gradient of this straight line, choose two points on the straight line graph, such as A and C.

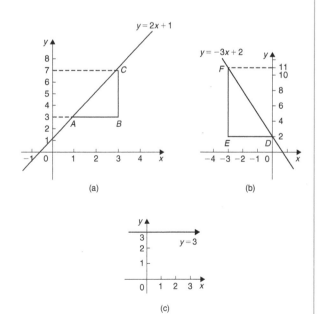

Figure 17.8

Then construct a right-angled triangle, such as ABC, where BC is vertical and AB is horizontal.

Then, **gradient of** $AC = \dfrac{\text{change in } y}{\text{change in } x} = \dfrac{CB}{BA}$

$$= \dfrac{7-3}{3-1} = \dfrac{4}{2} = \mathbf{2}$$

In Fig. 17.8(b), a straight line graph $y = -3x + 2$ is shown. To find the gradient of this straight line, choose two points on the straight line graph, such as D and F. Then construct a right-angled triangle, such as DEF, where EF is vertical and DE is horizontal.

Then, **gradient of** $DF = \dfrac{\text{change in } y}{\text{change in } x} = \dfrac{FE}{ED}$

$$= \dfrac{11-2}{-3-0} = \dfrac{9}{-3} = \mathbf{-3}$$

Fig. 17.8(c) shows a straight line graph $y = 3$. **Since the straight line is horizontal the gradient is zero.**

17.4.2 The y-axis intercept

The value of y when $x = 0$ is called the **y-axis intercept**. In Fig. 17.8(a) the y-axis intercept is 1 and in Fig. 17.8(b) the y-axis intercept is 2

17.4.3 The equation of a straight line graph

The general equation of a straight line graph is

$$y = mx + c$$

where m is the gradient and c is the y-axis intercept. Thus, as we have found in Fig. 17.8(a), $y = 2x + 1$ represents a straight line of gradient 2 and y-axis intercept 1. So, given the equation $y = 2x + 1$, we are able to state, on sight, that the gradient $= 2$ and the y-axis intercept $= 1$, without the need for any analysis.

Similarly, in Fig. 17.8(b), $y = -3x + 2$ represents a straight line of gradient -3 and y-axis intercept 2
In Fig. 17.8(c), $y = 3$ may be rewritten as $y = 0x + 3$ and therefore represents a straight line of gradient 0 and y-axis intercept 3
Here are some worked problems to help understanding of gradients, intercepts and equations of graphs.

Problem 2. Plot the following graphs on the same axes in the range $x = -4$ to $x = +4$ and determine the gradient of each.

(a) $y = x$ (b) $y = x + 2$

(c) $y = x + 5$ (d) $y = x - 3$

A table of co-ordinates is produced for each graph.

(a) $y = x$

x	-4	-3	-2	-1	0	1	2	3	4
y	-4	-3	-2	-1	0	1	2	3	4

(b) $y = x + 2$

x	-4	-3	-2	-1	0	1	2	3	4
y	-2	-1	0	1	2	3	4	5	6

(c) $y = x + 5$

x	-4	-3	-2	-1	0	1	2	3	4
y	1	2	3	4	5	6	7	8	9

(d) $y = x - 3$

x	-4	-3	-2	-1	0	1	2	3	4
y	-7	-6	-5	-4	-3	-2	-1	0	1

The co-ordinates are plotted and joined for each graph. The results are shown in Fig. 17.9. Each of the straight lines produced is parallel to the others; i.e. the slope or gradient is the same for each.
To find the gradient of any straight line, say, $y = x - 3$, a horizontal and vertical component needs to be constructed. In Fig. 17.9, AB is constructed vertically at $x = 4$ and BC is constructed horizontally at $y = -3$.

The gradient of $AC = \dfrac{AB}{BC} = \dfrac{1-(-3)}{4-0} = \dfrac{4}{4} = 1$

i.e. the gradient of the straight line $y = x - 3$ is 1, which could have been deduced 'on sight' since $y = 1x - 3$ represents a straight line graph with gradient 1 and y-axis intercept of -3
The actual positioning of AB and BC is unimportant because the gradient is also given by

$$\dfrac{DE}{EF} = \dfrac{-1-(-2)}{2-1} = \dfrac{1}{1} = 1$$

The slope or gradient of each of the straight lines in Fig. 17.9 is thus 1 since they are parallel to each other.

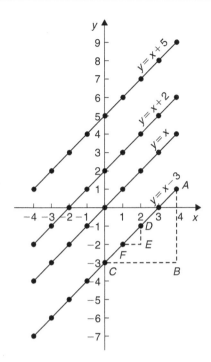

Figure 17.9

Problem 3.   Plot the following graphs on the same axes between the values $x = -3$ to $x = +3$ and determine the gradient and y-axis intercept of each.

(a) $y = 3x$ (b) $y = 3x + 7$

(c) $y = -4x + 4$ (d) $y = -4x - 5$

A table of co-ordinates is drawn up for each equation.

(a) $y = 3x$

x	-3	-2	-1	0	1	2	3
y	-9	-6	-3	0	3	6	9

(b) $y = 3x + 7$

x	-3	-2	-1	0	1	2	3
y	-2	1	4	7	10	13	16

(c) $y = -4x + 4$

x	-3	-2	-1	0	1	2	3
y	16	12	8	4	0	-4	-8

(d) $y = -4x - 5$

x	-3	-2	-1	0	1	2	3
y	7	3	-1	-5	-9	-13	-17

Each of the graphs is plotted as shown in Fig. 17.10 and each is a straight line. $y = 3x$ and $y = 3x + 7$ are parallel to each other and thus have the same gradient. The gradient of AC is given by

$$\frac{CB}{BA} = \frac{16 - 7}{3 - 0} = \frac{9}{3} = 3$$

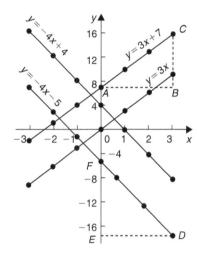

Figure 17.10

Hence, **the gradients of both $y = 3x$ and $y = 3x + 7$ are 3**, which could have been deduced 'on sight'.
$y = -4x + 4$ and $y = -4x - 5$ are parallel to each other and thus have the same gradient. The gradient of DF is given by

$$\frac{FE}{ED} = \frac{-5 - (-17)}{0 - 3} = \frac{12}{-3} = -4$$

Hence, **the gradient of both $y = -4x + 4$ and $y = -4x - 5$ is -4**, which, again, could have been deduced 'on sight'.
The y-axis intercept means the value of y where the straight line cuts the y-axis. From Fig. 17.10,

    $y = 3x$ cuts the y-axis at $y = 0$
    $y = 3x + 7$ cuts the y-axis at $y = +7$
    $y = -4x + 4$ cuts the y-axis at $y = +4$
    $y = -4x - 5$ cuts the y-axis at $y = -5$

Some general conclusions can be drawn from the graphs shown in Figs 17.9 and 17.10. When an equation is of the form $y = mx + c$, where m and c are constants, then

(a) a graph of y against x produces a straight line,

(b) m represents the slope or gradient of the line, and

(c) c represents the y-axis intercept.

Thus, given an equation such as $y = 3x + 7$, it may be deduced 'on sight' that its gradient is $+3$ and its y-axis intercept is $+7$, as shown in Fig. 17.10. Similarly, if $y = -4x - 5$, the gradient is -4 and the y-axis intercept is -5, as shown in Fig. 17.10.

When plotting a graph of the form $y = mx + c$, only two co-ordinates need be determined. When the co-ordinates are plotted a straight line is drawn between the two points. Normally, three co-ordinates are determined, the third one acting as a check.

Problem 4. Plot the graph $3x + y + 1 = 0$ and $2y - 5 = x$ on the same axes and find their point of intersection

Rearranging $3x + y + 1 = 0$ gives $y = -3x - 1$

Rearranging $2y - 5 = x$ gives $2y = x + 5$ and

$$y = \frac{1}{2}x + 2\frac{1}{2}$$

Since both equations are of the form $y = mx + c$, both are straight lines. Knowing an equation is a straight line means that only two co-ordinates need to be plotted and a straight line drawn through them. A third co-ordinate is usually determined to act as a check. A table of values is produced for each equation as shown below.

x	1	0	-1
$-3x - 1$	-4	-1	2

x	2	0	-3
$\frac{1}{2}x + 2\frac{1}{2}$	$3\frac{1}{2}$	$2\frac{1}{2}$	1

The graphs are plotted as shown in Fig. 17.11.
The **two straight lines are seen to intersect at $(-1, 2)$**

Problem 5. If graphs of y against x were to be plotted for each of the following, state (i) the gradient and (ii) the y-axis intercept.

(a) $y = 9x + 2$ (b) $y = -4x + 7$

(c) $y = 3x$ (d) $y = -5x - 3$

(e) $y = 6$ (f) $y = x$

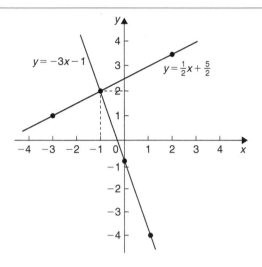

Figure 17.11

If $y = mx + c$ then $m = $ **gradient** and $c = y$-**axis intercept**.

(a) If $y = 9x + 2$, then   (i) **gradient = 9**

                      (ii) **y-axis intercept = 2**

(b) If $y = -4x + 7$, then   (i) **gradient = -4**

                      (ii) **y-axis intercept = 7**

(c) If $y = 3x$ i.e. $y = 3x + 0$,
then   (i) **gradient = 3**

                      (ii) **y-axis intercept = 0**

i.e. the straight line passes through the origin.

(d) If $y = -5x - 3$, then   (i) **gradient = -5**

                      (ii) **y-axis intercept = -3**

(e) If $y = 6$ i.e. $y = 0x + 6$,
then   (i) **gradient = 0**

                      (ii) **y-axis intercept = 6**

i.e. $y = 6$ is a straight horizontal line.

(f) If $y = x$ i.e. $y = 1x + 0$,
then   (i) **gradient = 1**

                      (ii) **y-axis intercept = 0**

Since $y = x$, as x increases, y increases by the same amount; i.e. **y is directly proportional to x.**

Problem 6. Without drawing graphs, determine the gradient and y-axis intercept for each of the following equations.

(a) $y + 4 = 3x$ (b) $2y + 8x = 6$ (c) $3x = 4y + 7$

If $y = mx + c$ then $m =$ **gradient** and $c = y$**-axis intercept**.

(a) Transposing $y + 4 = 3x$ gives $y = 3x - 4$

Hence, **gradient $= 3$** and **y-axis intercept $= -4$**

(b) Transposing $2y + 8x = 6$ gives $2y = -8x + 6$

Dividing both sides by 2 gives $y = -4x + 3$

Hence, **gradient $= -4$** and **y-axis intercept $= 3$**

(c) Transposing $3x = 4y + 7$ gives $3x - 7 = 4y$
or $\qquad\qquad\qquad\qquad 4y = 3x - 7$

Dividing both sides by 4 gives $y = \dfrac{3}{4}x - \dfrac{7}{4}$,

or $\qquad\qquad\qquad\qquad y = 0.75x - 1.75$

Hence, **gradient $= 0.75$** and
y-axis intercept $= -1.75$

Problem 7. Without plotting graphs, determine the gradient and y-axis intercept values of the following equations.

(a) $y = 7x - 3$ $\qquad$ (b) $3y = -6x + 2$

(c) $y - 2 = 4x + 9$ $\qquad$ (d) $\dfrac{y}{3} = \dfrac{x}{3} - \dfrac{1}{5}$

(e) $2x + 9y + 1 = 0$

(a) $y = 7x - 3$ is of the form $y = mx + c$

Hence, **gradient, $m = 7$** and **y-axis intercept, $c = -3$**

(b) Rearranging $3y = -6x + 2$ gives $y = -\dfrac{6x}{3} + \dfrac{2}{3}$,

i.e. $\qquad\qquad y = -2x + \dfrac{2}{3}$

which is of the form $y = mx + c$

Hence, **gradient $m = -2$** and **y-axis intercept, $c = \dfrac{2}{3}$**

(c) Rearranging $y - 2 = 4x + 9$ gives $y = 4x + 11$

Hence, **gradient $= 4$** and **y-axis intercept $= 11$**

(d) Rearranging $\dfrac{y}{3} = \dfrac{x}{2} - \dfrac{1}{5}$ gives $y = 3\left(\dfrac{x}{2} - \dfrac{1}{5}\right)$

$\qquad\qquad\qquad\qquad\qquad = \dfrac{3}{2}x - \dfrac{3}{5}$

Hence, **gradient $= \dfrac{3}{2}$** and
y-axis intercept $= -\dfrac{3}{5}$

(e) Rearranging $2x + 9y + 1 = 0$ gives $9y = -2x - 1$,

i.e. $\quad y = -\dfrac{2}{9}x - \dfrac{1}{9}$

Hence, **gradient $= -\dfrac{2}{9}$** and
y-axis intercept $= -\dfrac{1}{9}$

Problem 8. Determine for the straight line shown in Fig. 17.12 (a) the gradient and (b) the equation of the graph

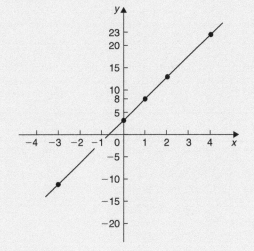

Figure 17.12

(a) A right-angled triangle ABC is constructed on the graph as shown in Fig. 17.13.

$$\text{Gradient} = \dfrac{AC}{CB} = \dfrac{23 - 8}{4 - 1} = \dfrac{15}{3} = 5$$

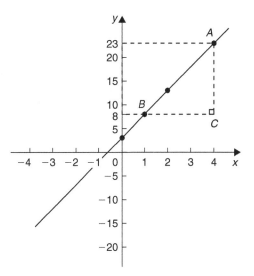

Figure 17.13

(b) The y-axis intercept at $x = 0$ is seen to be at $y = 3$

$y = mx + c$ is a straight line graph where $m = $ gradient and $c = y$-axis intercept.

From above, $m = 5$ and $c = 3$

Hence, the equation of the graph is $y = 5x + 3$

Problem 9. Determine the equation of the straight line shown in Fig. 17.14.

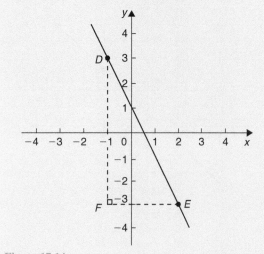

Figure 17.14

The triangle *DEF* is shown constructed in Fig. 17.14.

$$\text{Gradient of } DE = \frac{DF}{FE} = \frac{3 - (-3)}{-1 - 2} = \frac{6}{-3} = -2$$

and the *y*-axis intercept $= 1$

Hence, **the equation of the straight line is $y = mx + c$** **i.e. $y = -2x + 1$**

Problem 10. The velocity of a body was measured at various times and the results obtained were as follows:

Velocity v (m/s)	8	10.5	13	15.5	18	20.5	23
Time t (s)	1	2	3	4	5	6	7

Plot a graph of velocity (vertically) against time (horizontally) and determine the equation of the graph

Suitable scales are chosen and the co-ordinates (1, 8), (2, 10.5), (3, 13), and so on, are plotted as shown in Fig. 17.15.

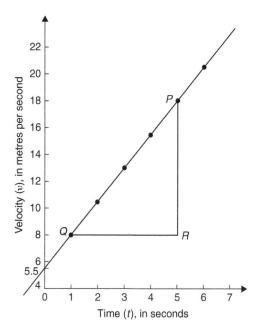

Figure 17.15

The right-angled triangle *PRQ* is constructed on the graph as shown in Fig. 17.15.

$$\textbf{Gradient of } PQ = \frac{PR}{RQ} = \frac{18 - 8}{5 - 1} = \frac{10}{4} = \textbf{2.5}$$

The **vertical axis intercept** is at $v = \textbf{5.5 m/s}$.
The equation of a straight line graph is $y = mx + c$. In this case, t corresponds to x and v corresponds to y. Hence, the equation of the graph shown in Fig. 17.15 is $v = mt + c$. But, from above, gradient, $m = 2.5$ and v-axis intercept, $c = 5.5$
Hence, **the equation of the graph is $v = 2.5t + 5.5$**

Problem 11. Determine the gradient of the straight line graph passing through the co-ordinates (a) $(-2, 5)$ and $(3, 4)$, and (b) $(-2, -3)$ and $(-1, 3)$

From Fig. 17.16, a straight line graph passing through co-ordinates (x_1, y_1) and (x_2, y_2) has a gradient given by

$$m = \frac{y_2 - y_1}{x_2 - x_1}$$

(a) A straight line passes through $(-2, 5)$ and $(3, 4)$, hence $x_1 = -2$, $y_1 = 5$, $x_2 = 3$ and $y_2 = 4$, hence,
gradient, $m = \dfrac{y_2 - y_1}{x_2 - x_1} = \dfrac{4 - 5}{3 - (-2)} = -\dfrac{1}{5}$

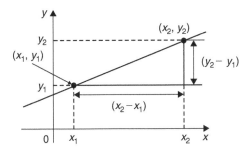

Figure 17.16

(b) A straight line passes through $(-2, -3)$ and $(-1, 3)$, hence $x_1 = -2$, $y_1 = -3$, $x_2 = -1$ and $y_2 = 3$, hence, **gradient,**

$$m = \frac{y_2 - y_1}{x_2 - x_1} = \frac{3 - (-3)}{-1 - (-2)} = \frac{3 + 3}{-1 + 2} = \frac{6}{1} = \mathbf{6}$$

Now try the following Practice Exercise

Practice Exercise 68   Gradients, intercepts and equations of graphs (answers on page 429)

1. The equation of a line is $4y = 2x + 5$. A table of corresponding values is produced and is shown below. Complete the table and plot a graph of y against x. Find the gradient of the graph.

x	-4	-3	-2	-1	0	1	2	3	4
y		-0.25			1.25				3.25

2. Determine the gradient and intercept on the y-axis for each of the following equations.
   - (a) $y = 4x - 2$
   - (b) $y = -x$
   - (c) $y = -3x - 4$
   - (d) $y = 4$

3. Find the gradient and intercept on the y-axis for each of the following equations.
   - (a) $2y - 1 = 4x$
   - (b) $6x - 2y = 5$
   - (c) $3(2y - 1) = \dfrac{x}{4}$

   Determine the gradient and y-axis intercept for each of the equations in Problems 4 and 5 and sketch the graphs.

4. (a) $y = 6x - 3$ (b) $y = -2x + 4$
   (c) $y = 3x$ (d) $y = 7$

5. (a) $2y + 1 = 4x$ (b) $2x + 3y + 5 = 0$
   (c) $3(2y - 4) = \dfrac{x}{3}$ (d) $5x - \dfrac{y}{2} - \dfrac{7}{3} = 0$

6. Determine the gradient of the straight line graphs passing through the co-ordinates:
   - (a) $(2, 7)$ and $(-3, 4)$
   - (b) $(-4, -1)$ and $(-5, 3)$
   - (c) $\left(\dfrac{1}{4}, -\dfrac{3}{4}\right)$ and $\left(-\dfrac{1}{2}, \dfrac{5}{8}\right)$

7. State which of the following equations will produce graphs which are parallel to one another.
   - (a) $y - 4 = 2x$ (b) $4x = -(y + 1)$
   - (c) $x = \dfrac{1}{2}(y + 5)$ (d) $1 + \dfrac{1}{2}y = \dfrac{3}{2}x$
   - (e) $2x = \dfrac{1}{2}(7 - y)$

8. Draw on the same axes the graphs of $y = 3x - 5$ and $3y + 2x = 7$. Find the co-ordinates of the point of intersection. Check the result obtained by solving the two simultaneous equations algebraically.

9. Plot the graphs $y = 2x + 3$ and $2y = 15 - 2x$ on the same axes and determine their point of intersection.

10. Draw on the same axes the graphs of $y = 3x - 1$ and $y + 2x = 4$. Find the co-ordinates of the point of intersection.

11. A piece of elastic is tied to a support so that it hangs vertically and a pan, on which weights can be placed, is attached to the free end. The length of the elastic is measured as various weights are added to the pan and the results obtained are as follows:

Load, W (N)	5	10	15	20	25
Length, l (cm)	60	72	84	96	108

   Plot a graph of load (horizontally) against length (vertically) and determine
   (a) the value of length when the load is 17 N

(b) the value of load when the length is 74 cm

(c) its gradient

(d) the equation of the graph.

12. The following table gives the effort P to lift a load W with a small lifting machine.

W (N)	10	20	30	40	50	60
P (N)	5.1	6.4	8.1	9.6	10.9	12.4

Plot W horizontally against P vertically and show that the values lie approximately on a straight line. Determine the probable relationship connecting P and W in the form $P = aW + b$.

13. In an experiment the speeds N rpm of a flywheel slowly coming to rest were recorded against the time t in minutes. Plot the results and show that N and t are connected by an equation of the form $N = at + b$. Find probable values of a and b.

t (min)	2	4	6	8	10	12	14
N (rev/min)	372	333	292	252	210	177	132

17.5 Practical problems involving straight line graphs

When a set of co-ordinate values are given or are obtained experimentally and it is believed that they follow a law of the form $y = mx + c$, if a straight line can be drawn reasonably close to most of the co-ordinate values when plotted, this verifies that a law of the form $y = mx + c$ exists. From the graph, constants m (i.e. gradient) and c (i.e. y-axis intercept) can be determined.

Here are some worked problems in which practical situations are featured.

Problem 12. The temperature in degrees Celsius* and the corresponding values in degrees Fahrenheit

*Who was Celsius? To find out more go to
www.routledge.com/cw/bird

are shown in the table below. Construct rectangular axes, choose suitable scales and plot a graph of degrees Celsius (on the horizontal axis) against degrees Fahrenheit (on the vertical scale).

°C	10	20	40	60	80	100
°F	50	68	104	140	176	212

From the graph find (a) the temperature in degrees Fahrenheit at 55°C, (b) the temperature in degrees Celsius at 167°F, (c) the Fahrenheit temperature at 0°C and (d) the Celsius temperature at 230°F

The co-ordinates (10, 50), (20, 68), (40, 104), and so on are plotted as shown in Fig. 17.17. When the co-ordinates are joined, a straight line is produced. Since a straight line results, there is a linear relationship between degrees Celsius and degrees Fahrenheit.

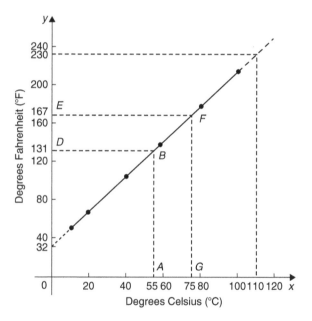

Figure 17.17

(a) To find the Fahrenheit temperature at 55°C, a vertical line AB is constructed from the horizontal axis to meet the straight line at B. The point where the horizontal line BD meets the vertical axis indicates the equivalent Fahrenheit temperature.

Hence, **55°C is equivalent to 131°F.**

This process of finding an equivalent value in between the given information in the above table is called **interpolation**.

(b) To find the Celsius temperature at 167°F, a horizontal line *EF* is constructed as shown in Fig. 17.17. The point where the vertical line *FG* cuts the horizontal axis indicates the equivalent Celsius temperature.

Hence, 167°F is equivalent to 75°C.

(c) If the graph is assumed to be linear even outside of the given data, the graph may be extended at both ends (shown by broken lines in Fig. 17.17).

From Fig. 17.17, **0°C corresponds to 32°F**.

(d) **230°F is seen to correspond to 110°C**.

The process of finding equivalent values outside of the given range is called **extrapolation**.

Problem 13.   In an experiment on Charles's law,* the value of the volume of gas, V m^3, was measured for various temperatures T°C. The results are shown below.

V m^3	25.0	25.8	26.6	27.4	28.2	29.0
T°C	60	65	70	75	80	85

Plot a graph of volume (vertical) against temperature (horizontal) and from it find (a) the temperature when the volume is 28.6 m^3 and (b) the volume when the temperature is 67°C

If a graph is plotted with both the scales starting at zero then the result is as shown in Fig. 17.18. All of the points lie in the top right-hand corner of the graph, making interpolation difficult. A more accurate graph is obtained if the temperature axis starts at 55°C and the volume axis starts at 24.5 m^3. The axes corresponding to these values are shown by the broken lines in Fig. 17.18 and are called **false axes**, since the origin is not now at zero. A magnified version of this relevant part of the graph is shown in Fig. 17.19. From the graph,

(a) When the volume is 28.6 m^3, the equivalent temperature is **82.5°C**.

(b) When the temperature is 67°C, the equivalent volume is **26.1 m^3**.

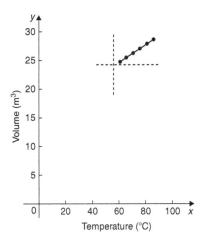

Figure 17.18

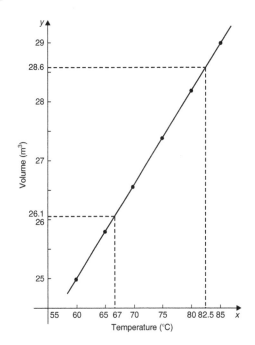

Figure 17.19

Problem 14.   In an experiment demonstrating Hooke's law*, the strain in an aluminium wire was measured for various stresses. The results were:

---

*Who was Charles? To find out more go to www.routledge.com/cw/bird

*Who was Hooke? To find out more go to www.routledge.com/cw/bird

Stress (N/mm^2)	4.9	8.7	15.0
Strain	0.00007	0.00013	0.00021

Stress (N/mm^2)	18.4	24.2	27.3
Strain	0.00027	0.00034	0.00039

Plot a graph of stress (vertically) against strain (horizontally). Find (a) Young's modulus of elasticity* for aluminium, which is given by the gradient of the graph, (b) the value of the strain at a stress of 20 N/mm^2 and (c) the value of the stress when the strain is 0.00020

The co-ordinates (0.00007, 4.9), (0.00013, 8.7), and so on, are plotted as shown in Fig. 17.20. The graph produced is the best straight line which can be drawn corresponding to these points. (With experimental results it is unlikely that all the points will lie exactly on a straight

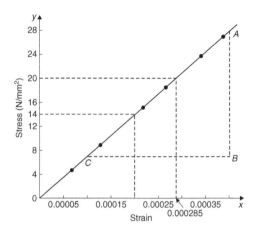

Figure 17.20

line.) The graph, and each of its axes, are labelled. Since the straight line passes through the origin, stress is directly proportional to strain for the given range of values.

(a) The gradient of the straight line AC is given by

$$\frac{AB}{BC} = \frac{28 - 7}{0.00040 - 0.00010} = \frac{21}{0.00030}$$

$$= \frac{21}{3 \times 10^{-4}} = \frac{7}{10^{-4}} = 7 \times 10^4$$

$$= 70\,000\,\text{N/mm}^2$$

Thus, **Young's modulus of elasticity for aluminium is 70 000 N/mm^2**. Since $1\,\text{m}^2 = 10^6\,\text{mm}^2$, $70\,000\,\text{N/mm}^2$ is equivalent to $70\,000 \times 10^6\,\text{N/m}^2$, i.e. **70 × 10^9 N/m^2 (or pascals)**.

From Fig. 17.20,

(b) The value of the strain at a stress of 20 N/mm^2 is **0.000285**

(c) The value of the stress when the strain is 0.00020 is **14 N/mm^2**

*Who was Young? **Thomas Young** (13 June 1773–10 May 1829) was an English polymath. He is famous for having partly deciphered Egyptian hieroglyphics (specifically the Rosetta Stone). Young made notable scientific contributions to the fields of vision, light, solid mechanics, energy, physiology, language, musical harmony and Egyptology. To find out more go to www.routledge.com/cw/bird

Problem 15. The following values of resistance R ohms and corresponding voltage V volts are obtained from a test on a filament lamp.

R ohms	30	48.5	73	107	128
V volts	16	29	52	76	94

Choose suitable scales and plot a graph with R representing the vertical axis and V the horizontal axis. Determine (a) the gradient of the graph, (b) the

R axis intercept value, (c) the equation of the graph, (d) the value of resistance when the voltage is 60 V and (e) the value of the voltage when the resistance is 40 ohms. (f) If the graph were to continue in the same manner, what value of resistance would be obtained at 110 V?

The co-ordinates (16, 30), (29, 48.5), and so on are shown plotted in Fig. 17.21, where the best straight line is drawn through the points.

(a) The slope or gradient of the straight line AC is given by

$$\frac{AB}{BC} = \frac{135 - 10}{100 - 0} = \frac{125}{100} = \mathbf{1.25}$$

(Note that the vertical line AB and the horizontal line BC may be constructed anywhere along the length of the straight line. However, calculations are made easier if the horizontal line BC is carefully chosen; in this case, 100)

(b) The R-axis intercept is at $R = \mathbf{10\,ohms}$ (by extrapolation).

(c) The equation of a straight line is $y = mx + c$, when y is plotted on the vertical axis and x on the horizontal axis. m represents the gradient and c the y-axis intercept. In this case, R corresponds to y,

V corresponds to x, $m = 1.25$ and $c = 10$. Hence, the equation of the graph is $\mathbf{R = (1.25\,V + 10)\,\Omega}$.

From Fig. 17.21,

(d) When the voltage is 60 V, the resistance is $\mathbf{85\,\Omega}$.

(e) When the resistance is 40 ohms, the voltage is $\mathbf{24V}$.

(f) By extrapolation, when the voltage is 110 V, the resistance is $\mathbf{147\,\Omega}$.

Problem 16. Experimental tests to determine the breaking stress σ of rolled copper at various temperatures t gave the following results.

Stress σ (N/cm^2)	8.46	8.04	7.78
Temperature $t(^\circ C)$	70	200	280

Stress σ (N/cm^2)	7.37	7.08	6.63
Temperature $t(^\circ C)$	410	500	640

Show that the values obey the law $\sigma = at + b$, where a and b are constants, and determine approximate values for a and b. Use the law to determine the stress at 250°C and the temperature when the stress is 7.54 N/cm^2

The co-ordinates (70, 8.46), (200, 8.04), and so on, are plotted as shown in Fig. 17.22. Since the graph is a straight line then the values obey the law $\sigma = at + b$,

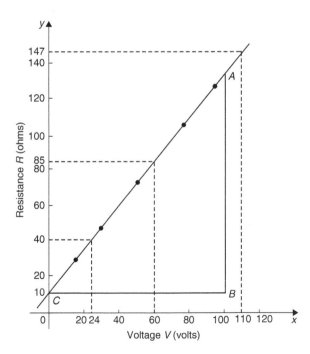

Figure 17.21

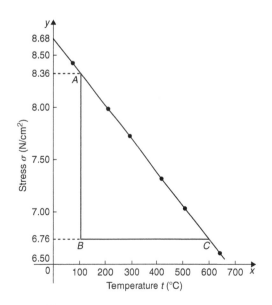

Figure 17.22

and the gradient of the straight line is

$$a = \frac{AB}{BC} = \frac{8.36 - 6.76}{100 - 600} = \frac{1.60}{-500} = -0.0032$$

Vertical axis intercept, $b = 8.68$

Hence, the law of the graph is $\sigma = 0.0032t + 8.68$

When the temperature is 250°C, stress σ is given by

$$\sigma = -0.0032(250) + 8.68 = 7.88\,\text{N/cm}^2$$

Rearranging $\sigma = -0.0032t + 8.68$ gives

$$0.0032t = 8.68 - \sigma, \quad \text{i.e.} \quad t = \frac{8.68 - \sigma}{0.0032}$$

Hence, when the stress, $\sigma = 7.54\,\text{N/cm}^2$,

$$\textbf{temperature, } t = \frac{8.68 - 7.54}{0.0032} = \textbf{356.3°C}$$

Now try the following Practice Exercise

Practice Exercise 69   Practical problems involving straight line graphs (answers on page 430)

1. The resistance R ohms of a copper winding is measured at various temperatures $t\,°\text{C}$ and the results are as follows:

R (ohms)	112	120	126	131	134
$t\,°\text{C}$	20	36	48	58	64

   Plot a graph of R (vertically) against t (horizontally) and find from it (a) the temperature when the resistance is $122\,\Omega$ and (b) the resistance when the temperature is 52°C.

2. The speed of a motor varies with armature voltage as shown by the following experimental results.

n (rev/min)	285	517	615
V (volts)	60	95	110

n (rev/min)	750	917	1050
V (volts)	130	155	175

   Plot a graph of speed (horizontally) against voltage (vertically) and draw the best straight line through the points. Find from the graph (a) the speed at a voltage of 145 V and (b) the voltage at a speed of 400 rev/min.

3. The following table gives the force F newtons which, when applied to a lifting machine, overcomes a corresponding load of L newtons.

Force F (newtons)	25	47	64
Load L (newtons)	50	140	210

Force F (newtons)	120	149	187
Load L (newtons)	430	550	700

   Choose suitable scales and plot a graph of F (vertically) against L (horizontally). Draw the best straight line through the points. Determine from the graph

   (a) the gradient,

   (b) the F-axis intercept,

   (c) the equation of the graph,

   (d) the force applied when the load is 310 N, and

   (e) the load that a force of 160 N will overcome.

   (f) If the graph were to continue in the same manner, what value of force will be needed to overcome a 800 N load?

4. The following table gives the results of tests carried out to determine the breaking stress σ of rolled copper at various temperatures, t.

Stress σ (N/cm²)	8.51	8.07	7.80
Temperature $t(°\text{C})$	75	220	310

Stress σ (N/cm²)	7.47	7.23	6.78
Temperature $t(°\text{C})$	420	500	650

   Plot a graph of stress (vertically) against temperature (horizontally). Draw the best straight line through the plotted co-ordinates. Determine the slope of the graph and the vertical axis intercept.

5. The velocity v of a body after varying time intervals t was measured as follows:

t (seconds)	2	5	8
v (m/s)	16.9	19.0	21.1

t (seconds)	11	15	18
v (m/s)	23.2	26.0	28.1

Plot v vertically and t horizontally and draw a graph of velocity against time. Determine from the graph (a) the velocity after 10 s, (b) the time at 20 m/s and (c) the equation of the graph.

6. The mass m of a steel joist varies with length L as follows:

mass, m (kg)	80	100	120	140	160
length, L (m)	3.00	3.74	4.48	5.23	5.97

Plot a graph of mass (vertically) against length (horizontally). Determine the equation of the graph.

7. The crushing strength of mortar varies with the percentage of water used in its preparation, as shown below.

Crushing strength, F (tonnes)	1.67	1.40	1.13
% of water used, w%	6	9	12

Crushing strength, F (tonnes)	0.86	0.59	0.32
% of water used, w%	15	18	21

Plot a graph of F (vertically) against w (horizontally).

(a) Interpolate and determine the crushing strength when 10% water is used.

(b) Assuming the graph continues in the same manner, extrapolate and determine the percentage of water used when the crushing strength is 0.15 tonnes.

(c) What is the equation of the graph?

8. In an experiment demonstrating Hooke's law, the strain in a copper wire was measured for various stresses. The results were

Stress (pascals)	10.6×10^6	18.2×10^6	24.0×10^6
Strain	0.00011	0.00019	0.00025

Stress (pascals)	30.7×10^6	39.4×10^6
Strain	0.00032	0.00041

Plot a graph of stress (vertically) against strain (horizontally). Determine

(a) Young's modulus of elasticity for copper, which is given by the gradient of the graph,

(b) the value of strain at a stress of 21×10^6 Pa,

(c) the value of stress when the strain is 0.00030

9. An experiment with a set of pulley blocks gave the following results.

Effort, E (newtons)	9.0	11.0	13.6
Load, L (newtons)	15	25	38

Effort, E (newtons)	17.4	20.8	23.6
Load, L (newtons)	57	74	88

Plot a graph of effort (vertically) against load (horizontally). Determine

(a) the gradient,

(b) the vertical axis intercept,

(c) the law of the graph,

(d) the effort when the load is 30 N,

(e) the load when the effort is 19 N.

10. The variation of pressure p in a vessel with temperature T is believed to follow a law of the form $p = aT + b$, where a and b are constants. Verify this law for the results given below and determine the approximate values of a and b. Hence, determine the pressures at temperatures of 285 K and 310 K and the temperature at a pressure of 250 kPa.

Pressure, p (kPa)	244	247	252
Temperature, T (K)	273	277	282

Pressure, p (kPa)	258	262	267
Temperature, T (K)	289	294	300

For fully worked solutions to each of the problems in Practice Exercises 67 to 69 in this chapter, go to the website:

www.routledge.com/cw/bird

Graphs reducing non-linear laws to linear form

Why it is important to understand: Graphs reducing non-linear laws to linear form

Graphs are important tools for analysing and displaying data between two experimental quantities. Many times situations occur in which the relationship between the variables is not linear. By manipulation, a straight line graph may be plotted to produce a law relating the two variables. Sometimes this involves using the laws of logarithms. The relationship between the resistance of wire and its diameter is not a linear one. Similarly, the periodic time of oscillations of a pendulum does not have a linear relationship with its length, and the head of pressure and the flow velocity are not linearly related. There are thus plenty of examples in engineering where determination of law is needed.

At the end of this chapter, you should be able to:

- understand what is meant by determination of law
- prepare co-ordinates for a non-linear relationship between two variables
- plot prepared co-ordinates and draw a straight line graph
- determine the gradient and vertical axis intercept of a straight line graph
- state the equation of a straight line graph
- plot straight line graphs involving practical engineering examples
- determine straight line laws involving logarithms: $y = ax^n$, $y = ab^x$ and $y = ae^{bx}$
- plot straight line graphs involving logarithms

18.1 Introduction

In Chapter 17 we discovered that the equation of a straight line graph is of the form $y = mx + c$, where m is the gradient and c is the y-axis intercept. This chapter explains how the law of a graph can still be determined even when it is not of the linear form $y = mx + c$. The method used is called **determination of law** and is explained in the following sections.

18.2 Determination of law

Frequently, the relationship between two variables, say x and y, is not a linear one; i.e. when x is plotted against y a curve results. In such cases the non-linear equation may be modified to the linear form, $y = mx + c$, so that the constants, and thus the law relating the variables, can be determined. This technique is called '**determination of law**'.

Some examples of the reduction of equations to linear form include

(i) $y = ax^2 + b$ compares with $Y = mX + c$, where $m = a$, $c = b$ and $X = x^2$
   Hence, y is plotted vertically against x^2 horizontally to produce a straight line graph of gradient a and y-axis intercept b.

(ii) $y = \dfrac{a}{x} + b$,   i.e. $y = a\left(\dfrac{1}{x}\right) + b$

   y is plotted vertically against $\dfrac{1}{x}$ horizontally to produce a straight line graph of gradient a and y-axis intercept b.

(iii) $y = ax^2 + bx$
   Dividing both sides by x gives $\dfrac{y}{x} = ax + b$

   Comparing with $Y = mX + c$ shows that $\dfrac{y}{x}$ is plotted vertically against x horizontally to produce a straight line graph of gradient a and $\dfrac{y}{x}$ axis intercept b.

Here are some worked problems to demonstrate determination of law.

Problem 1.   Experimental values of x and y, shown below, are believed to be related by the law $y = ax^2 + b$. By plotting a suitable graph, verify this law and determine approximate values of a and b

x	1	2	3	4	5
y	9.8	15.2	24.2	36.5	53.0

If y is plotted against x a curve results and it is not possible to determine the values of constants a and b from the curve.
Comparing $y = ax^2 + b$ with $Y = mX + c$ shows that y is to be plotted vertically against x^2 horizontally. A table of values is drawn up as shown below.

x	1	2	3	4	5
x^2	1	4	9	16	25
y	9.8	15.2	24.2	36.5	53.0

A graph of y against x^2 is shown in Fig. 18.1, with the best straight line drawn through the points. **Since a straight line graph results, the law is verified**.

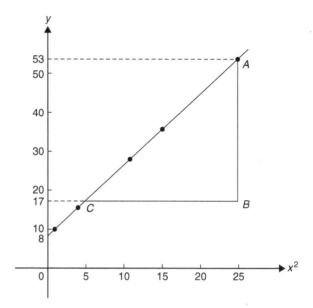

Figure 18.1

From the graph,   **gradient, $a = \dfrac{AB}{BC} = \dfrac{53 - 17}{25 - 5}$**

$$= \dfrac{36}{20} = \mathbf{1.8}$$

and the **y-axis intercept, $b = 8.0$**

Hence, the law of the graph is $y = \mathbf{1.8}x^2 + \mathbf{8.0}$

Problem 2.   Values of load L newtons and distance d metres obtained experimentally are shown in the following table.

Load, L(N)	32.3	29.6	27.0	23.2
Distance, d(m)	0.75	0.37	0.24	0.17

Load, L(N)	18.3	12.8	10.0	6.4
Distance, d(m)	0.12	0.09	0.08	0.07

(a)   Verify that load and distance are related by a law of the form $L = \dfrac{a}{d} + b$ and determine approximate values of a and b

(b)   Hence, calculate the load when the distance is 0.20 m and the distance when the load is 20 N

(a)   Comparing $L = \dfrac{a}{d} + b$ i.e. $L = a\left(\dfrac{1}{d}\right) + b$ with $Y = mX + c$ shows that L is to be plotted vertically against $\dfrac{1}{d}$ horizontally. Another table of values is drawn up as shown below.

L	32.3	29.6	27.0	23.2	18.3	12.8	10.0	6.4
d	0.75	0.37	0.24	0.17	0.12	0.09	0.08	0.07
$\dfrac{1}{d}$	1.33	2.70	4.17	5.88	8.33	11.11	12.50	14.29

A graph of L against $\dfrac{1}{d}$ is shown in Fig. 18.2. **A straight line can be drawn through the points, which verifies that load and distance are related by a law of the form $L = \dfrac{a}{d} + b$**

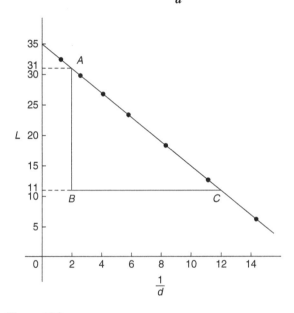

Figure 18.2

Gradient of straight line, $a = \dfrac{AB}{BC} = \dfrac{31 - 11}{2 - 12}$

$$= \dfrac{20}{-10} = -2$$

L-axis intercept, $b = 35$

Hence, the law of the graph is $L = -\dfrac{2}{d} + 35$

(b)   When the distance $d = 0.20\,$m,

load, $L = \dfrac{-2}{0.20} + 35 = 25.0\,$N

Rearranging $L = -\dfrac{2}{d} + 35$ gives

$\dfrac{2}{d} = 35 - L$ and $d = \dfrac{2}{35 - L}$

Hence, when the load $L = 20\,$N, **distance,**

$$d = \dfrac{2}{35 - 20} = \dfrac{2}{15} = \mathbf{0.13\,m}$$

Problem 3.   The solubility s of potassium chlorate is shown by the following table.

$t\,^\circ$C	10	20	30	40	50	60	80	100
s	4.9	7.6	11.1	15.4	20.4	26.4	40.6	58.0

The relationship between s and t is thought to be of the form $s = 3 + at + bt^2$. Plot a graph to test the supposition and use the graph to find approximate values of a and b. Hence, calculate the solubility of potassium chlorate at $70\,^\circ$C

Rearranging $s = 3 + at + bt^2$ gives

$$s - 3 = at + bt^2$$

and $\qquad \dfrac{s - 3}{t} = a + bt$

or $\qquad \dfrac{s - 3}{t} = bt + a$

which is of the form $Y = mX + c$

This shows that $\dfrac{s - 3}{t}$ is to be plotted vertically and t horizontally, with gradient b and vertical axis intercept a.

Another table of values is drawn up as shown below.

t	10	20	30	40	50	60	80	100
s	4.9	7.6	11.1	15.4	20.4	26.4	40.6	58.0
$\dfrac{s - 3}{t}$	0.19	0.23	0.27	0.31	0.35	0.39	0.47	0.55

A graph of $\dfrac{s - 3}{t}$ against t is shown plotted in Fig. 18.3. **A straight line fits the points, which shows that s and t are related by $s = 3 + at + bt^2$**

Gradient of straight line, $b = \dfrac{AB}{BC} = \dfrac{0.39 - 0.19}{60 - 10}$

$$= \dfrac{0.20}{50} = \mathbf{0.004}$$

Vertical axis intercept, $a = \mathbf{0.15}$
Hence, the law of the graph is $s = 3 + 0.15t + 0.004t^2$
The solubility of potassium chlorate at $70\,^\circ$C is given by

$$s = 3 + 0.15(70) + 0.004(70)^2$$

$$= 3 + 10.5 + 19.6 = \mathbf{33.1}$$

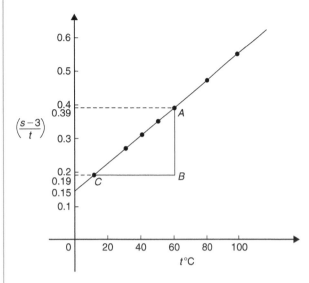

Figure 18.3

___

Now try the following Practice Exercise

Practice Exercise 70   Graphs reducing non-linear laws to linear form (answers on page 430)

In Problems 1 to 5, x and y are two related variables and all other letters denote constants. For the stated laws to be verified it is necessary to plot graphs of the variables in a modified form. State for each, (a) what should be plotted on the vertical axis, (b) what should be plotted on the horizontal axis, (c) the gradient and (d) the vertical axis intercept.

1. $y = d + cx^2$ 2. $y - a = b\sqrt{x}$

3. $y - e = \dfrac{f}{x}$ 4. $y - cx = bx^2$

5. $y = \dfrac{a}{x} + bx$

6.   In an experiment the resistance of wire is measured for wires of different diameters with the following results.

R (ohms)	1.64	1.14	0.89	0.76	0.63
d (mm)	1.10	1.42	1.75	2.04	2.56

It is thought that R is related to d by the law $R = \dfrac{a}{d^2} + b$, where a and b are constants. Verify this and find the approximate values for

a and b. Determine the cross-sectional area needed for a resistance reading of 0.50 ohms.

7.   Corresponding experimental values of two quantities x and y are given below.

x	1.5	3.0	4.5	6.0	7.5	9.0
y	11.5	25.0	47.5	79.0	119.5	169.0

By plotting a suitable graph, verify that y and x are connected by a law of the form $y = kx^2 + c$, where k and c are constants. Determine the law of the graph and hence find the value of x when y is 60.0

8.   Experimental results of the safe load L kN, applied to girders of varying spans, d m, are shown below.

Span, d (m)	2.0	2.8	3.6	4.2	4.8
Load, L (kN)	475	339	264	226	198

It is believed that the relationship between load and span is $L = c/d$, where c is a constant. Determine (a) the value of constant c and (b) the safe load for a span of 3.0 m.

9.   The following results give corresponding values of two quantities x and y which are believed to be related by a law of the form $y = ax^2 + bx$, where a and b are constants.

x	33.86	55.54	72.80	84.10	111.4	168.1
y	3.4	5.2	6.5	7.3	9.1	12.4

Verify the law and determine approximate values of a and b. Hence, determine (a) the value of y when x is 8.0 and (b) the value of x when y is 146.5

18.3   Revision of laws of logarithms

The laws of logarithms were stated in Chapter 15 as follows:

$$\log(A \times B) = \log A + \log B \qquad (1)$$

$$\log\left(\frac{A}{B}\right) = \log A - \log B \qquad (2)$$

$$\log A^n = n \times \log A \qquad (3)$$

Also, $\ln e = 1$ and if, say, $\lg x = 1.5$,
then $x = 10^{1.5} = 31.62$

Further, if $3^x = 7$ then $\lg 3^x = \lg 7$ and $x \lg 3 = \lg 7$, from which $x = \dfrac{\lg 7}{\lg 3} = 1.771$

These laws and techniques are used whenever non-linear laws of the form $y = ax^n$, $y = ab^x$ and $y = ae^{bx}$ are reduced to linear form with the values of a and b needing to be calculated. This is demonstrated in the following section.

18.4 Determination of laws involving logarithms

Examples of the reduction of equations to linear form involving logarithms include

(a) **$y = ax^n$**

Taking logarithms to **a base of 10** of both sides gives

$$\lg y = \lg(ax^n)$$
$$= \lg a + \lg x^n \quad \text{by law (1)}$$

i.e. $\lg y = n \lg x + \lg a$ by law (3)

which compares with $Y = mX + c$

and shows that **lg y is plotted vertically against lg x horizontally** to produce a straight line **graph of gradient n and lg y-axis intercept lg a**.
See worked Problems 4 and 5 to demonstrate how this law is determined.

(b) **$y = ab^x$**

Taking logarithms to **a base of 10** of both sides gives

$$\lg y = \lg(ab^x)$$

i.e. $\lg y = \lg a + \lg b^x$ by law (1)

$\lg y = \lg a + x \lg b$ by law (3)

i.e. $\lg y = x \lg b + \lg a$

or $\lg y = (\lg b)x + \lg a$

which compares with

$$Y = mX + c$$

and shows that **lg y is plotted vertically against x horizontally** to produce **a straight line graph of gradient lg b and lg y-axis intercept lg a**.

See worked Problem 6 to demonstrate how this law is determined.

(c) **$y = ae^{bx}$**

Taking logarithms to **a base of e** of both sides gives

$$\ln y = \ln(ae^{bx})$$

i.e. $\ln y = \ln a + \ln e^{bx}$ by law (1)

i.e. $\ln y = \ln a + bx \ln e$ by law (3)

i.e. $\ln y = bx + \ln a$ since $\ln e = 1$

which compares with

$$Y = mX + c$$

and shows that **ln y is plotted vertically against x horizontally** to produce **a straight line graph of gradient b and ln y-axis intercept ln a**.

See worked Problem 7 to demonstrate how this law is determined.

> **Problem 4.** The current flowing in, and the power dissipated by, a resistor are measured experimentally for various values and the results are as shown below.
>
> | Current, I (amperes) | 2.2 | 3.6 | 4.1 | 5.6 | 6.8 |
> |---|---|---|---|---|---|
> | Power, P (watts) | 116 | 311 | 403 | 753 | 1110 |
>
> Show that the law relating current and power is of the form $P = RI^n$, where R and n are constants, and determine the law

Taking logarithms to a base of 10 of both sides of $P = RI^n$ gives
$$\lg P = \lg(RI^n) = \lg R + \lg I^n = \lg R + n \lg I$$
by the laws of logarithms

i.e. $\lg P = n \lg I + \lg R$

which is of the form $Y = mX + c$, showing that $\lg P$ is to be plotted vertically against $\lg I$ horizontally.

A table of values for $\lg I$ and $\lg P$ is drawn up as shown below.

I	2.2	3.6	4.1	5.6	6.8
$\lg I$	0.342	0.556	0.613	0.748	0.833
P	116	311	403	753	1110
$\lg P$	2.064	2.493	2.605	2.877	3.045

A graph of $\lg P$ against $\lg I$ is shown in Fig. 18.4 and, **since a straight line results, the law $P = RI^n$ is verified**.

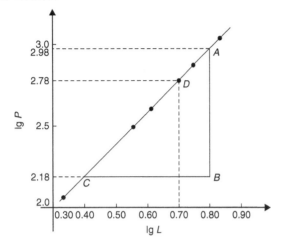

Figure 18.4

Gradient of straight line, $n = \dfrac{AB}{BC} = \dfrac{2.98 - 2.18}{0.8 - 0.4}$

$$= \frac{0.80}{0.4} = 2$$

It is not possible to determine the vertical axis intercept on sight since the horizontal axis scale does not start at zero. Selecting any point from the graph, say point D, where $\lg I = 0.70$ and $\lg P = 2.78$ and substituting values into

$$\lg P = n \lg I + \lg R$$

gives $2.78 = (2)(0.70) + \lg R$

from which, $\lg R = 2.78 - 1.40 = 1.38$

Hence, $R = \text{antilog } 1.38 = 10^{1.38} = \mathbf{24.0}$

Hence, **the law of the graph is $P = 24.0I^2$**

Problem 5. The periodic time, T, of oscillation of a pendulum is believed to be related to its length, L, by a law of the form $T = kL^n$, where k and n are constants. Values of T were measured for various lengths of the pendulum and the results are as shown below.

Periodic time, T(s)	1.0	1.3	1.5	1.8	2.0	2.3
Length, L(m)	0.25	0.42	0.56	0.81	1.0	1.32

Show that the law is true and determine the approximate values of k and n. Hence find the periodic time when the length of the pendulum is 0.75 m

From para (a), page 171, if $T = kL^n$

then $\lg T = n \lg L + \lg k$

and comparing with $Y = mX + c$

shows that $\lg T$ is plotted vertically against $\lg L$ horizontally, with gradient n and vertical axis intercept $\lg k$.

A table of values for $\lg T$ and $\lg L$ is drawn up as shown below.

T	1.0	1.3	1.5	1.8	2.0	2.3
$\lg T$	0	0.114	0.176	0.255	0.301	0.362
L	0.25	0.42	0.56	0.81	1.0	1.32
$\lg L$	-0.602	-0.377	-0.252	-0.092	0	0.121

A graph of $\lg T$ against $\lg L$ is shown in Fig. 18.5 and the **law $T = kL^n$ is true since a straight line results**.

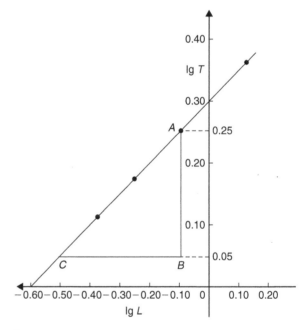

Figure 18.5

From the graph, gradient of straight line,

$$n = \frac{AB}{BC} = \frac{0.25 - 0.05}{-0.10 - (-0.50)} = \frac{0.20}{0.40} = \frac{1}{2}$$

Vertical axis intercept, $\lg k = 0.30$. Hence,
$k = \text{antilog } 0.30 = 10^{0.30} = \mathbf{2.0}$

Hence, **the law of the graph is $T = 2.0L^{1/2}$ or $T = 2.0\sqrt{L}$.**
When length $L = 0.75$ m, $\mathbf{T} = 2.0\sqrt{0.75} = \mathbf{1.73s}$

Problem 6. Quantities x and y are believed to be related by a law of the form $y = ab^x$, where a and b are constants. The values of x and corresponding values of y are

x	0	0.6	1.2	1.8	2.4	3.0
y	5.0	9.67	18.7	36.1	69.8	135.0

Verify the law and determine the approximate values of a and b. Hence determine (a) the value of y when x is 2.1 and (b) the value of x when y is 100

From para (b), page 171, if $\quad y = ab^x$

then $\qquad\qquad\qquad \lg y = (\lg b)x + \lg a$

and comparing with $\qquad\quad Y = mX + c$

shows that $\lg y$ is plotted vertically and x horizontally, with gradient $\lg b$ and vertical axis intercept $\lg a$. Another table is drawn up as shown below.

x	0	0.6	1.2	1.8	2.4	3.0
y	5.0	9.67	18.7	36.1	69.8	135.0
$\lg y$	0.70	0.99	1.27	1.56	1.84	2.13

A graph of $\lg y$ against x is shown in Fig. 18.6 and, **since a straight line results, the law $y = ab^x$ is verified**.

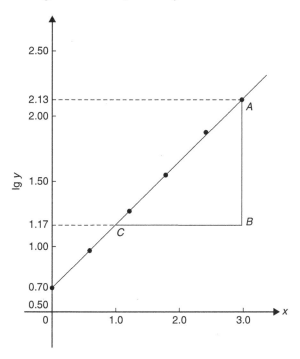

Figure 18.6

Gradient of straight line,

$$\lg b = \frac{AB}{BC} = \frac{2.13 - 1.17}{3.0 - 1.0} = \frac{0.96}{2.0} = 0.48$$

Hence, $b =$ antilog $0.48 = 10^{0.48} = \mathbf{3.0}$, correct to 2 significant figures.

Vertical axis intercept, $\quad \lg a = 0.70$, from which

$$a = \text{antilog } 0.70$$

$$= 10^{0.70} = \mathbf{5.0}, \text{ correct to}$$
$$2 \text{ significant figures.}$$

Hence, **the law of the graph is $y = 5.0(3.0)^x$**

(a) When $x = 2.1, y = 5.0(3.0)^{2.1} = \mathbf{50.2}$

(b) When $y = 100, 100 = 5.0(3.0)^x$, from which
$$100/5.0 = (3.0)^x$$

i.e. $\quad 20 = (3.0)^x$

Taking logarithms of both sides gives
$$\lg 20 = \lg(3.0)^x = x \lg 3.0$$

Hence, $x = \dfrac{\lg 20}{\lg 3.0} = \dfrac{1.3010}{0.4771} = \mathbf{2.73}$

Problem 7. The current i mA flowing in a capacitor which is being discharged varies with time t ms, as shown below.

i (mA)	203	61.14	22.49	6.13	2.49	0.615
t (ms)	100	160	210	275	320	390

Show that these results are related by a law of the form $i = Ie^{t/T}$, where I and T are constants. Determine the approximate values of I and T

Taking Napierian logarithms of both sides of
$$i = Ie^{t/T}$$

gives $\qquad \ln i = \ln(Ie^{t/T}) = \ln I + \ln e^{t/T}$

$$= \ln I + \frac{t}{T}\ln e$$

i.e. $\qquad \ln i = \ln I + \dfrac{t}{T}$ since $\ln e = 1$

or $\qquad \ln i = \left(\dfrac{1}{T}\right)t + \ln I$

which compares with $\quad y = mx + c$

showing that $\ln i$ is plotted vertically against t horizontally, with gradient $\dfrac{1}{T}$ and vertical axis intercept $\ln I$.

Another table of values is drawn up as shown below.

t	100	160	210	275	320	390
i	203	61.14	22.49	6.13	2.49	0.615
$\ln i$	5.31	4.11	3.11	1.81	0.91	−0.49

A graph of $\ln i$ against t is shown in Fig. 18.7 and, since a straight line results, **the law $i = Ie^{t/T}$ is verified**.

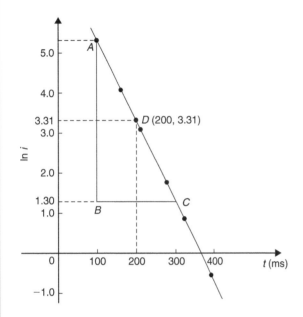

Figure 18.7

Gradient of straight line, $\dfrac{1}{T} = \dfrac{AB}{BC} = \dfrac{5.30 - 1.30}{100 - 300}$

$$= \dfrac{4.0}{-200} = -0.02$$

Hence, $\qquad T = \dfrac{1}{-0.02} = \mathbf{-50}$

Selecting any point on the graph, say point D, where $t = 200$ and $\ln i = 3.31$, and substituting

into $\qquad \ln i = \left(\dfrac{1}{T}\right) t + \ln I$

gives $\qquad 3.31 = -\dfrac{1}{50}(200) + \ln I$

from which, $\quad \ln I = 3.31 + 4.0 = 7.31$

and $\quad I = $ antilog $7.31 = e^{7.31} = 1495$ or **1500** correct to 3 significant figures.

Hence, **the law of the graph is $i = 1500e^{-t/50}$**

Now try the following Practice Exercise

Practice Exercise 71   Determination of laws involving logarithms (answers on page 430)

In Problems 1 to 3, x and y are two related variables and all other letters denote constants. For the stated laws to be verified it is necessary to plot graphs of the variables in a modified form. State for each, (a) what should be plotted on the vertical axis, (b) what should be plotted on the horizontal axis, (c) the gradient and (d) the vertical axis intercept.

1. $y = ba^x$ 2. $y = kx^L$ 3. $\dfrac{y}{m} = e^{nx}$

4. The luminosity I of a lamp varies with the applied voltage V, and the relationship between I and V is thought to be $I = kV^n$. Experimental results obtained are

I (candelas)	1.92	4.32	9.72
V (volts)	40	60	90

I (candelas)	15.87	23.52	30.72
V (volts)	115	140	160

Verify that the law is true and determine the law of the graph. Also determine the luminosity when 75 V is applied across the lamp.

5. The head of pressure h and the flow velocity v are measured and are believed to be connected by the law $v = ah^b$, where a and b are constants. The results are as shown below.

h	10.6	13.4	17.2	24.6	29.3
v	9.77	11.0	12.44	14.88	16.24

Verify that the law is true and determine values of a and b.

6. Experimental values of x and y are measured as follows.

x	0.4	0.9	1.2	2.3	3.8
y	8.35	13.47	17.94	51.32	215.20

The law relating x and y is believed to be of the form $y = ab^x$, where a and b are constants. Determine the approximate values of a and b. Hence, find the value of y when x is 2.0 and the value of x when y is 100.

7. The activity of a mixture of radioactive isotopes is believed to vary according to the law $R = R_0 t^{-c}$, where R_0 and c are constants. Experimental results are shown below.

R	9.72	2.65	1.15	0.47	0.32	0.23
t	2	5	9	17	22	28

Verify that the law is true and determine approximate values of R_0 and c.

8. Determine the law of the form $y = ae^{kx}$ which relates the following values.

y	0.0306	0.285	0.841	5.21	173.2	1181
x	−4.0	5.3	9.8	17.4	32.0	40.0

9. The tension T in a belt passing round a pulley wheel and in contact with the pulley over an angle of θ radians is given by $T = T_0 e^{\mu\theta}$, where T_0 and μ are constants. Experimental results obtained are

T (newtons)	47.9	52.8	60.3	70.1	80.9
θ (radians)	1.12	1.48	1.97	2.53	3.06

Determine approximate values of T_0 and μ. Hence, find the tension when θ is 2.25 radians and the value of θ when the tension is 50.0 newtons.

For fully worked solutions to each of the problems in Practice Exercises 70 and 71 in this chapter, go to the website:

www.routledge.com/cw/bird

Chapter 19

Graphical solution of equations

Why it is important to understand: Graphical solution of equations

It has been established in previous chapters that the solution of linear, quadratic, simultaneous and cubic equations occur often in engineering and science and may be solved using algebraic means. Being able to solve equations graphically provides another method to aid understanding and interpretation of equations. Engineers, including architects, surveyors and a variety of engineers in fields such as biomedical, chemical, electrical, mechanical and nuclear, all use equations which need solving by one means or another.

At the end of this chapter, you should be able to:

- solve two simultaneous equations graphically
- solve a quadratic equation graphically
- solve a linear and quadratic equation simultaneously by graphical means
- solve a cubic equation graphically

19.1 Graphical solution of simultaneous equations

Linear simultaneous equations in two unknowns may be solved graphically by

(a) plotting the two straight lines on the same axes, and

(b) noting their point of intersection.

The co-ordinates of the point of intersection give the required solution.

Here are some worked problems to demonstrate the graphical solution of simultaneous equations.

Problem 1. Solve graphically the simultaneous equations

$$2x - y = 4$$
$$x + y = 5$$

Rearranging each equation into $y = mx + c$ form gives

$$y = 2x - 4$$
$$y = -x + 5$$

Only three co-ordinates need be calculated for each graph since both are straight lines.

Basic Engineering Mathematics. 978-0-415-66278-9, © 2014 John Bird. Published by Taylor & Francis.

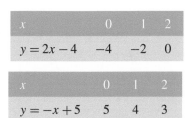

x	0	1	2
$y = 2x - 4$	-4	-2	0

x	0	1	2
$y = -x + 5$	5	4	3

Each of the graphs is plotted as shown in Fig. 19.1. The point of intersection is at (3, 2) and since this is the only point which lies simultaneously on both lines then $x = 3, y = 2$ is the solution of the simultaneous equations.

x	0	1	2
$y = 0.20x - 1.70$	-1.70	-1.50	-1.30

The two sets of co-ordinates are plotted as shown in Fig. 19.2. The point of intersection is (2.50, -1.20). Hence, the solution of the simultaneous equations is $x = 2.50, y = -1.20$

(It is sometimes useful to initially sketch the two straight lines to determine the region where the point of intersection is. Then, for greater accuracy, a graph having a smaller range of values can be drawn to 'magnify' the point of intersection.)

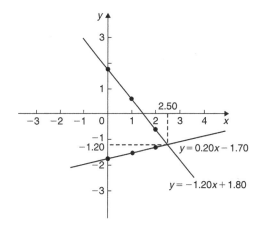

Figure 19.2

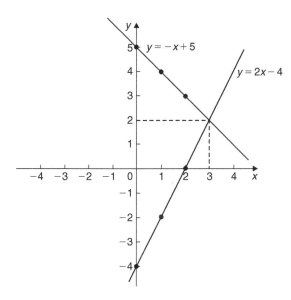

Figure 19.1

Now try the following Practice Exercise

Practice Exercise 72   Graphical solution of simultaneous equations (Answers on page 430)

Problem 2.   Solve graphically the equations

$$1.20x + y = 1.80$$
$$x - 5.0y = 8.50$$

Rearranging each equation into $y = mx + c$ form gives

$$y = -1.20x + 1.80 \qquad (1)$$
$$y = \frac{x}{5.0} - \frac{8.5}{5.0}$$

i.e. $\qquad y = 0.20x - 1.70 \qquad (2)$

Three co-ordinates are calculated for each equation as shown below.

x	0	1	2
$y = -1.20x + 1.80$	1.80	0.60	-0.60

In Problems 1 to 6, solve the simultaneous equations graphically.

1. $y = 3x - 2$
   $y = -x + 6$

2. $x + y = 2$
   $3y - 2x = 1$

3. $y = 5 - x$
   $x - y = 2$

4. $3x + 4y = 5$
   $2x - 5y + 12 = 0$

5. $1.4x - 7.06 = 3.2y$
   $2.1x - 6.7y = 12.87$

6. $3x - 2y = 0$
   $4x + y + 11 = 0$

7. The friction force F newtons and load L newtons are connected by a law of the form $F = aL + b$, where a and b are constants. When $F = 4$N, $L = 6$N and when $F = 2.4$N, $L = 2$N. Determine graphically the values of a and b.

19.2 Graphical solution of quadratic equations

A general **quadratic equation** is of the form $y = ax^2 + bx + c$, where a, b and c are constants and a is not equal to zero.

A graph of a quadratic equation always produces a shape called a **parabola**.

The gradients of the curves between 0 and A and between B and C in Fig. 19.3 are positive, whilst the gradient between A and B is negative. Points such as A and B are called **turning points**. At A the gradient is zero and, as x increases, the gradient of the curve changes from positive just before A to negative just after. Such a point is called a **maximum value**. At B the gradient is also zero and, as x increases, the gradient of the curve changes from negative just before B to positive just after. Such a point is called a **minimum value**.

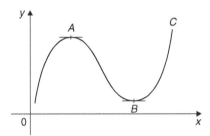

Figure 19.3

Following are three examples of solutions using **quadratic graphs.**

(a) $y = ax^2$

Graphs of $y = x^2$, $y = 3x^2$ and $y = \frac{1}{2}x^2$ are shown in Fig. 19.4. All have minimum values at the origin $(0, 0)$

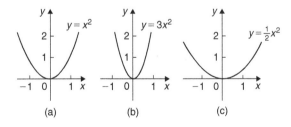

(a) (b) (c)

Figure 19.4

Graphs of $y = -x^2$, $y = -3x^2$ and $y = -\frac{1}{2}x^2$ are shown in Fig. 19.5. All have maximum values at the origin $(0, 0)$

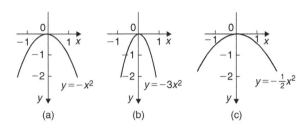

(a) (b) (c)

Figure 19.5

When $y = ax^2$,

(i) curves are symmetrical about the y-axis,
(ii) the magnitude of a affects the gradient of the curve, and
(iii) the sign of a determines whether it has a maximum or minimum value.

(b) $y = ax^2 + c$

Graphs of $y = x^2 + 3$, $y = x^2 - 2$, $y = -x^2 + 2$ and $y = -2x^2 - 1$ are shown in Fig. 19.6.

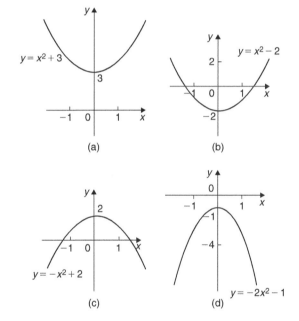

(a) (b)

(c) (d)

Figure 19.6

When $y = ax^2 + c$,

(i) curves are symmetrical about the y-axis,
(ii) the magnitude of a affects the gradient of the curve, and
(iii) the constant c is the y-axis intercept.

(c) $y = ax^2 + bx + c$

Whenever b has a value other than zero the curve is displaced to the right or left of the y-axis.

When b/a is positive, the curve is displaced $b/2a$ to the left of the y-axis, as shown in Fig. 19.7(a).

When b/a is negative, the curve is displaced $b/2a$ to the right of the y-axis, as shown in Fig. 19.7(b).

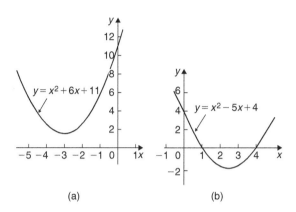

(a)   (b)

Figure 19.7

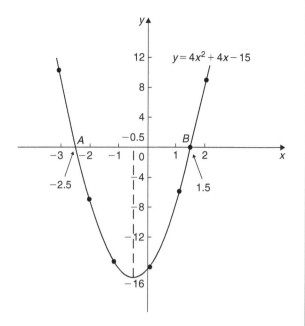

Figure 19.8

Quadratic equations of the form $ax^2 + bx + c = 0$ may be solved graphically by

(a)   plotting the graph $y = ax^2 + bx + c$, and

(b)   noting the points of intersection on the x-axis (i.e. where $y = 0$)

The x values of the points of intersection give the required solutions since at these points both $y = 0$ and $ax^2 + bx + c = 0$

The number of solutions, or roots, of a quadratic equation depends on how many times the curve cuts the x-axis. There can be no real roots, as in Fig. 19.7(a), one root, as in Figs 19.4 and 19.5, or two roots, as in Fig. 19.7(b).

Here are some worked problems to demonstrate the graphical solution of quadratic equations.

A graph of $y = 4x^2 + 4x - 15$ is shown in Fig. 19.8. The only points where $y = 4x^2 + 4x - 15$ and $y = 0$ are the points marked A and B. This occurs at $x = -2.5$ and $x = 1.5$ and these are the solutions of the quadratic equation $4x^2 + 4x - 15 = 0$
By substituting $x = -2.5$ and $x = 1.5$ into the original equation the solutions may be checked.
The curve has a turning point at $(-0.5, -16)$ and the nature of the point is a **minimum**.
An alternative graphical method of solving $4x^2 + 4x - 15 = 0$ is to rearrange the equation as $4x^2 = -4x + 15$ and then plot two separate graphs — in this case, $y = 4x^2$ and $y = -4x + 15$. Their points of intersection give the roots of the equation $4x^2 = -4x + 15$, i.e. $4x^2 + 4x - 15 = 0$. This is shown in Fig. 19.9, where the roots are $x = -2.5$ and $x = 1.5$, as before.

> **Problem 3.**   Solve the quadratic equation $4x^2 + 4x - 15 = 0$ graphically, given that the solutions lie in the range $x = -3$ to $x = 2$. Determine also the co-ordinates and nature of the turning point of the curve

Let $y = 4x^2 + 4x - 15$. A table of values is drawn up as shown below.

x			-3	-2	-1	0	1	2
$y = 4x^2 + 4x - 15$			9	-7	-15	-15	-7	9

> **Problem 4.**   Solve graphically the quadratic equation $-5x^2 + 9x + 7.2 = 0$ given that the solutions lie between $x = -1$ and $x = 3$. Determine also the co-ordinates of the turning point and state its nature

Let $y = -5x^2 + 9x + 7.2$. A table of values is drawn up as shown below.

x			-1	-0.5	0	1
$y = -5x^2 + 9x + 7.2$			-6.8	1.45	7.2	11.2

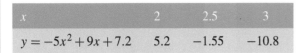

x		2	2.5	3
$y = -5x^2 + 9x + 7.2$		5.2	−1.55	−10.8

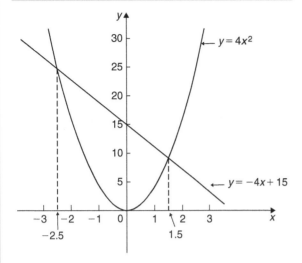

Figure 19.9

A graph of $y = -5x^2 + 9x + 7.2$ is shown plotted in Fig. 19.10. The graph crosses the x-axis (i.e. where $y = 0$) at $x = -0.6$ and $x = 2.4$ and these are the solutions of the quadratic equation $-5x^2 + 9x + 7.2 = 0$
The turning point is a **maximum**, having co-ordinates **(0.9, 11.25)**

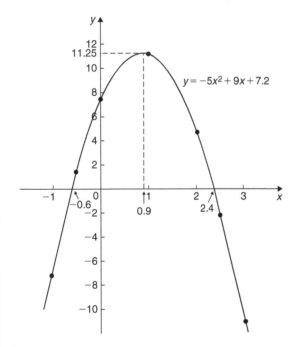

Figure 19.10

Problem 5. Plot a graph of $y = 2x^2$ and hence solve the equations
(a) $2x^2 - 8 = 0$ (b) $2x^2 - x - 3 = 0$

A graph of $y = 2x^2$ is shown in Fig. 19.11.

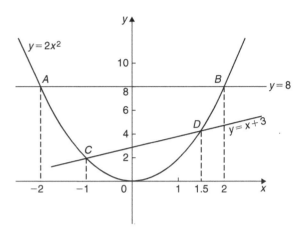

Figure 19.11

(a) Rearranging $2x^2 - 8 = 0$ gives $2x^2 = 8$ and the solution of this equation is obtained from the points of intersection of $y = 2x^2$ and $y = 8$; i.e. at co-ordinates $(-2, 8)$ and $(2, 8)$, shown as A and B, respectively, in Fig. 19.11. Hence, the solutions of $2x^2 - 8 = 0$ are $x = -2$ and $x = +2$

(b) Rearranging $2x^2 - x - 3 = 0$ gives $2x^2 = x + 3$ and the solution of this equation is obtained from the points of intersection of $y = 2x^2$ and $y = x + 3$; i.e. at C and D in Fig. 19.11. Hence, the solutions of $2x^2 - x - 3 = 0$ are $x = -1$ and $x = 1.5$

Problem 6. Plot the graph of $y = -2x^2 + 3x + 6$ for values of x from $x = -2$ to $x = 4$. Use the graph to find the roots of the following equations.
(a) $-2x^2 + 3x + 6 = 0$ (b) $-2x^2 + 3x + 2 = 0$
(c) $-2x^2 + 3x + 9 = 0$ (d) $-2x^2 + x + 5 = 0$

A table of values for $y = -2x^2 + 3x + 6$ is drawn up as shown below.

x	−2	−1	0	1	2	3	4
y	−8	1	6	7	4	−3	−14

A graph of $y = -2x^2 + 3x + 6$ is shown in Fig. 19.12.

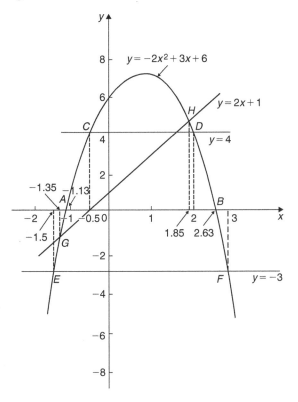

Figure 19.12

(a) The parabola $y = -2x^2 + 3x + 6$ and the straight line $y = 0$ intersect at A and B, where $x = -1.13$ and $x = 2.63$, and these are the roots of the equation $-2x^2 + 3x + 6 = 0$

(b) Comparing $y = -2x^2 + 3x + 6$ (1)

with $0 = -2x^2 + 3x + 2$ (2)

shows that, if 4 is added to both sides of equation (2), the RHS of both equations will be the same. Hence, $4 = -2x^2 + 3x + 6$. The solution of this equation is found from the points of intersection of the line $y = 4$ and the parabola $y = -2x^2 + 3x + 6$; i.e. points C and D in Fig. 19.12. Hence, the roots of $-2x^2 + 3x + 2 = 0$ are $x = -0.5$ and $x = 2$

(c) $-2x^2 + 3x + 9 = 0$ may be rearranged as $-2x^2 + 3x + 6 = -3$ and the solution of this equation is obtained from the points of intersection of the line $y = -3$ and the parabola $y = -2x^2 + 3x + 6$; i.e. at points E and F in Fig. 19.12. Hence, the roots of $-2x^2 + 3x + 9 = 0$ are $x = -1.5$ and $x = 3$

(d) Comparing $y = -2x^2 + 3x + 6$ (3)

with $0 = -2x^2 + x + 5$ (4)

shows that, if $2x + 1$ is added to both sides of equation (4), the RHS of both equations will be the same. Hence, equation (4) may be written as $2x + 1 = -2x^2 + 3x + 6$. The solution of this equation is found from the points of intersection of the line $y = 2x + 1$ and the parabola $y = -2x^2 + 3x + 6$; i.e. points G and H in Fig. 19.12. Hence, the roots of $-2x^2 + x + 5 = 0$ are $x = -1.35$ and $x = 1.85$

Now try the following Practice Exercise

Practice Exercise 73   Solving quadratic equations graphically (answers on page 430)

1. Sketch the following graphs and state the nature and co-ordinates of their respective turning points.
   (a) $y = 4x^2$ (b) $y = 2x^2 - 1$
   (c) $y = -x^2 + 3$ (d) $y = -\frac{1}{2}x^2 - 1$

Solve graphically the quadratic equations in Problems 2 to 5 by plotting the curves between the given limits. Give answers correct to 1 decimal place.

2. $4x^2 - x - 1 = 0$; $x = -1$ to $x = 1$

3. $x^2 - 3x = 27$; $x = -5$ to $x = 8$

4. $2x^2 - 6x - 9 = 0$; $x = -2$ to $x = 5$

5. $2x(5x - 2) = 39.6$; $x = -2$ to $x = 3$

6. Solve the quadratic equation $2x^2 + 7x + 6 = 0$ graphically, given that the solutions lie in the range $x = -3$ to $x = 1$. Determine also the nature and co-ordinates of its turning point.

7. Solve graphically the quadratic equation $10x^2 - 9x - 11.2 = 0$, given that the roots lie between $x = -1$ and $x = 2$

8. Plot a graph of $y = 3x^2$ and hence solve the following equations.
   (a) $3x^2 - 8 = 0$ (b) $3x^2 - 2x - 1 = 0$

9. Plot the graphs $y = 2x^2$ and $y = 3 - 4x$ on the same axes and find the co-ordinates of the points of intersection. Hence, determine the roots of the equation $2x^2 + 4x - 3 = 0$

10. Plot a graph of $y = 10x^2 - 13x - 30$ for values of x between $x = -2$ and $x = 3$. Solve the equation $10x^2 - 13x - 30 = 0$ and from the graph determine
    (a) the value of y when x is 1.3
    (b) the value of x when y is 10
    (c) the roots of the equation
    $10x^2 - 15x - 18 = 0$

19.3 Graphical solution of linear and quadratic equations simultaneously

The solution of **linear and quadratic equations simultaneously** may be achieved graphically by

(a) plotting the straight line and parabola on the same axes, and

(b) noting the points of intersection.

The co-ordinates of the points of intersection give the required solutions.
Here is a worked problem to demonstrate the simultaneous solution of a linear and quadratic equation.

Problem 7. Determine graphically the values of x and y which simultaneously satisfy the equations $y = 2x^2 - 3x - 4$ and $y = 2 - 4x$

$y = 2x^2 - 3x - 4$ is a parabola and a table of values is drawn up as shown below.

x	-2	-1	0	1	2	3
y	10	1	-4	-5	-2	5

$y = 2 - 4x$ is a straight line and only three co-ordinates need be calculated:

x	0	1	2
y	2	-2	-6

The two graphs are plotted in Fig. 19.13 and the points of intersection, shown as A and B, are at co-ordinates $(-2, 10)$ and $(1.5, -4)$. Hence, the simultaneous solutions occur when $x = -2$, $y = 10$ and when $x = 1.5$, $y = -4$
These solutions may be checked by substituting into each of the original equations.

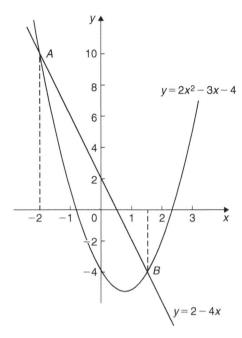

Figure 19.13

Now try the following Practice Exercise

Practice Exercise 74   Solving linear and quadratic equations simultaneously (answers on page 430)

1. Determine graphically the values of x and y which simultaneously satisfy the equations $y = 2(x^2 - 2x - 4)$ and $y + 4 = 3x$

2. Plot the graph of $y = 4x^2 - 8x - 21$ for values of x from -2 to $+4$. Use the graph to find the roots of the following equations.
   (a) $4x^2 - 8x - 21 = 0$
   (b) $4x^2 - 8x - 16 = 0$
   (c) $4x^2 - 6x - 18 = 0$

19.4 Graphical solution of cubic equations

A **cubic equation** of the form $ax^3 + bx^2 + cx + d = 0$ may be solved graphically by

(a) plotting the graph $y = ax^3 + bx^2 + cx + d$, and

(b) noting the points of intersection on the x-axis (i.e. where $y = 0$)

The x-values of the points of intersection give the required solution since at these points both $y = 0$ and $ax^3 + bx^2 + cx + d = 0$.

The number of solutions, or roots, of a cubic equation depends on how many times the curve cuts the x-axis and there can be one, two or three possible roots, as shown in Fig. 19.14.

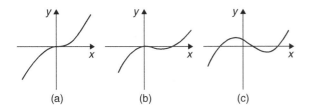

Figure 19.14

Here are some worked problems to demonstrate the graphical solution of cubic equations.

Problem 8. Solve graphically the cubic equation $4x^3 - 8x^2 - 15x + 9 = 0$, given that the roots lie between $x = -2$ and $x = 3$. Determine also the co-ordinates of the turning points and distinguish between them

Let $y = 4x^3 - 8x^2 - 15x + 9$. A table of values is drawn up as shown below.

x	-2	-1	0	1	2	3
y	-25	12	9	-10	-21	0

A graph of $y = 4x^3 - 8x^2 - 15x + 9$ is shown in Fig. 19.15.

The graph crosses the x-axis (where $y = 0$) at $x = -1.5$, $x = 0.5$ and $x = 3$ and these are the solutions to the cubic equation $4x^3 - 8x^2 - 15x + 9 = 0$

The turning points occur at $(-0.6, 14.2)$, which is a **maximum**, and $(2, -21)$, which is a **minimum**.

Problem 9. Plot the graph of $y = 2x^3 - 7x^2 + 4x + 4$ for values of x between $x = -1$ and $x = 3$. Hence, determine the roots of the equation $2x^3 - 7x^2 + 4x + 4 = 0$

A table of values is drawn up as shown below.

x	-1	0	1	2	3
y	-9	4	3	0	7

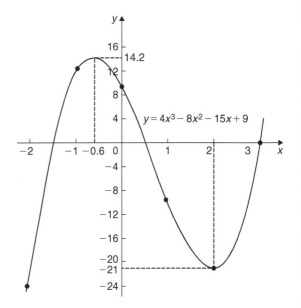

Figure 19.15

A graph of $y = 2x^3 - 7x^2 + 4x + 4$ is shown in Fig. 19.16. The graph crosses the x-axis at $x = -0.5$ and touches the x-axis at $x = 2$

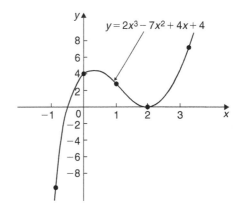

Figure 19.16

Hence the solutions of the equation $2x^3 - 7x^2 + 4x + 4 = 0$ are $x = -0.5$ and $x = 2$

Now try the following Practice Exercise

Practice Exercise 75   Solving cubic equations (answers on page 430)

1. Plot the graph $y = 4x^3 + 4x^2 - 11x - 6$ between $x = -3$ and $x = 2$ and use

the graph to solve the cubic equation $4x^3 + 4x^2 - 11x - 6 = 0$

2. By plotting a graph of $y = x^3 - 2x^2 - 5x + 6$ between $x = -3$ and $x = 4$, solve the equation $x^3 - 2x^2 - 5x + 6 = 0$. Determine also the co-ordinates of the turning points and distinguish between them.

In Problems 3 to 6, solve graphically the cubic equations given, each correct to 2 significant figures.

3. $x^3 - 1 = 0$

4. $x^3 - x^2 - 5x + 2 = 0$

5. $x^3 - 2x^2 = 2x - 2$

6. $2x^3 - x^2 - 9.08x + 8.28 = 0$

7. Show that the cubic equation $8x^3 + 36x^2 + 54x + 27 = 0$ has only one real root and determine its value.

For fully worked solutions to each of the problems in Practice Exercises 72 to 75 in this chapter, go to the website:
www.routledge.com/cw/bird

This assignment covers the material contained in Chapters 17–19. *The marks available are shown in brackets at the end of each question.*

1. Determine the value of P in the following table of values.

x	0	1	4
$y = 3x - 5$	-5	-2	P

(2)

2. Assuming graph paper measuring 20 cm by 20 cm is available, suggest suitable scales for the following ranges of values.

   Horizontal axis: 5 N to 70 N; vertical axis: 20 mm to 190 mm. (2)

3. Corresponding values obtained experimentally for two quantities are:

x	-5	-3	-1	0	2	4
y	-17	-11	-5	-2	4	10

   Plot a graph of y (vertically) against x (horizontally) to scales of 1 cm = 1 for the horizontal x-axis and 1 cm = 2 for the vertical y-axis. From the graph, find
   (a) the value of y when $x = 3$
   (b) the value of y when $x = -4$
   (c) the value of x when $y = 1$
   (d) the value of x when $y = -20$ (8)

4. If graphs of y against x were to be plotted for each of the following, state (i) the gradient, and (ii) the y-axis intercept.
   (a) $y = -5x + 3$ (b) $y = 7x$
   (c) $2y + 4 = 5x$ (d) $5x + 2y = 6$
   (e) $2x - \dfrac{y}{3} = \dfrac{7}{6}$ (10)

5. The resistance R ohms of a copper winding is measured at various temperatures $t°C$ and the results are as follows.

R (Ω)	38	47	55	62	72
t (°C)	16	34	50	64	84

   Plot a graph of R (vertically) against t (horizontally) and find from it
   (a) the temperature when the resistance is 50 Ω
   (b) the resistance when the temperature is 72°C
   (c) the gradient
   (d) the equation of the graph. (10)

6. x and y are two related variables and all other letters denote constants. For the stated laws to be verified it is necessary to plot graphs of the variables in a modified form. State for each
   (a) what should be plotted on the vertical axis,
   (b) what should be plotted on the horizontal axis,
   (c) the gradient,
   (d) the vertical axis intercept.
   (i) $y = p + rx^2$ (ii) $y = \dfrac{a}{x} + bx$ (4)

7. The following results give corresponding values of two quantities x and y which are believed to be related by a law of the form $y = ax^2 + bx$ where a and b are constants.

y	33.9	55.5	72.8	84.1	111.4	168.1
x	3.4	5.2	6.5	7.3	9.1	12.4

   Verify the law and determine approximate values of a and b.
   Hence determine (i) the value of y when x is 8.0 and (ii) the value of x when y is 146.5 (18)

8. By taking logarithms of both sides of $y = kx^n$, show that lg y needs to be plotted vertically and lg x needs to be plotted horizontally to produce a straight line graph. Also, state the gradient and vertical axis intercept. (6)

9. By taking logarithms of both sides of $y = ae^{kx}$ show that ln y needs to be plotted vertically and x needs to be plotted horizontally to produce a straight line graph. Also, state the gradient and vertical axis intercept. (6)

10. Show from the following results of voltage V and admittance Y of an electrical circuit that the law connecting the quantities is of the form $V = kY^n$ and determine the values of k and n.

Voltage, V (volts)	2.88	2.05	1.60	1.22	0.96
Admittance, Y (siemens)	0.52	0.73	0.94	1.23	1.57

(12)

11. The current i flowing in a discharging capacitor varies with time t as shown.

i (mA)	50.0	17.0	5.8	1.7	0.58	0.24
t (ms)	200	255	310	375	425	475

Show that these results are connected by the law of the form $i = I e^{\frac{t}{T}}$ where I is the initial current flowing and T is a constant. Determine approximate values of constants I and T. (15)

12. Solve, correct to 1 decimal place, the quadratic equation $2x^2 - 6x - 9 = 0$ by plotting values of x from $x = -2$ to $x = 5$ (8)

13. Plot the graph of $y = x^3 + 4x^2 + x - 6$ for values of x between $x = -4$ and $x = 2$. Hence determine the roots of the equation $x^3 + 4x^2 + x - 6 = 0$ (9)

14. Plot a graph of $y = 2x^2$ from $x = -3$ to $x = +3$ and hence solve the following equations.
    (a) $2x^2 - 8 = 0$
    (b) $2x^2 - 4x - 6 = 0$ (10)

Chapter 20

Angles and triangles

Why it is important to understand: Angles and triangles

Knowledge of angles and triangles is very important in engineering. Trigonometry is needed in surveying and architecture, for building structures/systems, designing bridges and solving scientific problems. Trigonometry is also used in electrical engineering; the functions that relate angles and side lengths in right-angled triangles are useful in expressing how a.c. electric current varies with time. Engineers use triangles to determine how much force it will take to move along an incline, GPS satellite receivers use triangles to determine exactly where they are in relation to satellites orbiting hundreds of miles away. Whether you want to build a skateboard ramp, a stairway, or a bridge, you can't escape trigonometry.

At the end of this chapter, you should be able to:

- define an angle: acute, right, obtuse, reflex, complementary, supplementary
- define parallel lines, transversal, vertically opposite angles, corresponding angles, alternate angles, interior angles
- define degrees, minutes, seconds, radians
- add and subtract angles
- state types of triangles – acute, right, obtuse, equilateral, isosceles, scalene
- define hypotenuse, adjacent and opposite sides with reference to an angle in a right-angled triangle
- recognise congruent triangles
- recognise similar triangles
- construct triangles, given certain sides and angles

20.1 Introduction

Trigonometry is a subject that involves the measurement of sides and angles of triangles and their relationship to each other. This chapter involves the measurement of angles and introduces types of triangle.

20.2 Angular measurement

An **angle** is the amount of rotation between two straight lines. Angles may be measured either in **degrees** or in **radians**.

If a circle is divided into 360 equal parts, then each part is called **1 degree** and is written as **1°**

i.e.   **1 revolution = 360°**

or   **1 degree is $\frac{1}{360}$th of a revolution**

Some angles are given **special names**.

- Any angle between 0° and 90° is called an **acute angle**.
- An angle equal to 90° is called a **right angle**.
- Any angle between 90° and 180° is called an **obtuse angle**.

- Any angle greater than 180° and less than 360° is called a **reflex angle**.
- An angle of 180° lies on a **straight line**.
- If two angles add up to 90° they are called **complementary angles**.
- If two angles add up to 180° they are called **supplementary angles**.
- **Parallel lines** are straight lines which are in the same plane and never meet. Such lines are denoted by arrows, as in Fig. 20.1.
- A straight line which crosses two parallel lines is called a **transversal** (see *MN* in Fig. 20.1).

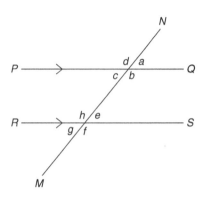

Figure 20.1

With reference to Fig. 20.1,

(a) $a = c$, $b = d$, $e = g$ and $f = h$. Such pairs of angles are called **vertically opposite angles**.

(b) $a = e$, $b = f$, $c = g$ and $d = h$. Such pairs of angles are called **corresponding angles**.

(c) $c = e$ and $b = h$. Such pairs of angles are called **alternate angles**.

(d) $b + e = 180°$ and $c + h = 180°$. Such pairs of angles are called **interior angles**.

20.2.1 Minutes and seconds

One degree may be subdivided into 60 parts, called **minutes**.

i.e.   **1 degree = 60 minutes**

which is written as   **1° = 60′**

41 degrees and 29 minutes is written as 41°29′. 41°29′ is equivalent to $41\dfrac{29°}{60} = 41.483°$ as a decimal, correct to 3 decimal places by calculator.

1 minute further subdivides into 60 seconds,

i.e.   **1 minute = 60 seconds**

which is written as   **1′ = 60″**

(Notice that for minutes, 1 dash is used and for seconds, 2 dashes are used.)
For example, 56 degrees, 36 minutes and 13 seconds is written as 56°36′13″

20.2.2 Radians and degrees

One radian is defined as the angle subtended at the centre of a circle by an arc equal in length to the radius. (For more on circles, see Chapter 26.)
With reference to Fig. 20.2, for arc length s,
θ radians $= \dfrac{s}{r}$

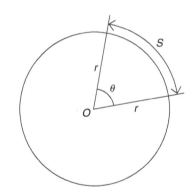

Figure 20.2

When s is the whole circumference, i.e. when $s = 2\pi r$,

$$\theta = \frac{s}{r} = \frac{2\pi r}{r} = 2\pi$$

In one revolution, $\theta = 360°$. Hence, the relationship between **degrees and radians** is

$$360° = 2\pi \text{ radians or } 180° = \pi \text{ rad}$$

i.e.   **1 rad** $= \dfrac{180°}{\pi} \approx 57.30°$

Here are some worked examples on angular measurement.

Problem 1.   Evaluate $43°29′ + 27°43′$

$$
\begin{array}{r}
43°\ 29' \\
+\ \underline{27°\ 43'} \\
\underline{71°\ 12'} \\
1°
\end{array}
$$

(i) $29' + 43' = 72'$

(ii)   Since $60' = 1°, 72' = 1°12'$

(iii)   The $12'$ is placed in the minutes column and $1°$ is carried in the degrees column.

(iv) $43° + 27° + 1°$ (carried) $= 71°$. Place $71°$ in the degrees column.

This answer can be obtained using the **calculator** as follows.

1. Enter 43   2. Press $°$ $'$ $''$ 3. Enter 29
4. Press $°$ $'$ $''$ 5. Press $+$ 6. Enter 27
7. Press $°$ $'$ $''$ 8. Enter 43   9. Press $°$ $'$ $''$
10. Press $=$ Answer $= $ **71°12'**

Thus, **$43°29' + 27°43' = 71°12'$**

Problem 2.   Evaluate $84°13' - 56°39'$

$$\begin{array}{r} 84° \ 13' \\ - \ 56° \ 39' \\ \hline 27° \ 34' \end{array}$$

(i) $13' - 39'$ cannot be done.

(ii) $1°$ or $60'$ is 'borrowed' from the degrees column, which leaves $83°$ in that column.

(iii) $(60' + 13') - 39' = 34'$, which is placed in the minutes column.

(iv) $83° - 56° = 27°$, which is placed in the degrees column.

This answer can be obtained using the **calculator** as follows.

1. Enter 84   2. Press $°$ $'$ $''$ 3. Enter 13
4. Press $°$ $'$ $''$ 5. Press $-$ 6. Enter 56
7. Press $°$ $'$ $''$ 8. Enter 39   9. Press $°$ $'$ $''$
10. Press $=$ Answer $= $ **27°34'**

Thus, **$84°13' - 56°39' = 27°34'$**

Problem 3.   Evaluate $19° 51'47'' + 63°27'34''$

$$\begin{array}{r} 19° \ 51' \ 47'' \\ + \ 63° \ 27' \ 34'' \\ \hline 83° \ 19' \ 21'' \\ 1° \ \ 1' \end{array}$$

(i) $47'' + 34'' = 81''$

(ii)   Since $60'' = 1', 81'' = 1'21''$

(iii)   The $21''$ is placed in the seconds column and $1'$ is carried in the minutes column.

(iv) $51' + 27' + 1' = 79'$

(v)   Since $60' = 1°, 79' = 1°19'$

(vi)   The $19'$ is placed in the minutes column and $1°$ is carried in the degrees column.

(vii) $19° + 63° + 1°$ (carried) $= 83°$. Place $83°$ in the degrees column.

This answer can be obtained using the **calculator** as follows.

1. Enter 19   2. Press $°$ $'$ $''$ 3. Enter 51
4. Press $°$ $'$ $''$ 5. Enter 47   6. Press $°$ $'$ $''$
7. Press $+$ 8. Enter 63   9. Press $°$ $'$ $''$
10. Enter 27   11. Press $°$ $'$ $''$
12. Enter 34   13. Press $°$ $'$ $''$
14. Press $=$ Answer $= $ **83°19'21''**

Thus, **$19°51'47'' + 63°27'34'' = 83°19'21''$**

Problem 4.   Convert $39° 27'$ to degrees in decimal form

$$39°27' = 39\frac{27°}{60}$$

$$\frac{27°}{60} = 0.45° \quad \text{by calculator}$$

Hence,   **$39°27' = 39\dfrac{27°}{60} = 39.45°$**

This answer can be obtained using the **calculator** as follows.

1. Enter 39   2. Press $°$ $'$ $''$ 3. Enter 27
4. Press $°$ $'$ $''$ 5. Press $=$ 6. Press $°$ $'$ $''$
Answer $= $ **39.45°**

Problem 5.   Convert $63° 26' 51''$ to degrees in decimal form, correct to 3 decimal places

$$63° 26'51'' = 63° 26\frac{51'}{60} = 63° 26.85'$$

$$63°26.85' = 63\frac{26.85°}{60} = 63.4475°$$

Hence,   **$63° 26'51'' = 63.448°$**   correct to 3 decimal places.

This answer can be obtained using the **calculator** as follows.

1. Enter 63   2. Press $°$ $'$ $''$ 3. Enter 26
4. Press $°$ $'$ $''$ 5. Enter 51   6. Press $°$ $'$ $''$
7. Press $=$ 8. Press $°$ $'$ $''$ Answer $= $ **63.4475°**

Problem 6. Convert $53.753°$ to degrees, minutes and seconds

$$0.753° = 0.753 \times 60' = 45.18'$$
$$0.18' = 0.18 \times 60'' = 11''$$
to the nearest second

Hence, $\quad$ **$53.753° = 53°45'11''$**

This answer can be obtained using the **calculator** as follows.

1. Enter 53.753 $\quad$ 2. Press $=$
3. Press $°\ '\ ''$ $\quad$ Answer $=$ **$53°45'10.8''$**

Now try the following Practice Exercise

Practice Exercise 76 $\quad$ **Angular measurement (answers on page 430)**

1. Evaluate $52°39' + 29°48'$

2. Evaluate $76°31' - 48°37'$

3. Evaluate $77°22' + 41°36' - 67°47'$

4. Evaluate $41°37'16'' + 58°29'36''$

5. Evaluate $54°37'42'' - 38°53'25''$

6. Evaluate $79°26'19'' - 45°58'56'' + 53°21'38''$

7. Convert $72°33'$ to degrees in decimal form.

8. Convert $27°45'15''$ to degrees correct to 3 decimal places.

9. Convert $37.952°$ to degrees and minutes.

10. Convert $58.381°$ to degrees, minutes and seconds.

Here are some further worked examples on angular measurement.

Problem 7. State the general name given to the following angles: (a) $157°$ (b) $49°$ (c) $90°$ (d) $245°$

(a) Any angle between $90°$ and $180°$ is called an obtuse angle.

   Thus, **$157°$ is an obtuse angle**.

(b) Any angle between $0°$ and $90°$ is called an acute angle.

   Thus, **$49°$ is an acute angle**.

(c) An angle equal to $90°$ is called a **right angle**.

(d) Any angle greater than $180°$ and less than $360°$ is called a reflex angle.

   Thus, **$245°$ is a reflex angle**.

Problem 8. Find the angle complementary to $48°39'$

If two angles add up to $90°$ they are called **complementary angles**. Hence, **the angle complementary to $48°39'$ is**

$$90° - 48°39' = \mathbf{41°21'}$$

Problem 9. Find the angle supplementary to $74°25'$

If two angles add up to $180°$ they are called **supplementary angles**. Hence, **the angle supplementary to $74°25'$ is**

$$180° - 74°25' = \mathbf{105°35'}$$

Problem 10. Evaluate angle θ in each of the diagrams shown in Fig. 20.3

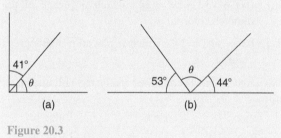

Figure 20.3

(a) The symbol shown in Fig. 20.4 is called a **right angle** and it equals $90°$

Figure 20.4

Hence, from Fig. 20.3(a),

$$\theta + 41° = 90°$$
from which, $\qquad \theta = 90° - 41° = \mathbf{49°}$

(b) An angle of 180° lies on a straight line. Hence, from Fig. 20.3(b),

$$180° = 53° + \theta + 44°$$

from which, $\theta = 180° - 53° - 44° = \mathbf{83°}$

Problem 11. Evaluate angle θ in the diagram shown in Fig. 20.5

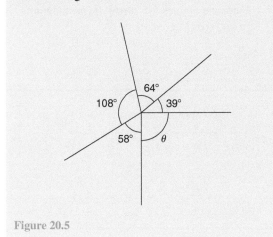

Figure 20.5

There are 360° in a complete revolution of a circle. Thus, $360° = 58° + 108° + 64° + 39° + \theta$ from which, $\theta = 360° - 58° - 108° - 64° - 39° = \mathbf{91°}$

Problem 12. Two straight lines AB and CD intersect at 0. If $\angle AOC$ is 43°, find $\angle AOD$, $\angle DOB$ and $\angle BOC$

From Fig. 20.6, $\angle AOD$ is supplementary to $\angle AOC$. Hence, $\angle AOD = 180° - 43° = \mathbf{137°}$
When two straight lines intersect, the vertically opposite angles are equal.
Hence, $\angle DOB = \mathbf{43°}$ and $\angle BOC$ $\mathbf{137°}$

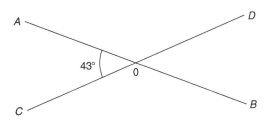

Figure 20.6

Problem 13. Determine angle β in Fig. 20.7

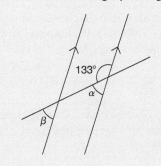

Figure 20.7

$\alpha = 180° - 133° = 47°$ (i.e. supplementary angles). $\alpha = \beta = \mathbf{47°}$ (corresponding angles between parallel lines).

Problem 14. Determine the value of angle θ in Fig. 20.8

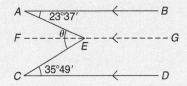

Figure 20.8

Let a straight line FG be drawn through E such that FG is parallel to AB and CD.
$\angle BAE = \angle AEF$ (alternate angles between parallel lines AB and FG), hence $\angle AEF = 23°37'$
$\angle ECD = \angle FEC$ (alternate angles between parallel lines FG and CD), hence $\angle FEC = 35°49'$

Angle $\theta = \angle AEF + \angle FEC = 23°37' + 35°49'$

$$= \mathbf{59°26'}$$

Problem 15. Determine angles c and d in Fig. 20.9

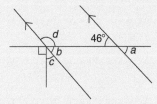

Figure 20.9

$a = b = 46°$ (corresponding angles between parallel lines).

Also, $b + c + 90° = 180°$ (angles on a straight line).

Hence, $46° + c + 90° = 180°$, from which, $c = 44°$

b and d are supplementary, hence $d = 180° - 46°$
$$= 134°$$

Alternatively, $90° + c = d$ (vertically opposite angles).

Problem 16. Convert the following angles to radians, correct to 3 decimal places.
(a) $73°$ (b) $25°37'$

Although we may be more familiar with degrees, radians is the SI unit of angular measurement in engineering (1 radian $\approx 57.3°$).

(a) Since $180° = \pi$ rad then $1° = \dfrac{\pi}{180}$ rad.

Hence, $\mathbf{73°} = 73 \times \dfrac{\pi}{180}$ rad $= \mathbf{1.274}$ **rad**.

(b) $25°37' = 25\dfrac{37°}{60} = 25.616666\ldots$

. Hence, $\mathbf{25°37'} = 25.616666\ldots°$

$$= 25.616666\ldots \times \frac{\pi}{180} \text{rad}$$

$$= \mathbf{0.447} \textbf{ rad}.$$

Problem 17. Convert 0.743 rad to degrees and minutes

Since $180° = \pi$ rad then 1 rad $= \dfrac{180°}{\pi}$

Hence, **0.743 rad** $= 0.743 \times \dfrac{180°}{\pi} = 42.57076\ldots°$

$$= \mathbf{42°34'}$$

Since π rad $= 180°$, then $\dfrac{\pi}{2}$ rad $= 90°$, $\dfrac{\pi}{4}$ rad $= 45°$,

$\dfrac{\pi}{3}$ rad $= 60°$ and $\dfrac{\pi}{6}$ rad $= 30°$

Now try the following Practice Exercise

Practice Exercise 77   Further angular measurement (answers on page 430)

1. State the general name given to an angle of $197°$

2. State the general name given to an angle of $136°$

3. State the general name given to an angle of $49°$

4. State the general name given to an angle of $90°$

5. Determine the angles complementary to the following.
   (a) $69°$ (b) $27°37'$ (c) $41°3'43''$

6. Determine the angles supplementary to
   (a) $78°$ (b) $15°$ (c) $169°41'11''$

7. Find the values of angle θ in diagrams (a) to (i) of Fig. 20.10.

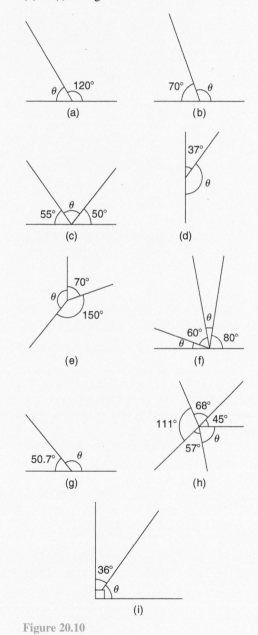

Figure 20.10

8. With reference to Fig. 20.11, what is the name given to the line *XY*? Give examples of each of the following.

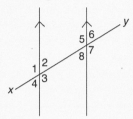

Figure 20.11

   (a) vertically opposite angles
   (b) supplementary angles
   (c) corresponding angles
   (d) alternate angles

9. In Fig. 20.12, find angle α.

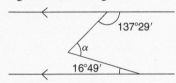

Figure 20.12

10. In Fig. 20.13, find angles a, b and c.

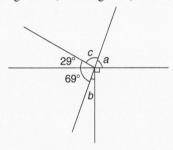

Figure 20.13

11. Find angle β in Fig. 20.14.

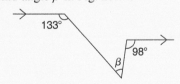

Figure 20.14

12. Convert $76°$ to radians, correct to 3 decimal places.

13. Convert $34°40'$ to radians, correct to 3 decimal places.

14. Convert 0.714 rad to degrees and minutes.

20.3 Triangles

A **triangle** is a figure enclosed by three straight lines. **The sum of the three angles of a triangle is equal to 180°**

20.3.1 Types of triangle

An **acute-angled triangle** is one in which all the angles are acute; i.e. all the angles are less than 90°. An example is shown in triangle *ABC* in Fig. 20.15(a).

A **right-angled triangle** is one which contains a right angle; i.e. one in which one of the angles is 90°. An example is shown in triangle *DEF* in Fig. 20.15(b).

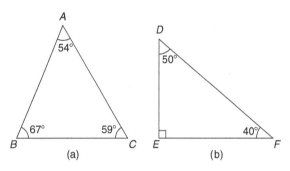

Figure 20.15

An **obtuse-angled triangle** is one which contains an obtuse angle; i.e. one angle which lies between 90° and 180°. An example is shown in triangle *PQR* in Fig. 20.16(a).

An **equilateral triangle** is one in which all the sides and all the angles are equal; i.e. each is 60°. An example is shown in triangle *ABC* in Fig. 20.16(b).

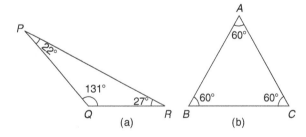

Figure 20.16

An **isosceles triangle** is one in which two angles and two sides are equal. An example is shown in triangle *EFG* in Fig. 20.17(a).

A **scalene triangle** is one with unequal angles and therefore unequal sides. An example of an acute-angled scalene triangle is shown in triangle *ABC* in Fig. 20.17(b).

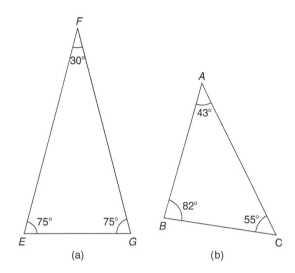

Figure 20.17

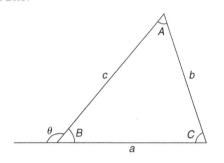

Figure 20.18

With reference to Fig. 20.18,

(a) Angles A, B and C are called **interior angles** of the triangle.

(b) Angle θ is called an **exterior angle** of the triangle and is equal to the sum of the two opposite interior angles; i.e. $\theta = A + C$.

(c) $a + b + c$ is called the **perimeter** of the triangle.

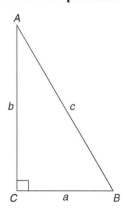

Figure 20.19

A right-angled triangle ABC is shown in Fig. 20.19. The point of intersection of two lines is called a vertex (plural **vertices**); the three vertices of the triangle are labelled as A, B and C, respectively. The right angle is angle C. The side opposite the right angle is given the special name of the **hypotenuse**. The hypotenuse, length AB in Fig. 20.19, is always the longest side of a right-angled triangle. With reference to angle B, AC is the **opposite** side and BC is called the **adjacent** side. With reference to angle A, BC is the **opposite** side and AC is the **adjacent** side.

Often sides of a triangle are labelled with lower case letters, a being the side opposite angle A, b being the side opposite angle B and c being the side opposite angle C. So, in the triangle ABC, length $AB = c$, length $BC = a$ and length $AC = b$. Thus, c is the hypotenuse in the triangle ABC.

$\angle$ is the symbol used for 'angle'. For example, in the triangle shown, $\angle C = 90°$. Another way of indicating an angle is to use all three letters. For example, $\angle ABC$ actually means $\angle B$; i.e. we take the middle letter as the angle. Similarly, $\angle BAC$ means $\angle A$ and $\angle ACB$ means $\angle C$.

Here are some worked examples to help us understand more about triangles.

Problem 18. Name the types of triangle shown in Fig. 20.20

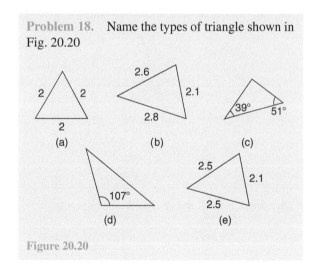

Figure 20.20

(a) **Equilateral triangle** (since all three sides are equal).

(b) **Acute-angled scalene triangle** (since all the angles are less than $90°$).

(c) **Right-angled triangle** ($39° + 51° = 90°$; hence, the third angle must be $90°$, since there are $180°$ in a triangle).

(d) **Obtuse-angled scalene triangle** (since one of the angles lies between 90° and 180°).

(e) **Isosceles triangle** (since two sides are equal).

> **Problem 19.** In the triangle *ABC* shown in Fig. 20.21, with reference to angle θ, which side is the adjacent?
>
>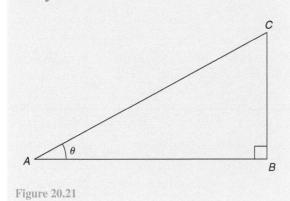
>
> Figure 20.21

The triangle is right-angled; thus, side *AC* is the hypotenuse. With reference to angle θ, the opposite side is *BC*. The remaining side, **AB**, **is the adjacent side**.

> **Problem 20.** In the triangle shown in Fig. 20.22, determine angle θ
>
>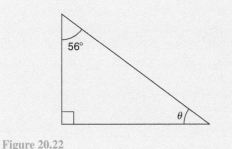
>
> Figure 20.22

The sum of the three angles of a triangle is equal to 180°

The triangle is right-angled. Hence,

$$90° + 56° + \angle\theta = 180°$$

from which, $\angle\theta = 180° - 90° - 56° = \mathbf{34°}$

> **Problem 21.** Determine the value of θ and α in Fig. 20.23

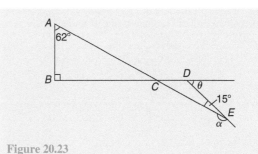

Figure 20.23

In triangle *ABC*, $\angle A + \angle B + \angle C = 180°$ (the angles in a triangle add up to 180°)

Hence, $\angle C = 180° - 90° - 62° = 28°$. Thus, $\angle DCE = 28°$ (vertically opposite angles).

$\theta = \angle DCE + \angle DEC$ (the exterior angle of a triangle is equal to the sum of the two opposite interior angles).

Hence, $\angle\theta = 28° + 15° = \mathbf{43°}$

$\angle\alpha$ and $\angle DEC$ are supplementary; thus,

$\alpha = 180° - 15° = \mathbf{165°}$

> **Problem 22.** *ABC* is an isosceles triangle in which the unequal angle *BAC* is 56°. *AB* is extended to *D* as shown in Fig. 20.24. Find, for the triangle, $\angle ABC$ and $\angle ACB$. Also, calculate $\angle DBC$
>
>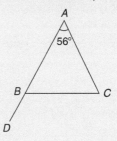
>
> Figure 20.24

Since triangle *ABC* is isosceles, two sides – i.e. *AB* and *AC* – are equal and two angles – i.e. $\angle ABC$ and $\angle ACB$ – are equal.

The sum of the three angles of a triangle is equal to 180°. Hence, $\angle ABC + \angle ACB = 180° - 56° = 124°$

Since $\angle ABC = \angle ACB$ then

$$\angle\mathbf{ABC} = \angle\mathbf{ACB} = \frac{124°}{2} = \mathbf{62°}$$

An angle of 180° lies on a straight line; hence, $\angle ABC + \angle DBC = 180°$ from which,

$\angle\mathbf{DBC} = 180° - \angle ABC = 180° - 62° = \mathbf{118°}$

Alternatively, $\angle DBC = \angle A + \angle C$ (exterior angle equals sum of two interior opposite angles),

i.e. $\angle\mathbf{DBC} = 56° + 62° = \mathbf{118°}$

Problem 23. Find angles a, b, c, d and e in Fig. 20.25

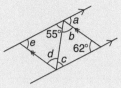

Figure 20.25

$a = 62°$ and $c = 55°$ (alternate angles between parallel lines).
$55° + b + 62° = 180°$ (angles in a triangle add up to $180°$); hence, $b = 180° - 55° - 62° = 63°$
$b = d = 63°$ (alternate angles between parallel lines).
$e + 55° + 63° = 180°$ (angles in a triangle add up to $180°$); hence, $e = 180° - 55° - 63° = 62°$
Check: $e = a = 62°$ (corresponding angles between parallel lines).

Now try the following Practice Exercise

Practice Exercise 78   Triangles (answers on page 431)

1. Name the types of triangle shown in diagrams (a) to (f) in Fig. 20.26.

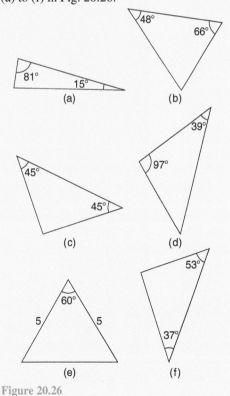

Figure 20.26

2. Find the angles a to f in Fig. 20.27.

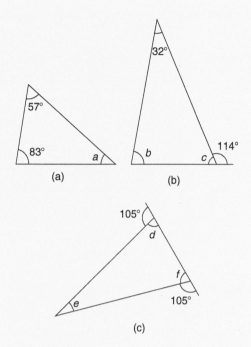

Figure 20.27

3. In the triangle DEF of Fig. 20.28, which side is the hypotenuse? With reference to angle D, which side is the adjacent?

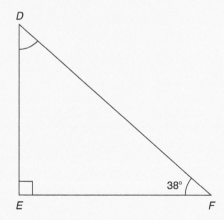

Figure 20.28

4. In triangle DEF of Fig. 20.28, determine angle D.

5. MNO is an isosceles triangle in which the unequal angle is $65°$ as shown in Fig. 20.29. Calculate angle θ.

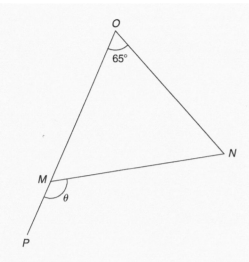

Figure 20.29

6. Determine $\angle\phi$ and $\angle x$ in Fig. 20.30.

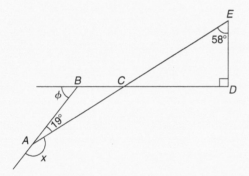

Figure 20.30

7. In Fig. 20.31(a) and (b), find angles w, x, y and z. What is the name given to the types of triangle shown in (a) and (b)?

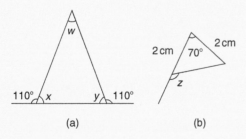

Figure 20.31

8. Find the values of angles a to g in Fig. 20.32(a) and (b).

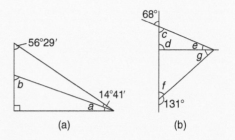

Figure 20.32

9. Find the unknown angles a to k in Fig. 20.33.

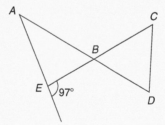

Figure 20.33

10. Triangle ABC has a right angle at B and $\angle BAC$ is $34°$. BC is produced to D. If the bisectors of $\angle ABC$ and $\angle ACD$ meet at E, determine $\angle BEC$.

11. If in Fig. 20.34 triangle BCD is equilateral, find the interior angles of triangle ABE.

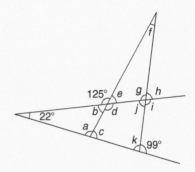

Figure 20.34

20.4 Congruent triangles

Two triangles are said to be **congruent** if they are equal in all respects; i.e. three angles and three sides in one triangle are equal to three angles and three sides in the other triangle. Two triangles are congruent if

(a) the three sides of one are equal to the three sides of the other (SSS),

(b) two sides of one are equal to two sides of the other and the angles included by these sides are equal (SAS),

(c) two angles of the one are equal to two angles of the other and any side of the first is equal to the corresponding side of the other (ASA), or

(d) their hypotenuses are equal and one other side of one is equal to the corresponding side of the other (RHS).

Problem 24. State which of the pairs of triangles shown in Fig. 20.35 are congruent and name their sequence

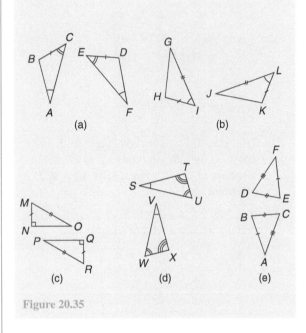

Figure 20.35

(a) Congruent *ABC*, *FDE* (angle, side, angle; i.e. ASA).

(b) Congruent *GIH*, *JLK* (side, angle, side; i.e. SAS).

(c) Congruent *MNO*, *RQP* (right angle, hypotenuse, side; i.e. RHS).

(d) Not necessarily congruent. It is not indicated that any side coincides.

(e) Congruent *ABC*, *FED* (side, side, side; i.e. SSS).

Problem 25. In Fig. 20.36, triangle *PQR* is isosceles with *Z*, the mid-point of *PQ*. Prove that triangles *PXZ* and *QYZ* are congruent and that triangles *RXZ* and *RYZ* are congruent. Determine the values of angles *RPZ* and *RXZ*

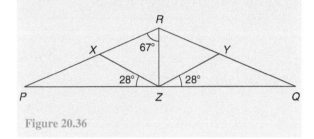

Figure 20.36

Since triangle *PQR* is isosceles, $PR = RQ$ and thus $\angle QPR = \angle RQP$.

$\angle RXZ = \angle QPR + 28°$ and $\angle RYZ = \angle RQP + 28°$ (exterior angles of a triangle equal the sum of the two interior opposite angles). Hence, $\angle RXZ = \angle RYZ$.

$\angle PXZ = 180° - \angle RXZ$ and $\angle QYZ = 180° - \angle RYZ$. Thus, $\angle PXZ = \angle QYZ$.

Triangles *PXZ* and *QYZ* are congruent since $\angle XPZ = \angle YQZ, PZ = ZQ$ and $\angle XZP = \angle YZQ$ (ASA). Hence, $XZ = YZ$.

Triangles *PRZ* and *QRZ* are congruent since $PR = RQ$, $\angle RPZ = \angle RQZ$ and $PZ = ZQ$ (SAS). Hence, $\angle RZX = \angle RZY$.

Triangles *RXZ* and *RYZ* are congruent since $\angle RXZ = \angle RYZ$, $XZ = YZ$ and $\angle RZX = \angle RZY$ (ASA).

$\angle QRZ = 67°$ and thus $\angle PRQ = 67° + 67° = 134°$

Hence, $\angle RPZ = \angle RQZ = \dfrac{180° - 134°}{2} = \mathbf{23°}$

$\angle RXZ = 23° + 28° = \mathbf{51°}$ (external angle of a triangle equals the sum of the two interior opposite angles).

Now try the following Practice Exercise

Practice Exercise 79   Congruent triangles (answers on page 431)

1. State which of the pairs of triangles in Fig. 20.37 are congruent and name their sequences.

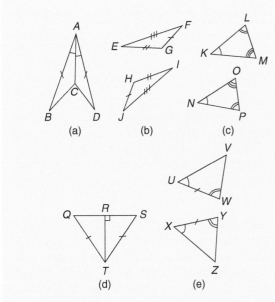

Figure 20.37

2. In a triangle ABC, $AB = BC$ and D and E are points on AB and BC, respectively, such that $AD = CE$. Show that triangles AEB and CDB are congruent.

20.5 Similar triangles

Two triangles are said to be **similar** if the angles of one triangle are equal to the angles of the other triangle. With reference to Fig. 20.38, triangles ABC and PQR are similar and the corresponding sides are in proportion to each other,

i.e.
$$\frac{p}{a} = \frac{q}{b} = \frac{r}{c}$$

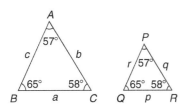

Figure 20.38

Problem 26. In Fig. 20.39, find the length of side a

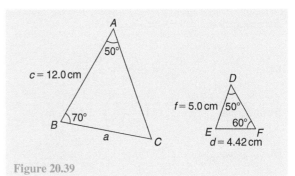

Figure 20.39

In triangle ABC, $50° + 70° + \angle C = 180°$, from which $\angle C = 60°$

In triangle DEF, $\angle E = 180° - 50° - 60° = 70°$ Hence, triangles ABC and DEF are similar, since their angles are the same. Since corresponding sides are in proportion to each other,

$$\frac{a}{d} = \frac{c}{f} \quad \text{i.e.} \quad \frac{a}{4.42} = \frac{12.0}{5.0}$$

Hence, **side,** $a = \dfrac{12.0}{5.0}(4.42) = \mathbf{10.61\,cm}$.

Problem 27. In Fig. 20.40, find the dimensions marked r and p

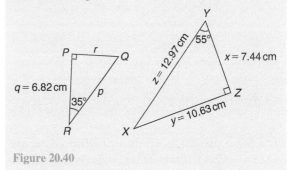

Figure 20.40

In triangle PQR, $\angle Q = 180° - 90° - 35° = 55°$
In triangle XYZ, $\angle X = 180° - 90° - 55° = 35°$
Hence, triangles PQR and ZYX are similar since their angles are the same. The triangles may he redrawn as shown in Fig. 20.41.

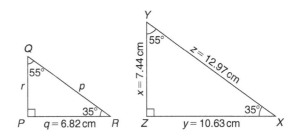

Figure 20.41

By proportion: $\dfrac{p}{z} = \dfrac{r}{x} = \dfrac{q}{y}$

i.e. $\dfrac{p}{12.97} = \dfrac{r}{7.44} = \dfrac{6.82}{10.63}$

from which, $r = 7.44\left(\dfrac{6.82}{10.63}\right) = \mathbf{4.77\,cm}$

By proportion: $\dfrac{p}{z} = \dfrac{q}{y}$ i.e. $\dfrac{p}{12.97} = \dfrac{6.82}{10.63}$

Hence, $p = 12.97\left(\dfrac{6.82}{10.63}\right) = \mathbf{8.32\,cm}$

Problem 28. In Fig. 20.42, show that triangles *CBD* and *CAE* are similar and hence find the length of *CD* and *BD*

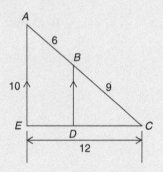

Figure 20.42

Since *BD* is parallel to *AE* then $\angle CBD = \angle CAE$ and $\angle CDB = \angle CEA$ (corresponding angles between parallel lines). Also, $\angle C$ is common to triangles *CBD* and *CAE*.

Since the angles in triangle *CBD* are the same as in triangle *CAE*, the triangles are similar. Hence,

by proportion: $\dfrac{CB}{CA} = \dfrac{CD}{CE}\left(= \dfrac{BD}{AE}\right)$

i.e. $\dfrac{9}{6+9} = \dfrac{CD}{12}$, from which

$$CD = 12\left(\dfrac{9}{15}\right) = \mathbf{7.2\,cm}$$

Also, $\dfrac{9}{15} = \dfrac{BD}{10}$, from which

$$BD = 10\left(\dfrac{9}{15}\right) = \mathbf{6\,cm}$$

Problem 29. A rectangular shed 2 m wide and 3 m high stands against a perpendicular building of height 5.5 m. A ladder is used to gain access to the roof of the building. Determine the minimum distance between the bottom of the ladder and the shed

A side view is shown in Fig. 20.43, where *AF* is the minimum length of the ladder. Since *BD* and *CF* are parallel, $\angle ADB = \angle DFE$ (corresponding angles between parallel lines). Hence, triangles *BAD* and *EDF* are similar since their angles are the same.

$$AB = AC - BC = AC - DE = 5.5 - 3 = 2.5\,m$$

By proportion: $\dfrac{AB}{DE} = \dfrac{BD}{EF}$ i.e. $\dfrac{2.5}{3} = \dfrac{2}{EF}$

Hence, $EF = 2\left(\dfrac{3}{2.5}\right) = \mathbf{2.4\,m = minimum\ distance}$

from bottom of ladder to the shed.

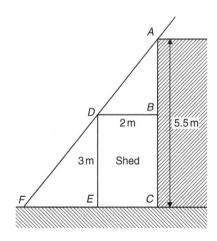

Figure 20.43

Now try the following Practice Exercise

1. In Fig. 20.44, find the lengths x and y.

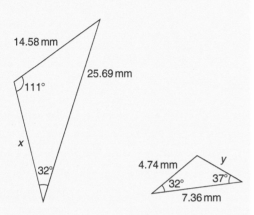

Figure 20.44

2. *PQR* is an equilateral triangle of side 4 cm. When *PQ* and *PR* are produced to *S* and *T*, respectively, *ST* is found to be parallel with *QR*. If *PS* is 9 cm, find the length of *ST*. X is a point on *ST* between *S* and *T* such that the line *PX* is the bisector of $\angle SPT$. Find the length of *PX*.

3. In Fig. 20.45, find
    (a) the length of *BC* when *AB* = 6 cm, *DE* = 8 cm and *DC* = 3 cm,
    (b) the length of *DE* when *EC* = 2 cm, *AC* = 5 cm and *AB* = 10 cm.

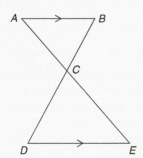

Figure 20.45

4. In Fig. 20.46, *AF* = 8 m, *AB* = 5 m and *BC* = 3 m. Find the length of *BD*.

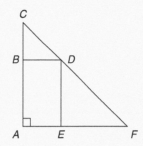

Figure 20.46

20.6 Construction of triangles

To construct any triangle, the following drawing instruments are needed:

(a) ruler and/or straight edge

(b) compass

(c) protractor

(d) pencil.

Here are some worked problems to demonstrate triangle construction.

> **Problem 30.**   Construct a triangle whose sides are 6 cm, 5 cm and 3 cm

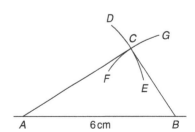

Figure 20.47

With reference to Fig. 20.47:

(i) Draw a straight line of any length and, with a pair of compasses, mark out 6 cm length and label it *AB*.

(ii) Set compass to 5 cm and with centre at *A* describe arc *DE*.

(iii) Set compass to 3 cm and with centre at *B* describe arc *FG*.

(iv) The intersection of the two curves at *C* is the vertex of the required triangle. Join *AC* and *BC* by straight lines.

It may be proved by measurement that the ratio of the angles of a triangle is not equal to the ratio of the sides (i.e. in this problem, the angle opposite the 3 cm side is not equal to half the angle opposite the 6 cm side).

> **Problem 31.** Construct a triangle *ABC* such that $a = 6$ cm, $b = 3$ cm and $\angle C = 60°$

With reference to Fig. 20.48:

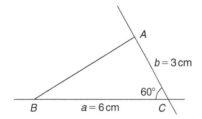

Figure 20.48

(i) Draw a line *BC*, 6 cm long.

(ii) Using a protractor centred at *C*, make an angle of 60° to *BC*.

(iii) From *C* measure a length of 3 cm and label *A*.

(iv) Join *B* to *A* by a straight line.

> **Problem 32.** Construct a triangle *PQR* given that $QR = 5$ cm, $\angle Q = 70°$ and $\angle R = 44°$

With reference to Fig. 20.49:

(i) Draw a straight line 5 cm long and label it *QR*.

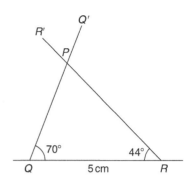

Figure 20.49

(ii) Use a protractor centred at *Q* and make an angle of 70°. Draw *QQ'*.

(iii) Use a protractor centred at *R* and make an angle of 44°. Draw *RR'*.

(iv) The intersection of *QQ'* and *RR'* forms the vertex *P* of the triangle.

> **Problem 33.** Construct a triangle *XYZ* given that $XY = 5$ cm, the hypotenuse $YZ = 6.5$ cm and $\angle X = 90°$

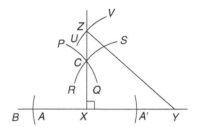

Figure 20.50

With reference to Fig. 20.50:

(i) Draw a straight line 5 cm long and label it *XY*.

(ii) Produce *XY* any distance to *B*. With compass centred at *X* make an arc at *A* and *A'*. (The length *XA* and *XA'* is arbitrary.) With compass centred at *A* draw the arc *PQ*. With the same compass setting and centred at *A'*, draw the arc *RS*. Join the intersection of the arcs, *C* to *X*, and a right angle to *XY* is produced at *X*. (Alternatively, a protractor can be used to construct a 90° angle.)

(iii) The hypotenuse is always opposite the right angle. Thus, *YZ* is opposite $\angle X$. Using a compass centred at *Y* and set to 6.5 cm, describe the arc *UV*.

(iv) The intersection of the arc UV with XC produced, forms the vertex Z of the required triangle. Join YZ with a straight line.

Now try the following Practice Exercise

Practice Exercise 81   Construction of triangles (answers on page 431)

In the following, construct the triangles ABC for the given sides/angles.

1. $a = 8$ cm, $b = 6$ cm and $c = 5$ cm.

2. $a = 40$ mm, $b = 60$ mm and $C = 60°$.

3. $a = 6$ cm, $C = 45°$ and $B = 75°$.

4. $c = 4$ cm, $A = 130°$ and $C = 15°$.

5. $a = 90$ mm, $B = 90°$, hypotenuse $= 105$ mm.

For fully worked solutions to each of the problems in Practice Exercises 76 to 81 in this chapter, go to the website:
www.routledge.com/cw/bird

Chapter 21

Introduction to trigonometry

> **Why it is important to understand: Introduction to trigonometry**
>
> There are an enormous number of uses of trigonometry and trigonometric functions. Fields that use trigonometry or trigonometric functions include astronomy (especially for locating apparent positions of celestial objects, in which spherical trigonometry is essential) and hence navigation (on the oceans, in aircraft, and in space), music theory, acoustics, optics, analysis of financial markets, electronics, probability theory, statistics, biology, medical imaging (CAT scans and ultrasound), pharmacy, chemistry, number theory (and hence cryptology), seismology, meteorology, oceanography, many physical sciences, land surveying and geodesy (a branch of earth sciences), architecture, phonetics, economics, electrical engineering, mechanical engineering, civil engineering, computer graphics, cartography, crystallography and game development. It is clear that a good knowledge of trigonometry is essential in many fields of engineering.

At the end of this chapter, you should be able to:

- state the theorem of Pythagoras and use it to find the unknown side of a right-angled triangle
- define sine, cosine and tangent of an angle in a right-angled triangle
- evaluate trigonometric ratios of angles
- solve right-angled triangles
- understand angles of elevation and depression

21.1 Introduction

Trigonometry is a subject that involves the measurement of sides and angles of triangles and their relationship to each other.

The theorem of Pythagoras and trigonometric ratios are used with right-angled triangles only. However, there are many practical examples in engineering where knowledge of right-angled triangles is very important.

In this chapter, three trigonometric ratios – i.e. sine, cosine and tangent – are defined and then evaluated using a calculator. Finally, solving right-angled triangle problems using Pythagoras and trigonometric ratios is demonstrated, together with some practical examples involving angles of elevation and depression.

21.2 The theorem of Pythagoras

The **theorem of Pythagoras*** states:

In any right-angled triangle, the square of the hypotenuse is equal to the sum of the squares of the other two sides.

---

*Who was Pythagoras? To find out more go to www.routledge.com/cw/bird

In the right-angled triangle ABC shown in Fig. 21.1, this means

$$b^2 = a^2 + c^2 \qquad (1)$$

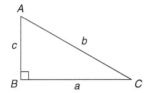

Figure 21.1

If the lengths of any two sides of a right-angled triangle are known, the length of the third side may be calculated by Pythagoras' theorem.

From equation (1): $b = \sqrt{a^2 + c^2}$

Transposing equation (1) for a gives $a^2 = b^2 - c^2$, from which $a = \sqrt{b^2 - c^2}$
Transposing equation (1) for c gives $c^2 = b^2 - a^2$, from which $c = \sqrt{b^2 - a^2}$
Here are some worked problems to demonstrate the theorem of Pythagoras.

Problem 1.   In Fig. 21.2, find the length of BC

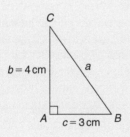

Figure 21.2

From Pythagoras, $a^2 = b^2 + c^2$

i.e. $a^2 = 4^2 + 3^2$

                    $= 16 + 9 = 25$

Hence, $a = \sqrt{25} = 5\,\text{cm}.$

$\sqrt{25} = \pm 5$ but in a practical example like this an answer of $a = -5\,\text{cm}$ has no meaning, so we take only the positive answer.

Thus $a = BC = 5\,\text{cm}.$

ABC is a 3, 4, 5 triangle. There are not many right-angled triangles which have integer values (i.e. whole numbers) for all three sides.

Problem 2.   In Fig. 21.3, find the length of EF

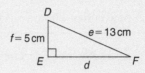

Figure 21.3

By Pythagoras' theorem, $e^2 = d^2 + f^2$

Hence, $13^2 = d^2 + 5^2$

                    $169 = d^2 + 25$

                    $d^2 = 169 - 25 = 144$

Thus, $d = \sqrt{144} = 12\,\text{cm}$

i.e. $d = EF = 12\,\text{cm}$

DEF is a 5, 12, 13 triangle, another right-angled triangle which has integer values for all three sides.

Problem 3.   Two aircraft leave an airfield at the same time. One travels due north at an average speed of 300 km/h and the other due west at an average speed of 220 km/h. Calculate their distance apart after 4 hours

After 4 hours, the first aircraft has travelled

$$4 \times 300 = 1200\,\text{km due north}$$

and the second aircraft has travelled

$$4 \times 220 = 880\,\text{km due west,}$$

as shown in Fig. 21.4. The distance apart after 4 hours = BC.

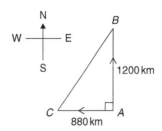

Figure 21.4

From Pythagoras' theorem,

$$BC^2 = 1200^2 + 880^2$$

$$= 1440000 + 774400 = 2214400$$

and $BC = \sqrt{2214400} = 1488\,\text{km}.$

Hence, distance apart after 4 hours = 1488 km.

Now try the following Practice Exercise

1. Find the length of side x in Fig. 21.5.

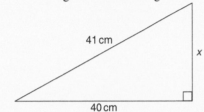

41 cm

x

40 cm

Figure 21.5

2. Find the length of side x in Fig. 21.6(a).

3. Find the length of side x in Fig. 21.6(b), correct to 3 significant figures.

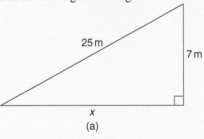

25 m

7 m

x

(a)

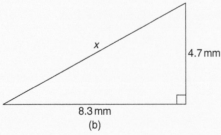

x

4.7 mm

8.3 mm

(b)

Figure 21.6

4. In a triangle ABC, $AB = 17\,\text{cm}$, $BC = 12\,\text{cm}$ and $\angle ABC = 90°$. Determine the length of AC, correct to 2 decimal places.

5. A tent peg is 4.0 m away from a 6.0 m high tent. What length of rope, correct to the nearest centimetre, runs from the top of the tent to the peg?

6. In a triangle ABC, $\angle B$ is a right angle, $AB = 6.92\,\text{cm}$ and $BC = 8.78\,\text{cm}$. Find the length of the hypotenuse.

7. In a triangle CDE, $D = 90°$, $CD = 14.83\,\text{mm}$ and $CE = 28.31\,\text{mm}$. Determine the length of DE.

8. Show that if a triangle has sides of 8, 15 and 17 cm it is right-angled.

9. Triangle PQR is isosceles, Q being a right angle. If the hypotenuse is 38.46 cm find (a) the lengths of sides PQ and QR and (b) the value of $\angle QPR$.

10. A man cycles 24 km due south and then 20 km due east. Another man, starting at the same time as the first man, cycles 32 km due east and then 7 km due south. Find the distance between the two men.

11. A ladder 3.5 m long is placed against a perpendicular wall with its foot 1.0 m from the wall. How far up the wall (to the nearest centimetre) does the ladder reach? If the foot of the ladder is now moved 30 cm further away from the wall, how far does the top of the ladder fall?

12. Two ships leave a port at the same time. One travels due west at 18.4 knots and the other due south at 27.6 knots. If 1 knot = 1 nautical mile per hour, calculate how far apart the two ships are after 4 hours.

13. Fig. 21.7 shows a bolt rounded off at one end. Determine the dimension h.

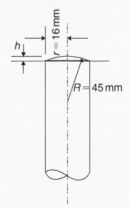

h

$r = 16$ mm

$R = 45$ mm

Figure 21.7

14. Fig. 21.8 shows a cross-section of a component that is to be made from a round bar. If the diameter of the bar is 74 mm, calculate the dimension x.

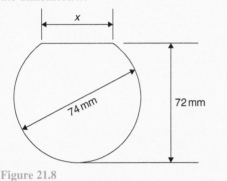

Figure 21.8

21.3 Sines, cosines and tangents

With reference to angle θ in the right-angled triangle ABC shown in Fig. 21.9,

$$\textbf{sine}\,\theta = \frac{\textbf{opposite side}}{\textbf{hypotenuse}}$$

'Sine' is abbreviated to 'sin', thus $\textbf{sin}\,\theta = \dfrac{BC}{AC}$

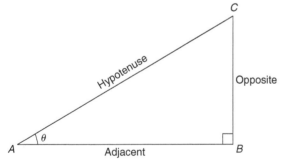

Figure 21.9

Also, $\textbf{cosine}\,\theta = \dfrac{\textbf{adjacent side}}{\textbf{hypotenuse}}$

'Cosine' is abbreviated to 'cos', thus $\textbf{cos}\,\theta = \dfrac{AB}{AC}$

Finally, $\textbf{tangent}\,\theta = \dfrac{\textbf{opposite side}}{\textbf{adjacent side}}$

'Tangent' is abbreviated to 'tan', thus $\textbf{tan}\,\theta = \dfrac{BC}{AB}$

These three trigonometric ratios only apply to right-angled triangles. Remembering these three equations is very important and the mnemonic '**SOH CAH TOA**' is one way of remembering them.

SOH indicates $\underline{\text{s}}\text{in} = \underline{\text{o}}\text{pposite} \div \underline{\text{h}}\text{ypotenuse}$

CAH indicates $\underline{\text{c}}\text{os} = \underline{\text{a}}\text{djacent} \div \underline{\text{h}}\text{ypotenuse}$

TOA indicates $\underline{\text{t}}\text{an} = \underline{\text{o}}\text{pposite} \div \underline{\text{a}}\text{djacent}$

Here are some worked problems to help familiarise ourselves with trigonometric ratios.

Problem 4. In triangle PQR shown in Fig. 21.10, determine $\sin\theta$, $\cos\theta$ and $\tan\theta$

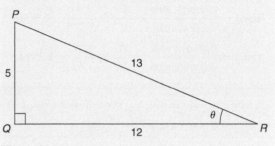

Figure 21.10

$$\textbf{sin}\,\theta = \frac{\text{opposite side}}{\text{hypotenuse}} = \frac{PQ}{PR} = \frac{5}{13} = \textbf{0.3846}$$

$$\textbf{cos}\,\theta = \frac{\text{adjacent side}}{\text{hypotenuse}} = \frac{QR}{PR} = \frac{12}{13} = \textbf{0.9231}$$

$$\textbf{tan}\,\theta = \frac{\text{opposite side}}{\text{adjacent side}} = \frac{PQ}{QR} = \frac{5}{12} = \textbf{0.4167}$$

Problem 5. In triangle ABC of Fig. 21.11, determine length AC, $\sin C$, $\cos C$, $\tan C$, $\sin A$, $\cos A$ and $\tan A$

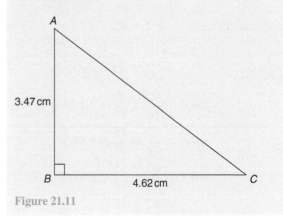

Figure 21.11

By Pythagoras, $\quad AC^2 = AB^2 + BC^2$

i.e. $\quad\quad\quad\quad AC^2 = 3.47^2 + 4.62^2$

from which $\quad AC = \sqrt{3.47^2 + 4.62^2} = \textbf{5.778 cm}$

$$\sin C = \frac{\text{opposite side}}{\text{hypotenuse}} = \frac{AB}{AC} = \frac{3.47}{5.778} = \mathbf{0.6006}$$

$$\cos C = \frac{\text{adjacent side}}{\text{hypotenuse}} = \frac{BC}{AC} = \frac{4.62}{5.778} = \mathbf{0.7996}$$

$$\tan C = \frac{\text{opposite side}}{\text{adjacent side}} = \frac{AB}{BC} = \frac{3.47}{4.62} = \mathbf{0.7511}$$

$$\sin A = \frac{\text{opposite side}}{\text{hypotenuse}} = \frac{BC}{AC} = \frac{4.62}{5.778} = \mathbf{0.7996}$$

$$\cos A = \frac{\text{adjacent side}}{\text{hypotenuse}} = \frac{AB}{AC} = \frac{3.47}{5.778} = \mathbf{0.6006}$$

$$\tan A = \frac{\text{opposite side}}{\text{adjacent side}} = \frac{BC}{AB} = \frac{4.62}{3.47} = \mathbf{1.3314}$$

Problem 6. If $\tan B = \dfrac{8}{15}$, determine the value of $\sin B, \cos B, \sin A$ and $\tan A$

A right-angled triangle ABC is shown in Fig. 21.12. If $\tan B = \dfrac{8}{15}$, then $AC = 8$ and $BC = 15$

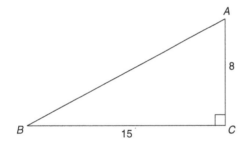

Figure 21.12

By Pythagoras, $AB^2 = AC^2 + BC^2$

i.e. $AB^2 = 8^2 + 15^2$

from which $AB = \sqrt{8^2 + 15^2} = 17$

$$\mathbf{\sin B} = \frac{AC}{AB} = \frac{8}{17} \text{ or } \mathbf{0.4706}$$

$$\mathbf{\cos B} = \frac{BC}{AB} = \frac{15}{17} \text{ or } \mathbf{0.8824}$$

$$\mathbf{\sin A} = \frac{BC}{AB} = \frac{15}{17} \text{ or } \mathbf{0.8824}$$

$$\mathbf{\tan A} = \frac{BC}{AC} = \frac{15}{8} \text{ or } \mathbf{1.8750}$$

Problem 7. Point A lies at co-ordinate $(2, 3)$ and point B at $(8, 7)$. Determine (a) the distance AB and (b) the gradient of the straight line AB

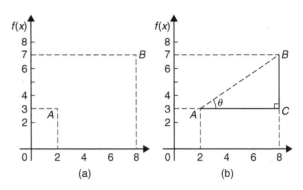

Figure 21.13

(a) Points A and B are shown in Fig. 21.13(a).

In Fig. 21.13(b), the horizontal and vertical lines AC and BC are constructed. Since ABC is a right-angled triangle, and $AC = (8 - 2) = 6$ and $BC = (7 - 3) = 4$, by Pythagoras' theorem,

$$AB^2 = AC^2 + BC^2 = 6^2 + 4^2$$

and $AB = \sqrt{6^2 + 4^2} = \sqrt{52}$

$$= \mathbf{7.211} \text{ correct to 3 decimal places.}$$

(b) The gradient of AB is given by $\tan \theta$, i.e.

$$\mathbf{gradient} = \tan \theta = \frac{BC}{AC} = \frac{4}{6} = \frac{2}{3}$$

Now try the following Practice Exercise

Practice Exercise 83   Trigonometric ratios (answers on page 431)

1. Sketch a triangle XYZ such that $\angle Y = 90°$, $XY = 9$ cm and $YZ = 40$ cm. Determine $\sin Z, \cos Z, \tan X$ and $\cos X$.

2. In triangle ABC shown in Fig. 21.14, find $\sin A, \cos A, \tan A, \sin B, \cos B$ and $\tan B$.

Figure 21.14

3. If $\cos A = \dfrac{15}{17}$, find $\sin A$ and $\tan A$, in fraction form.

4. If $\tan X = \dfrac{15}{112}$, find $\sin X$ and $\cos X$, in fraction form.

5. For the right-angled triangle shown in Fig. 21.15, find (a) $\sin\alpha$ (b) $\cos\theta$ (c) $\tan\theta$.

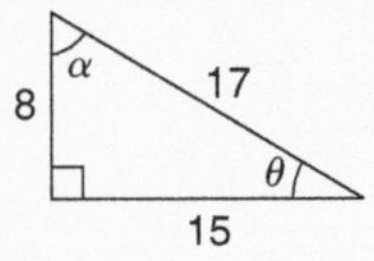

Figure 21.15

6. If $\tan\theta = \dfrac{7}{24}$, find (a) $\sin\theta$ and (b) $\cos\theta$ in fraction form.

7. Point P lies at co-ordinate $(-3, 1)$ and point Q at $(5, -4)$. Determine
   (a)   the distance PQ
   (b)   the gradient of the straight line PQ

21.4   Evaluating trigonometric ratios of acute angles

The easiest way to evaluate trigonometric ratios of any angle is to use a calculator. Use a calculator to check the following (each correct to 4 decimal places).

$$\sin 29^\circ = \mathbf{0.4848} \quad \sin 53.62^\circ = \mathbf{0.8051}$$

$$\cos 67^\circ = \mathbf{0.3907} \quad \cos 83.57^\circ = \mathbf{0.1120}$$

$$\tan 34^\circ = \mathbf{0.6745} \quad \tan 67.83^\circ = \mathbf{2.4541}$$

$$\sin 67^\circ 43' = \sin 67\frac{43^\circ}{60} = \sin 67.7166666\ldots^\circ = \mathbf{0.9253}$$

$$\cos 13^\circ 28' = \cos 13\frac{28^\circ}{60} = \cos 13.466666\ldots^\circ = \mathbf{0.9725}$$

$$\tan 56^\circ 54' = \tan 56\frac{54^\circ}{60} = \tan 56.90^\circ = \mathbf{1.5340}$$

If we know the value of a trigonometric ratio and need to find the angle we use the **inverse function** on our calculators. For example, using shift and sin on our calculator gives $\sin^{-1}($

If, for example, we know the sine of an angle is 0.5 then the value of the angle is given by

$$\sin^{-1} 0.5 = \mathbf{30^\circ} \text{ (Check that } \sin 30^\circ = 0.5)$$

Similarly, if

$$\cos\theta = 0.4371 \text{ then } \theta = \cos^{-1} 0.4371 = \mathbf{64.08^\circ}$$

and if

$$\tan A = 3.5984 \text{ then } A = \tan^{-1} 3.5984 = \mathbf{74.47^\circ}$$

each correct to 2 decimal places.
Use your calculator to check the following worked problems.

Problem 8.   Determine, correct to 4 decimal places, $\sin 43^\circ 39'$

$$\sin 43^\circ 39' = \sin 43\frac{39^\circ}{60} = \sin 43.65^\circ = \mathbf{0.6903}$$

This answer can be obtained using the **calculator** as follows:

1. Press sin   2. Enter 43   3. Press° ′′′
4. Enter 39   5. Press° ′′′   6. Press)
7. Press =   Answer = **0.6902512**…

Problem 9.   Determine, correct to 3 decimal places, $6\cos 62^\circ 12'$

$$6\cos 62^\circ 12' = 6\cos 62\frac{12^\circ}{60} = 6\cos 62.20^\circ = \mathbf{2.798}$$

This answer can be obtained using the **calculator** as follows:

1. Enter 6   2. Press cos   3. Enter 62
4. Press° ′′′   5. Enter 12   6. Press° ′′′
7. Press)   8. Press =   Answer = **2.798319**…

Problem 10.   Evaluate $\sin 1.481$, correct to 4 significant figures

$\sin 1.481$ means the sine of 1.481 **radians**. (If there is no degrees sign, i.e. °, then radians are assumed). Therefore a calculator needs to be on the radian function.
Hence, $\sin 1.481 = \mathbf{0.9960}$

Problem 11.   Evaluate $\cos(3\pi/5)$, correct to 4 significant figures

As in Problem 10, $3\pi/5$ is in radians.
Hence, $\mathbf{\cos(3\pi/5)} = \cos 1.884955\ldots = \mathbf{-0.3090}$
Since, from page 188, π radians $= 180^\circ$,
$$3\pi/5 \text{ rad} = \frac{3}{5} \times 180^\circ = 108^\circ$$
i.e. $3\pi/5$ rad $= 108^\circ$. Check with your calculator that $\mathbf{\cos 108^\circ = -0.3090}$

Problem 12. Evaluate tan 2.93, correct to 4 significant figures

Again, since there is no degrees sign, 2.93 means 2.93 radians.
Hence, **tan 2.93 = −0.2148**
It is important to know when to have your calculator on either degrees mode or radian mode. A lot of mistakes can arise from this if we are not careful.

Problem 13. Find the acute angle $\sin^{-1} 0.4128$ in degrees, correct to 2 decimal places

$\sin^{-1} 0.4128$ means 'the angle whose sine is 0.4128'. Using a calculator,

1. Press shift   2. Press sin   3. Enter 0.4128

4. Press)   5. Press =

The answer 24.380848... is displayed.
Hence, $\sin^{-1} 0.4128 = 24.38°$ correct to 2 decimal places.

Problem 14. Find the acute angle $\cos^{-1} 0.2437$ in degrees and minutes

$\cos^{-1} 0.2437$ means 'the angle whose cosine is 0.2437'. Using a calculator,

1. Press shift   2. Press cos   3. Enter 0.2437

4. Press)   5. Press =

The answer 75.894979... is displayed.
6. Press ° '" and 75° 53′ 41.93″ is displayed.
Hence, $\cos^{-1} 0.2437 = 75.89° = 77°54′$ correct to the nearest minute.

Problem 15. Find the acute angle $\tan^{-1} 7.4523$ in degrees and minutes

$\tan^{-1} 7.4523$ means 'the angle whose tangent is 7.4523'. Using a calculator,

1. Press shift   2. Press tan   3. Enter 7.4523

4. Press)   5. Press =

The answer 82.357318... is displayed.
6. Press ° '" and 82° 21′ 26.35″ is displayed.
Hence, $\tan^{-1} 7.4523 = 82.36° = 82°21′$ correct to the nearest minute.

Problem 16. In triangle *EFG* in Fig. 21.16, calculate angle *G*

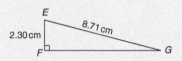

Figure 21.16

With reference to $\angle G$, the two sides of the triangle given are the opposite side *EF* and the hypotenuse *EG*; hence, sine is used,

i.e. $\sin G = \dfrac{2.30}{8.71} = 0.26406429\ldots$

from which, $G = \sin^{-1} 0.26406429\ldots$

i.e. $G = 15.311360°$

Hence, $\angle G = 15.31°$ or $15°19′$

Problem 17. Evaluate the following expression, correct to 3 significant figures

$$\frac{4.2 \tan 49°26′ - 3.7 \sin 66°1′}{7.1 \cos 29°34′}$$

By calculator:

$$\tan 49°26′ = \tan \left(49\frac{26}{60}\right)° = 1.1681,$$

$$\sin 66°1′ = 0.9137 \text{ and } \cos 29°34′ = 0.8698$$

Hence, $\dfrac{4.2 \tan 49°26′ - 3.7 \sin 66°1′}{7.1 \cos 29°34′}$

$$= \frac{(4.2 \times 1.1681) - (3.7 \times 0.9137)}{(7.1 \times 0.8698)}$$

$$= \frac{4.9060 - 3.3807}{6.1756} = \frac{1.5253}{6.1756}$$

$$= 0.2470 = \textbf{0.247},$$

correct to 3 significant figures.

Now try the following Practice Exercise

1. Determine, correct to 4 decimal places,
    $3 \sin 66°41'$

2. Determine, correct to 3 decimal places,
    $5 \cos 14°15'$

3. Determine, correct to 4 significant figures,
    $7 \tan 79°9'$

4. Determine

    (a) sine $\dfrac{2\pi}{3}$ (b) $\cos 1.681$ (c) $\tan 3.672$

5. Find the acute angle $\sin^{-1} 0.6734$ in degrees,
    correct to 2 decimal places.

6. Find the acute angle $\cos^{-1} 0.9648$ in degrees,
    correct to 2 decimal places.

7. Find the acute angle $\tan^{-1} 3.4385$ in degrees,
    correct to 2 decimal places.

8. Find the acute angle $\sin^{-1} 0.1381$ in degrees
    and minutes.

9. Find the acute angle $\cos^{-1} 0.8539$ in degrees
    and minutes.

10. Find the acute angle $\tan^{-1} 0.8971$ in degrees
    and minutes.

11. In the triangle shown in Fig. 21.17,
    determine angle θ, correct to 2 decimal
    places.

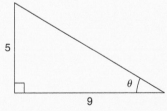

Figure 21.17

12. In the triangle shown in Fig. 21.18, determine
    angle θ in degrees and minutes.

Figure 21.18

13. Evaluate, correct to 4 decimal places,
    $$\frac{4.5 \cos 67°34' - \sin 90°}{2 \tan 45°}$$

14. Evaluate, correct to 4 significant figures,
    $$\frac{(3 \sin 37.83°)(2.5 \tan 57.48°)}{4.1 \cos 12.56°}$$

15. For the supported beam AB shown in Fig.
    21.19, determine (a) the angle the support-
    ing stay CD makes with the beam, i.e.
    θ, correct to the nearest degree, (b) the
    length of the stay, CD, correct to the nearest
    centimetre.

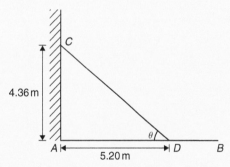

Figure 21.19

21.5 Solving right-angled triangles

'Solving a right-angled triangle' means 'finding the
unknown sides and angles'. This is achieved using

(a) the theorem of Pythagoras and/or

(b) trigonometric ratios.

Six pieces of information describe a triangle completely; i.e. three sides and three angles. As long as at least three facts are known, the other three can usually be calculated.

Here are some worked problems to demonstrate the solution of right-angled triangles.

Problem 18. In the triangle ABC shown in Fig. 21.20, find the lengths AC and AB

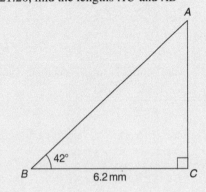

Figure 21.20

There is usually more than one way to solve such a triangle.

In triangle ABC,

$$\tan 42° = \frac{AC}{BC} = \frac{AC}{6.2}$$

(Remember SOH CAH TOA)

Transposing gives
$$AC = 6.2 \tan 42° = \mathbf{5.583\,mm}$$

$$\cos 42° = \frac{BC}{AB} = \frac{6.2}{AB}, \text{ from which}$$

$$AB = \frac{6.2}{\cos 42°} = \mathbf{8.343\,mm}$$

Alternatively, by Pythagoras, $AB^2 = AC^2 + BC^2$ from which $\mathbf{AB} = \sqrt{AC^2 + BC^2} = \sqrt{5.583^2 + 6.2^2}$
$$= \sqrt{69.609889} = \mathbf{8.343\,mm}.$$

Problem 19. Sketch a right-angled triangle ABC such that $B = 90°$, $AB = 5$ cm and $BC = 12$ cm. Determine the length of AC and hence evaluate $\sin A$, $\cos C$ and $\tan A$

Triangle ABC is shown in Fig. 21.21.

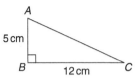

Figure 21.21

By Pythagoras' theorem, $AC = \sqrt{5^2 + 12^2} = 13$

By definition: $\sin A = \dfrac{\text{opposite side}}{\text{hypotenuse}} = \dfrac{12}{13}$ or **0.9231**

(Remember SOH CAH TOA)

$$\cos C = \frac{\text{adjacent side}}{\text{hypotenuse}} = \frac{12}{13} \text{ or } \mathbf{0.9231}$$

and $\tan A = \dfrac{\text{opposite side}}{\text{adjacent side}} = \dfrac{12}{5}$ or **2.400**

Problem 20. In triangle PQR shown in Fig. 21.22, find the lengths of PQ and PR

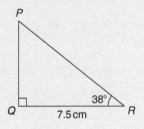

Figure 21.22

$\tan 38° = \dfrac{PQ}{QR} = \dfrac{PQ}{7.5}$, hence

$$PQ = 7.5 \tan 38° = 7.5(0.7813) = \mathbf{5.860\,cm}$$

$\cos 38° = \dfrac{QR}{PR} = \dfrac{7.5}{PR}$, hence

$$PR = \frac{7.5}{\cos 38°} = \frac{7.5}{0.7880} = \mathbf{9.518\,cm}$$

Check: using Pythagoras' theorem,
$(7.5)^2 + (5.860)^2 = 90.59 = (9.518)^2$

Problem 21. Solve the triangle ABC shown in Fig. 21.23

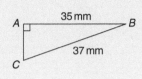

Figure 21.23

To 'solve the triangle ABC' means 'to find the length AC and angles B and C'

$$\sin C = \frac{35}{37} = 0.94595, \text{ hence}$$

$$C = \sin^{-1} 0.94595 = \mathbf{71.08°}$$

$B = 180° - 90° - 71.08° = \mathbf{18.92°}$ (since the angles in a triangle add up to $180°$)

$$\sin B = \frac{AC}{37}, \text{ hence}$$

$$AC = 37 \sin 18.92° = 37(0.3242) = \mathbf{12.0\,mm}$$

or, using Pythagoras' theorem, $37^2 = 35^2 + AC^2$, from which $AC = \sqrt{(37^2 - 35^2)} = \mathbf{12.0\,mm}$.

Problem 22. Solve triangle XYZ given $\angle X = 90°$, $\angle Y = 23°17'$ and $YZ = 20.0\,mm$

It is always advisable to make a reasonably accurate sketch so as to visualise the expected magnitudes of unknown sides and angles. Such a sketch is shown in Fig. 21.24.

$$\angle Z = 180° - 90° - 23°17' = \mathbf{66°43'}$$

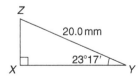

Figure 21.24

$$\sin 23°17' = \frac{XZ}{20.0}, \text{ hence } XZ = 20.0 \sin 23°17'$$

$$= 20.0(0.3953) = \mathbf{7.906\,mm}$$

$$\cos 23°17' = \frac{XY}{20.0}, \text{ hence } XY = 20.0 \cos 23°17'$$

$$= 20.0(0.9186) = \mathbf{18.37\,mm}$$

Check: using Pythagoras' theorem,
$(18.37)^2 + (7.906)^2 = 400.0 = (20.0)^2$

Now try the following Practice Exercise

Practice Exercise 85   Solving right-angled triangles (answers on page 431)

1. Calculate the dimensions shown as x in Figs 21.25(a) to (f), each correct to 4 significant figures.

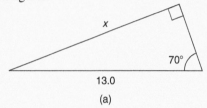

(a)

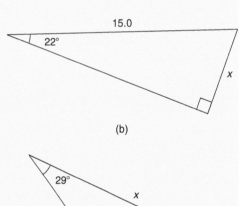

(b)

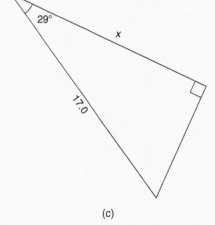

(c)

Figure 21.25

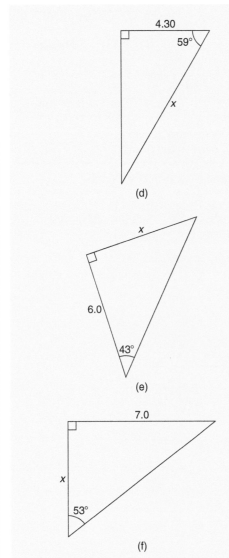

(d)

(e)

(f)

Figure 21.25

2. Find the unknown sides and angles in the right-angled triangles shown in Fig. 21.26. The dimensions shown are in centimetres.

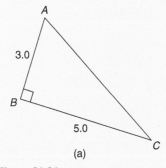

(a)

Figure 21.26

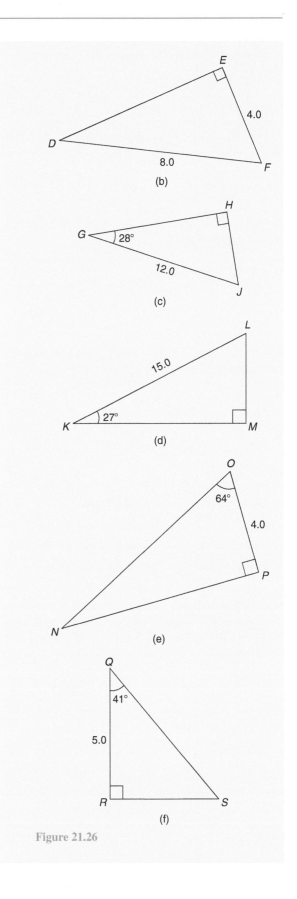

(b)

(c)

(d)

(e)

(f)

Figure 21.26

3. A ladder rests against the top of the perpendicular wall of a building and makes an angle of 73° with the ground. If the foot of the ladder is 2 m from the wall, calculate the height of the building.

4. Determine the length x in Fig. 21.27.

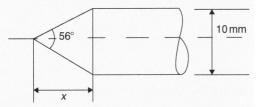

Figure 21.27

5. A symmetrical part of a bridge lattice is shown in Fig. 21.28. If $AB = 6$ m, angle $BAD = 56°$ and E is the mid-point of $ABCD$, determine the height h, correct to the nearest centimetre.

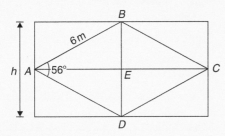

Figure 21.28

21.6 Angles of elevation and depression

If, in Fig. 21.29, BC represents horizontal ground and AB a vertical flagpole, the **angle of elevation** of the top of the flagpole, A, from the point C is the angle that the imaginary straight line AC must be raised (or elevated) from the horizontal CB; i.e. angle θ

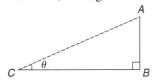

Figure 21.29

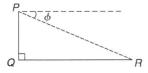

Figure 21.30

If, in Fig. 21.30, PQ represents a vertical cliff and R a ship at sea, the **angle of depression** of the ship from point P is the angle through which the imaginary straight line PR must be lowered (or depressed) from the horizontal to the ship; i.e. angle ϕ. (Note, $\angle PRQ$ is also ϕ − **alternate angles** between parallel lines.)

Problem 23. An electricity pylon stands on horizontal ground. At a point 80 m from the base of the pylon, the angle of elevation of the top of the pylon is 23°. Calculate the height of the pylon to the nearest metre

Fig. 21.31 shows the pylon AB and the angle of elevation of A from point C is 23°

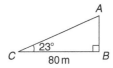

Figure 21.31

$$\tan 23° = \frac{AB}{BC} = \frac{AB}{80}$$

Hence, height of pylon $AB = 80 \tan 23°$
$$= 80(0.4245) = 33.96 \text{ m}$$
$$= \textbf{34 m to the nearest metre}.$$

Problem 24. A surveyor measures the angle of elevation of the top of a perpendicular building as 19°. He moves 120 m nearer to the building and finds the angle of elevation is now 47°. Determine the height of the building

The building PQ and the angles of elevation are shown in Fig. 21.32.

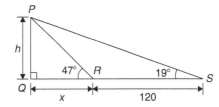

Figure 21.32

In triangle PQS, $\tan 19° = \dfrac{h}{x + 120}$

Hence, $h = \tan 19°(x + 120)$

i.e. $h = 0.3443(x + 120)$ (1)

In triangle PQR, $\tan 47° = \dfrac{h}{x}$

Hence, $h = \tan 47°(x)$ i.e. $h = 1.0724x$ (2)

Equating equations (1) and (2) gives

$$0.3443(x + 120) = 1.0724x$$
$$0.3443x + (0.3443)(120) = 1.0724x$$
$$(0.3443)(120) = (1.0724 - 0.3443)x$$
$$41.316 = 0.7281x$$
$$x = \dfrac{41.316}{0.7281} = 56.74 \,\text{m}$$

From equation (2), **height of building,**
$h = 1.0724x = 1.0724(56.74) = $ **60.85 m.**

> **Problem 25.** The angle of depression of a ship viewed at a particular instant from the top of a 75 m vertical cliff is 30°. Find the distance of the ship from the base of the cliff at this instant. The ship is sailing away from the cliff at constant speed and 1 minute later its angle of depression from the top of the cliff is 20°. Determine the speed of the ship in km/h

Fig. 21.33 shows the cliff AB, the initial position of the ship at C and the final position at D. Since the angle of depression is initially 30°, $\angle ACB = 30°$ (alternate angles between parallel lines).

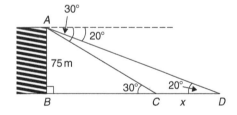

Figure 21.33

$\tan 30° = \dfrac{AB}{BC} = \dfrac{75}{BC}$ hence,

$$BC = \dfrac{75}{\tan 30°}$$

$$= \textbf{129.9 m} = \textbf{initial position}$$
$$\textbf{of ship from base of cliff}$$

In triangle ABD,

$$\tan 20° = \dfrac{AB}{BD} = \dfrac{75}{BC + CD} = \dfrac{75}{129.9 + x}$$

Hence, $\qquad 129.9 + x = \dfrac{75}{\tan 20°} = 206.06\,\text{m}$

from which $\qquad x = 206.06 - 129.9 = 76.16\,\text{m}$

Thus, the ship sails 76.16 m in 1 minute; i.e. 60 s,

Hence, **speed of ship** $= \dfrac{\text{distance}}{\text{time}} = \dfrac{76.16}{60}\,\text{m/s}$

$$= \dfrac{76.16 \times 60 \times 60}{60 \times 1000}\,\text{km/h} = \textbf{4.57 km/h.}$$

Now try the following Practice Exercise

Practice Exercise 86   **Angles of elevation and depression (answers on page 432)**

1. A vertical tower stands on level ground. At a point 105 m from the foot of the tower the angle of elevation of the top is 19°. Find the height of the tower.

2. If the angle of elevation of the top of a vertical 30 m high aerial is 32°, how far is it to the aerial?

3. From the top of a vertical cliff 90.0 m high the angle of depression of a boat is 19°50′. Determine the distance of the boat from the cliff.

4. From the top of a vertical cliff 80.0 m high the angles of depression of two buoys lying due west of the cliff are 23° and 15°, respectively. How far apart are the buoys?

5. From a point on horizontal ground a surveyor measures the angle of elevation of the top of a flagpole as 18°40′. He moves 50 m nearer to the flagpole and measures the angle of elevation as 26°22′. Determine the height of the flagpole.

6. A flagpole stands on the edge of the top of a building. At a point 200 m from the building the angles of elevation of the top and bottom of the pole are 32° and 30° respectively. Calculate the height of the flagpole.

7. From a ship at sea, the angles of elevation of the top and bottom of a vertical lighthouse standing on the edge of a vertical cliff are 31° and 26°, respectively. If the lighthouse is 25.0 m high, calculate the height of the cliff.

8. From a window 4.2 m above horizontal ground the angle of depression of the foot of a building across the road is 24° and the angle of elevation of the top of the building is 34°. Determine, correct to the nearest centimetre, the width of the road and the height of the building.

9. The elevation of a tower from two points, one due west of the tower and the other due east of it are 20° and 24°, respectively, and the two points of observation are 300 m apart. Find the height of the tower to the nearest metre.

For fully worked solutions to each of the problems in Practice Exercises 82 to 86 in this chapter, go to the website:
www.routledge.com/cw/bird

Revision Test 8: Angles, triangles and trigonometry

This assignment covers the material contained in Chapters 20 and 21. *The marks available are shown in brackets at the end of each question.*

1. Determine $38°48' + 17°23'$ (2)

2. Determine $47°43'12'' - 58°35'53'' + 26°17'29''$ (3)

3. Change $42.683°$ to degrees and minutes. (2)

4. Convert $77°42'34''$ to degrees correct to 3 decimal places. (3)

5. Determine angle θ in Fig. RT8.1. (3)

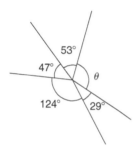

Figure RT8.1

6. Determine angle θ in the triangle in Fig. RT8.2. (2)

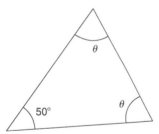

Figure RT8.2

7. Determine angle θ in the triangle in Fig. RT8.3. (2)

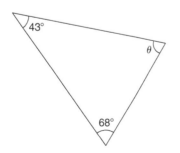

Figure RT8.3

8. In Fig. RT8.4, if triangle ABC is equilateral, determine $\angle CDE$. (3)

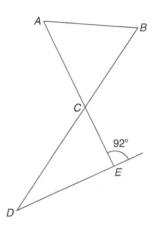

Figure RT8.4

9. Find angle J in Fig. RT8.5. (2)

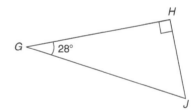

Figure RT8.5

10. Determine angle θ in Fig. RT8.6. (3)

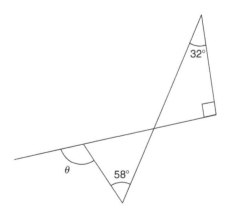

Figure RT8.6

11. State the angle (a) supplementary to $49°$ (b) complementary to $49°$ (2)

12. In Fig. RT8.7, determine angles x, y and z. (3)

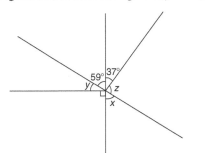

Figure RT8.7

13. In Fig. RT8.8, determine angles a to e. (5)

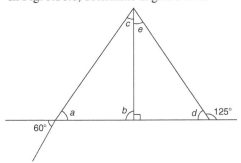

Figure RT8.8

14. In Fig. RT8.9, determine the length of AC. (4)

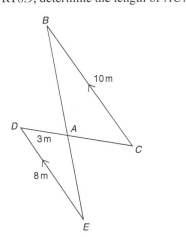

Figure RT8.9

15. In triangle JKL in Fig. RT8.10, find
    (a) the length KJ correct to 3 significant figures.
    (b) $\sin L$ and $\tan K$, each correct to 3 decimal places. (4)

16. Two ships leave a port at the same time. Ship X travels due west at 30 km/h and ship Y travels due

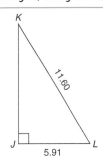

Figure RT8.10

north. After 4 hours the two ships are 130 km apart. Calculate the velocity of ship Y. (4)

17. If $\sin A = \dfrac{12}{37}$, find $\tan A$ in fraction form. (3)

18. Evaluate $5 \tan 62°11'$ correct to 3 significant figures. (2)

19. Determine the acute angle $\cos^{-1} 0.3649$ in degrees and minutes. (2)

20. In triangle PQR in Fig. RT8.11, find angle P in decimal form, correct to 2 decimal places. (2)

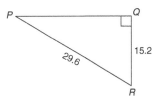

Figure RT8.11

21. Evaluate, correct to 3 significant figures, $3 \tan 81.27° - 5 \cos 7.32° - 6 \sin 54.81°$ (2)

22. In triangle ABC in Fig. RT8.12, find lengths AB and AC, correct to 2 decimal places. (4)

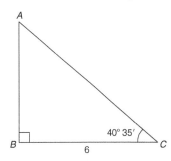

Figure RT8.12

23. From a point P, the angle of elevation of a 40 m high electricity pylon is $20°$. How far is point P from the base of the pylon, correct to the nearest metre? (3)

For lecturers/instructors/teachers, fully worked solutions to each of the problems in Revision Test 8, together with a full marking scheme, are available at the website:

www.routledge.com/cw/bird

Chapter 22

Trigonometric waveforms

Why it is important to understand: Trigonometric waveforms

Trigonometric graphs are commonly used in all areas of science and engineering for modelling many different natural and mechanical phenomena such as waves, engines, acoustics, electronics, populations, UV intensity, growth of plants and animals, and so on. Periodic trigonometric graphs mean that the shape repeats itself exactly after a certain amount of time. Anything that has a regular cycle, like the tides, temperatures, rotation of the earth, and so on, can be modelled using a sine or cosine curve. The most common periodic signal waveform that is used in electrical and electronic engineering is the sinusoidal waveform. However, an alternating a.c. waveform may not always take the shape of a smooth shape based around the sine and cosine function; a.c. waveforms can also take the shape of square or triangular waves, i.e. complex waves. In engineering, it is therefore important to have some clear understanding of sine and cosine waveforms.

At the end of this chapter, you should be able to:

- sketch sine, cosine and tangent waveforms
- determine angles of any magnitude
- understand cycle, amplitude, period, periodic time, frequency, lagging/leading angles with reference to sine and cosine waves
- perform calculations involving sinusoidal form $A \sin(\omega t \pm \alpha)$

22.1 Graphs of trigonometric functions

By drawing up tables of values from $0°$ to $360°$, graphs of $y = \sin A$, $y = \cos A$ and $y = \tan A$ may be plotted. Values obtained with a calculator (correct to 3 decimal places – which is more than sufficient for plotting graphs), using $30°$ intervals, are shown below, with the respective graphs shown in Fig. 22.1.

(a) $y = \sin A$

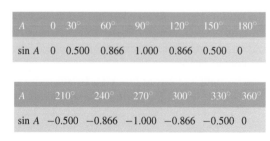

A	0	30°	60°	90°	120°	150°	180°
sin A	0	0.500	0.866	1.000	0.866	0.500	0

A	210°	240°	270°	300°	330°	360°
sin A	−0.500	−0.866	−1.000	−0.866	−0.500	0

(b) $y = \cos A$

A	0	30°	60°	90°	120°	150°	180°
cos A	1.000	0.866	0.500	0	−0.500	−0.866	−1.000

A	210°	240°	270°	300°	330°	360°
cos A	−0.866	−0.500	0	0.500	0.866	1.000

(c) $y = \tan A$

A	0	30°	60°	90°	120°	150°	180°
tan A	0	0.577	1.732	∞	−1.732	−0.577	0

A	210°	240°	270°	300°	330°	360°
tan A	0.577	1.732	∞	−1.732	−0.577	0

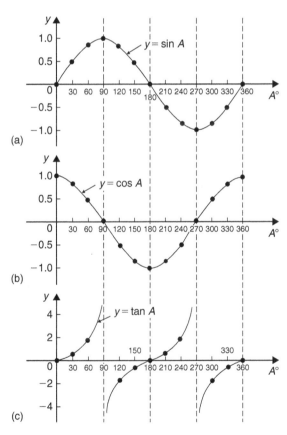

Figure 22.1

From Fig. 22.1 it is seen that

(a) Sine and cosine graphs oscillate between peak values of ±1

(b) The cosine curve is the same shape as the sine curve but displaced by 90°

(c) The sine and cosine curves are continuous and they repeat at intervals of 360°, and the tangent curve appears to be discontinuous and repeats at intervals of 180°

22.2 Angles of any magnitude

Fig. 22.2 shows rectangular axes XX' and YY' intersecting at origin 0. As with graphical work, measurements made to the right and above 0 are positive, while those to the left and downwards are negative.

Let $0A$ be free to rotate about 0. By convention, when $0A$ moves anticlockwise angular measurement is considered positive, and vice versa.

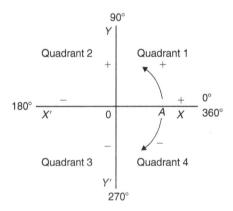

Figure 22.2

Let $0A$ be rotated anticlockwise so that θ_1 is any angle in the first quadrant and let perpendicular AB be constructed to form the right-angled triangle $0AB$ in Fig. 22.3. Since all three sides of the triangle are positive, the trigonometric ratios sine, cosine and tangent will all be positive in the first quadrant. (Note: $0A$ is always positive since it is the radius of a circle.)

Let $0A$ be further rotated so that θ_2 is any angle in the second quadrant and let AC be constructed to form the right-angled triangle $0AC$. Then,

$$\sin\theta_2 = \frac{+}{+} = + \qquad \cos\theta_2 = \frac{-}{+} = -$$

$$\tan\theta_2 = \frac{+}{-} = -$$

Let $0A$ be further rotated so that θ_3 is any angle in the third quadrant and let AD be constructed to form the

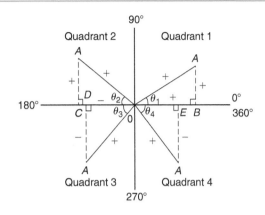

Figure 22.3

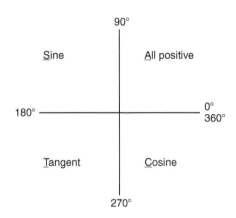

Figure 22.4

right-angled triangle 0AD. Then,

$$\sin\theta_3 = \frac{-}{+} = - \qquad \cos\theta_3 = \frac{-}{+} = -$$

$$\tan\theta_3 = \frac{-}{-} = +$$

Let 0A be further rotated so that θ_4 is any angle in the fourth quadrant and let AE be constructed to form the right-angled triangle 0AE. Then,

$$\sin\theta_4 = \frac{-}{+} = - \qquad \cos\theta_4 = \frac{+}{+} = +$$

$$\tan\theta_4 = \frac{-}{+} = -$$

The above results are summarised in Fig. 22.4, in which all three trigonometric ratios are positive in the first quadrant, only sine is positive in the second quadrant, only tangent is positive in the third quadrant and only cosine is positive in the fourth quadrant.

The underlined letters in Fig. 22.4 spell the word CAST when starting in the fourth quadrant and moving in an anticlockwise direction.

It is seen that, in the first quadrant of Fig. 22.1, all of the curves have positive values; in the second only sine is positive; in the third only tangent is positive; and in the fourth only cosine is positive – exactly as summarised in Fig. 22.4.

A knowledge of angles of any magnitude is needed when finding, for example, all the angles between $0°$ and $360°$ whose sine is, say, 0.3261. If 0.3261 is entered into a calculator and then the inverse sine key pressed (or $\sin^{-1}$ key) the answer $19.03°$ appears. However, there is a second angle between $0°$ and $360°$ which the calculator does not give. Sine is also positive in the second quadrant (either from CAST or from Fig. 22.1(a)). The other angle is shown in Fig. 22.5 as angle θ, where $\theta = 180° - 19.03° = 160.97°$. Thus,

$19.03°$ **and** $160.97°$ are the angles between $0°$ and $360°$ whose sine is 0.3261 (check that $\sin 160.97° = 0.3261$ on your calculator).

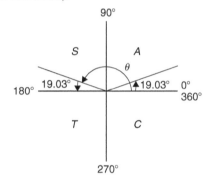

Figure 22.5

Be careful! Your calculator only gives you one of these answers. The second answer needs to be deduced from a knowledge of angles of any magnitude, as shown in the following worked problems.

Problem 1. Determine all of the angles between $0°$ and $360°$ whose sine is -0.4638

The angles whose sine is -0.4638 occur in the third and fourth quadrants since sine is negative in these quadrants – see Fig. 22.6.

From Fig. 22.7, $\theta = \sin^{-1} 0.4638 = 27.63°$. Measured from $0°$, the two angles between $0°$ and $360°$ whose sine is -0.4638 are $180° + 27.63°$ i.e. **207.63°** and $360° - 27.63°$, i.e. **332.37°**. (Note that if a calculator is used to determine $\sin^{-1}(-0.4638)$ it only gives one answer: $-27.632588°$)

Problem 2. Determine all of the angles between $0°$ and $360°$ whose tangent is 1.7629

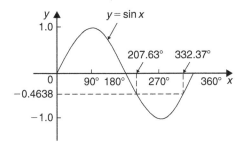

Figure 22.6

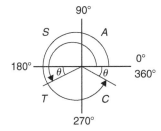

Figure 22.7

A tangent is positive in the first and third quadrants – see Fig. 22.8.

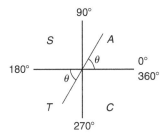

Figure 22.8

From Fig. 22.9, $\theta = \tan^{-1} 1.7629 = 60.44°$. Measured from $0°$, the two angles between $0°$ and $360°$ whose tangent is 1.7629 are **60.44°** and $180° + 60.44°$, i.e. **240.44°**

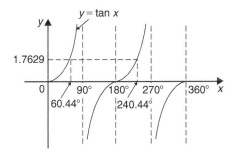

Figure 22.9

Problem 3. Solve the equation $\cos^{-1}(-0.2348) = \alpha$ for angles of α between $0°$ and $360°$

Cosine is positive in the first and fourth quadrants and thus negative in the second and third quadrants – see Fig. 22.10 or from Fig. 22.1(b).

In Fig. 22.10, angle $\theta = \cos^{-1}(0.2348) = 76.42°$. Measured from $0°$, the two angles whose cosine is -0.2348 are $\alpha = 180° - 76.42°$, i.e. **103.58°** and $\alpha = 180° + 76.42°$, i.e. **256.42°**

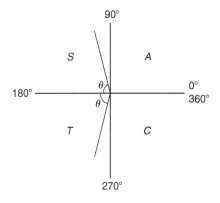

Figure 22.10

Now try the following Practice Exercise

1. Determine all of the angles between $0°$ and $360°$ whose sine is
    (a) 0.6792   (b) -0.1483

2. Solve the following equations for values of x between $0°$ and $360°$.
    (a) $x = \cos^{-1} 0.8739$

    (b) $x = \cos^{-1}(-0.5572)$

3. Find the angles between $0°$ to $360°$ whose tangent is
    (a) 0.9728   (b) -2.3420

In Problems 4 to 6, solve the given equations in the range $0°$ to $360°$, giving the answers in degrees and minutes.

4. $\cos^{-1}(-0.5316) = t$

5. $\sin^{-1}(-0.6250) = \alpha$

6. $\tan^{-1} 0.8314 = \theta$

22.3 The production of sine and cosine waves

In Fig. 22.11, let *OR* be a vector 1 unit long and free to rotate anticlockwise about 0. In one revolution a circle is produced and is shown with 15° sectors. Each radius arm has a vertical and a horizontal component. For example, at 30°, the vertical component is *TS* and the horizontal component is *OS*.

From triangle *OST*,

$$\sin 30° = \frac{TS}{TO} = \frac{TS}{1} \quad \text{i.e.} \quad TS = \sin 30°$$

$$\text{and} \quad \cos 30° = \frac{OS}{TO} = \frac{OS}{1} \quad \text{i.e.} \quad OS = \cos 30°$$

22.3.1 Sine waves

The vertical component *TS* may be projected across to *T'S'*, which is the corresponding value of 30° on the graph of *y* against angle *x*°. If all such vertical components as *TS* are projected on to the graph, a **sine wave** is produced as shown in Fig. 22.11.

22.3.2 Cosine waves

If all horizontal components such as *OS* are projected on to a graph of *y* against angle *x*°, a **cosine wave** is produced. It is easier to visualise these projections by redrawing the circle with the radius arm *OR* initially in a vertical position as shown in Fig. 22.12.

It is seen from Figs 22.11 and 22.12 that a cosine curve is of the same form as the sine curve but is displaced by 90° (or π/2 radians). Both sine and cosine waves repeat every 360°

22.4 Terminology involved with sine and cosine waves

Sine waves are extremely important in engineering, with examples occurring with alternating currents and voltages – the mains supply is a sine wave – and with simple harmonic motion.

22.4.1 Cycle

When a sine wave has passed through a complete series of values, both positive and negative, it is said to have completed one **cycle**. One cycle of a sine wave is shown in Fig. 22.1(a) on page 221 and in Fig. 22.11.

22.4.2 Amplitude

The amplitude is the maximum value reached in a half cycle by a sine wave. Another name for **amplitude** is **peak value** or **maximum value**.

A sine wave $y = 5 \sin x$ has an amplitude of 5, a sine wave $v = 200 \sin 314t$ has an amplitude of 200 and the sine wave $y = \sin x$ shown in Fig. 22.11 has an amplitude of 1.

22.4.3 Period

The waveforms $y = \sin x$ and $y = \cos x$ repeat themselves every 360°. Thus, for each, the **period** is 360°. A waveform of $y = \tan x$ has a period of 180° (from Fig. 22.1(c)).

A graph of $y = 3 \sin 2A$, as shown in Fig. 22.13, has an **amplitude of 3** and **period 180°**

A graph of $y = \sin 3A$, as shown in Fig. 22.14, has an **amplitude of 1** and **period of 120°**

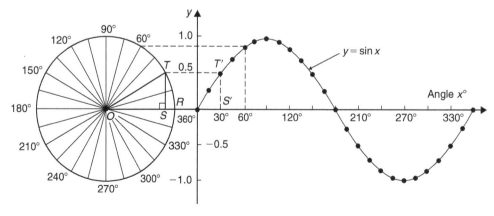

Figure 22.11

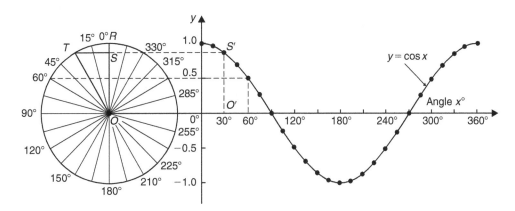

Figure 22.12

A graph of $y = 4\cos 2x$, as shown in Fig. 22.15, has an amplitude of 4 and a period of $180°$

In general, **if** $y = A\sin px$ **or** $y = A\cos px$,

$$\textbf{amplitude} = A \textbf{ and period} = \frac{360°}{p}$$

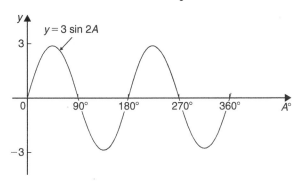

Figure 22.13

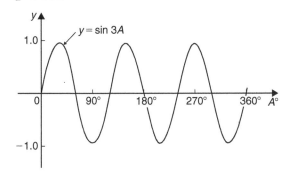

Figure 22.14

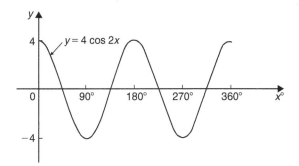

Figure 22.15

A sketch of $y = 2\sin\dfrac{3}{5}A$ is shown in Fig. 22.16.

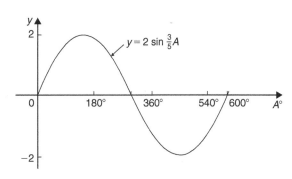

Figure 22.16

Problem 4.   Sketch $y = 2\sin\dfrac{3}{5}A$ over one cycle

$$\textbf{Amplitude} = 2; \ \textbf{period} = \frac{360°}{\frac{3}{5}} = \frac{360° \times 5}{3} = \textbf{600°}$$

22.4.4   Periodic time

In practice, the horizontal axis of a sine wave will be time. The time taken for a sine wave to complete one cycle is called the **periodic time, T**.

In the sine wave of voltage v (volts) against time t (milliseconds) shown in Fig. 22.17, the amplitude is $10\,\text{V}$ and the periodic time is $20\,\text{ms}$; i.e. $T = 20\,\text{ms}$.

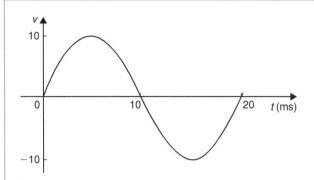

Figure 22.17

22.4.5 Frequency

The number of cycles completed in one second is called the **frequency** f and is measured in **hertz, Hz**.

$$f = \frac{1}{T} \text{ or } T = \frac{1}{f}$$

Problem 5. Determine the frequency of the sine wave shown in Fig. 22.17

In the sine wave shown in Fig. 22.17, $T = 20$ ms, hence

$$\textbf{frequency, } f = \frac{1}{T} = \frac{1}{20 \times 10^{-3}} = \textbf{50 Hz}$$

Problem 6. If a waveform has a frequency of 200 kHz, determine the periodic time

If a waveform has a frequency of 200 kHz, the periodic time T is given by

$$\textbf{periodic time, } T = \frac{1}{f} = \frac{1}{200 \times 10^3}$$

$$= 5 \times 10^{-6}\text{s} = \textbf{5 } \mu\textbf{s}$$

22.4.6 Lagging and leading angles

A sine or cosine curve may not always start at $0°$. To show this, a periodic function is represented by $y = A\sin(x \pm \alpha)$ where α is a phase displacement compared with $y = A\sin x$. For example, $y = \sin A$ is shown by the broken line in Fig. 22.18 and, on the same axes, $y = \sin(A - 60°)$ is shown. **The graph $y = \sin(A - 60°)$ is said to lag $y = \sin A$ by $60°$** In another example, $y = \cos A$ is shown by the broken line in Fig. 22.19 and, on the same axes, $y = \cos(A + 45°)$ is shown. **The graph $y = \cos(A + 45°)$ is said to lead $y = \cos A$ by $45°$**

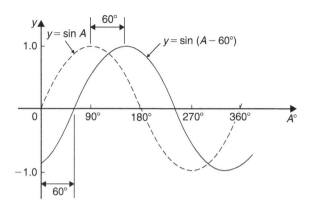

Figure 22.18

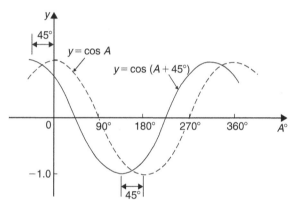

Figure 22.19

Problem 7. Sketch $y = 5\sin(A + 30°)$ from $A = 0°$ to $A = 360°$

Amplitude $= 5$ and **period $= 360°/1 = 360°$**

$5\sin(A + 30°)$ **leads** $5\sin A$ by $30°$ (i.e. starts $30°$ earlier).
A sketch of $y = 5\sin(A + 30°)$ is shown in Fig. 22.20.

Problem 8. Sketch $y = 7\sin(2A - \pi/3)$ in the range $0 \leq A \leq 360°$

Amplitude $= 7$ and **period $= 2\pi/2 = \pi$ radians**

In general, $\textbf{\textit{y} = sin(\textit{pt} - \boldsymbol{\alpha})}$ **lags** $\textbf{\textit{y} = sin \textit{pt}}$ by α/p, hence $7\sin(2A - \pi/3)$ lags $7\sin 2A$ by $(\pi/3)/2$, i.e. $\pi/6$ rad or $30°$
A sketch of $y = 7\sin(2A - \pi/3)$ is shown in Fig. 22.21.

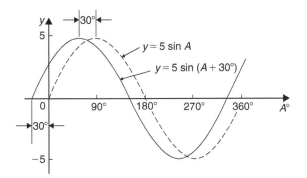

Figure 22.20

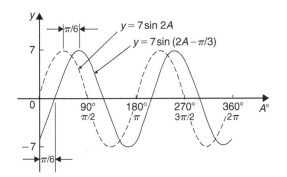

Figure 22.21

Problem 9.   Sketch $y = 2\cos(\omega t - 3\pi/10)$ over one cycle

Amplitude = 2 and **period = $2\pi/\omega$ rad**

$2\cos(\omega t - 3\pi/10)$ **lags** $2\cos\omega t$ by $3\pi/10\omega$ seconds. A sketch of $y = 2\cos(\omega t - 3\pi/10)$ is shown in Fig. 22.22.

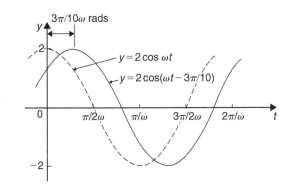

Figure 22.22

Now try the following Practice Exercise

Practice Exercise 88   Trigonometric waveforms (answers on page 432)

1. A sine wave is given by $y = 5\sin 3x$. State its peak value.

2. A sine wave is given by $y = 4\sin 2x$. State its period in degrees.

3. A periodic function is given by $y = 30\cos 5x$. State its maximum value.

4. A periodic function is given by $y = 25\cos 3x$. State its period in degrees.

In Problems 5 to 11, state the amplitude and period of the waveform and sketch the curve between $0°$ and $360°$.

5. $y = \cos 3A$

6. $y = 2\sin\dfrac{5x}{2}$

7. $y = 3\sin 4t$

8. $y = 5\cos\dfrac{\theta}{2}$

9. $y = \dfrac{7}{2}\sin\dfrac{3x}{8}$

10. $y = 6\sin(t - 45°)$

11. $y = 4\cos(2\theta + 30°)$

12. The frequency of a sine wave is $200\,\text{Hz}$. Calculate the periodic time.

13. Calculate the frequency of a sine wave that has a periodic time of $25\,\text{ms}$.

14. Calculate the periodic time for a sine wave having a frequency of $10\,\text{kHz}$.

15. An alternating current completes 15 cycles in $24\,\text{ms}$. Determine its frequency.

16. Graphs of $y_1 = 2\sin x$ and $y_2 = 3\sin(x + 50°)$ are drawn on the same axes. Is y_2 lagging or leading y_1?

17. Graphs of $y_1 = 6\sin x$ and $y_2 = 5\sin(x - 70°)$ are drawn on the same axes. Is y_1 lagging or leading y_2?

22.5   Sinusoidal form: $A\sin(\omega t \pm \alpha)$

If a sine wave is expressed in the form $y = A\sin(\omega t \pm \alpha)$ then

(a) $A = $ amplitude

(b) $\omega = $ angular velocity $= 2\pi f$ rad/s

(c) frequency, $f = \dfrac{\omega}{2\pi}$ hertz

(d) periodic time, $T = \dfrac{2\pi}{\omega}$ seconds $\left(\text{i.e. } T = \dfrac{1}{f}\right)$

(e) $\alpha =$ angle of lead or lag (compared with $y = A\sin\omega t$)

Here are some worked problems involving the sinusoidal form $A\sin(\omega t \pm \alpha)$

> **Problem 10.** An alternating current is given by $i = 30\sin(100\pi t + 0.35)$ amperes. Find the (a) amplitude, (b) frequency, (c) periodic time and (d) phase angle (in degrees and minutes)

(a) $i = 30\sin(100\pi t + 0.35)A$; hence, **amplitude = 30 A**

(b) Angular velocity, $\omega = 100\pi$, rad/s, hence

$$\textbf{frequency}, f = \frac{\omega}{2\pi} = \frac{100\pi}{2\pi} = \textbf{50 Hz}$$

(c) **Periodic time,** $T = \dfrac{1}{f} = \dfrac{1}{50} = \textbf{0.02 s or 20 ms}.$

(d) 0.35 is the angle in **radians**. The relationship between radians and degrees is

$$360° = 2\pi \text{ radians or } \mathbf{180° = \pi \, radians}$$

from which,

$$\mathbf{1° = \frac{\pi}{180} rad} \text{ and } \mathbf{1\,rad} = \frac{180°}{\pi} \ (\approx 57.30°)$$

Hence, **phase angle,** $\alpha = 0.35$ rad
$$= \left(0.35 \times \frac{180}{\pi}\right)^° = \textbf{20.05° or 20°3′ leading}$$
$$i = 30\sin(100\pi t)$$

> **Problem 11.** An oscillating mechanism has a maximum displacement of 2.5 m and a frequency of 60 Hz. At time $t = 0$ the displacement is 90 cm. Express the displacement in the general form $A\sin(\omega t \pm \alpha)$

Amplitude = maximum displacement = 2.5 m.

Angular velocity, $\omega = 2\pi f = 2\pi (60) = 120\pi$ rad/s.

Hence, **displacement $= 2.5\sin(120\pi t + \alpha)$ m.**

When $t = 0$, displacement $= 90$ cm $= 0.90$ m

Hence, $0.90 = 2.5\sin(0 + \alpha)$

i.e. $\sin\alpha = \dfrac{0.90}{2.5} = 0.36$

Hence, $\alpha = \sin^{-1} 0.36 = 21.10°$

$$= 21°6′ = 0.368 \text{ rad}.$$

Thus, **displacement $= 2.5\sin(120\pi t + 0.368)$ m**

> **Problem 12.** The instantaneous value of voltage in an a.c. circuit at any time t seconds is given by $v = 340\sin(50\pi t - 0.541)$ volts. Determine the (a) amplitude, frequency, periodic time and phase angle (in degrees), (b) value of the voltage when $t = 0$, (c) value of the voltage when $t = 10$ ms, (d) time when the voltage first reaches 200 V and (e) time when the voltage is a maximum. Also, (f) sketch one cycle of the waveform

(a) **Amplitude = 340 V**

   Angular velocity, $\omega = 50\pi$

   Frequency, $f = \dfrac{\omega}{2\pi} = \dfrac{50\pi}{2\pi} = \textbf{25 Hz}$

   Periodic time, $T = \dfrac{1}{f} = \dfrac{1}{25} = \textbf{0.04 s or 40 ms}$

   Phase angle $= 0.541$ rad $= \left(0.541 \times \dfrac{180}{\pi}\right)^°$
   $$= \textbf{31° lagging } v = 340\sin(50\pi t)$$

(b) **When $t = 0$,**
   $$v = 340\sin(0 - 0.541)$$
   $$= 340\sin(-31°) = \textbf{-175.1 V}$$

(c) **When $t = 10$ ms,**
   $$v = 340\sin(50\pi \times 10 \times 10^{-3} - 0.541)$$
   $$= 340\sin(1.0298)$$
   $$= 340\sin 59° = \textbf{291.4 volts}$$

(d)   When $v = 200$ volts,

$$200 = 340 \sin(50\pi t - 0.541)$$

$$\frac{200}{340} = \sin(50\pi t - 0.541)$$

Hence, $(50\pi t - 0.541) = \sin^{-1} \dfrac{200}{340}$

$$= 36.03° \text{ or } 0.628875 \text{ rad}$$

$$50\pi t = 0.628875 + 0.541$$

$$= 1.169875$$

Hence, when $v = 200$ V,

time, $t = \dfrac{1.169875}{50\pi} = \mathbf{7.448\,ms}$

(e)   When the voltage is a maximum, $v = 340$ V.

Hence, $340 = 340 \sin(50\pi t - 0.541)$

$$1 = \sin(50\pi t - 0.541)$$

$$50\pi t - 0.541 = \sin^{-1} 1 = 90° \text{ or } 1.5708 \text{ rad}$$

$$50\pi t = 1.5708 + 0.541 = 2.1118$$

Hence, **time,** $t = \dfrac{2.1118}{50\pi} = \mathbf{13.44\,ms}$

(f)   A sketch of $v = 340 \sin(50\pi t - 0.541)$ volts is shown in Fig. 22.23.

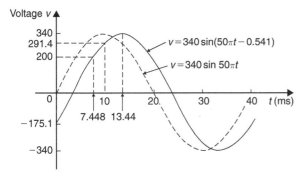

Figure 22.23

Now try the following Practice Exercise

Practice Exercise 89   Sinusoidal form
$A \sin(\omega t \pm \alpha)$ **(answers on page 432)**

In Problems 1 to 3 find the (a) amplitude, (b) frequency, (c) periodic time and (d) phase angle (stating whether it is leading or lagging $\sin \omega t$) of the alternating quantities given.

1. $i = 40 \sin(50\pi t + 0.29)\,\text{mA}$

2. $y = 75 \sin(40t - 0.54)\,\text{cm}$

3. $v = 300 \sin(200\pi t - 0.412)\,\text{V}$

4.   A sinusoidal voltage has a maximum value of 120 V and a frequency of 50 Hz. At time $t = 0$, the voltage is (a) zero and (b) 50 V. Express the instantaneous voltage v in the form $v = A \sin(\omega t \pm \alpha)$.

5.   An alternating current has a periodic time of 25 ms and a maximum value of 20 A. When time $= 0$, current $i = -10$ amperes. Express the current i in the form $i = A \sin(\omega t \pm \alpha)$.

6.   An oscillating mechanism has a maximum displacement of 3.2 m and a frequency of 50 Hz. At time $t = 0$ the displacement is 150 cm. Express the displacement in the general form $A \sin(\omega t \pm \alpha)$

7.   The current in an a.c. circuit at any time t seconds is given by

$$i = 5 \sin(100\pi t - 0.432) \text{ amperes}$$

Determine the
(a)   amplitude, frequency, periodic time and phase angle (in degrees),
(b)   value of current at $t = 0$,
(c)   value of current at $t = 8$ ms,
(d)   time when the current is first a maximum,
(e)   time when the current first reaches 3A.
Also,
(f)   sketch one cycle of the waveform showing relevant points.

For fully worked solutions to each of the problems in Practice Exercises 87 to 89 in this chapter, go to the website:
www.routledge.com/cw/bird

Chapter 23

Non-right-angled triangles and some practical applications

Why it is important to understand: Non-right-angled triangles and some practical applications

As was mentioned earlier, fields that use trigonometry include astronomy, navigation, music theory, acoustics, optics, electronics, probability theory, statistics, biology, medical imaging (CAT scans and ultrasound), pharmacy, chemistry, seismology, meteorology, oceanography, many physical sciences, land surveying, architecture, economics, electrical engineering, mechanical engineering, civil engineering, computer graphics, cartography and crystallography. There are so many examples where triangles are involved in engineering, and the ability to solve such triangles is of great importance.

At the end of this chapter, you should be able to:

- state and use the sine rule
- state and use the cosine rule
- use various formulae to determine the area of any triangle
- apply the sine and cosine rules to solving practical trigonometric problems

23.1 The sine and cosine rules

To 'solve a triangle' means 'to find the values of unknown sides and angles'. If a triangle is **right-angled**, trigonometric ratios and the theorem of Pythagoras* may be used for its solution, as shown in Chapter 21. However, for a **non-right-angled triangle**,

trigonometric ratios and Pythagoras' theorem cannot be used. Instead, two rules, called the **sine rule** and the **cosine rule**, are used.

23.1.1 The sine rule

With reference to triangle ABC of Fig. 23.1, the **sine rule** states

$$\frac{a}{\sin A} = \frac{b}{\sin B} = \frac{c}{\sin C}$$

The rule may be used only when

*Who was Pythagoras? To find out more go to www.routledge.com/cw/bird

Basic Engineering Mathematics. 978-0-415-66278-9, © 2014 John Bird. Published by Taylor & Francis. All rights reserved.

Figure 23.1

(a) 1 side and any 2 angles are initially given, or
(b) 2 sides and an angle (not the included angle) are initially given.

23.1.2 The cosine rule

With reference to triangle ABC of Fig. 23.1, the **cosine rule** states

$$a^2 = b^2 + c^2 - 2bc \cos A$$

or $$b^2 = a^2 + c^2 - 2ac \cos B$$

or $$c^2 = a^2 + b^2 - 2ab \cos C$$

The rule may be used only when

(a) 2 sides and the included angle are initially given, or

(b) 3 sides are initially given.

23.2 Area of any triangle

The **area of any triangle** such as ABC of Fig. 23.1 is given by

(a) $\dfrac{1}{2}$ × base × perpendicular height

or (b) $\dfrac{1}{2}ab \sin C$ or $\dfrac{1}{2}ac \sin B$ or $\dfrac{1}{2}bc \sin A$

or (c) $\sqrt{[s(s-a)(s-b)(s-c)]}$ where $s = \dfrac{a+b+c}{2}$

23.3 Worked problems on the solution of triangles and their areas

Problem 1. In a triangle XYZ, $\angle X = 51°$, $\angle Y = 67°$ and $YZ = 15.2\,cm$. Solve the triangle and find its area

The triangle XYZ is shown in Fig. 23.2. Solving the triangle means finding $\angle Z$ and sides XZ and XY.
Since the angles in a triangle add up to $180°$, $Z = 180° - 51° - 67° = \mathbf{62°}$

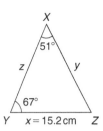

Figure 23.2

Applying the sine rule, $\dfrac{15.2}{\sin 51°} = \dfrac{y}{\sin 67°}$

$$= \dfrac{z}{\sin 62°}$$

Using $\dfrac{15.2}{\sin 51°} = \dfrac{y}{\sin 67°}$

and transposing gives $y = \dfrac{15.2 \sin 67°}{\sin 51°}$

$$= \mathbf{18.00\,cm} = XZ$$

Using $\dfrac{15.2}{\sin 51°} = \dfrac{z}{\sin 62°}$

and transposing gives $z = \dfrac{15.2 \sin 62°}{\sin 51°}$

$$= \mathbf{17.27\,cm} = XY$$

Area of triangle $XYZ = \dfrac{1}{2}xy \sin Z$

$$= \dfrac{1}{2}(15.2)(18.00) \sin 62° = \mathbf{120.8\,cm^2}$$

(or area $= \dfrac{1}{2}xz \sin Y = \dfrac{1}{2}(15.2)(17.27) \sin 67°$

$$= \mathbf{120.8\,cm^2})$$

It is always worth checking with triangle problems that the longest side is opposite the largest angle and vice versa. In this problem, Y is the largest angle and XZ is the longest of the three sides.

Problem 2. Solve the triangle ABC given $B = 78°51'$, $AC = 22.31\,mm$ and $AB = 17.92\,mm$. Also find its area

Triangle ABC is shown in Fig. 23.3. Solving the triangle means finding angles A and C and side BC.

Applying the sine rule, $\dfrac{22.31}{\sin 78°51'} = \dfrac{17.92}{\sin C}$

from which $\sin C = \dfrac{17.92 \sin 78°51'}{22.31} = 0.7881$

Hence, $C = \sin^{-1} 0.7881 = 52°0'$ or $128°0'$

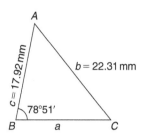

Figure 23.3

Since $B = 78°51'$, C cannot be $128°0'$, since $128°0' + 78°51'$ is greater than $180°$. Thus, only $C = 52°0'$ is valid.

Angle $A = 180° - 78°51' - 52°0' = 49°9'$

Applying the sine rule, $\dfrac{a}{\sin 49°9'} = \dfrac{22.31}{\sin 78°51'}$

from which $a = \dfrac{22.31 \sin 49°9'}{\sin 78°51'} = 17.20 \text{ mm}$

Hence, $A = 49°9'$, $C = 52°0'$ and $BC = 17.20 \text{ mm}$.

Area of triangle $ABC = \dfrac{1}{2} ac \sin B$

$= \dfrac{1}{2}(17.20)(17.92) \sin 78°51'$

$= 151.2 \text{ mm}^2$

> **Problem 3.** Solve the triangle PQR and find its area given that $QR = 36.5 \text{ mm}, PR = 29.6 \text{ mm}$ and $\angle Q = 36°$

Triangle PQR is shown in Fig. 23.4.

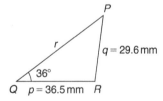

Figure 23.4

Applying the sine rule, $\dfrac{29.6}{\sin 36°} = \dfrac{36.5}{\sin P}$

from which $\sin P = \dfrac{36.5 \sin 36°}{29.6} = 0.7248$

Hence, $P = \sin^{-1} 0.7248 = 46.45°$ or $133.55°$

When $P = 46.45°$ and $Q = 36°$ then
$R = 180° - 46.45° - 36° = 97.55°$

When $P = 133.55°$ and $Q = 36°$ then
$R = 180° - 133.55° - 36° = 10.45°$

Thus, in this problem, there are **two** separate sets of results and both are feasible solutions. Such a situation is called the **ambiguous case**.

Case 1. $P = 46.45°$, $Q = 36°$, $R = 97.55°$, $p = 36.5 \text{ mm}$ and $q = 29.6 \text{ mm}$

From the sine rule, $\dfrac{r}{\sin 97.55°} = \dfrac{29.6}{\sin 36°}$

from which $r = \dfrac{29.6 \sin 97.55°}{\sin 36°} = 49.92 \text{ mm} = PQ$

Area of $PQR = \dfrac{1}{2} pq \sin R = \dfrac{1}{2}(36.5)(29.6) \sin 97.55°$

$= 535.5 \text{ mm}^2$

Case 2. $P = 133.55°$, $Q = 36°$, $R = 10.45°$, $p = 36.5 \text{ mm}$ and $q = 29.6 \text{ mm}$

From the sine rule, $\dfrac{r}{\sin 10.45°} = \dfrac{29.6}{\sin 36°}$

from which $r = \dfrac{29.6 \sin 10.45°}{\sin 36°} = 9.134 \text{ mm} = PQ$

Area of $PQR = \dfrac{1}{2} pq \sin R = \dfrac{1}{2}(36.5)(29.6) \sin 10.45°$

$= 97.98 \text{ mm}^2$

The triangle PQR for case 2 is shown in Fig. 23.5.

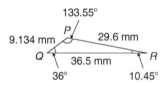

Figure 23.5

Now try the following Practice Exercise

Practice Exercise 90   Solution of triangles and their areas (answers on page 432)

In Problems 1 and 2, use the sine rule to solve the triangles ABC and find their areas.

1. $A = 29°$, $B = 68°$, $b = 27 \text{ mm}$

2. $B = 71°26'$, $C = 56°32'$, $b = 8.60 \text{ cm}$

In Problems 3 and 4, use the sine rule to solve the triangles DEF and find their areas.

3. $d = 17 \text{ cm}$, $f = 22 \text{ cm}$, $F = 26°$

4. $d = 32.6 \text{ mm}$, $e = 25.4 \text{ mm}$, $D = 104°22'$

In Problems 5 and 6, use the sine rule to solve the triangles JKL and find their areas.

5. $j = 3.85\,\text{cm}, k = 3.23\,\text{cm}, K = 36°$

6. $k = 46\,\text{mm}, l = 36\,\text{mm}, L = 35°$

23.4 Further worked problems on the solution of triangles and their areas

Problem 4. Solve triangle DEF and find its area given that $EF = 35.0\,\text{mm}, DE = 25.0\,\text{mm}$ and $\angle E = 64°$

Triangle DEF is shown in Fig. 23.6. Solving the triangle means finding angles D and F and side DF. Since two sides and the angle in between the two sides are given, the cosine rule needs to be used.

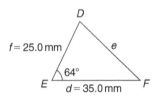

Figure 23.6

Applying the cosine rule, $e^2 = d^2 + f^2 - 2df\cos E$

i.e. $e^2 = (35.0)^2 + (25.0)^2 - [2(35.0)(25.0)\cos 64°]$

$= 1225 + 625 - 767.15$

$= 1082.85$

from which $e = \sqrt{1082.85}$

$= \textbf{32.91\,mm} = \textbf{DF}$

Applying the sine rule, $\dfrac{32.91}{\sin 64°} = \dfrac{25.0}{\sin F}$

from which $\sin F = \dfrac{25.0\sin 64°}{32.91} = 0.6828$

Thus, $\angle F = \sin^{-1} 0.6828 = 43°4'$ or $136°56'$

$F = 136°56'$ is not possible in this case since $136°56' + 64°$ is greater than $180°$. Thus, only $\textbf{F = 43°4'}$ is valid. Then $\angle D = 180° - 64° - 43°4' = \textbf{72°56'}$

Area of triangle DEF $= \dfrac{1}{2} df \sin E$

$= \dfrac{1}{2}(35.0)(25.0)\sin 64° = \textbf{393.2\,mm}^2$

Problem 5. A triangle ABC has sides $a = 9.0\,\text{cm}, b = 7.5\,\text{cm}$ and $c = 6.5\,\text{cm}$. Determine its three angles and its area

Triangle ABC is shown in Fig. 23.7. It is usual first to calculate the largest angle to determine whether the triangle is acute or obtuse. In this case the largest angle is A (i.e. opposite the longest side).

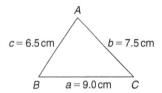

Figure 23.7

Applying the cosine rule, $a^2 = b^2 + c^2 - 2bc\cos A$

from which $2bc\cos A = b^2 + c^2 - a^2$

and $\cos A = \dfrac{b^2 + c^2 - a^2}{2bc} = \dfrac{7.5^2 + 6.5^2 - 9.0^2}{2(7.5)(6.5)}$

$= 0.1795$

Hence, $A = \cos^{-1} 0.1795 = \textbf{79.67°}$
(or $280.33°$, which is clearly impossible)

The triangle is thus acute angled since $\cos A$ is positive. (If $\cos A$ had been negative, angle A would be obtuse; i.e. would lie between $90°$ and $180°$)

Applying the sine rule, $\dfrac{9.0}{\sin 79.67°} = \dfrac{7.5}{\sin B}$

from which $\sin B = \dfrac{7.5\sin 79.67°}{9.0} = 0.8198$

Hence, $B = \sin^{-1} 0.8198 = \textbf{55.07°}$

and $C = 180° - 79.67° - 55.07° = \textbf{45.26°}$

Area $= \sqrt{[s(s-a)(s-b)(s-c)]}$, where

$s = \dfrac{a+b+c}{2} = \dfrac{9.0 + 7.5 + 6.5}{2} = 11.5\,\text{cm}$

Hence,

$\textbf{area} = \sqrt{[11.5(11.5 - 9.0)(11.5 - 7.5)(11.5 - 6.5)]}$

$= \sqrt{[11.5(2.5)(4.0)(5.0)]} = \textbf{23.98\,cm}^2$

Alternatively, **area** $= \dfrac{1}{2} ac \sin B$

$= \dfrac{1}{2}(9.0)(6.5)\sin 55.07° = \textbf{23.98\,cm}^2$

Problem 6. Solve triangle XYZ, shown in Fig. 23.8, and find its area given that $Y = 128°, XY = 7.2$ cm and $YZ = 4.5$ cm

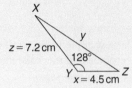

Figure 23.8

Applying the cosine rule,

$$y^2 = x^2 + z^2 - 2xz \cos Y$$

$$= 4.5^2 + 7.2^2 - [2(4.5)(7.2)\cos 128°]$$

$$= 20.25 + 51.84 - [-39.89]$$

$$= 20.25 + 51.84 + 39.89 = 112.0$$

$$y = \sqrt{112.0} = \mathbf{10.58\,cm = XZ}$$

Applying the sine rule, $\dfrac{10.58}{\sin 128°} = \dfrac{7.2}{\sin Z}$

from which $\sin Z = \dfrac{7.2 \sin 128°}{10.58} = 0.5363$

Hence, $\qquad Z = \sin^{-1} 0.5363 = \mathbf{32.43°}$
$\qquad\qquad$ (or $147.57°$ which is not possible)

Thus, $X = 180° - 128° - 32.43° = \mathbf{19.57°}$

Area of XYZ $= \dfrac{1}{2} xz \sin Y = \dfrac{1}{2}(4.5)(7.2)\sin 128°$

$$= \mathbf{12.77\,cm^2}$$

Now try the following Practice Exercise

In Problems 1 and 2, use the cosine and sine rules to solve the triangles PQR and find their areas.

1. $q = 12$ cm, $r = 16$ cm, $P = 54°$

2. $q = 3.25$ m, $r = 4.42$ m, $P = 105°$

In Problems 3 and 4, use the cosine and sine rules to solve the triangles XYZ and find their areas.

3. $x = 10.0$ cm, $y = 8.0$ cm, $z = 7.0$ cm

4. $x = 21$ mm, $y = 34$ mm, $z = 42$ mm

23.5 Practical situations involving trigonometry

There are a number of **practical situations** in which the use of trigonometry is needed to find unknown sides and angles of triangles. This is demonstrated in the following worked problems.

Problem 7. A room 8.0 m wide has a span roof which slopes at 33° on one side and 40° on the other. Find the length of the roof slopes, correct to the nearest centimetre

A section of the roof is shown in Fig. 23.9.

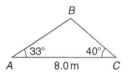

Figure 23.9

Angle at ridge, $B = 180° - 33° - 40° = 107°$

From the sine rule, $\qquad \dfrac{8.0}{\sin 107°} = \dfrac{a}{\sin 33°}$

from which $\qquad a = \dfrac{8.0 \sin 33°}{\sin 107°} = \mathbf{4.556\,m = BC}$

Also from the sine rule, $\qquad \dfrac{8.0}{\sin 107°} = \dfrac{c}{\sin 40°}$

from which $\qquad c = \dfrac{8.0 \sin 40°}{\sin 107°} = \mathbf{5.377\,m = AB}$

Hence, **the roof slopes are 4.56 m and 5.38 m**, correct to the nearest centimetre.

Problem 8. A man leaves a point walking at 6.5 km/h in a direction E 20° N (i.e. a bearing of 70°). A cyclist leaves the same point at the same time in a direction E 40° S (i.e. a bearing of 130°) travelling at a constant speed. Find the average speed of the cyclist if the walker and cyclist are 80 km apart after 5 hours

After 5 hours the walker has travelled $5 \times 6.5 = 32.5$ km (shown as AB in Fig. 23.10). If AC is the distance the cyclist travels in 5 hours then $BC = 80$ km.

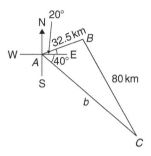

Figure 23.10

Applying the sine rule, $\dfrac{80}{\sin 60°} = \dfrac{32.5}{\sin C}$

from which $\sin C = \dfrac{32.5 \sin 60°}{80} = 0.3518$

Hence, $C = \sin^{-1} 0.3518 = 20.60°$

    (or $159.40°$, which is not possible)

and $B = 180° - 60° - 20.60° = 99.40°$

Applying the sine rule again, $\dfrac{80}{\sin 60°} = \dfrac{b}{\sin 99.40°}$

from which $b = \dfrac{80 \sin 99.40°}{\sin 60°} = 91.14 \, \text{km}$

Since the cyclist travels 91.14 km in 5 hours,

average speed $= \dfrac{\text{distance}}{\text{time}} = \dfrac{91.14}{5} = \mathbf{18.23 \, km/h}$

Problem 9. Two voltage phasors are shown in Fig. 23.11. If $V_1 = 40\,\text{V}$ and $V_2 = 100\,\text{V}$, determine the value of their resultant (i.e. length OA) and the angle the resultant makes with V_1

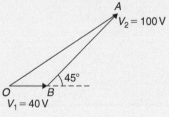

Figure 23.11

Angle $OBA = 180° - 45° = 135°$

Applying the cosine rule,

$OA^2 = V_1^2 + V_2^2 - 2V_1 V_2 \cos OBA$

$= 40^2 + 100^2 - \{2(40)(100)\cos 135°\}$

$= 1600 + 10000 - \{-5657\}$

$= 1600 + 10000 + 5657 = 17257$

Thus,   **resultant**, $OA = \sqrt{17257} = \mathbf{131.4\,V}$

Applying the sine rule $\dfrac{131.4}{\sin 135°} = \dfrac{100}{\sin AOB}$

from which $\sin AOB = \dfrac{100 \sin 135°}{131.4} = 0.5381$

Hence, angle $AOB = \sin^{-1} 0.5381 = 32.55°$ (or $147.45°$, which is not possible)

Hence, **the resultant voltage is 131.4 volts at 32.55° to** V_1

Problem 10. In Fig. 23.12, PR represents the inclined jib of a crane and is 10.0 m long. PQ is 4.0 m long. Determine the inclination of the jib to the vertical and the length of tie QR.

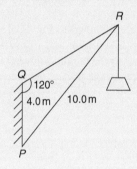

Figure 23.12

Applying the sine rule, $\dfrac{PR}{\sin 120°} = \dfrac{PQ}{\sin R}$

from which $\sin R = \dfrac{PQ \sin 120°}{PR} = \dfrac{(4.0)\sin 120°}{10.0}$

$= 0.3464$

Hence, $\angle R = \sin^{-1} 0.3464 = 20.27°$ (or $159.73°$, which is not possible)

$\angle P = 180° - 120° - 20.27° = \mathbf{39.73°}$, **which is the inclination of the jib to the vertical**

Applying the sine rule, $\dfrac{10.0}{\sin 120°} = \dfrac{QR}{\sin 39.73°}$

from which   **length of tie,** $QR = \dfrac{10.0 \sin 39.73°}{\sin 120°}$

$= \mathbf{7.38\,m}$

Now try the following Practice Exercise

Practice Exercise 92   Practical situations
involving trigonometry (answers on
page 432)

1. A ship P sails at a steady speed of 45 km/h in a direction of W 32° N (i.e. a bearing of 302°) from a port. At the same time another ship Q leaves the port at a steady speed of 35 km/h in a direction N 15° E (i.e. a bearing of 015°). Determine their distance apart after 4 hours.

2. Two sides of a triangular plot of land are 52.0 m and 34.0 m, respectively. If the area of the plot is 620 m², find (a) the length of fencing required to enclose the plot and (b) the angles of the triangular plot.

3. A jib crane is shown in Fig. 23.13. If the tie rod PR is 8.0 m long and PQ is 4.5 m long, determine (a) the length of jib RQ and (b) the angle between the jib and the tie rod.

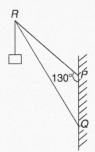

Figure 23.13

4. A building site is in the form of a quadrilateral, as shown in Fig. 23.14, and its area is 1510 m². Determine the length of the perimeter of the site.

Figure 23.14

5. Determine the length of members BF and EB in the roof truss shown in Fig. 23.15.

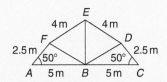

Figure 23.15

6. A laboratory 9.0 m wide has a span roof which slopes at 36° on one side and 44° on the other. Determine the lengths of the roof slopes.

7. PQ and QR are the phasors representing the alternating currents in two branches of a circuit. Phasor PQ is 20.0 A and is horizontal. Phasor QR (which is joined to the end of PQ to form triangle PQR) is 14.0 A and is at an angle of 35° to the horizontal. Determine the resultant phasor PR and the angle it makes with phasor PQ.

23.6 Further practical situations involving trigonometry

Problem 11. A vertical aerial stands on horizontal ground. A surveyor positioned due east of the aerial measures the elevation of the top as 48°. He moves due south 30.0 m and measures the elevation as 44°. Determine the height of the aerial

In Fig. 23.16, DC represents the aerial, A is the initial position of the surveyor and B his final position.

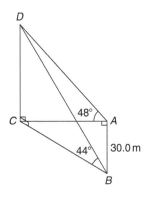

Figure 23.16

From triangle ACD, $\tan 48° = \dfrac{DC}{AC}$ from which

$$AC = \frac{DC}{\tan 48°}$$

Similarly, from triangle BCD, $BC = \dfrac{DC}{\tan 44°}$

For triangle ABC, using Pythagoras' theorem,

$$BC^2 = AB^2 + AC^2$$

$$\left(\frac{DC}{\tan 44°}\right)^2 = (30.0)^2 + \left(\frac{DC}{\tan 48°}\right)^2$$

$$DC^2\left(\frac{1}{\tan^2 44°} - \frac{1}{\tan^2 48°}\right) = 30.0^2$$

$$DC^2(1.072323 - 0.810727) = 30.0^2$$

$$DC^2 = \frac{30.0^2}{0.261596} = 3440.4$$

Hence, **height of aerial, $DC = \sqrt{3340.4} = 58.65\,\text{m}$**.

Problem 12.   A crank mechanism of a petrol engine is shown in Fig. 23.17. Arm OA is 10.0 cm long and rotates clockwise about O. The connecting rod AB is 30.0 cm long and end B is constrained to move horizontally.

(a)   For the position shown in Fig. 23.17, determine the angle between the connecting rod AB and the horizontal, and the length of OB.

(b)   How far does B move when angle AOB changes from 50° to 120°?

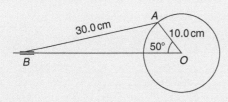

Figure 23.17

(a)   Applying the sine rule, $\dfrac{AB}{\sin 50°} = \dfrac{AO}{\sin B}$

from which $\sin B = \dfrac{AO \sin 50°}{AB} = \dfrac{10.0 \sin 50°}{30.0}$

$$= 0.2553$$

Hence, $B = \sin^{-1} 0.2553 = 14.79°$ (or 165.21°, which is not possible)

Hence, **the connecting rod AB makes an angle of 14.79° with the horizontal**.

Angle $OAB = 180° - 50° - 14.79° = 115.21°$

Applying the sine rule: $\dfrac{30.0}{\sin 50°} = \dfrac{OB}{\sin 115.21°}$

from which

$$OB = \frac{30.0 \sin 115.21°}{\sin 50°}$$

$$= \mathbf{35.43\,cm}$$

(b)   Fig. 23.18 shows the initial and final positions of the crank mechanism.

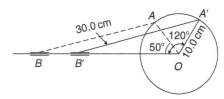

Figure 23.18

In triangle $OA'B'$, applying the sine rule,

$$\frac{30.0}{\sin 120°} = \frac{10.0}{\sin A'B'O}$$

from which $\sin A'B'O = \dfrac{10.0 \sin 120°}{30.0} = 0.28868$

Hence, $A'B'O = \sin^{-1} 0.28868 = 16.78°$ (or 163.22°, which is not possible)

Angle $OA'B' = 180° - 120° - 16.78° = 43.22°$

Applying the sine rule, $\dfrac{30.0}{\sin 120°} = \dfrac{OB'}{\sin 43.22°}$

from which $OB' = \dfrac{30.0 \sin 43.22°}{\sin 120°} = 23.72\,\text{cm}$

Since $OB = 35.43\,\text{cm}$ and $OB' = 23.72\,\text{cm}$, $BB' = 35.43 - 23.72 = 11.71\,\text{cm}$

Hence, **B moves 11.71 cm when angle AOB changes from 50° to 120°**

Problem 13.   The area of a field is in the form of a quadrilateral $ABCD$ as shown in Fig. 23.19. Determine its area

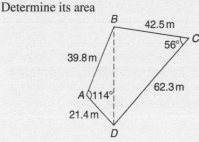

Figure 23.19

A diagonal drawn from B to D divides the quadrilateral into two triangles.

Area of quadrilateral $ABCD$

$$= \text{area of triangle } ABD$$
$$+ \text{area of triangle } BCD$$

$$= \frac{1}{2}(39.8)(21.4)\sin 114°$$

$$+ \frac{1}{2}(42.5)(62.3)\sin 56°$$

$$= 389.04 + 1097.5$$

$$= \mathbf{1487\,m^2}$$

Now try the following Practice Exercise

Practice Exercise 93   More practical situations involving trigonometry (answers on page 432)

1. Three forces acting on a fixed point are represented by the sides of a triangle of dimensions 7.2 cm, 9.6 cm and 11.0 cm. Determine the angles between the lines of action and the three forces.

2. A vertical aerial AB, 9.60 m high, stands on ground which is inclined 12° to the horizontal. A stay connects the top of the aerial A to a point C on the ground 10.0 m downhill from B, the foot of the aerial. Determine (a) the length of the stay and (b) the angle the stay makes with the ground.

3. A reciprocating engine mechanism is shown in Fig. 23.20. The crank AB is 12.0 cm long and the connecting rod BC is 32.0 cm long. For the position shown determine the length of AC and the angle between the crank and the connecting rod.

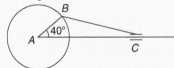

Figure 23.20

4. From Fig. 23.20, determine how far C moves, correct to the nearest millimetre, when angle CAB changes from 40° to 160°, B moving in an anticlockwise direction.

5. A surveyor standing W 25°S of a tower measures the angle of elevation of the top of the tower as 46°30′. From a position E 23°S from the tower the elevation of the top is 37°15′. Determine the height of the tower if the distance between the two observations is 75 m.

6. Calculate, correct to 3 significant figures, the co-ordinates x and y to locate the hole centre at P shown in Fig. 23.21.

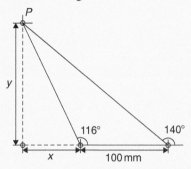

Figure 23.21

7. An idler gear, 30 mm in diameter, has to be fitted between a 70 mm diameter driving gear and a 90 mm diameter driven gear, as shown in Fig. 23.22. Determine the value of angle θ between the centre lines.

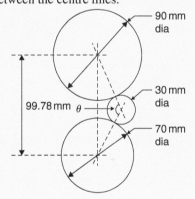

Figure 23.22

8. 16 holes are equally spaced on a pitch circle of 70 mm diameter. Determine the length of the chord joining the centres of two adjacent holes.

For fully worked solutions to each of the problems in Practice Exercises 90 to 93 in this chapter, go to the website:
www.routledge.com/cw/bird

Chapter 24

Cartesian and polar co-ordinates

Why it is important to understand: Cartesian and polar co-ordinates

Applications where polar co-ordinates would be used include terrestrial navigation with sonar-like devices, and those in engineering and science involving energy radiation patterns. Applications where Cartesian co-ordinates would be used include any navigation on a grid and anything involving raster graphics (i.e. bitmap – a dot matrix data structure representing a generally rectangular grid of pixels). The ability to change from Cartesian to polar co-ordinates is vitally important when using complex numbers and their use in a.c. electrical circuit theory and with vector geometry.

At the end of this chapter, you should be able to:

- change from Cartesian to polar co-ordinates
- change from polar to Cartesian co-ordinates
- use a scientific notation calculator to change from Cartesian to polar co-ordinates and vice-versa

24.1 Introduction

There are two ways in which the position of a point in a plane can be represented. These are

(a) Cartesian co-ordinates, (named after Descartes*), i.e. (x, y)

(b) Polar co-ordinates, i.e. (r, θ), where r is a radius from a fixed point and θ is an angle from a fixed point.

---

*Who was Descartes? To find out more go to
www.routledge.com/cw/bird

24.2 Changing from Cartesian to polar co-ordinates

In Fig. 24.1, if lengths x and y are known then the length of r can be obtained from Pythagoras' theorem (see Chapter 21) since OPQ is a right-angled triangle.

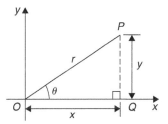

Figure 24.1

Hence, $r^2 = (x^2 + y^2)$, from which $r = \sqrt{x^2 + y^2}$

From trigonometric ratios (see Chapter 21), $\tan\theta = \dfrac{y}{x}$

from which $\theta = \tan^{-1}\dfrac{y}{x}$

$r = \sqrt{x^2 + y^2}$ and $\theta = \tan^{-1}\dfrac{y}{x}$ are the two formulae we need to change from Cartesian to polar co-ordinates. The angle θ, which may be expressed in degrees or radians, must **always** be measured from the positive x-axis; i.e. measured from the line OQ in Fig. 24.1. It is suggested that when changing from Cartesian to polar co-ordinates a diagram should always be sketched.

Problem 1.   Change the Cartesian co-ordinates (3, 4) into polar co-ordinates

A diagram representing the point (3, 4) is shown in Fig. 24.2.

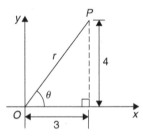

Figure 24.2

From Pythagoras' theorem, $r = \sqrt{3^2 + 4^2} = 5$ (note that -5 has no meaning in this context).

By trigonometric ratios, $\theta = \tan^{-1}\dfrac{4}{3} = 53.13°$ or 0.927 rad.

Note that $53.13° = 53.13 \times \dfrac{\pi}{180}$ rad $= 0.927$ rad.

Hence, **(3, 4) in Cartesian co-ordinates corresponds to (5, 53.13°) or (5, 0.927 rad) in polar co-ordinates**.

Problem 2.   Express in polar co-ordinates the position (−4, 3)

A diagram representing the point using the Cartesian co-ordinates (−4, 3) is shown in Fig. 24.3.

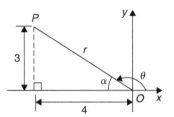

Figure 24.3

From Pythagoras' theorem, $r = \sqrt{4^2 + 3^2} = 5$

By trigonometric ratios, $\alpha = \tan^{-1}\dfrac{3}{4} = 36.87°$ or 0.644 rad

Hence, $\theta = 180° - 36.87° = 143.13°$

or $\theta = \pi - 0.644 = 2.498$ rad

Hence, **the position of point P in polar co-ordinate form is (5, 143.13°) or (5, 2.498 rad).**

Problem 3.   Express (−5, −12) in polar co-ordinates

A sketch showing the position (−5, −12) is shown in Fig. 24.4.

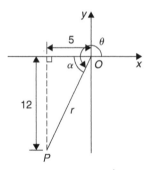

Figure 24.4

$r = \sqrt{5^2 + 12^2} = 13$ and $\alpha = \tan^{-1}\dfrac{12}{5} = 67.38°$ or 1.176 rad

Hence, $\theta = 180° + 67.38° = 247.38°$

or $\theta = \pi + 1.176 = 4.318$ rad.

Thus, **(−5, −12) in Cartesian co-ordinates corresponds to (13, 247.38°) or (13, 4.318 rad) in polar co-ordinates**.

Problem 4.   Express (2, −5) in polar co-ordinates

A sketch showing the position (2, −5) is shown in Fig. 24.5.

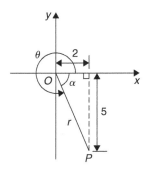

Figure 24.5

$r = \sqrt{2^2 + 5^2} = \sqrt{29} = 5.385$, correct to 3 decimal places

$\alpha = \tan^{-1} \dfrac{5}{2} = 68.20°$ or 1.190 rad

Hence, $\theta = 360° - 68.20° = 291.80°$
or $\theta = \quad 2\pi - 1.190 = 5.093$ rad.

Thus, **(2, −5) in Cartesian co-ordinates corresponds to (5.385, 291.80°) or (5.385, 5.093 rad) in polar co-ordinates**.

Now try the following Practice Exercise

Practice Exercise 94   Changing from Cartesian to polar co-ordinates (answers on page 432)

In Problems 1 to 8, express the given Cartesian co-ordinates as polar co-ordinates, correct to 2 decimal places, in both degrees and radians.

1. $(3, 5)$ 2. $(6.18, 2.35)$

3. $(-2, 4)$ 4. $(-5.4, 3.7)$

5. $(-7, -3)$ 6. $(-2.4, -3.6)$

7. $(5, -3)$ 8. $(9.6, -12.4)$

24.3 Changing from polar to Cartesian co-ordinates

From the right-angled triangle OPQ in Fig. 24.6,

$\cos\theta = \dfrac{x}{r}$ and $\sin\theta = \dfrac{y}{r}$ from trigonometric ratios

Hence, $x = r\cos\theta$ and $y = r\sin\theta$

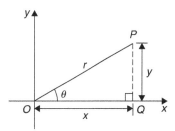

Figure 24.6

If lengths r and angle θ are known then $x = r\cos\theta$ and $y = r\sin\theta$ are the two formulae we need to change from polar to Cartesian co-ordinates.

Problem 5.   Change $(4, 32°)$ into Cartesian co-ordinates

A sketch showing the position $(4, 32°)$ is shown in Fig. 24.7.

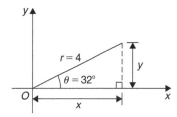

Figure 24.7

Now $x = r\cos\theta = 4\cos 32° = 3.39$

and $y = r\sin\theta = 4\sin 32° = 2.12$

Hence, **$(4, 32°)$ in polar co-ordinates corresponds to $(3.39, 2.12)$ in Cartesian co-ordinates**.

Problem 6.   Express $(6, 137°)$ in Cartesian co-ordinates

A sketch showing the position $(6, 137°)$ is shown in Fig. 24.8.

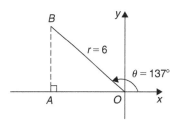

Figure 24.8

$$x = r\cos\theta = 6\cos 137° = -4.388$$

which corresponds to length OA in Fig. 24.8.

$$y = r\sin\theta = 6\sin 137° = 4.092$$

which corresponds to length AB in Fig. 24.8.

Thus, **(6, 137°) in polar co-ordinates corresponds to (−4.388, 4.092) in Cartesian co-ordinates**.

(Note that when changing from polar to Cartesian co-ordinates it is not quite so essential to draw a sketch. Use of $x = r\cos\theta$ and $y = r\sin\theta$ automatically produces the correct values and signs.)

> **Problem 7.** Express (4.5, 5.16 rad) in Cartesian co-ordinates

A sketch showing the position (4.5, 5.16 rad) is shown in Fig. 24.9.

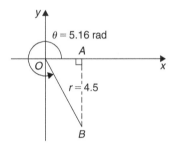

Figure 24.9

$$x = r\cos\theta = 4.5\cos 5.16 = 1.948$$

which corresponds to length OA in Fig. 24.9.

$$y = r\sin\theta = 4.5\sin 5.16 = -4.057$$

which corresponds to length AB in Fig. 24.9.

Thus, **(1.948, −4.057) in Cartesian co-ordinates corresponds to (4.5, 5.16 rad) in polar co-ordinates**.

Now try the following Practice Exercise

> **Practice Exercise 95   Changing polar to Cartesian co-ordinates (answers on page 433)**

In Problems 1 to 8, express the given polar co-ordinates as Cartesian co-ordinates, correct to 3 decimal places.

1. (5, 75°)
2. (4.4, 1.12 rad)
3. (7, 140°)
4. (3.6, 2.5 rad)
5. (10.8, 210°)
6. (4, 4 rad)
7. (1.5, 300°)
8. (6, 5.5 rad)

9. Fig. 24.10 shows 5 equally spaced holes on an 80 mm pitch circle diameter. Calculate their co-ordinates relative to axes Ox and Oy in (a) polar form, (b) Cartesian form.

10. In Fig. 24.10, calculate the shortest distance between the centres of two adjacent holes.

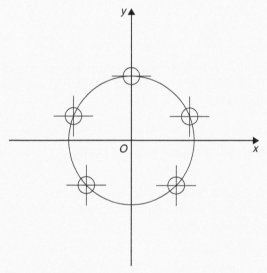

Figure 24.10

24.4 Use of Pol/Rec functions on calculators

Another name for Cartesian co-ordinates is **rectangular** co-ordinates. Many scientific notation calculators have **Pol** and **Rec** functions. 'Rec' is an abbreviation of 'rectangular' (i.e. Cartesian) and 'Pol' is an abbreviation of 'polar'. Check the operation manual for your particular calculator to determine how to use these two functions. They make changing from Cartesian to polar co-ordinates, and vice-versa, so much quicker and easier. For example, with the Casio fx-991ES PLUS calculator, or similar, to change the Cartesian number (3, 4) into polar form, the following procedure is adopted.

1. Press 'shift'   2. Press 'Pol'   3. Enter 3

4. Enter 'comma' (obtained by 'shift' then))

5. Enter 4   6. Press)   7. Press =

The answer is $r = 5, \theta = 53.13°$

Hence, **(3, 4) in Cartesian form is the same as (5, 53.13°) in polar form**.

If the angle is required in **radians**, then before repeating the above procedure press 'shift', 'mode' and then 4 to change your calculator to radian mode.

Similarly, to change the polar form number (7, 126°) into Cartesian or rectangular form, adopt the following procedure.

1. Press 'shift'   2. Press 'Rec'

3. Enter 7   4. Enter 'comma'

5. Enter 126 (assuming your calculator is in degrees mode)

6. Press)   7. Press =

The answer is $X = -4.11$ and, scrolling across, $Y = 5.66$, correct to 2 decimal places.

Hence, **(7, 126°) in polar form is the same as (-4.11, 5.66) in rectangular or Cartesian form**.

Now return to Practice Exercises 94 and 95 in this chapter and use your calculator to determine the answers, and see how much more quickly they may be obtained.

For fully worked solutions to each of the problems in Practice Exercises 94 and 95 in this chapter, go to the website:

www.routledge.com/cw/bird

This assignment covers the material contained in Chapters 22–24. *The marks available are shown in brackets at the end of each question.*

1. A sine wave is given by $y = 8\sin 4x$. State its peak value and its period, in degrees. (2)

2. A periodic function is given by $y = 15\tan 2x$. State its period in degrees. (2)

3. The frequency of a sine wave is 800 Hz. Calculate the periodic time in milliseconds. (2)

4. Calculate the frequency of a sine wave that has a periodic time of $40\,\mu$s. (2)

5. Calculate the periodic time for a sine wave having a frequency of 20 kHz. (2)

6. An alternating current completes 12 cycles in 16 ms. What is its frequency? (3)

7. A sinusoidal voltage is given by $e = 150\sin(500\pi t - 0.25)$ volts. Determine the
   (a) amplitude,
   (b) frequency,
   (c) periodic time,
   (d) phase angle (stating whether it is leading or lagging $150\sin 500\pi t$). (4)

8. Determine the acute angles in degrees, degrees and minutes, and radians.
   (a) $\sin^{-1} 0.4721$ (b) $\cos^{-1} 0.8457$
   (c) $\tan^{-1} 1.3472$ (9)

9. Sketch the following curves, labelling relevant points.
   (a) $y = 4\cos(\theta + 45°)$ (b) $y = 5\sin(2t - 60°)$ (8)

10. The current in an alternating current circuit at any time t seconds is given by $i = 120\sin(100\pi t + 0.274)$ amperes. Determine
    (a) the amplitude, frequency, periodic time and phase angle (with reference to $120\sin 100\pi t$),
    (b) the value of current when $t = 0$,
    (c) the value of current when $t = 6$ ms.
    Sketch one cycle of the oscillation. (16)

11. A triangular plot of land ABC is shown in Fig. RT9.1. Solve the triangle and determine its area. (10)

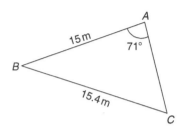

Figure RT9.1

12. A car is travelling 20 m above sea level. It then travels 500 m up a steady slope of $17°$. Determine, correct to the nearest metre, how high the car is now above sea level. (3)

13. Fig. RT9.2 shows a roof truss PQR with rafter $PQ = 3$ m. Calculate the length of
    (a) the roof rise PP',
    (b) rafter PR,
    (c) the roof span QR.
    Find also (d) the cross-sectional area of the roof truss. (11)

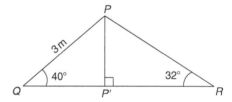

Figure RT9.2

14. Solve triangle ABC given $b = 10$ cm, $c = 15$ cm and $\angle A = 60°$. (10)

15. Change the following Cartesian co-ordinates into polar co-ordinates, correct to 2 decimal places, in both degrees and in radians.
    (a) $(-2.3, 5.4)$ (b) $(7.6, -9.2)$ (10)

16. Change the following polar co-ordinates into Cartesian co-ordinates, correct to 3 decimal places.
    (a) $(6.5, 132°)$ (b) $(3, 3\,\text{rad})$ (6)

Multiple choice questions Test 3
Graphs and trigonometry
This test covers the material in Chapters 17 to 24

All questions have only one correct answer (Answers on page 440).

1. In the right-angled triangle ABC shown in Fig. M3.1, sine A is given by:
   (a) b/a (b) c/b (c) b/c (d) a/b

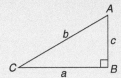

Figure M3.1

2. In the right-angled triangle ABC shown in Fig. M3.1, cosine C is given by:
   (a) a/b (b) c/b (c) a/c (d) b/a

3. In the right-angled triangle shown in Fig. M3.1, tangent A is given by:
   (a) b/c (b) a/c (c) a/b (d) c/a

4. A graph of resistance against voltage for an electrical circuit is shown in Fig. M3.2. The equation relating resistance R and voltage V is:
   (a) $R = 1.45\,V + 40$
   (b) $R = 0.8\,V + 20$
   (c) $R = 1.45\,V + 20$
   (d) $R = 1.25\,V + 20$

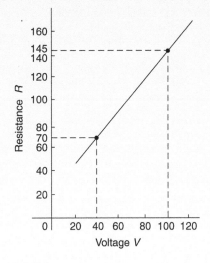

Figure M3.2

5. $\dfrac{3\pi}{4}$ radians is equivalent to:
   (a) $135°$ (b) $270°$ (c) $45°$ (d) $67.5°$

6. In the triangular template ABC shown in Fig. M3.3, the length AC is:
   (a) 6.17 cm   (b) 11.17 cm
   (c) 9.22 cm   (d) 12.40 cm

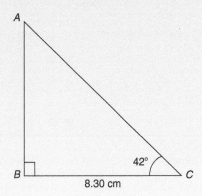

Figure M3.3

7. A graph of y against x, two engineering quantities, produces a straight line.
   A table of values is shown below:

x	2	-1	p
y	9	3	5

   The value of p is:
   (a) $-\dfrac{1}{2}$ (b) -2 (c) 3   (d) 0

8. $(-4, 3)$ in polar co-ordinates is:
   (a) $(5, 2.498 \text{ rad})$ (b) $(7,\ 36.87°)$
   (c) $(5, 36.87°)$ (d) $(5, 323.13°)$

9. Correct to 3 decimal places, $\sin(-2.6 \text{ rad})$ is:
   (a) 0.516 (b) -0.045 (c) -0.516 (d) 0.045

10. Which of the straight lines shown in Fig. M3.4 has the equation $y + 4 = 2x$?
    (a) (i)   (b) (ii)   (c) (iii)   (d) (iv)

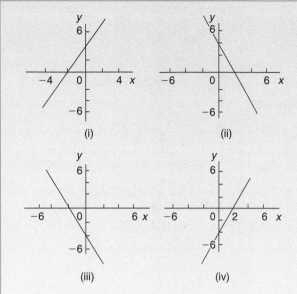

(i)   (ii)

(iii)   (iv)

Figure M3.4

11. For the right-angled triangle PQR shown in Fig. M3.5, angle R is equal to:

   (a) 41.41°   (b) 48.59°   (c) 36.87°   (d) 53.13°

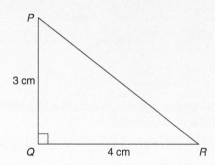

Figure M3.5

12. If $\cos A = \dfrac{12}{13}$, then $\sin A$ is equal to:

   (a) $\dfrac{5}{13}$ (b) $\dfrac{13}{12}$ (c) $\dfrac{5}{12}$ (d) $\dfrac{12}{5}$

13. The area of triangle XYZ in Fig. M3.6 is:
   (a) 24.22 cm^2 (b) 19.35 cm^2
   (c) 38.72 cm^2 (d) 32.16 cm^2

Questions 14 to 17 relate to the following information:

x and y are two related engineering variables and p and q are constants.

For the law $y - p = \dfrac{q}{x}$ to be verified it is necessary to plot a graph of the variables.

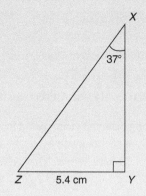

Figure M3.6

14. On the vertical axis is plotted:

   (a) y (b) p (c) q (d) x

15. On the horizontal axis is plotted:

   (a) x (b) $\dfrac{q}{x}$ (c) $\dfrac{1}{x}$ (d) p

16. The gradient of the graph is:

   (a) y (b) p (c) q (d) x

17. The vertical axis intercept is:

   (a) y (b) p (c) q (d) x

18. The value, correct to 3 decimal places, of $\cos\left(\dfrac{-3\pi}{4}\right)$ is:

   (a) 0.999   (b) 0.707   (c) -0.999 (d) -0.707

19. A triangle has sides $a = 9.0$ cm, $b = 8.0$ cm and $c = 6.0$ cm. Angle A is equal to:

   (a) 82.42°   (b) 56.49°   (c) 78.58°   (d) 79.87°

20. An alternating current is given by: $i = 15\sin(100\pi t - 0.25)$ amperes. When time $t = 5$ ms, the current i has a value of:

   (a) 0.35 A   (b) 14.53 A   (c) 15 A   (d) 0.41 A

21. A graph relating effort E (plotted vertically) against load L (plotted horizontally) for a set of pulleys is given by $L + 30 = 6E$. The gradient of the graph is:

   (a) $\dfrac{1}{6}$ (b) 5   (c) 6   (d) $\dfrac{1}{5}$

22. The displacement x metres of a mass from a fixed point about which it is oscillating is given by

$x = 3\cos\omega t - 4\sin\omega t$, where t is the time in seconds. x may be expressed as:

(a) $-\sin(\omega t - 2.50)$ metres

(b) $7\sin(\omega t - 36.87°)$ metres

(c) $5\sin\omega t$ metres

(d) $5\sin(\omega t + 2.50)$ metres

23. A sinusoidal current is given by: $i = R\sin(\omega t + \alpha)$. Which of the following statements is incorrect?

(a) R is the average value of the current

(b) frequency $= \dfrac{\omega}{2\pi}$Hz

(c) ω = angular velocity

(d) periodic time $= \dfrac{2\pi}{\omega}$s

24. A vertical tower stands on level ground. At a point 100 m from the foot of the tower the angle of elevation of the top is 20°. The height of the tower is:

(a) 274.7 m   (b) 36.4 m   (c) 34.3 m   (d) 94.0 m

Questions 25 to 28 relate to the following information:

x and y are two related engineering variables and a and b are constants.
For the law $y - bx = ax^2$ to be verified it is necessary to plot a graph of the variables.

25. On the vertical axis is plotted:

(a) y (b) a (c) $\dfrac{y}{x}$ (d) x

26. On the horizontal axis is plotted:

(a) x (b) $\dfrac{a}{x}$ (c) $\dfrac{1}{x}$ (d) $\dfrac{b}{x}$

27. The gradient of the graph is:

(a) y (b) a (c) b (d) x

28. The vertical axis intercept is:

(a) y (b) a (c) b (d) x

29. $(7, 141°)$ in Cartesian co-ordinates is:

(a) $(5.44, -4.41)$ (b) $(-5.44, -4.41)$

(c) $(5.44, 4.41)$ (d) $(-5.44, 4.41)$

30. The angles between 0° and 360° whose tangent is -1.7624 are:

(a) 60.43° and 240.43°

(b) 119.57° and 299.57°

(c) 119.57° and 240.43°

(d) 150.43° and 299.57°

31. In the triangular template DEF shown in Fig. M3.7, angle F is equal to:

(a) 43.5°   (b) 28.6°   (c) 116.4°   (d) 101.5°

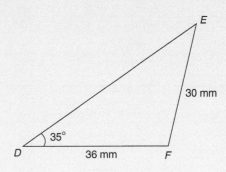

Figure M3.7

32. The area of the triangular template DEF shown in Fig. M3.7 is:
(a) 529.2 mm^2 (b) 258.5 mm^2
(c) 483.7 mm^2 (d) 371.7 mm^2

33. Here are four equations in x and y. When x is plotted against y, in each case a straight line results.

(i) $y + 3 = 3x$ (ii) $y + 3x = 3$ (iii) $\dfrac{y}{2} - \dfrac{3}{2} = x$
(iv) $\dfrac{y}{3} = x + \dfrac{2}{3}$

Which of these equations are parallel to each other?

(a) (i) and (ii)   (b) (i) and (iv)

(c) (ii) and (iii)   (d) (ii) and (iv)

34. An alternating voltage v is given by $v = 100\sin\left(100\pi t + \dfrac{\pi}{4}\right)$ volts. When $v = 50$ volts, the time t is equal to:

(a) 0.093 s   (b) -0.908 ms

(c) -0.833 ms   (d) -0.162 s

35. The equation of the graph shown in Fig. M3.8 is:

(a) $x(x + 1) = \dfrac{15}{4}$

(b) $4x^2 - 4x - 15 = 0$

(c) $x^2 - 4x - 5 = 0$

(d) $4x^2 + 4x - 15 = 0$

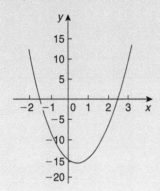

Figure M3.8

36. The area of triangle PQR is given by:

(a) $\dfrac{1}{2}pr \cos Q$

(b) $\sqrt{[(s-p)(s-q)(s-r)]}$ where
$s = \dfrac{p+q+r}{2}$

(c) $\dfrac{1}{2}rq \sin P$

(d) $\dfrac{1}{2}pq \sin Q$

37. The relationship between two related engineering variables x and y is $y - cx = bx^2$ where b and c are constants. To produce a straight line graph it is necessary to plot:

(a) x vertically against y horizontally

(b) y vertically against x^2 horizontally

(c) $\dfrac{y}{x}$ vertically against x^2 horizontally

(d) y vertically against x horizontally

38. In triangle ABC in Fig. M3.9, length AC is:
(a) 14.90 cm   (b) 18.15 cm

(c) 13.16 cm   (d) 14.04 cm

39. In an experiment demonstrating Hooke's law, the strain in a copper wire was measured for various stresses. The results included

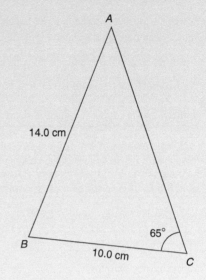

Figure M3.9

Stress (megapascals)	18.24	24.00	39.36
Strain	0.00019	0.00025	0.00041

When stress is plotted vertically against strain horizontally a straight line graph results. Young's modulus of elasticity for copper, which is given by the gradient of the graph, is:

(a) 96×10^9 Pa   (b) 1.04×10^{-11} Pa

(c) 96 Pa   (d) 96000 Pa

40. In triangle ABC in Fig. M3.10, the length AC is:
(a) 18.79 cm   (b) 70.89 cm

(c) 22.89 cm   (d) 16.10 cm

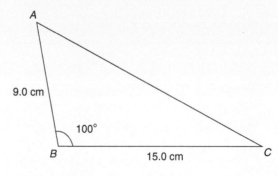

Figure M3.10

The companion website for this book contains the above multiple-choice test. If you prefer to attempt the test online then visit:

www.routledge.com/cw/bird

For a copy of this multiple choice test, go to:

www.routledge.com/cw/bird

Chapter 25

Areas of common shapes

Why it is important to understand: Areas of common shapes

To paint, wallpaper or panel a wall, you must know the total area of the wall so you can buy the appropriate amount of finish. When designing a new building, or seeking planning permission, it is often necessary to specify the total floor area of the building. In construction, calculating the area of a gable end of a building is important when determining the number of bricks and amount of mortar to order. When using a bolt, the most important thing is that it is long enough for your particular application and it may also be necessary to calculate the shear area of the bolt connection. Ridge vents allow a home to properly vent, while disallowing rain or other forms of precipitation to leak into the attic or crawlspace underneath the roof. Equal amounts of cool air and warm air flowing through the vents is paramount for proper heat exchange. Calculating how much surface area is available on the roof aids in determining how long the ridge vent should run. Arches are everywhere, from sculptures and monuments to pieces of architecture and strings on musical instruments; finding the height of an arch or its cross-sectional area is often required. Determining the cross-sectional areas of beam structures is vitally important in design engineering. There are thus a large number of situations in engineering where determining area is important.

At the end of this chapter, you should be able to:

- state the SI unit of area
- identify common polygons – triangle, quadrilateral, pentagon, hexagon, heptagon and octagon
- identify common quadrilaterals – rectangle, square, parallelogram, rhombus and trapezium
- calculate areas of quadrilaterals and circles
- appreciate that areas of similar shapes are proportional to the squares of the corresponding linear dimensions

25.1 Introduction

Area is a measure of the size or extent of a plane surface. Area is measured in **square units** such as mm^2, cm^2 and m^2. This chapter deals with finding the areas of common shapes.

In engineering it is often important to be able to calculate simple areas of various shapes. In everyday life its important to be able to measure area to, say, lay a carpet, order sufficient paint for a decorating job or order sufficient bricks for a new wall.

On completing this chapter you will be able to recognise common shapes and be able to find the areas of rectangles, squares, parallelograms, triangles, trapeziums and circles.

25.2 Common shapes

25.2.1 Polygons

A polygon is a closed plane figure bounded by straight lines. A polygon which has

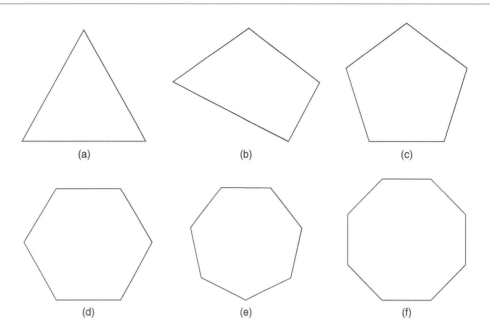

Figure 25.1

3 sides is called a **triangle** – see Fig. 25.1(a)
4 sides is called a **quadrilateral** – see Fig. 25.1(b)
5 sides is called a **pentagon** – see Fig. 25.1(c)
6 sides is called a **hexagon** – see Fig. 25.1(d)
7 sides is called a **heptagon** – see Fig. 25.1(e)
8 sides is called an **octagon** – see Fig. 25.1(f)

25.2.2 Quadrilaterals

There are five types of quadrilateral, these being rectangle, square, parallelogram, rhombus and trapezium. If the opposite corners of any quadrilateral are joined by a straight line, two triangles are produced. Since the sum of the angles of a triangle is 180°, the sum of the angles of a quadrilateral is 360°

Rectangle

In the rectangle *ABCD* shown in Fig. 25.2,

(a) all four angles are right angles,

(b) the opposite sides are parallel and equal in length, and

(c) diagonals *AC* and *BD* are equal in length and bisect one another.

Square

In the square *PQRS* shown in Fig. 25.3,

(a) all four angles are right angles,

(b) the opposite sides are parallel,

(c) all four sides are equal in length, and

(d) diagonals *PR* and *QS* are equal in length and bisect one another at right angles.

Parallelogram

In the parallelogram *WXYZ* shown in Fig. 25.4,

(a) opposite angles are equal,

(b) opposite sides are parallel and equal in length, and

(c) diagonals *WY* and *XZ* bisect one another.

Rhombus

In the rhombus *ABCD* shown in Fig. 25.5,

(a) opposite angles are equal,

(b) opposite angles are bisected by a diagonal,

(c) opposite sides are parallel,

(d) all four sides are equal in length, and

(e) diagonals *AC* and *BD* bisect one another at right angles.

Trapezium

In the trapezium *EFGH* shown in Fig. 25.6,

(a) only one pair of sides is parallel.

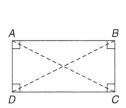

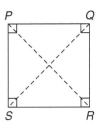

Figure 25.2   **Figure 25.3**

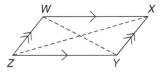

Figure 25.4

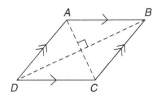

Figure 25.5   **Figure 25.6**

Problem 1. State the types of quadrilateral shown in Fig. 25.7 and determine the angles marked a to l

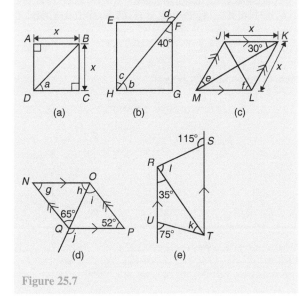

Figure 25.7

(a) **ABCD is a square**
The diagonals of a square bisect each of the right angles, hence

$$a = \frac{90°}{2} = \mathbf{45°}$$

(b) **EFGH is a rectangle**
In triangle FGH, $40° + 90° + b = 180°$, since the angles in a triangle add up to $180°$, from which $\mathbf{b = 50°}$. Also, $\mathbf{c = 40°}$ (alternate angles between parallel lines EF and HG). (Alternatively, b and c are complementary; i.e. add up to $90°$)
$d = 90° + c$ (external angle of a triangle equals the sum of the interior opposite angles), hence $\mathbf{d = 90° + 40° = 130°}$ (or $\angle EFH = 50°$ and $d = 180° - 50° = 130°$)

(c) **JKLM is a rhombus**
The diagonals of a rhombus bisect the interior angles and the opposite internal angles are equal. Thus, $\angle JKM = \angle MKL = \angle JMK = \angle LMK = 30°$, hence, $\mathbf{e = 30°}$
In triangle KLM, $30° + \angle KLM + 30° = 180°$ (the angles in a triangle add up to $180°$), hence, $\angle KLM = 120°$. The diagonal JL bisects $\angle KLM$, hence, $\mathbf{f = \dfrac{120°}{2} = 60°}$

(d) **NOPQ is a parallelogram**
$\mathbf{g = 52°}$ since the opposite interior angles of a parallelogram are equal.
In triangle NOQ, $g + h + 65° = 180°$ (the angles in a triangle add up to $180°$), from which $\mathbf{h = 180° - 65° - 52° = 63°}$
$\mathbf{i = 65°}$ (alternate angles between parallel lines NQ and OP).
$\mathbf{j = 52° + i = 52° + 65° = 117°}$ (the external angle of a triangle equals the sum of the interior opposite angles). (Alternatively, $\angle PQO = h = 63°$; hence, $j = 180° - 63° = 117°$)

(e) **RSTU is a trapezium**
$35° + k = 75°$ (external angle of a triangle equals the sum of the interior opposite angles), hence, $\mathbf{k = 40°}$
$\angle STR = 35°$ (alternate angles between parallel lines RU and ST). $l + 35° = 115°$ (external angle of a triangle equals the sum of the interior opposite angles), hence, $\mathbf{l = 115° - 35° = 80°}$

Now try the following Practice Exercise

Practice Exercise 96   Common shapes (answers on page 433)

1. Find the angles p and q in Fig. 25.8(a).

2. Find the angles r and s in Fig. 25.8(b).

3.   Find the angle t in Fig. 25.8(c).

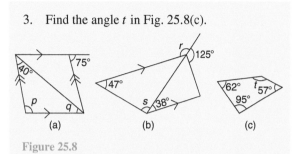

Figure 25.8

25.3   Areas of common shapes

The formulae for the areas of common shapes are shown in Table 25.1.

Here are some worked problems to demonstrate how the formulae are used to determine the area of common shapes.

Problem 2.   Calculate the area and length of the perimeter of the square shown in Fig. 25.9

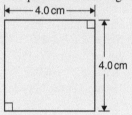

Figure 25.9

$$\textbf{Area of square} = x^2 = (4.0)^2 = 4.0\,\text{cm} \times 4.0\,\text{cm}$$
$$= \textbf{16.0\,cm}^2$$

(Note the unit of area is cm × cm = cm²; i.e. square centimetres or centimetres squared.)

$$\textbf{Perimeter of square} = 4.0\,\text{cm} + 4.0\,\text{cm} + 4.0\,\text{cm}$$
$$+ 4.0\,\text{cm} = \textbf{16.0\,cm}$$

Problem 3.   Calculate the area and length of the perimeter of the rectangle shown in Fig. 25.10

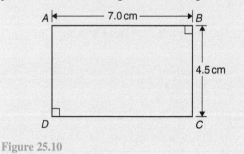

Figure 25.10

$$\textbf{Area of rectangle} = l \times b = 7.0 \times 4.5$$
$$= \textbf{31.5\,cm}^2$$

$$\textbf{Perimeter of rectangle} = 7.0\,\text{cm} + 4.5\,\text{cm}$$
$$+ 7.0\,\text{cm} + 4.5\,\text{cm}$$
$$= \textbf{23.0\,cm}$$

Problem 4.   Calculate the area of the parallelogram shown in Fig. 25.11

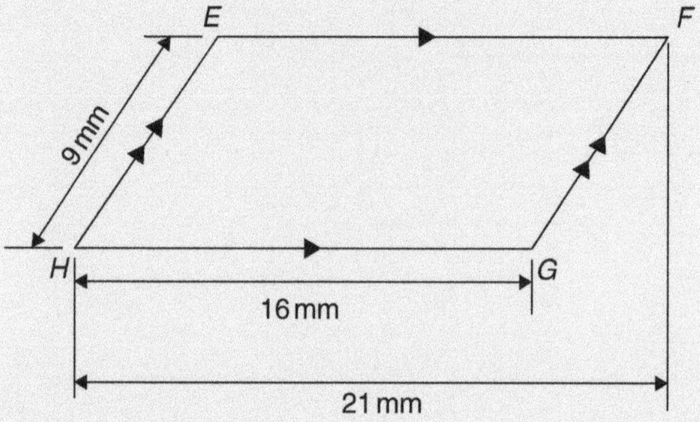

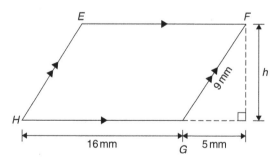

Figure 25.11

Area of a parallelogram = base × perpendicular height

The perpendicular height h is not shown in Fig. 25.11 but may be found using Pythagoras' theorem (see Chapter 21).

From Fig. 25.12, $9^2 = 5^2 + h^2$, from which $h^2 = 9^2 - 5^2 = 81 - 25 = 56$

Hence, perpendicular height,

$$h = \sqrt{56} = 7.48\,\text{mm}.$$

Figure 25.12

Hence, **area of parallelogram EFGH**

$$= 16\,\text{mm} \times 7.48\,\text{mm}$$
$$= \textbf{120\,mm}^2$$

Table 25.1 **Formulae for the areas of common shapes**

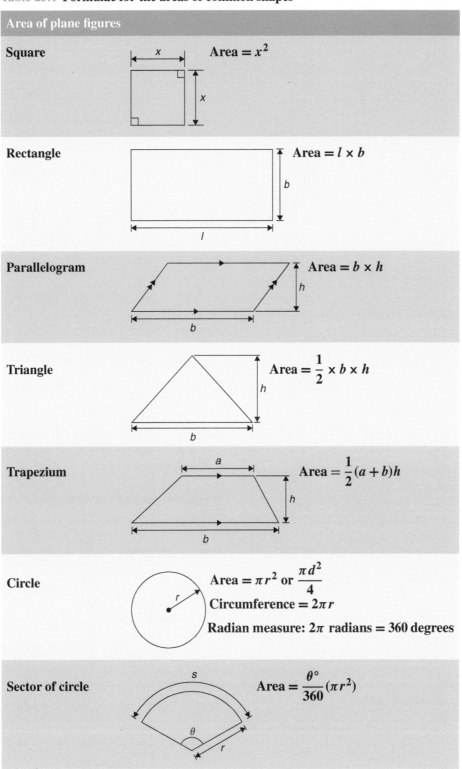

Area of plane figures		
Square		Area $= x^2$
Rectangle		Area $= l \times b$
Parallelogram		Area $= b \times h$
Triangle		Area $= \dfrac{1}{2} \times b \times h$
Trapezium		Area $= \dfrac{1}{2}(a + b)h$
Circle		Area $= \pi r^2$ or $\dfrac{\pi d^2}{4}$ Circumference $= 2\pi r$ Radian measure: 2π radians $= 360$ degrees
Sector of circle		Area $= \dfrac{\theta°}{360}(\pi r^2)$

Problem 5. Calculate the area of the triangle shown in Fig. 25.13

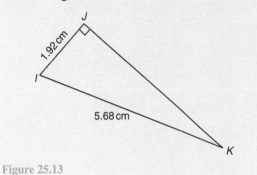

Figure 25.13

Area of triangle $IJK = \dfrac{1}{2} \times$ base $\times$ perpendicular height

$$= \dfrac{1}{2} \times IJ \times JK$$

To find JK, Pythagoras' theorem is used; i.e.

$$5.68^2 = 1.92^2 + JK^2, \text{ from which}$$

$$JK = \sqrt{5.68^2 - 1.92^2} = 5.346 \text{cm}$$

Hence, **area of triangle $IJK = \dfrac{1}{2} \times 1.92 \times 5.346$**

$$= \textbf{5.132cm}^2$$

Problem 6. Calculate the area of the trapezium shown in Fig. 25.14

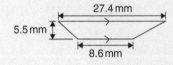

Figure 25.14

Area of a trapezium $= \dfrac{1}{2} \times$ (sum of parallel sides)

$\times$ (perpendicular distance between the parallel sides)

Hence, **area of trapezium $LMNO$**

$$= \dfrac{1}{2} \times (27.4 + 8.6) \times 5.5$$

$$= \dfrac{1}{2} \times 36 \times 5.5 = \textbf{99mm}^2$$

Problem 7. A rectangular tray is 820mm long and 400mm wide. Find its area in (a) mm² (b) cm² (c) m²

(a) **Area of tray** = length × width = 820 × 400

$$= \textbf{328000 mm}^2$$

(b) Since 1cm = 10mm, 1cm² = 1cm × 1cm

$$= 10\text{mm} \times 10\text{mm} = 100\text{mm}^2, \text{ or}$$

$$1\text{mm}^2 = \dfrac{1}{100}\text{cm}^2 = 0.01\text{cm}^2$$

Hence, **328000 mm²** = 328000 × 0.01 cm²

$$= \textbf{3280cm}^2$$

(c) Since 1m = 100cm, 1m² = 1m × 1m

$$= 100\text{cm} \times 100\text{cm} = 10000\text{cm}^2, \text{ or}$$

$$1\text{cm}^2 = \dfrac{1}{10000}\text{m}^2 = 0.0001\text{m}^2$$

Hence, **3280 cm²** = 3280 × 0.0001 m²

$$= \textbf{0.3280 m}^2$$

Problem 8. The outside measurements of a picture frame are 100cm by 50cm. If the frame is 4cm wide, find the area of the wood used to make the frame

A sketch of the frame is shown shaded in Fig. 25.15.

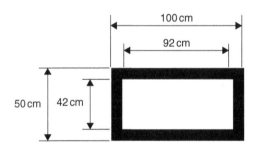

Figure 25.15

Area of wood = area of large rectangle − area of small rectangle

$$= (100 \times 50) - (92 \times 42)$$

$$= 5000 - 3864$$

$$= \textbf{1136cm}^2$$

Problem 9. Find the cross-sectional area of the girder shown in Fig. 25.16

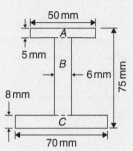

Figure 25.16

The girder may be divided into three separate rectangles, as shown.

$$\text{Area of rectangle } A = 50 \times 5 = 250\,\text{mm}^2$$

$$\text{Area of rectangle } B = (75 - 8 - 5) \times 6$$

$$= 62 \times 6 = 372\,\text{mm}^2$$

$$\text{Area of rectangle } C = 70 \times 8 = 560\,\text{mm}^2$$

Total area of girder $= 250 + 372 + 560$

$$= \textbf{1182}\,\textbf{mm}^2 \text{ or } \textbf{11.82}\,\textbf{cm}^2$$

Problem 10. Fig. 25.17 shows the gable end of a building. Determine the area of brickwork in the gable end

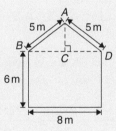

Figure 25.17

The shape is that of a rectangle and a triangle.

$$\text{Area of rectangle } = 6 \times 8 = 48\,\text{m}^2$$

$$\text{Area of triangle } = \frac{1}{2} \times \text{base} \times \text{height}$$

$CD = 4\,$m and $AD = 5\,$m, hence $AC = 3\,$m (since it is a 3, 4, 5 triangle – or by Pythagoras).

Hence, area of triangle $ABD = \frac{1}{2} \times 8 \times 3 = 12\,\text{m}^2$

Total area of brickwork $= 48 + 12$

$$= \textbf{60}\,\textbf{m}^2$$

Now try the following Practice Exercise

1. Name the types of quadrilateral shown in Fig. 25.18(i) to (iv) and determine for each (a) the area and (b) the perimeter.

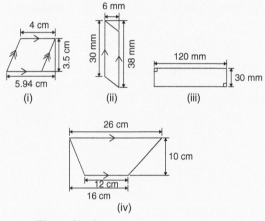

Figure 25.18

2. A rectangular plate is 85 mm long and 42 mm wide. Find its area in square centimetres.

3. A rectangular field has an area of 1.2 hectares and a length of 150 m. If 1 hectare = 10000 m^2, find (a) the field's width and (b) the length of a diagonal.

4. Find the area of a triangle whose base is 8.5 cm and perpendicular height is 6.4 cm.

5. A square has an area of 162 cm^2. Determine the length of a diagonal.

6. A rectangular picture has an area of 0.96 m^2. If one of the sides has a length of 800 mm, calculate, in millimetres, the length of the other side.

7. Determine the area of each of the angle iron sections shown in Fig. 25.19.

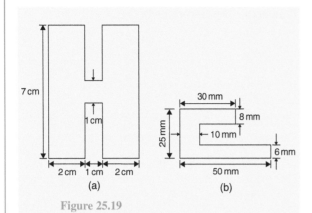

Figure 25.19

8. Fig. 25.20 shows a 4 m wide path around the outside of a 41 m by 37 m garden. Calculate the area of the path.

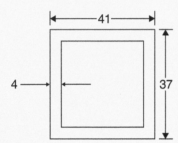

Figure 25.20

9. The area of a trapezium is 13.5 cm^2 and the perpendicular distance between its parallel sides is 3 cm. If the length of one of the parallel sides is 5.6 cm, find the length of the other parallel side.

10. Calculate the area of the steel plate shown in Fig. 25.21.

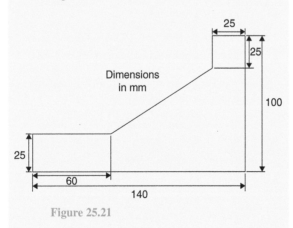

Figure 25.21

11. Determine the area of an equilateral triangle of side 10.0 cm.

12. If paving slabs are produced in 250 mm by 250 mm squares, determine the number of slabs required to cover an area of 2 m^2.

13. Fig. 25.22 shows a plan view of an office block to be built. The walls will have a height of 8 m, and it is necessary to make an estimate of the number of bricks required to build the walls. Assuming that any doors and windows are ignored in the calculation and that 48 bricks are required to build 1m^2 of wall, calculate the number of external bricks required.

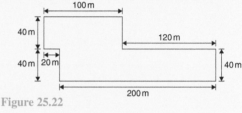

Figure 25.22

Here are some further worked problems on finding the areas of common shapes, using the formulae in Table 25.1, page 253.

Problem 11. Find the area of a circle having a radius of 5 cm

$$\textbf{Area of circle} = \pi r^2 = \pi (5)^2 = 25\pi = \textbf{78.54 cm}^2$$

Problem 12. Find the area of a circle having a diameter of 15 mm

$$\textbf{Area of circle} = \frac{\pi d^2}{4} = \frac{\pi (15)^2}{4} = \frac{225\pi}{4} = \textbf{176.7 mm}^2$$

Problem 13. Find the area of a circle having a circumference of 70 mm

Circumference, $c = 2\pi r$, hence

$$\text{radius}, r = \frac{c}{2\pi} = \frac{70}{2\pi} = \frac{35}{\pi} \text{ mm}$$

$$\textbf{Area of circle} = \pi r^2 = \pi \left(\frac{35}{\pi}\right)^2 = \frac{35^2}{\pi}$$

$$= \textbf{389.9 mm}^2 \text{ or } \textbf{3.899 cm}^2$$

Problem 14. Calculate the area of the sector of a circle having radius 6 cm with angle subtended at centre 50°

$$\text{Area of sector} = \frac{\theta}{360}(\pi r^2) = \frac{50}{360}(\pi 6^2)$$

$$= \frac{50 \times \pi \times 36}{360} = \textbf{15.71 cm}^2$$

Problem 15. Calculate the area of the sector of a circle having diameter 80 mm with angle subtended at centre 107°42′

If diameter = 80 mm then radius, $r = 40$ mm, and

$$\text{area of sector} = \frac{107°42′}{360}(\pi 40^2) = \frac{107\frac{42}{60}}{360}(\pi 40^2)$$

$$= \frac{107.7}{360}(\pi 40^2)$$

$$= \textbf{1504 mm}^2 \text{ or } \textbf{15.04 cm}^2$$

Problem 16. A hollow shaft has an outside diameter of 5.45 cm and an inside diameter of 2.25 cm. Calculate the cross-sectional area of the shaft

The cross-sectional area of the shaft is shown by the shaded part in Fig. 25.23 (often called an **annulus**).

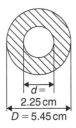

Figure 25.23

Area of shaded part = area of large circle − area of
small circle

$$= \frac{\pi D^2}{4} - \frac{\pi d^2}{4} = \frac{\pi}{4}(D^2 - d^2)$$

$$= \frac{\pi}{4}(5.45^2 - 2.25^2)$$

$$= \textbf{19.35 cm}^2$$

Now try the following Practice Exercise

1. A rectangular garden measures 40 m by 15 m. A 1 m flower border is made round the two shorter sides and one long side. A circular swimming pool of diameter 8 m is constructed in the middle of the garden. Find, correct to the nearest square metre, the area remaining.

2. Determine the area of circles having (a) a radius of 4 cm (b) a diameter of 30 mm (c) a circumference of 200 mm.

3. An annulus has an outside diameter of 60 mm and an inside diameter of 20 mm. Determine its area.

4. If the area of a circle is 320 mm², find (a) its diameter and (b) its circumference.

5. Calculate the areas of the following sectors of circles.

    (a) radius 9 cm, angle subtended at centre 75°
    (b) diameter 35 mm, angle subtended at centre 48°37′

6. Determine the shaded area of the template shown in Fig. 25.24.

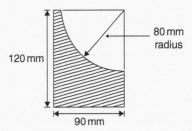

Figure 25.24

7. An archway consists of a rectangular opening topped by a semicircular arch, as shown in Fig. 25.25. Determine the area of the opening if the width is 1 m and the greatest height is 2 m.

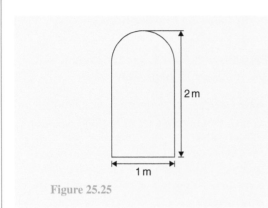

Figure 25.25

Here are some further worked problems of common shapes.

Problem 17. Calculate the area of a regular octagon if each side is 5 cm and the width across the flats is 12 cm

An octagon is an 8-sided polygon. If radii are drawn from the centre of the polygon to the vertices then 8 equal triangles are produced, as shown in Fig. 25.26.

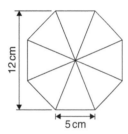

Figure 25.26

$$\text{Area of one triangle} = \frac{1}{2} \times \text{base} \times \text{height}$$

$$= \frac{1}{2} \times 5 \times \frac{12}{2} = 15 \, \text{cm}^2$$

Area of octagon $= 8 \times 15 = \mathbf{120 \, cm^2}$

Problem 18. Determine the area of a regular hexagon which has sides 8 cm long

A hexagon is a 6-sided polygon which may be divided into 6 equal triangles as shown in Fig. 25.27. The angle subtended at the centre of each triangle is $360° \div 6 = 60°$. The other two angles in the triangle add up to $120°$ and are equal to each other. Hence, each of the triangles is equilateral with each angle $60°$ and each side 8 cm.

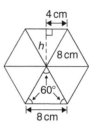

Figure 25.27

$$\text{Area of one triangle} = \frac{1}{2} \times \text{base} \times \text{height}$$

$$= \frac{1}{2} \times 8 \times h$$

h is calculated using Pythagoras' theorem:

$$8^2 = h^2 + 4^2$$

from which $\quad h = \sqrt{8^2 - 4^2} = 6.928 \, \text{cm}$

Hence,

$$\text{Area of one triangle} = \frac{1}{2} \times 8 \times 6.928 = 27.71 \, \text{cm}^2$$

Area of hexagon $= 6 \times 27.71$

$$= \mathbf{166.3 \, cm^2}$$

Problem 19. Fig. 25.28 shows a plan of a floor of a building which is to be carpeted. Calculate the area of the floor in square metres. Calculate the cost, correct to the nearest pound, of carpeting the floor with carpet costing £16.80 per m², assuming 30% extra carpet is required due to wastage in fitting

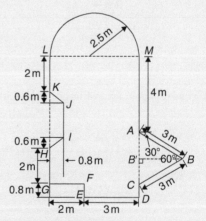

Figure 25.28

Area of floor plan

  = area of triangle ABC + area of semicircle

    + area of rectangle $CGLM$

    + area of rectangle $CDEF$

    − area of trapezium $HIJK$

Triangle ABC is equilateral since $AB = BC = 3$ m and, hence, angle $B'CB = 60°$

$$\sin B'CB = BB'/3$$

i.e. $$BB' = 3\sin 60° = 2.598\,\text{m}.$$

$$\text{Area of triangle } ABC = \frac{1}{2}(AC)(BB')$$

$$= \frac{1}{2}(3)(2.598) = 3.897\,\text{m}^2$$

$$\text{Area of semicircle } = \frac{1}{2}\pi r^2 = \frac{1}{2}\pi (2.5)^2$$

$$= 9.817\,\text{m}^2$$

$$\text{Area of } CGLM = 5 \times 7 = 35\,\text{m}^2$$

$$\text{Area of } CDEF = 0.8 \times 3 = 2.4\,\text{m}^2$$

$$\text{Area of } HIJK = \frac{1}{2}(KH + IJ)(0.8)$$

Since $MC = 7$ m then $LG = 7$ m, hence $JI = 7 - 5.2 = 1.8$ m. Hence,

$$\text{Area of } HIJK = \frac{1}{2}(3 + 1.8)(0.8) = 1.92\,\text{m}^2$$

Total floor area $= 3.897 + 9.817 + 35 + 2.4 - 1.92$

$$= 49.194\,\text{m}^2$$

To allow for 30% wastage, amount of carpet required $= 1.3 \times 49.194 = 63.95\,\text{m}^2$

Cost of carpet at £16.80 per m^2
$= 63.95 \times 16.80 = $ **£1074**, correct to the nearest pound.

Now try the following Practice Exercise

Practice Exercise 99   Areas of common shapes (answers on page 433)

1. Calculate the area of a regular octagon if each side is 20 mm and the width across the flats is 48.3 mm.

2. Determine the area of a regular hexagon which has sides 25 mm.

3. A plot of land is in the shape shown in Fig. 25.29. Determine

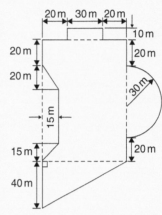

Figure 25.29

(a) its area in hectares (1 ha $= 10^4\,\text{m}^2$)

(b) the length of fencing required, to the nearest metre, to completely enclose the plot of land.

25.4 Areas of similar shapes

Fig. 25.30 shows two squares, one of which has sides three times as long as the other.

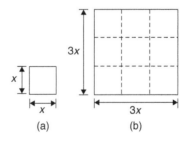

Figure 25.30

Area of Fig. 25.30(a) $= (x)(x) = x^2$

Area of Fig. 25.30(b) $= (3x)(3x) = 9x^2$

Hence, Fig. 25.30(b) has an area $(3)^2$; i.e. 9 times the area of Fig. 25.30(a).

In summary, **the areas of similar shapes are proportional to the squares of corresponding linear dimensions**.

Problem 20.   A rectangular garage is shown on a building plan having dimensions 10 mm by 20 mm. If the plan is drawn to a scale of 1 to 250, determine the true area of the garage in square metres

$$\text{Area of garage on the plan} = 10\,\text{mm} \times 20\,\text{mm}$$
$$= 200\,\text{mm}^2$$

Since the areas of similar shapes are proportional to the squares of corresponding dimensions,

$$\text{True area of garage} = 200 \times (250)^2$$
$$= 12.5 \times 10^6\,\text{mm}^2$$
$$= \frac{12.5 \times 10^6}{10^6}\,\text{m}^2$$
$$\text{since }\; 1\,\text{m}^2 = 10^6\,\text{mm}^2$$
$$= \mathbf{12.5\,m^2}$$

Now try the following Practice Exercise

Practice Exercise 100   Areas of similar shapes (answers on page 433)

1. The area of a park on a map is 500 mm². If the scale of the map is 1 to 40 000, determine the true area of the park in hectares (1 hectare = 10^4 m²).

2. A model of a boiler is made having an overall height of 75 mm corresponding to an overall height of the actual boiler of 6 m. If the area of metal required for the model is 12 500 mm², determine, in square metres, the area of metal required for the actual boiler.

3. The scale of an Ordnance Survey map is 1 : 2500. A circular sports field has a diameter of 8 cm on the map. Calculate its area in hectares, giving your answer correct to 3 significant figures. (1 hectare = 10^4 m²)

For fully worked solutions to each of the problems in Practice Exercises 96 to 100 in this chapter, go to the website:
www.routledge.com/cw/bird

Chapter 26

The circle and its properties

Why it is important to understand: The circle and its properties

A circle is one of the fundamental shapes of geometry; it consists of all the points that are equidistant from a central point. Knowledge of calculations involving circles is needed with crank mechanisms, with determinations of latitude and longitude, with pendulums, and even in the design of paper clips. The floodlit area at a football ground, the area an automatic garden sprayer sprays and the angle of lap of a belt drive all rely on calculations involving the arc of a circle. The ability to handle calculations involving circles and its properties is clearly essential in several branches of engineering design.

At the end of this chapter, you should be able to:

- define a circle
- state some properties of a circle – including radius, circumference, diameter, semicircle, quadrant, tangent, sector, chord, segment and arc
- appreciate the angle in a semicircle is a right angle
- define a radian, and change radians to degrees, and vice versa
- determine arc length, area of a circle and area of a sector of a circle
- state the equation of a circle
- sketch a circle given its equation

26.1 Introduction

A **circle** is a plain figure enclosed by a curved line, every point on which is equidistant from a point within, called the **centre**.

In Chapter 25, worked problems on the areas of circles and sectors were demonstrated. In this chapter, properties of circles are listed and arc lengths are calculated, together with more practical examples on the areas of sectors of circles. Finally, the equation of a circle is explained.

26.2 Properties of circles

(a)   The distance from the centre to the curve is called the **radius**, r, of the circle (see OP in Fig. 26.1).

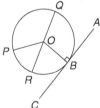

Figure 26.1

(b) The boundary of a circle is called the **circumference**, **c**

(c) Any straight line passing through the centre and touching the circumference at each end is called the **diameter**, **d** (see QR in Fig. 26.1). Thus, **d = 2r**

(d) The ratio $\dfrac{\text{circumference}}{\text{diameter}}$ is a constant for any circle. This constant is denoted by the Greek letter π (pronounced 'pie'), where $\pi = 3.14159$, correct to 5 decimal places (check with your calculator). Hence, $\dfrac{c}{d} = \pi$ or $c = \pi d$ or $c = 2\pi r$

(e) A **semicircle** is one half of a whole circle.

(f) A **quadrant** is one quarter of a whole circle.

(g) A **tangent** to a circle is a straight line which meets the circle at one point only and does not cut the circle when produced. AC in Fig. 26.1 is a tangent to the circle since it touches the curve at point B only. If radius OB is drawn, **angle ABO is a right angle**.

(h) The **sector** of a circle is the part of a circle between radii (for example, the portion OXY of Fig. 26.2 is a sector). If a sector is less than a semicircle it is called a **minor sector**; if greater than a semicircle it is called a **major sector**.

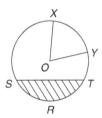

Figure 26.2

(i) The **chord** of a circle is any straight line which divides the circle into two parts and is terminated at each end by the circumference. ST, in Fig. 26.2, is a chord.

(j) **Segment** is the name given to the parts into which a circle is divided by a chord. If the segment is less than a semicircle it is called a **minor segment** (see shaded area in Fig. 26.2). If the segment is greater than a semicircle it is called a **major segment** (see the un-shaded area in Fig. 26.2).

(k) An **arc** is a portion of the circumference of a circle. The distance SRT in Fig. 26.2 is called a **minor arc** and the distance $SXYT$ is called a **major arc**.

(l) The angle at the centre of a circle, subtended by an arc, is double the angle at the circumference subtended by the same arc. With reference to Fig. 26.3,

Angle $AOC = 2 \times$ angle ABC

Figure 26.3

(m) The angle in a semicircle is a right angle (see angle BQP in Fig. 26.3).

Problem 1. Find the circumference of a circle of radius 12.0 cm

Circumference, $c = 2 \times \pi \times \text{radius} = 2\pi r = 2\pi (12.0)$
$$= \mathbf{75.40\,cm}$$

Problem 2. If the diameter of a circle is 75 mm, find its circumference

Circumference, $c = \pi \times \text{diameter} = \pi d = \pi (75)$
$$= \mathbf{235.6\,mm}$$

Problem 3. Determine the radius of a circular pond if its perimeter is 112 m

Perimeter $=$ circumference, $c = 2\pi r$

Hence, **radius of pond, $r = \dfrac{c}{2\pi} = \dfrac{112}{2\pi} = \mathbf{17.83\,cm}$**

Problem 4. In Fig. 26.4, AB is a tangent to the circle at B. If the circle radius is 40 mm and $AB = 150$ mm, calculate the length AO

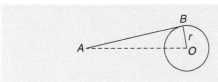

Figure 26.4

A tangent to a circle is at right angles to a radius drawn from the point of contact; i.e. **ABO = 90°**. Hence, using Pythagoras' theorem,

$$AO^2 = AB^2 + OB^2$$

from which, $AO = \sqrt{AB^2 + OB^2}$

$$= \sqrt{150^2 + 40^2} = \textbf{155.2 mm}$$

Now try the following Practice Exercise

Practice Exercise 101   Properties of a circle (answers on page 433)

1. Calculate the length of the circumference of a circle of radius 7.2 cm.

2. If the diameter of a circle is 82.6 mm, calculate the circumference of the circle.

3. Determine the radius of a circle whose circumference is 16.52 cm.

4. Find the diameter of a circle whose perimeter is 149.8 cm.

5. A crank mechanism is shown in Fig. 26.5, where XY is a tangent to the circle at point X. If the circle radius OX is 10 cm and length OY is 40 cm, determine the length of the connecting rod XY.

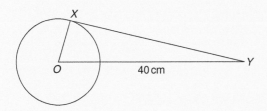

Figure 26.5

6. If the circumference of the earth is 40 000 km at the equator, calculate its diameter.

7. Calculate the length of wire in the paper clip shown in Fig. 26.6. The dimensions are in millimetres.

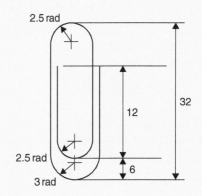

Figure 26.6

26.3 Radians and degrees

One **radian** is defined as the angle subtended at the centre of a circle by an arc equal in length to the radius. With reference to Fig. 26.7, for arc length s,

$$\theta \text{ radians} = \frac{s}{r}$$

Figure 26.7

When $s =$ whole circumference $(= 2\pi r)$ then

$$\theta = \frac{s}{r} = \frac{2\pi r}{r} = 2\pi$$

i.e. **2π radians = 360°** or **π radians = 180°**

Thus, **$1 \text{ rad} = \dfrac{180°}{\pi} = 57.30°$**, correct to 2 decimal places.

Since π rad $= 180°$, then $\dfrac{\pi}{2} = 90°$, $\dfrac{\pi}{3} = 60°$, $\dfrac{\pi}{4} = 45°$, and so on.

Problem 5.   Convert to radians: (a) 125° (b) 69°47′

(a) Since $180° = \pi$ rad, $1° = \dfrac{\pi}{180}$ rad, therefore

$$125° = 125\left(\dfrac{\pi}{180}\right) \text{rad} = \textbf{2.182 radians}.$$

(b) $69°47' = 69\dfrac{47°}{60} = 69.783°$ (or, with your calculator, enter $69°47'$ using $°\,'\,'\,'$ function, press $=$ and press $°\,'\,'\,'$ again).

and $69.783° = 69.783\left(\dfrac{\pi}{180}\right)$ rad

$$= \textbf{1.218 radians}.$$

Problem 6. Convert to degrees and minutes:
(a) 0.749 radians (b) $3\pi/4$ radians

(a) Since π rad $= 180°$, 1 rad $= \dfrac{180°}{\pi}$

therefore $0.749\,\text{rad} = 0.749\left(\dfrac{180}{\pi}\right)^° = 42.915°$

$0.915° = (0.915 \times 60)' = 55'$, correct to the nearest minute,

Hence, **0.749 radians $= 42°55'$**

(b) Since 1 rad $= \left(\dfrac{180}{\pi}\right)^°$ then

$$\dfrac{3\pi}{4}\text{ rad} = \dfrac{3\pi}{4}\left(\dfrac{180}{\pi}\right)^° = \dfrac{3}{4}(180)° = \textbf{135°}$$

Problem 7. Express in radians, in terms of π,
(a) 150° (b) 270° (c) 37.5°

Since $180° = \pi$ rad, $1° = \dfrac{\pi}{180}$ rad

(a) $150° = 150\left(\dfrac{\pi}{180}\right)$ rad $= \dfrac{5\pi}{6}$ **rad**

(b) $270° = 270\left(\dfrac{\pi}{180}\right)$ rad $= \dfrac{3\pi}{2}$ **rad**

(c) $37.5° = 37.5\left(\dfrac{\pi}{180}\right)$ rad $= \dfrac{75\pi}{360}$ rad $= \dfrac{5\pi}{24}$ **rad**

Now try the following Practice Exercise

Practice Exercise 102   Radians and degrees (answers on page 433)

1. Convert to radians in terms of π:
   (a) 30° (b) 75° (c) 225°

2. Convert to radians, correct to 3 decimal places:
   (a) 48° (b) 84°51' (c) 232°15'

3. Convert to degrees:
   (a) $\dfrac{7\pi}{6}$ rad (b) $\dfrac{4\pi}{9}$ rad (c) $\dfrac{7\pi}{12}$ rad

4. Convert to degrees and minutes:
   (a) 0.0125 rad   (b) 2.69 rad
   (c) 7.241 rad

5. A car engine speed is 1000 rev/min. Convert this speed into rad/s.

26.4 Arc length and area of circles and sectors

26.4.1 Arc length

From the definition of the radian in the previous section and Fig. 26.7,

$$\textbf{arc length}, s = r\theta \qquad \text{where } \theta \text{ is in radians}$$

26.4.2 Area of a circle

From Chapter 25, for any circle, area $= \pi \times (\text{radius})^2$

i.e. $$\textbf{area} = \pi r^2$$

Since $r = \dfrac{d}{2}$, $$\textbf{area} = \pi r^2 \text{ or } \dfrac{\pi d^2}{4}$$

26.4.3 Area of a sector

$$\textbf{Area of a sector} = \dfrac{\theta}{360}(\pi r^2) \text{ when } \theta \text{ is in degrees}$$

$$= \dfrac{\theta}{2\pi}(\pi r^2)$$

$$= \dfrac{1}{2}r^2\theta \qquad \text{when } \theta \text{ is in radians}$$

Problem 8. A hockey pitch has a semicircle of radius 14.63 m around each goal net. Find the area enclosed by the semicircle, correct to the nearest square metre

---

Area of a semicircle $= \dfrac{1}{2}\pi r^2$

When $r = 14.63\,\text{m}$, area $= \dfrac{1}{2}\pi(14.63)^2$

i.e. **area of semicircle $= 336\,\text{m}^2$**

Problem 9.   Find the area of a circular metal plate having a diameter of 35.0 mm, correct to the nearest square millimetre

Area of a circle $= \pi r^2 = \dfrac{\pi d^2}{4}$

When $d = 35.0$ mm,   area $= \dfrac{\pi (35.0)^2}{4}$

i.e.   **area of circular plate $= 962$ mm^2**

Problem 10.   Find the area of a circle having a circumference of 60.0 mm

Circumference, $c = 2\pi r$

from which radius, $r = \dfrac{c}{2\pi} = \dfrac{60.0}{2\pi} = \dfrac{30.0}{\pi}$

Area of a circle $= \pi r^2$

i.e.   **area $= \pi \left(\dfrac{30.0}{\pi}\right)^2 = 286.5$ mm^2**

Problem 11.   Find the length of the arc of a circle of radius 5.5 cm when the angle subtended at the centre is 1.20 radians

Length of arc, $s = r\theta$, where θ is in radians.
Hence, **arc length, $s = (5.5)(1.20) = 6.60$ cm.**

Problem 12.   Determine the diameter and circumference of a circle if an arc of length 4.75 cm subtends an angle of 0.91 radians

Since arc length, $s = r\theta$ then
radius, $r = \dfrac{s}{\theta} = \dfrac{4.75}{0.91} = 5.22$ cm
Diameter $= 2 \times$ radius $= 2 \times 5.22 = \textbf{10.44 cm}$
Circumference, $c = \pi d = \pi (10.44) = \textbf{32.80 cm}$

Problem 13.   If an angle of 125° is subtended by an arc of a circle of radius 8.4 cm, find the length of (a) the minor arc and (b) the major arc, correct to 3 significant figures

Since $180° = \pi$ rad then $1° = \left(\dfrac{\pi}{180}\right)$ rad and
$125° = 125\left(\dfrac{\pi}{180}\right)$ rad

(a)   Length of minor arc,

$$s = r\theta = (8.4)(125)\left(\dfrac{\pi}{180}\right) = \textbf{18.3 cm},$$

correct to 3 significant figures

(b)   Length of major arc $=$ (circumference $-$ minor arc) $= 2\pi(8.4) - 18.3 = \textbf{34.5 cm}$, correct to 3 significant figures.
(Alternatively, major arc $= r\theta$
$$= 8.4(360 - 125)\left(\dfrac{\pi}{180}\right) = \textbf{34.5 cm})$$

Problem 14.   Determine the angle, in degrees and minutes, subtended at the centre of a circle of diameter 42 mm by an arc of length 36 mm. Calculate also the area of the minor sector formed

Since length of arc, $s = r\theta$ then $\theta = \dfrac{s}{r}$

Radius, $r = \dfrac{\text{diameter}}{2} = \dfrac{42}{2} = 21$ mm,

hence $\theta = \dfrac{s}{r} = \dfrac{36}{21} = 1.7143$ radians.

1.7143 rad $= 1.7143 \times \left(\dfrac{180}{\pi}\right)° = 98.22° = \textbf{98°13}' =$
angle subtended at centre of circle.

From page 264,

$$\textbf{area of sector} = \dfrac{1}{2}r^2\theta = \dfrac{1}{2}(21)^2(1.7143)$$
$$= \textbf{378 mm}^2$$

Problem 15.   A football stadium floodlight can spread its illumination over an angle of 45° to a distance of 55 m. Determine the maximum area that is floodlit.

Floodlit area $=$ area of sector $= \dfrac{1}{2}r^2\theta$
$$= \dfrac{1}{2}(55)^2\left(45 \times \dfrac{\pi}{180}\right)$$
$$= \textbf{1188 m}^2$$

Problem 16.   An automatic garden sprayer produces spray to a distance of 1.8 m and revolves through an angle α which may be varied. If the desired spray catchment area is to be 2.5 m^2, to what should angle α be set, correct to the nearest degree?

Area of sector $= \dfrac{1}{2}r^2\theta$, hence $2.5 = \dfrac{1}{2}(1.8)^2\alpha$

from which,

$$\alpha = \frac{2.5 \times 2}{1.8^2} = 1.5432 \text{ radians}$$

$$1.5432 \,\text{rad} = \left(1.5432 \times \frac{180}{\pi}\right)^{\circ} = 88.42^{\circ}$$

Hence, **angle $\alpha = 88°$**, correct to the nearest degree.

Problem 17. The angle of a tapered groove is checked using a 20 mm diameter roller as shown in Fig. 26.8. If the roller lies 2.12 mm below the top of the groove, determine the value of angle θ

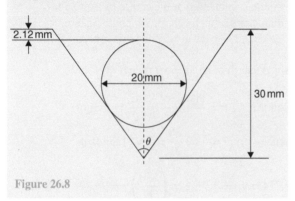

Figure 26.8

In Fig. 26.9, triangle ABC is right-angled at C (see property (g) in Section 26.2).

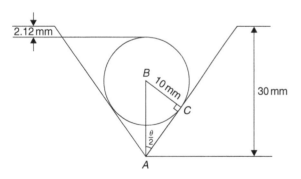

Figure 26.9

Length $BC = 10$ mm (i.e. the radius of the circle), and $AB = 30 - 10 - 2.12 = 17.88$ mm, from Fig. 26.9.

Hence, $\sin \dfrac{\theta}{2} = \dfrac{10}{17.88}$ and $\dfrac{\theta}{2} = \sin^{-1}\left(\dfrac{10}{17.88}\right) = 34^{\circ}$

and **angle $\theta = 68°$**.

Now try the following Practice Exercise

1. Calculate the area of a circle of radius 6.0 cm, correct to the nearest square centimetre.

2. The diameter of a circle is 55.0 mm. Determine its area, correct to the nearest square millimetre.

3. The perimeter of a circle is 150 mm. Find its area, correct to the nearest square millimetre.

4. Find the area of the sector, correct to the nearest square millimetre, of a circle having a radius of 35 mm with angle subtended at centre of 75°.

5. An annulus has an outside diameter of 49.0 mm and an inside diameter of 15.0 mm. Find its area correct to 4 significant figures.

6. Find the area, correct to the nearest square metre, of a 2 m wide path surrounding a circular plot of land 200 m in diameter.

7. A rectangular park measures 50 m by 40 m. A 3 m flower bed is made round the two longer sides and one short side. A circular fish pond of diameter 8.0 m is constructed in the centre of the park. It is planned to grass the remaining area. Find, correct to the nearest square metre, the area of grass.

8. With reference to Fig. 26.10, determine (a) the perimeter and (b) the area.

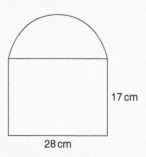

Figure 26.10

9. Find the area of the shaded portion of Fig. 26.11.

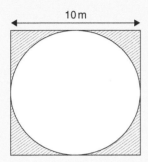

Figure 26.11

10. Find the length of an arc of a circle of radius 8.32 cm when the angle subtended at the centre is 2.14 radians. Calculate also the area of the minor sector formed.

11. If the angle subtended at the centre of a circle of diameter 82 mm is 1.46 rad, find the lengths of the (a) minor arc and (b) major arc.

12. A pendulum of length 1.5 m swings through an angle of 10° in a single swing. Find, in centimetres, the length of the arc traced by the pendulum bob.

13. Determine the shaded area of the section shown in Fig. 26.12

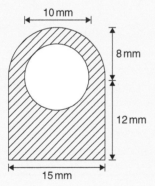

Figure 26.12

14. Determine the length of the radius and circumference of a circle if an arc length of 32.6 cm subtends an angle of 3.76 radians.

15. Determine the angle of lap, in degrees and minutes, if 180 mm of a belt drive are in contact with a pulley of diameter 250 mm.

16. Determine the number of complete revolutions a motorcycle wheel will make in travelling 2 km if the wheel's diameter is 85.1 cm.

17. A floodlight at a sports ground spread its illumination over an angle of 40° to a distance of 48 m. Determine (a) the angle in radians and (b) the maximum area that is floodlit.

18. Find the area swept out in 50 minutes by the minute hand of a large floral clock if the hand is 2 m long.

19. Determine (a) the shaded area in Fig. 26.13 and (b) the percentage of the whole sector that the shaded area represents.

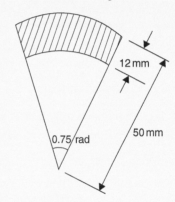

Figure 26.13

20. Determine the length of steel strip required to make the clip shown in Fig. 26.14.

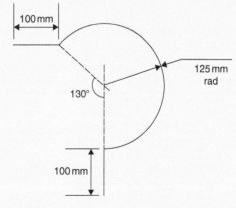

Figure 26.14

21. A 50° tapered hole is checked with a 40 mm diameter ball as shown in Fig. 26.15. Determine the length shown as x.

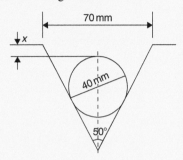

Figure 26.15

26.5 The equation of a circle

The simplest equation of a circle, centre at the origin and radius r, is given by

$$x^2 + y^2 = r^2$$

For example, Fig. 26.16 shows a circle $x^2 + y^2 = 9$

More generally, the equation of a circle, centre (a, b) and radius r, is given by

$$(x - a)^2 + (y - b)^2 = r^2 \qquad (1)$$

Fig. 26.17 shows a circle $(x - 2)^2 + (y - 3)^2 = 4$

The general equation of a circle is

$$x^2 + y^2 + 2ex + 2fy + c = 0 \qquad (2)$$

Multiplying out the bracketed terms in equation (1) gives

$$x^2 - 2ax + a^2 + y^2 - 2by + b^2 = r^2$$

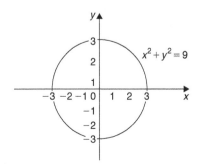

Figure 26.16

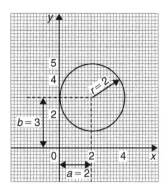

Figure 26.17

Comparing this with equation (2) gives

$$2e = -2a, \text{ i.e. } \boldsymbol{a = -\dfrac{2e}{2}}$$

and $$2f = -2b, \text{ i.e. } \boldsymbol{b = -\dfrac{2f}{2}}$$

and $$c = a^2 + b^2 - r^2, \text{ i.e. } \boldsymbol{r = \sqrt{a^2 + b^2 - c}}$$

Thus, for example, the equation

$$x^2 + y^2 - 4x - 6y + 9 = 0$$

represents a circle with centre,

$$a = -\left(\frac{-4}{2}\right), b = -\left(\frac{-6}{2}\right) \text{ i.e. at } (2, 3) \text{ and}$$

radius, $$r = \sqrt{2^2 + 3^2 - 9} = 2$$

Hence, $x^2 + y^2 - 4x - 6y + 9 = 0$ is the circle shown in Fig. 26.17 (which may be checked by multiplying out the brackets in the equation $(x - 2)^2 + (y - 3)^2 = 4$)

Problem 18. Determine (a) the radius and (b) the co-ordinates of the centre of the circle given by the equation $x^2 + y^2 + 8x - 2y + 8 = 0$

$x^2 + y^2 + 8x - 2y + 8 = 0$ is of the form shown in equation (2),

where $$a = -\left(\frac{8}{2}\right) = -4, b = -\left(\frac{-2}{2}\right) = 1$$

and $$r = \sqrt{(-4)^2 + 1^2 - 8} = \sqrt{9} = 3$$

Hence, $x^2 + y^2 + 8x - 2y + 8 = 0$ represents a circle **centre (−4, 1)** and **radius 3**, as shown in Fig. 26.18.

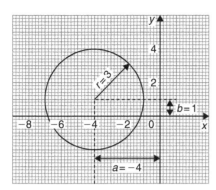

Figure 26.18

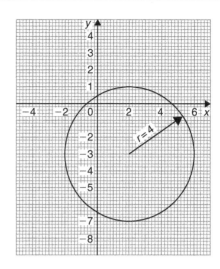

Figure 26.19

Alternatively, $x^2 + y^2 + 8x - 2y + 8 = 0$ may be re-arranged as

$$(x+4)^2 + (y-1)^2 - 9 = 0$$

i.e. $$(x+4)^2 + (y-1)^2 = 3^2$$

which represents a circle, **centre (−4, 1)** and **radius 3**, as stated above.

> **Problem 19.** Sketch the circle given by the equation $x^2 + y^2 - 4x + 6y - 3 = 0$

The equation of a circle, centre (a, b), radius r is given by

$$(x-a)^2 + (y-b)^2 = r^2$$

The general equation of a circle is
$x^2 + y^2 + 2ex + 2fy + c = 0$

From above $a = -\dfrac{2e}{2}, b = -\dfrac{2f}{2}$ and $r = \sqrt{a^2 + b^2 - c}$

Hence, if $x^2 + y^2 - 4x + 6y - 3 = 0$

then $a = -\left(\dfrac{-4}{2}\right) = 2, b = -\left(\dfrac{6}{2}\right) = -3$ and

$r = \sqrt{2^2 + (-3)^2 - (-3)} = \sqrt{16} = 4$

Thus, the circle has **centre (2, −3)** and **radius 4**, as shown in Fig. 26.19.

Alternatively, $x^2 + y^2 - 4x + 6y - 3 = 0$ may be re-arranged as

$$(x-2)^2 + (y+3)^2 - 3 - 13 = 0$$

i.e. $$(x-2)^2 + (y+3)^2 = 4^2$$

which represents a circle, **centre (2, −3)** and **radius 4**, as stated above.

Now try the following Practice Exercise

> **Practice Exercise 104   The equation of a circle (answers on page 433)**
>
> 1. Determine (a) the radius and (b) the co-ordinates of the centre of the circle given by the equation $x^2 + y^2 - 6x + 8y + 21 = 0$
>
> 2. Sketch the circle given by the equation $x^2 + y^2 - 6x + 4y - 3 = 0$
>
> 3. Sketch the curve $x^2 + (y-1)^2 - 25 = 0$
>
> 4. Sketch the curve $x = 6\sqrt{\left[1 - \left(\dfrac{y}{6}\right)^2\right]}$

For fully worked solutions to each of the problems in Practice Exercises 101 to 104 in this chapter, go to the website:
www.routledge.com/cw/bird

This assignment covers the material contained in Chapters 25 and 26. *The marks available are shown in brackets at the end of each question.*

1. A rectangular metal plate has an area of $9600\,\text{cm}^2$. If the length of the plate is $1.2\,\text{m}$, calculate the width, in centimetres. (3)

2. Calculate the cross-sectional area of the angle iron section shown in Fig. RT10.1, the dimensions being in millimetres. (4)

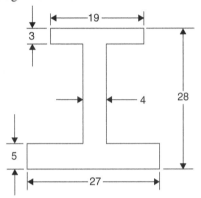

Figure RT10.1

3. Find the area of the trapezium $MNOP$ shown in Fig. RT10.2 when $a = 6.3\,\text{cm}, b = 11.7\,\text{cm}$ and $h = 5.5\,\text{cm}$. (3)

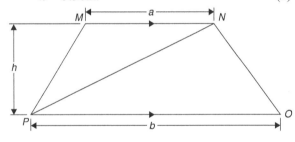

Figure RT10.2

4. Find the area of the triangle DEF in Fig. RT10.3, correct to 2 decimal places. (4)

5. A rectangular park measures $150\,\text{m}$ by $70\,\text{m}$. A $2\,\text{m}$ flower border is constructed round the two longer sides and one short side. A circular fish pond of diameter $15\,\text{m}$ is in the centre of the park and the remainder of the park is grass. Calculate, correct to the nearest square metre, the area of (a) the fish pond, (b) the flower borders and (c) the grass. (6)

6. A swimming pool is $55\,\text{m}$ long and $10\,\text{m}$ wide. The perpendicular depth at the deep end is $5\,\text{m}$ and at

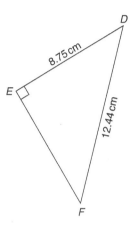

Figure RT10.3

the shallow end is $1.5\,\text{m}$, the slope from one end to the other being uniform. The inside of the pool needs two coats of a protective paint before it is filled with water. Determine how many litres of paint will be needed if 1 litre covers $10\,\text{m}^2$ (7)

7. Find the area of an equilateral triangle of side $20.0\,\text{cm}$. (4)

8. A steel template is of the shape shown in Fig. RT10.4, the circular area being removed. Determine the area of the template, in square centimetres, correct to 1 decimal place. (8)

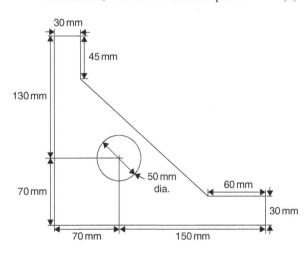

Figure RT10.4

9. The area of a plot of land on a map is $400\,\text{mm}^2$. If the scale of the map is 1 to 50000,

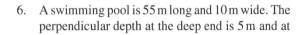

determine the true area of the land in hectares (1 hectare $= 10^4 \, \text{m}^2$) (4)

10. Determine the shaded area in Fig. RT10.5, correct to the nearest square centimetre. (3)

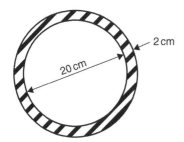

Figure RT10.5

11. Determine the diameter of a circle, correct to the nearest millimetre, whose circumference is $178.4 \, \text{cm}$. (2)

12. Calculate the area of a circle of radius $6.84 \, \text{cm}$, correct to 1 decimal place. (2)

13. The circumference of a circle is $250 \, \text{mm}$. Find its area, correct to the nearest square millimetre. (4)

14. Find the area of the sector of a circle having a radius of $50.0 \, \text{mm}$, with angle subtended at centre of $120°$. (3)

15. Determine the total area of the shape shown in Fig. RT10.6, correct to 1 decimal place. (7)

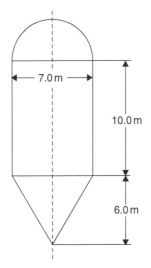

Figure RT10.6

16. The radius of a circular cricket ground is $75 \, \text{m}$. The boundary is painted with white paint and 1 tin of paint will paint a line $22.5 \, \text{m}$ long. How many tins of paint are needed? (3)

17. Find the area of a $1.5 \, \text{m}$ wide path surrounding a circular plot of land $100 \, \text{m}$ in diameter. (3)

18. A cyclometer shows 2530 revolutions in a distance of $3.7 \, \text{km}$. Find the diameter of the wheel in centimetres, correct to 2 decimal places. (4)

19. The minute hand of a wall clock is $10.5 \, \text{cm}$ long. How far does the tip travel in the course of 24 hours? (4)

20. Convert
    (a) $125°47'$ to radians.
    (b) 1.724 radians to degrees and minutes. (4)

21. Calculate the length of metal strip needed to make the clip shown in Fig. RT10.7. (7)

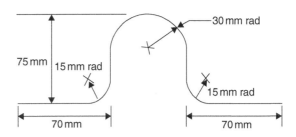

Figure RT10.7

22. A lorry has wheels of radius $50 \, \text{cm}$. Calculate the number of complete revolutions a wheel makes (correct to the nearest revolution) when travelling 3 miles (assume 1 mile $= 1.6 \, \text{km}$). (4)

23. The equation of a circle is
    $x^2 + y^2 + 12x - 4y + 4 = 0$. Determine
    (a) the diameter of the circle.
    (b) the co-ordinates of the centre of the circle. (7)

For lecturers/instructors/teachers, fully worked solutions to each of the problems in Revision Test 10, together with a full marking scheme, are available at the website:

www.routledge.com/cw/bird

Chapter 27

Volumes and surface areas of common solids

Why it is important to understand: Volumes and surface areas of common solids

There are many practical applications where volumes and surface areas of common solids are required. Examples include determining capacities of oil, water, petrol and fish tanks, ventilation shafts and cooling towers, determining volumes of blocks of metal, ball-bearings, boilers and buoys, and calculating the cubic metres of concrete needed for a path. Finding the surface areas of loudspeaker diaphragms and lampshades provide further practical examples. Understanding these calculations is essential for the many practical applications in engineering, construction, architecture and science.

At the end of this chapter, you should be able to:

- state the SI unit of volume
- calculate the volumes and surface areas of cuboids, cylinders, prisms, pyramids, cones and spheres
- calculate volumes and surface areas of frusta of pyramids and cones
- appreciate that volumes of similar bodies are proportional to the cubes of the corresponding linear dimensions

27.1 Introduction

The **volume** of any solid is a measure of the space occupied by the solid. Volume is measured in **cubic units** such as mm^3, cm^3 and m^3.

This chapter deals with finding volumes of common solids; in engineering it is often important to be able to calculate volume or capacity to estimate, say, the amount of liquid, such as water, oil or petrol, in different shaped containers.

A **prism** is a solid with a constant cross-section and with two ends parallel. The shape of the end is used to describe the prism. For example, there are rectangular prisms (called cuboids), triangular prisms and circular prisms (called cylinders).

On completing this chapter you will be able to calculate the volumes and surface areas of rectangular and other prisms, cylinders, pyramids, cones and spheres, together with frusta of pyramids and cones. Volumes of similar shapes are also considered.

27.2 Volumes and surface areas of common shapes

27.2.1 Cuboids or rectangular prisms

A cuboid is a solid figure bounded by six rectangular faces; all angles are right angles and opposite faces are equal. A typical cuboid is shown in Fig. 27.1 with length l, breadth b and height h.

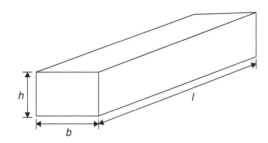

Figure 27.1

$$\textbf{Volume of cuboid} = \textbf{\textit{l}} \times \textbf{\textit{b}} \times \textbf{\textit{h}}$$

and

$$\textbf{surface area} = \textbf{2\textit{bh}} + \textbf{2\textit{hl}} + \textbf{2\textit{lb}} = \textbf{2}(\textbf{\textit{bh}} + \textbf{\textit{hl}} + \textbf{\textit{lb}})$$

A **cube** is a square prism. If all the sides of a cube are x then

$$\textbf{Volume} = \textbf{\textit{x}}^3 \text{ and } \textbf{surface area} = \textbf{6\textit{x}}^2$$

Problem 1. A cuboid has dimensions of 12 cm by 4 cm by 3 cm. Determine (a) its volume and (b) its total surface area

The cuboid is similar to that in Fig. 27.1, with $l = 12$ cm, $b = 4$ cm and $h = 3$ cm.

(a) **Volume of cuboid** $= l \times b \times h = 12 \times 4 \times 3$
$$= \textbf{144 cm}^3$$

(b) **Surface area** $= 2(bh + hl + lb)$
$$= 2(4 \times 3 + 3 \times 12 + 12 \times 4)$$
$$= 2(12 + 36 + 48)$$
$$= 2 \times 96 = \textbf{192 cm}^2$$

Problem 2. An oil tank is the shape of a cube, each edge being of length 1.5 m. Determine (a) the maximum capacity of the tank in m³ and litres and (b) its total surface area ignoring input and output orifices

(a) **Volume of oil tank** = volume of cube
$$= 1.5\,\text{m} \times 1.5\,\text{m} \times 1.5\,\text{m}$$
$$= 1.5^3\,\text{m}^3 = \textbf{3.375 m}^3$$

$1\,\text{m}^3 = 100\,\text{cm} \times 100\,\text{cm} \times 100\,\text{cm} = 10^6\,\text{cm}^3$. Hence,
$$\text{volume of tank} = 3.375 \times 10^6\,\text{cm}^3$$

$1\,\text{litre} = 1000\,\text{cm}^3$, hence **oil tank capacity**
$$= \frac{3.375 \times 10^6}{1000}\,\text{litres} = \textbf{3375 litres}$$

(b) Surface area of one side $= 1.5\,\text{m} \times 1.5\,\text{m}$
$$= 2.25\,\text{m}^2$$

A cube has six identical sides, hence

$$\textbf{total surface area of oil tank} = \textbf{6} \times \textbf{2.25}$$
$$= \textbf{13.5 m}^2$$

Problem 3. A water tank is the shape of a rectangular prism having length 2 m, breadth 75 cm and height 500 mm. Determine the capacity of the tank in (a) m³ (b) cm³ (c) litres

Capacity means volume; when dealing with liquids, the word capacity is usually used.
The water tank is similar in shape to that in Fig. 27.1, with $l = 2$ m, $b = 75$ cm and $h = 500$ mm.

(a) Capacity of water tank $= l \times b \times h$. To use this formula, all dimensions **must** be in the same units. Thus, $l = 2\,\text{m}, b = 0.75\,\text{m}$ and $h = 0.5\,\text{m}$ (since $1\,\text{m} = 100\,\text{cm} = 1000\,\text{mm}$). Hence,

$$\textbf{capacity of tank} = 2 \times 0.75 \times 0.5 = \textbf{0.75 m}^3$$

(b) $1\,\text{m}^3 = 1\,\text{m} \times 1\,\text{m} \times 1\,\text{m}$
$$= 100\,\text{cm} \times 100\,\text{cm} \times 100\,\text{cm}$$
i.e. $\textbf{1 m}^3 = \textbf{1 000 000} = \textbf{10}^6\,\textbf{cm}^3$. Hence,
$$\textbf{capacity} = 0.75\,\text{m}^3 = 0.75 \times 10^6\,\text{cm}^3$$
$$= \textbf{750 000 cm}^3$$

(c) $1\,\text{litre} = 1000\,\text{cm}^3$. Hence,

$$\textbf{750 000 cm}^3 = \frac{750,000}{1000} = \textbf{750 litres}$$

27.2.2 Cylinders

A cylinder is a circular prism. A cylinder of radius r and height h is shown in Fig. 27.2.

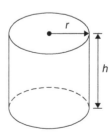

Figure 27.2

$$\textbf{Volume} = \pi r^2 h$$

$$\textbf{Curved surface area} = 2\pi rh$$

$$\textbf{Total surface area} = 2\pi rh + 2\pi r^2$$

Total surface area means the curved surface area plus the area of the two circular ends.

> **Problem 4.** A solid cylinder has a base diameter of 12 cm and a perpendicular height of 20 cm. Calculate (a) the volume and (b) the total surface area

(a) $\textbf{Volume} = \pi r^2 h = \pi \times \left(\dfrac{12}{2}\right)^2 \times 20$

$$= 720\pi = \textbf{2262 cm}^3$$

(b) **Total surface area**

$$= 2\pi rh + 2\pi r^2$$

$$= (2 \times \pi \times 6 \times 20) + (2 \times \pi \times 6^2)$$

$$= 240\pi + 72\pi = 312\pi = \textbf{980 cm}^2$$

> **Problem 5.** A copper pipe has the dimensions shown in Fig. 27.3. Calculate the volume of copper in the pipe, in cubic metres.

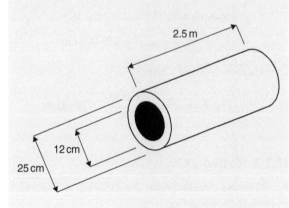

Figure 27.3

Outer diameter, $D = 25$ cm $= 0.25$ m and inner diameter, $d = 12$ cm $= 0.12$ m.

Area of cross-section of copper

$$= \frac{\pi D^2}{4} - \frac{\pi d^2}{4} = \frac{\pi (0.25)^2}{4} - \frac{\pi (0.12)^2}{4}$$

$$= 0.0491 - 0.0113 = 0.0378 \, \text{m}^2$$

Hence, **volume of copper**

$$= (\text{cross-sectional area}) \times \text{length of pipe}$$

$$= 0.0378 \times 2.5 = \textbf{0.0945 m}^3$$

27.2.3 More prisms

A right-angled triangular prism is shown in Fig. 27.4 with dimensions b, h and l.

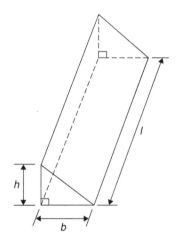

Figure 27.4

$$\textbf{Volume} = \frac{1}{2}bhl$$

and

$$\textbf{surface area} = \textbf{area of each end}$$

$$+ \textbf{area of three sides}$$

Notice that the volume is given by the area of the end (i.e. area of triangle $= \frac{1}{2}bh$) multiplied by the length l. In fact, the volume of any shaped prism is given by the area of an end multiplied by the length.

> **Problem 6.** Determine the volume (in cm^3) of the shape shown in Fig. 27.5

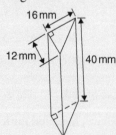

Figure 27.5

The solid shown in Fig. 27.5 is a triangular prism. The volume V of any prism is given by $V = Ah$, where A is the cross-sectional area and h is the perpendicular height. Hence,

$$\textbf{volume} = \frac{1}{2} \times 16 \times 12 \times 40 = 3840\,\text{mm}^3$$

$$= \textbf{3.840 cm}^3$$

$$(\text{since } 1\,\text{cm}^3 = 1000\,\text{mm}^3)$$

Problem 7. Calculate the volume of the right-angled triangular prism shown in Fig. 27.6. Also, determine its total surface area

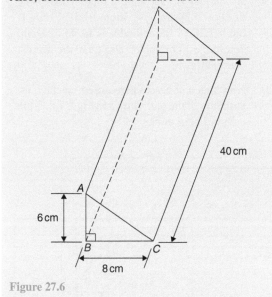

Figure 27.6

Volume of right-angled triangular prism

$$= \frac{1}{2}bhl = \frac{1}{2} \times 8 \times 6 \times 40$$

i.e.   **volume = 960 cm³**

Total surface area = area of each end + area of three sides.

In triangle ABC, $AC^2 = AB^2 + BC^2$

from which, $AC = \sqrt{AB^2 + BC^2} = \sqrt{6^2 + 8^2}$

$$= 10\,\text{cm}$$

Hence, total surface area

$$= 2\left(\frac{1}{2}bh\right) + (AC \times 40) + (BC \times 40) + (AB \times 40)$$

$$= (8 \times 6) + (10 \times 40) + (8 \times 40) + (6 \times 40)$$

$$= 48 + 400 + 320 + 240$$

i.e. **total surface area = 1008 cm²**

Problem 8. Calculate the volume and total surface area of the solid prism shown in Fig. 27.7

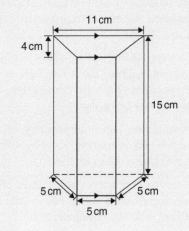

Figure 27.7

The solid shown in Fig. 27.7 is a **trapezoidal prism**.

Volume of prism = cross-sectional area × height

$$= \frac{1}{2}(11 + 5)4 \times 15 = 32 \times 15$$

$$= \textbf{480 cm}^3$$

Surface area of prism

= sum of two trapeziums + 4 rectangles

$$= (2 \times 32) + (5 \times 15) + (11 \times 15) + 2(5 \times 15)$$

$$= 64 + 75 + 165 + 150 = \textbf{454 cm}^2$$

Now try the following Practice Exercise

Practice Exercise 105   Volumes and surface areas of common shapes (answers on page 433)

1. Change a volume of $1\,200\,000\,\text{cm}^3$ to cubic metres.

2. Change a volume of $5000\,\text{mm}^3$ to cubic centimetres.

3. A metal cube has a surface area of $24\,\text{cm}^2$. Determine its volume.

4. A rectangular block of wood has dimensions of $40\,\text{mm}$ by $12\,\text{mm}$ by $8\,\text{mm}$. Determine

   (a) its volume, in cubic millimetres

    (b) its total surface area in square millimetres.

5. Determine the capacity, in litres, of a fish tank measuring 90 cm by 60 cm by 1.8 m, given 1 litre = 1000 cm^3.

6. A rectangular block of metal has dimensions of 40 mm by 25 mm by 15 mm. Determine its volume in cm^3. Find also its mass if the metal has a density of 9 g/cm^3.

7. Determine the maximum capacity, in litres, of a fish tank measuring 50 cm by 40 cm by 2.5 m (1 litre = 1000 cm^3).

8. Determine how many cubic metres of concrete are required for a 120 m long path, 150 mm wide and 80 mm deep.

9. A cylinder has a diameter 30 mm and height 50 mm. Calculate

    (a) its volume in cubic centimetres, correct to 1 decimal place

    (b) the total surface area in square centimetres, correct to 1 decimal place.

10. Find (a) the volume and (b) the total surface area of a right-angled triangular prism of length 80 cm and whose triangular end has a base of 12 cm and perpendicular height 5 cm.

11. A steel ingot whose volume is 2 m^2 is rolled out into a plate which is 30 mm thick and 1.80 m wide. Calculate the length of the plate in metres.

12. The volume of a cylinder is 75 cm^3. If its height is 9.0 cm, find its radius.

13. Calculate the volume of a metal tube whose outside diameter is 8 cm and whose inside diameter is 6 cm, if the length of the tube is 4 m.

14. The volume of a cylinder is 400 cm^3. If its radius is 5.20 cm, find its height. Also determine its curved surface area.

15. A cylinder is cast from a rectangular piece of alloy 5 cm by 7 cm by 12 cm. If the length of the cylinder is to be 60 cm, find its diameter.

16. Find the volume and the total surface area of a regular hexagonal bar of metal of length 3 m if each side of the hexagon is 6 cm.

17. A block of lead 1.5 m by 90 cm by 750 mm is hammered out to make a square sheet 15 mm thick. Determine the dimensions of the square sheet, correct to the nearest centimetre.

18. How long will it take a tap dripping at a rate of 800 mm^3/s to fill a 3-litre can?

19. A cylinder is cast from a rectangular piece of alloy 5.20 cm by 6.50 cm by 19.33 cm. If the height of the cylinder is to be 52.0 cm, determine its diameter, correct to the nearest centimetre.

20. How much concrete is required for the construction of the path shown in Fig. 27.8, if the path is 12 cm thick?

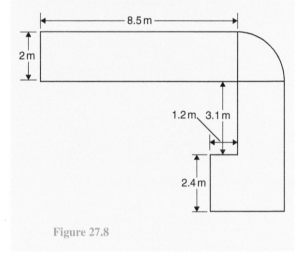

Figure 27.8

27.2.4 Pyramids

Volume of any pyramid

$$= \frac{1}{3} \times \textbf{area of base} \times \textbf{perpendicular height}$$

A square-based pyramid is shown in Fig. 27.9 with base dimension x by x and the perpendicular height of the pyramid h. For the square-base pyramid shown,

$$\textbf{volume} = \frac{1}{3}x^2 h$$

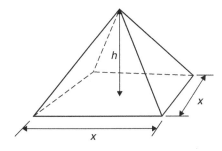

Figure 27.9

Problem 9. A square pyramid has a perpendicular height of 16 cm. If a side of the base is 6 cm, determine the volume of a pyramid

Volume of pyramid

$$= \frac{1}{3} \times \text{area of base} \times \text{perpendicular height}$$

$$= \frac{1}{3} \times (6 \times 6) \times 16$$

$$= \mathbf{192\,cm^3}$$

Problem 10. Determine the volume and the total surface area of the square pyramid shown in Fig. 27.10 if its perpendicular height is 12 cm.

Volume of pyramid

$$= \frac{1}{3}(\text{area of base}) \times \text{perpendicular height}$$

$$= \frac{1}{3}(5 \times 5) \times 12$$

$$= \mathbf{100\,cm^3}$$

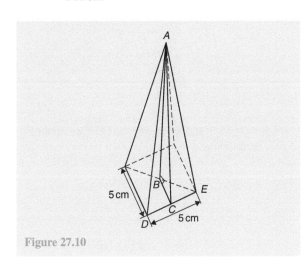

Figure 27.10

The total surface area consists of a square base and 4 equal triangles.

Area of triangle ADE

$$= \frac{1}{2} \times \text{base} \times \text{perpendicular height}$$

$$= \frac{1}{2} \times 5 \times AC$$

The length AC may be calculated using Pythagoras' theorem on triangle ABC, where $AB = 12$ cm and $BC = \frac{1}{2} \times 5 = 2.5$ cm.

$$AC = \sqrt{AB^2 + BC^2} = \sqrt{12^2 + 2.5^2} = 12.26\,\text{cm}$$

Hence,

$$\text{area of triangle } ADE = \frac{1}{2} \times 5 \times 12.26 = 30.65\,\text{cm}^2$$

Total surface area of pyramid $= (5 \times 5) + 4(30.65)$

$$= \mathbf{147.6\,cm^2}$$

Problem 11. A rectangular prism of metal having dimensions of 5 cm by 6 cm by 18 cm is melted down and recast into a pyramid having a rectangular base measuring 6 cm by 10 cm. Calculate the perpendicular height of the pyramid, assuming no waste of metal

Volume of rectangular prism $= 5 \times 6 \times 18 = 540\,\text{cm}^3$
Volume of pyramid

$$= \frac{1}{3} \times \text{area of base} \times \text{perpendicular height}$$

Hence,　$540 = \frac{1}{3} \times (6 \times 10) \times h$

from which,　$h = \dfrac{3 \times 540}{6 \times 10} = 27\,\text{cm}$

i.e. **perpendicular height of pyramid = 27 cm**

27.2.5 Cones

A cone is a circular-based pyramid. A cone of base radius r and perpendicular height h is shown in Fig. 27.11.

$$\text{Volume} = \frac{1}{3} \times \text{area of base} \times \text{perpendicular height}$$

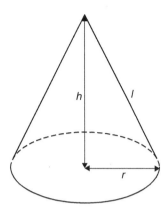

Figure 27.11

i.e.

$$\text{Volume} = \frac{1}{3}\pi r^2 h$$

$$\text{Curved surface area} = \pi r l$$

$$\text{Total surface area} = \pi r l + \pi r^2$$

Problem 12. Calculate the volume, in cubic centimetres, of a cone of radius 30 mm and perpendicular height 80 mm

Volume of cone $= \frac{1}{3}\pi r^2 h = \frac{1}{3} \times \pi \times 30^2 \times 80$

$$= 75398.2236\ldots\text{mm}^3$$

$1\,\text{cm} = 10\,\text{mm}$ and
$1\,\text{cm}^3 = 10\,\text{mm} \times 10\,\text{mm} \times 10\,\text{mm} = 10^3\,\text{mm}^3$, or

$$\mathbf{1\,mm^3 = 10^{-3}\,cm^3}$$

Hence, $75398.2236\ldots\text{mm}^3$
$$= 75398.2236\ldots \times 10^{-3}\,\text{cm}^3$$
i.e.

$$\textbf{volume} = \mathbf{75.40\,cm^3}$$

Alternatively, from the question, $r = 30\,\text{mm} = 3\,\text{cm}$ and $h = 80\,\text{mm} = 8\,\text{cm}$. Hence,

$$\textbf{volume} = \frac{1}{3}\pi r^2 h = \frac{1}{3} \times \pi \times 3^2 \times 8 = \mathbf{75.40\,cm^3}$$

Problem 13. Determine the volume and total surface area of a cone of radius 5 cm and perpendicular height 12 cm

The cone is shown in Fig. 27.12.

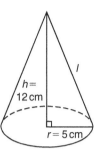

Figure 27.12

$$\textbf{Volume of cone} = \frac{1}{3}\pi r^2 h = \frac{1}{3} \times \pi \times 5^2 \times 12$$

$$= \mathbf{314.2\,cm^3}$$

Total surface area = curved surface area + area of base
$$= \pi r l + \pi r^2$$

From Fig. 27.12, slant height l may be calculated using Pythagoras' theorem:

$$l = \sqrt{12^2 + 5^2} = 13\,\text{cm}$$

Hence, **total surface area** $= (\pi \times 5 \times 13) + (\pi \times 5^2)$

$$= \mathbf{282.7\,cm^2}$$

27.2.6 Spheres

For the sphere shown in Fig. 27.13:

$$\textbf{Volume} = \frac{4}{3}\pi r^3 \quad \text{and} \quad \textbf{surface area} = 4\pi r^2$$

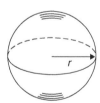

Figure 27.13

Problem 14. Find the volume and surface area of a sphere of diameter 10 cm

Since diameter $= 10\,\text{cm}$, radius, $r = 5\,\text{cm}$.

$$\textbf{Volume of sphere} = \frac{4}{3}\pi r^3 = \frac{4}{3} \times \pi \times 5^3$$

$$= \mathbf{523.6\,cm^3}$$

Surface area of sphere $= 4\pi r^2 = 4 \times \pi \times 5^2$

$$= \mathbf{314.2\,cm^2}$$

Problem 15. The surface area of a sphere is $201.1\,cm^2$. Find the diameter of the sphere and hence its volume

Surface area of sphere $= 4\pi r^2$

Hence, $201.1\,cm^2 = 4 \times \pi \times r^2$

from which $\qquad r^2 = \dfrac{201.1}{4 \times \pi} = 16.0$

and $\qquad$ radius, $r = \sqrt{16.0} = 4.0\,cm$

from which, **diameter** $= 2 \times r = 2 \times 4.0 = \mathbf{8.0\,cm}$

$$\mathbf{Volume\ of\ sphere} = \frac{4}{3}\pi r^3 = \frac{4}{3} \times \pi \times (4.0)^3$$

$$= \mathbf{268.1\,cm^3}$$

Now try the following Practice Exercise

Practice Exercise 106   Volumes and surface areas of common shapes (answers on page 433)

1. If a cone has a diameter of 80 mm and a perpendicular height of 120 mm, calculate its volume in cm^3 and its curved surface area.

2. A square pyramid has a perpendicular height of 4 cm. If a side of the base is 2.4 cm long, find the volume and total surface area of the pyramid.

3. A sphere has a diameter of 6 cm. Determine its volume and surface area.

4. If the volume of a sphere is $566\,cm^3$, find its radius.

5. A pyramid having a square base has a perpendicular height of 25 cm and a volume of $75\,cm^3$. Determine, in centimetres, the length of each side of the base.

6. A cone has a base diameter of 16 mm and a perpendicular height of 40 mm. Find its volume correct to the nearest cubic millimetre.

7. Determine (a) the volume and (b) the surface area of a sphere of radius 40 mm.

8. The volume of a sphere is $325\,cm^3$. Determine its diameter.

9. Given the radius of the earth is 6380 km, calculate, in engineering notation

    (a) its surface area in km^2

    (b) its volume in km^3

10. An ingot whose volume is $1.5\,m^3$ is to be made into ball bearings whose radii are 8.0 cm. How many bearings will be produced from the ingot, assuming 5% wastage?

11. A spherical chemical storage tank has an internal diameter of 5.6 m. Calculate the storage capacity of the tank, correct to the nearest cubic metre. If 1 litre = 1000 cm^3, determine the tank capacity in litres.

27.3 Summary of volumes and surface areas of common solids

A summary of volumes and surface areas of regular solids is shown in Table 27.1 on page 280.

27.4 More complex volumes and surface areas

Here are some worked problems involving more complex and composite solids.

Problem 16. A wooden section is shown in Fig. 27.14. Find (a) its volume in m^3 and (b) its total surface area

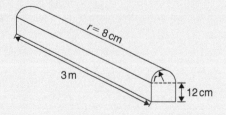

Figure 27.14

Table 27.1 Volumes and surface areas of regular solids

Rectangular prism (or cuboid)

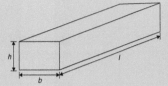

Volume $= l \times b \times h$
Surface area $= 2(bh + hl + lb)$

Cylinder

Volume $= \pi r^2 h$
Total surface area $= 2\pi rh + 2\pi r^2$

Triangular prism

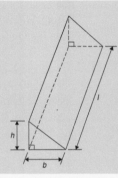

Volume $= \dfrac{1}{2}bhl$
Surface area $=$ area of each end $+$
    area of three sides

Pyramid

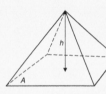

Volume $= \dfrac{1}{3} \times A \times h$
Total surface area $=$
sum of areas of triangles
forming sides $+$ area of base

Cone

Volume $= \dfrac{1}{3}\pi r^2 h$
Curved surface area $= \pi rl$
Total surface area $= \pi rl + \pi r^2$

Sphere

Volume $= \dfrac{4}{3}\pi r^3$
Curved surface area $= 4\pi r^2$

(a) The section of wood is a prism whose end comprises a rectangle and a semicircle. Since the radius of the semicircle is 8 cm, the diameter is 16 cm. Hence, the rectangle has dimensions 12 cm by 16 cm.

Area of end $= (12 \times 16) + \frac{1}{2}\pi 8^2 = 292.5\,\text{cm}^2$

Volume of wooden section

$= \text{area of end} \times \text{perpendicular height}$

$= 292.5 \times 300 = \mathbf{87\,750\,cm^3}$

$= \dfrac{87750}{10^6}\,\text{m}^3, \text{since } 1\,\text{m}^3 = 10^6\,\text{cm}^3$

$= \mathbf{0.08775\,m^3}$

(b) The total surface area comprises the two ends (each of area 292.5 cm²), three rectangles and a curved surface (which is half a cylinder). Hence,

total surface area

$= (2 \times 292.5) + 2(12 \times 300)$

$+ (16 \times 300) + \frac{1}{2}(2\pi \times 8 \times 300)$

$= 585 + 7200 + 4800 + 2400\pi$

$= \mathbf{20\,125\,cm^2} \quad \text{or} \quad \mathbf{2.0125\,m^2}$

Problem 17. A pyramid has a rectangular base 3.60 cm by 5.40 cm. Determine the volume and total surface area of the pyramid if each of its sloping edges is 15.0 cm

---

The pyramid is shown in Fig. 27.15. To calculate the volume of the pyramid, the perpendicular height EF is required. Diagonal BD is calculated using Pythagoras' theorem,

i.e. $BD = \sqrt{[3.60^2 + 5.40^2]} = 6.490\,\text{cm}$

Hence, $EB = \frac{1}{2}BD = \dfrac{6.490}{2} = 3.245\,\text{cm}$

Using Pythagoras' theorem on triangle BEF gives

$$BF^2 = EB^2 + EF^2$$

from which $EF = \sqrt{(BF^2 - EB^2)}$

$= \sqrt{15.0^2 - 3.245^2} = 14.64\,\text{cm}$

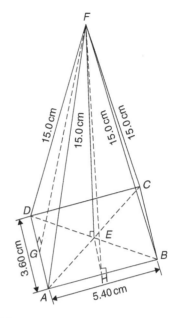

Figure 27.15

Volume of pyramid

$= \frac{1}{3}(\text{area of base})(\text{perpendicular height})$

$= \frac{1}{3}(3.60 \times 5.40)(14.64) = \mathbf{94.87\,cm^3}$

Area of triangle ADF (which equals triangle BCF) $= \frac{1}{2}(AD)(FG)$, where G is the mid-point of AD. Using Pythagoras' theorem on triangle FGA gives

$$FG = \sqrt{[15.0^2 - 1.80^2]} = 14.89\,\text{cm}$$

Hence, area of triangle $ADF = \frac{1}{2}(3.60)(14.89)$

$= 26.80\,\text{cm}^2$

Similarly, if H is the mid-point of AB,

$$FH = \sqrt{15.0^2 - 2.70^2} = 14.75\,\text{cm}$$

Hence, area of triangle ABF (which equals triangle CDF) $= \frac{1}{2}(5.40)(14.75) = 39.83\,\text{cm}^2$

Total surface area of pyramid

$= 2(26.80) + 2(39.83) + (3.60)(5.40)$

$= 53.60 + 79.66 + 19.44$

$= \mathbf{152.7\,cm^2}$

Problem 18. Calculate the volume and total surface area of a hemisphere of diameter 5.0 cm

Volume of hemisphere $= \dfrac{1}{2}$(volume of sphere)

$$= \dfrac{2}{3}\pi r^3 = \dfrac{2}{3}\pi\left(\dfrac{5.0}{2}\right)^3$$

$$= \mathbf{32.7\,cm^3}$$

Total surface area

$$= \text{curved surface area} + \text{area of circle}$$

$$= \dfrac{1}{2}(\text{surface area of sphere}) + \pi r^2$$

$$= \dfrac{1}{2}(4\pi r^2) + \pi r^2$$

$$= 2\pi r^2 + \pi r^2 = 3\pi r^2 = 3\pi\left(\dfrac{5.0}{2}\right)^2$$

$$= \mathbf{58.9\,cm^2}$$

Problem 19. A rectangular piece of metal having dimensions 4 cm by 3 cm by 12 cm is melted down and recast into a pyramid having a rectangular base measuring 2.5 cm by 5 cm. Calculate the perpendicular height of the pyramid

Volume of rectangular prism of metal $= 4 \times 3 \times 12$

$$= 144\,cm^3$$

Volume of pyramid

$$= \dfrac{1}{3}(\text{area of base})(\text{perpendicular height})$$

Assuming no waste of metal,

$$144 = \dfrac{1}{3}(2.5 \times 5)(\text{height})$$

i.e. **perpendicular height of pyramid** $= \dfrac{144 \times 3}{2.5 \times 5}$

$$= \mathbf{34.56\,cm}$$

Problem 20. A rivet consists of a cylindrical head, of diameter 1 cm and depth 2 mm, and a shaft of diameter 2 mm and length 1.5 cm. Determine the volume of metal in 2000 such rivets

Radius of cylindrical head $= \dfrac{1}{2}$ cm $= 0.5$ cm and height of cylindrical head $= 2$ mm $= 0.2$ cm.

Hence, volume of cylindrical head

$$= \pi r^2 h = \pi (0.5)^2 (0.2) = 0.1571\,cm^3$$

Volume of cylindrical shaft

$$= \pi r^2 h = \pi \left(\dfrac{0.2}{2}\right)^2 (1.5) = 0.0471\,cm^3$$

Total volume of 1 rivet $= 0.1571 + 0.0471$

$$= 0.2042\,cm^3$$

Volume of metal in 2000 such rivets

$$= 2000 \times 0.2042 = \mathbf{408.4\,cm^3}$$

Problem 21. A solid metal cylinder of radius 6 cm and height 15 cm is melted down and recast into a shape comprising a hemisphere surmounted by a cone. Assuming that 8% of the metal is wasted in the process, determine the height of the conical portion if its diameter is to be 12 cm

Volume of cylinder $= \pi r^2 h = \pi \times 6^2 \times 15$

$$= 540\pi\,cm^3$$

If 8% of metal is lost then 92% of 540π gives the volume of the new shape, shown in Fig. 27.16.

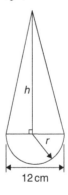

Figure 27.16

Hence, the volume of (hemisphere + cone)

$$= 0.92 \times 540\pi\,cm^3$$

i.e. $\dfrac{1}{2}\left(\dfrac{4}{3}\pi r^3\right) + \dfrac{1}{3}\pi r^2 h = 0.92 \times 540\pi$

Dividing throughout by π gives

$$\frac{2}{3}r^3 + \frac{1}{3}r^2 h = 0.92 \times 540$$

Since the diameter of the new shape is to be 12 cm, radius $r = 6$ cm,

then $$\frac{2}{3}(6)^3 + \frac{1}{3}(6)^2 h = 0.92 \times 540$$

$$144 + 12h = 496.8$$

i.e. **height of conical portion,**

$$h = \frac{496.8 - 144}{12} = \textbf{29.4 cm}$$

Problem 22.   A block of copper having a mass of 50 kg is drawn out to make 500 m of wire of uniform cross-section. Given that the density of copper is 8.91 g/cm^3, calculate (a) the volume of copper, (b) the cross-sectional area of the wire and (c) the diameter of the cross-section of the wire

(a)   A density of 8.91 g/cm^3 means that 8.91 g of copper has a volume of 1 cm^3, or 1 g of copper has a volume of $(1 \div 8.91)$ cm^3

$$\text{Density} = \frac{\text{mass}}{\text{volume}}$$

from which $$\text{volume} = \frac{\text{mass}}{\text{density}}$$

Hence, 50 kg, i.e. 50 000 g, has a

$$\textbf{volume} = \frac{\text{mass}}{\text{density}} = \frac{50000}{8.91} \text{ cm}^3 = \textbf{5612 cm}^3$$

(b)   Volume of wire = area of circular cross-section
$$\times \text{ length of wire.}$$

Hence, $5612 \text{ cm}^3 = \text{area} \times (500 \times 100 \text{ cm})$

from which, **area** $= \dfrac{5612}{500 \times 100} \text{ cm}^2$

$$= \textbf{0.1122 cm}^2$$

(c)   Area of circle $= \pi r^2$ or $\dfrac{\pi d^2}{4}$

hence, $0.1122 = \dfrac{\pi d^2}{4}$

from which, $d = \sqrt{\left(\dfrac{4 \times 0.1122}{\pi}\right)} = 0.3780$ cm

i.e. **diameter of cross-section is 3.780 mm**.

Problem 23.   A boiler consists of a cylindrical section of length 8 m and diameter 6 m, on one end of which is surmounted a hemispherical section of diameter 6 m and on the other end a conical section of height 4 m and base diameter 6 m. Calculate the volume of the boiler and the total surface area

The boiler is shown in Fig. 27.17.

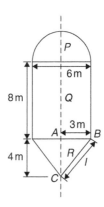

Figure 27.17

Volume of hemisphere, $P = \dfrac{2}{3}\pi r^3$

$$= \frac{2}{3} \times \pi \times 3^3 = 18\pi \text{ m}^3$$

Volume of cylinder, $Q = \pi r^2 h = \pi \times 3^2 \times 8$

$$= 72\pi \text{ m}^3$$

Volume of cone, $R = \dfrac{1}{3}\pi r^2 h = \dfrac{1}{3} \times \pi \times 3^2 \times 4$

$$= 12\pi \text{ m}^3$$

Total volume of boiler $= 18\pi + 72\pi + 12\pi$

$$= 102\pi = \textbf{320.4 m}^3$$

Surface area of hemisphere, $P = \dfrac{1}{2}(4\pi r^2)$

$$= 2 \times \pi \times 3^2 = 18\pi \text{ m}^2$$

Curved surface area of cylinder, $Q = 2\pi r h$

$$= 2 \times \pi \times 3 \times 8$$

$$= 48\pi \text{ m}^2$$

The slant height of the cone, l, is obtained by Pythagoras' theorem on triangle ABC, i.e.

$$l = \sqrt{(4^2 + 3^2)} = 5$$

Curved surface area of cone,

$$R = \pi r l = \pi \times 3 \times 5 = 15\pi \, \text{m}^2$$

Total surface area of boiler $= 18\pi + 48\pi + 15\pi$

$$= 81\pi = \textbf{254.5 m}^2$$

Now try the following Practice Exercise

Practice Exercise 107   More complex
volumes and surface areas (answers on
page 434)

1. Find the total surface area of a hemisphere of diameter 50 mm.

2. Find (a) the volume and (b) the total surface area of a hemisphere of diameter 6 cm.

3. Determine the mass of a hemispherical copper container whose external and internal radii are 12 cm and 10 cm, assuming that 1 cm³ of copper weighs 8.9 g.

4. A metal plumb bob comprises a hemisphere surmounted by a cone. If the diameter of the hemisphere and cone are each 4 cm and the total length is 5 cm, find its total volume.

5. A marquee is in the form of a cylinder surmounted by a cone. The total height is 6 m and the cylindrical portion has a height of 3.5 m with a diameter of 15 m. Calculate the surface area of material needed to make the marquee assuming 12% of the material is wasted in the process.

6. Determine (a) the volume and (b) the total surface area of the following solids.

    (i) a cone of radius 8.0 cm and perpendicular height 10 cm.

    (ii) a sphere of diameter 7.0 cm.

    (iii) a hemisphere of radius 3.0 cm.

    (iv) a 2.5 cm by 2.5 cm square pyramid of perpendicular height 5.0 cm.

    (v) a 4.0 cm by 6.0 cm rectangular pyramid of perpendicular height 12.0 cm.

    (vi) a 4.2 cm by 4.2 cm square pyramid whose sloping edges are each 15.0 cm

    (vii) a pyramid having an octagonal base of side 5.0 cm and perpendicular height 20 cm.

7. A metal sphere weighing 24 kg is melted down and recast into a solid cone of base radius 8.0 cm. If the density of the metal is 8000 kg/m³ determine

    (a) the diameter of the metal sphere.

    (b) the perpendicular height of the cone, assuming that 15% of the metal is lost in the process.

8. Find the volume of a regular hexagonal pyramid if the perpendicular height is 16.0 cm and the side of the base is 3.0 cm.

9. A buoy consists of a hemisphere surmounted by a cone. The diameter of the cone and hemisphere is 2.5 m and the slant height of the cone is 4.0 m. Determine the volume and surface area of the buoy.

10. A petrol container is in the form of a central cylindrical portion 5.0 m long with a hemispherical section surmounted on each end. If the diameters of the hemisphere and cylinder are both 1.2 m, determine the capacity of the tank in litres (1 litre = 1000 cm³).

11. Fig. 27.18 shows a metal rod section. Determine its volume and total surface area.

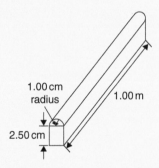

Figure 27.18

12. Find the volume (in cm³) of the die-casting shown in Fig. 27.19. The dimensions are in millimetres.

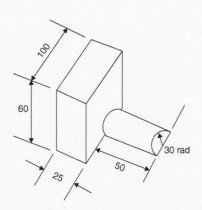

Figure 27.19

13. The cross-section of part of a circular ventilation shaft is shown in Fig. 27.20, ends *AB* and *CD* being open. Calculate

(a) the volume of the air, correct to the nearest litre, contained in the part of the system shown, neglecting the sheet metal thickness (given 1 litre $= 1000\,\text{cm}^3$)

(b) the cross-sectional area of the sheet metal used to make the system, in square metres

(c) the cost of the sheet metal if the material costs £11.50 per square metre, assuming that 25% extra metal is required due to wastage.

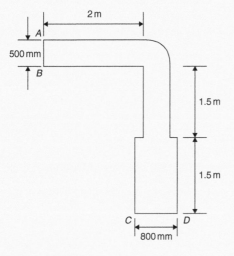

Figure 27.20

27.5 Volumes and surface areas of frusta of pyramids and cones

The **frustum** of a pyramid or cone is the portion remaining when a part containing the vertex is cut off by a plane parallel to the base.

The **volume of a frustum of a pyramid or cone** is given by the volume of the whole pyramid or cone minus the volume of the small pyramid or cone cut off.

The **surface area of the sides of a frustum of a pyramid or cone** is given by the surface area of the whole pyramid or cone minus the surface area of the small pyramid or cone cut off. This gives the lateral surface area of the frustum. If the total surface area of the frustum is required then the surface area of the two parallel ends are added to the lateral surface area.

There is an alternative method for finding the volume and surface area of a **frustum of a cone**. With reference to Fig. 27.21,

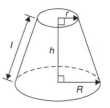

Figure 27.21

$$\text{Volume} = \frac{1}{3}\pi h(R^2 + Rr + r^2)$$

$$\text{Curved surface area} = \pi l(R + r)$$

$$\text{Total surface area} = \pi l(R + r) + \pi r^2 + \pi R^2$$

Problem 24. Determine the volume of a frustum of a cone if the diameter of the ends are 6.0 cm and 4.0 cm and its perpendicular height is 3.6 cm

(i) Method 1

A section through the vertex of a complete cone is shown in Fig. 27.22.

Using similar triangles, $\dfrac{AP}{DP} = \dfrac{DR}{BR}$

Hence, $\dfrac{AP}{2.0} = \dfrac{3.6}{1.0}$

from which $AP = \dfrac{(2.0)(3.6)}{1.0}$

$= 7.2\,\text{cm}$

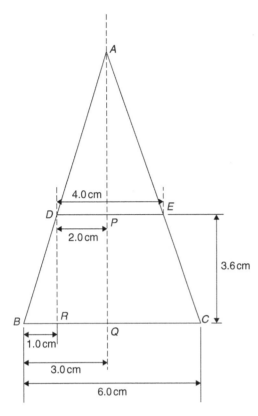

Figure 27.22

The height of the large cone $= 3.6 + 7.2$

$$= 10.8\,\text{cm}$$

Volume of frustum of cone

$=$ volume of large cone

$\quad\quad\quad\quad$ $-$ volume of small cone cut off

$$= \frac{1}{3}\pi(3.0)^2(10.8) - \frac{1}{3}\pi(2.0)^2(7.2)$$

$$= 101.79 - 30.16 = \textbf{71.6}\,\textbf{cm}^\textbf{3}$$

(ii) Method 2

From above, volume of the frustum of a cone

$$= \frac{1}{3}\pi h(R^2 + Rr + r^2)$$

where $\quad R = 3.0\,\text{cm}$,

$$r = 2.0\,\text{cm} \quad\text{and}\quad h = 3.6\,\text{cm}$$

Hence, **volume of frustum**

$$= \frac{1}{3}\pi(3.6)\left[(3.0)^2 + (3.0)(2.0) + (2.0)^2\right]$$

$$= \frac{1}{3}\pi(3.6)(19.0) = \textbf{71.6}\,\textbf{cm}^\textbf{3}$$

Problem 25. Find the total surface area of the frustum of the cone in Problem 24.

(i) Method 1

Curved surface area of frustum = curved surface area of large cone $-$ curved surface area of small cone cut off.

From Fig. 27.22, using Pythagoras' theorem,

$$AB^2 = AQ^2 + BQ^2$$

from which $\quad AB = \sqrt{\left[10.8^2 + 3.0^2\right]} = 11.21\,\text{cm}$

and $\quad\quad\quad AD^2 = AP^2 + DP^2$

from which $\quad AD = \sqrt{\left[7.2^2 + 2.0^2\right]} = 7.47\,\text{cm}$

Curved surface area of large cone $= \pi r l$

$$= \pi(BQ)(AB) = \pi(3.0)(11.21)$$

$$= 105.65\,\text{cm}^2$$

and curved surface area of small cone

$$= \pi(DP)(AD) = \pi(2.0)(7.47) = 46.94\,\text{cm}^2$$

Hence, curved surface area of frustum

$$= 105.65 - 46.94$$

$$= 58.71\,\text{cm}^2$$

Total surface area of frustum

$=$ curved surface area

$\quad\quad\quad\quad\quad$ $+$ area of two circular ends

$$= 58.71 + \pi(2.0)^2 + \pi(3.0)^2$$

$$= 58.71 + 12.57 + 28.27 = \textbf{99.6}\,\textbf{cm}^\textbf{2}$$

(ii) Method 2

From page 285, total surface area of frustum

$$= \pi l(R + r) + \pi r^2 + \pi R^2$$

where $l = BD = 11.21 - 7.47 = 3.74\,\text{cm}$, $R = 3.0\,\text{cm}$ and $r = 2.0\,\text{cm}$.

Hence, total surface area of frustum

$$= \pi(3.74)(3.0 + 2.0) + \pi(2.0)^2 + \pi(3.0)^2$$

$$= \textbf{99.6}\,\textbf{cm}^\textbf{2}$$

Problem 26. A storage hopper is in the shape of a frustum of a pyramid. Determine its volume if the ends of the frustum are squares of sides 8.0 m and

4.6 m, respectively, and the perpendicular height between its ends is 3.6 m

The frustum is shown shaded in Fig. 27.23(a) as part of a complete pyramid. A section perpendicular to the base through the vertex is shown in Fig. 27.23(b).

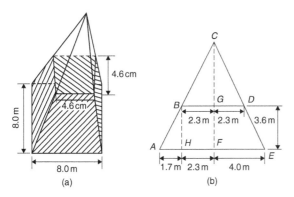

Figure 27.23

By similar triangles $\dfrac{CG}{BG} = \dfrac{BH}{AH}$

From which, height

$$CG = BG\left(\frac{BH}{AH}\right) = \frac{(2.3)(3.6)}{1.7} = 4.87\,\text{m}$$

Height of complete pyramid $= 3.6 + 4.87 = 8.47\,\text{m}$

Volume of large pyramid $= \dfrac{1}{3}(8.0)^2(8.47)$

$$= 180.69\,\text{m}^3$$

Volume of small pyramid cut off $= \dfrac{1}{3}(4.6)^2(4.87)$

$$= 34.35\,\text{m}^3$$

Hence, **volume of storage hopper** $= 180.69 - 34.35$

$$= \textbf{146.3}\,\textbf{m}^3$$

Problem 27. Determine the lateral surface area of the storage hopper in Problem 26

The lateral surface area of the storage hopper consists of four equal trapeziums. From Fig. 27.24,

$$\text{Area of trapezium } PRSU = \frac{1}{2}(PR + SU)(QT)$$

$OT = 1.7\,\text{m}$ (same as AH in Fig. 27.23(b) and $OQ = 3.6\,\text{m}$. By Pythagoras' theorem,

$$QT = \sqrt{(OQ^2 + OT^2)} = \sqrt{[3.6^2 + 1.7^2]} = 3.98\,\text{m}$$

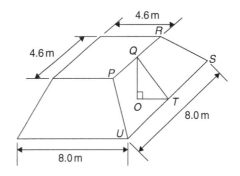

Figure 27.24

Area of trapezium $PRSU$

$$= \frac{1}{2}(4.6 + 8.0)(3.98) = 25.07\,\text{m}^2$$

Lateral surface area of hopper $= 4(25.07)$

$$= \textbf{100.3}\,\textbf{m}^2$$

Problem 28. A lampshade is in the shape of a frustum of a cone. The vertical height of the shade is 25.0 cm and the diameters of the ends are 20.0 cm and 10.0 cm, respectively. Determine the area of the material needed to form the lampshade, correct to 3 significant figures

The curved surface area of a frustum of a cone $= \pi l(R + r)$ from page 285. Since the diameters of the ends of the frustum are 20.0 cm and 10.0 cm, from Fig. 27.25,

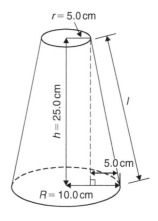

Figure 27.25

$$r = 5.0\,\text{cm}, R = 10.0\,\text{cm}$$

and $\quad l = \sqrt{[25.0^2 + 5.0^2]} = 25.50\,\text{cm}$

from Pythagoras' theorem.
Hence, curved surface area
$$= \pi(25.50)(10.0+5.0) = 1201.7\,\text{cm}^2$$
i.e. the area of material needed to form the lampshade is **1200 cm²**, correct to 3 significant figures.

Problem 29. A cooling tower is in the form of a cylinder surmounted by a frustum of a cone, as shown in Fig. 27.26. Determine the volume of air space in the tower if 40% of the space is used for pipes and other structures

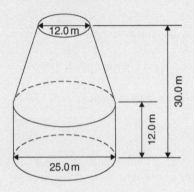

Figure 27.26

Volume of cylindrical portion $= \pi r^2 h$

$$= \pi\left(\frac{25.0}{2}\right)^2 (12.0)$$

$$= 5890\,\text{m}^3$$

Volume of frustum of cone $= \dfrac{1}{3}\pi h(R^2 + Rr + r^2)$

where $\qquad h = 30.0 - 12.0 = 18.0\,\text{m}$,

$$R = 25.0 \div 2 = 12.5\,\text{m}$$

and $\qquad r = 12.0 \div 2 = 6.0\,\text{m}.$

Hence, volume of frustum of cone

$$= \frac{1}{3}\pi(18.0)\left[(12.5)^2 + (12.5)(6.0) + (6.0)^2\right]$$

$$= 5038\,\text{m}^3$$

Total volume of cooling tower $= 5890 + 5038$

$$= 10\,928\,\text{m}^3$$

If 40% of space is occupied then

volume of air space $= 0.6 \times 10928 = \mathbf{6557\,m^3}$

Now try the following Practice Exercise

Practice Exercise 108   Volumes and surface areas of frusta of pyramids and cones (answers on page 434)

1. The radii of the faces of a frustum of a cone are 2.0 cm and 4.0 cm and the thickness of the frustum is 5.0 cm. Determine its volume and total surface area.

2. A frustum of a pyramid has square ends, the squares having sides 9.0 cm and 5.0 cm, respectively. Calculate the volume and total surface area of the frustum if the perpendicular distance between its ends is 8.0 cm.

3. A cooling tower is in the form of a frustum of a cone. The base has a diameter of 32.0 m, the top has a diameter of 14.0 m and the vertical height is 24.0 m. Calculate the volume of the tower and the curved surface area.

4. A loudspeaker diaphragm is in the form of a frustum of a cone. If the end diameters are 28.0 cm and 6.00 cm and the vertical distance between the ends is 30.0 cm, find the area of material needed to cover the curved surface of the speaker.

5. A rectangular prism of metal having dimensions 4.3 cm by 7.2 cm by 12.4 cm is melted down and recast into a frustum of a square pyramid, 10% of the metal being lost in the process. If the ends of the frustum are squares of side 3 cm and 8 cm respectively, find the thickness of the frustum.

6. Determine the volume and total surface area of a bucket consisting of an inverted frustum of a cone, of slant height 36.0 cm and end diameters 55.0 cm and 35.0 cm.

7. A cylindrical tank of diameter 2.0 m and perpendicular height 3.0 m is to be replaced by a tank of the same capacity but in the form of a frustum of a cone. If the diameters of the ends of the frustum are 1.0 m and 2.0 m, respectively, determine the vertical height required.

27.6   Volumes of similar shapes

Fig. 27.27 shows two cubes, one of which has sides three times as long as those of the other.

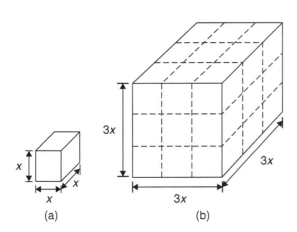

Figure 27.27

Volume of Fig. 27.27(a) $= (x)(x)(x) = x^3$

Volume of Fig. 27.27(b) $= (3x)(3x)(3x) = 27x^3$

Hence, Fig. 27.27(b) has a volume $(3)^3$, i.e. 27, times the volume of Fig. 27.27(a).

Summarising, **the volumes of similar bodies are proportional to the cubes of corresponding linear dimensions.**

$$\frac{\text{Volume of model}}{\text{Volume of car}} = \left(\frac{1}{50}\right)^3$$

since the volume of similar bodies are proportional to the cube of corresponding dimensions.

Mass = density × volume and, since both car and model are made of the same material,

$$\frac{\text{Mass of model}}{\text{Mass of car}} = \left(\frac{1}{50}\right)^3$$

Hence, mass of model

$$= (\text{mass of car})\left(\frac{1}{50}\right)^3 = \frac{1000}{50^3} = \textbf{0.008 kg or 8 g}$$

Now try the following Practice Exercise

Practice Exercise 109   Volumes of similar shapes (answers on page 434)

1. The diameter of two spherical bearings are in the ratio 2 : 5. What is the ratio of their volumes?

2. An engineering component has a mass of 400 g. If each of its dimensions are reduced by 30%, determine its new mass.

For fully worked solutions to each of the problems in Practice Exercises 105 to 109 in this chapter, go to the website:
www.routledge.com/cw/bird

Chapter 28

Irregular areas and volumes and mean values

Why it is important to understand: Irregular areas and volumes and mean values of waveforms

Surveyors, farmers and landscapers often need to determine the area of irregularly shaped pieces of land to work with the land properly. There are many applications in business, economics and the sciences, including all aspects of engineering, where finding the areas of irregular shapes, the volumes of solids, and the lengths of irregular shaped curves are important applications. Typical earthworks include roads, railway beds, causeways, dams and canals. Other common earthworks are land grading to reconfigure the topography of a site, or to stabilise slopes. Engineers need to concern themselves with issues of geotechnical engineering (such as soil density and strength) and with quantity estimation to ensure that soil volumes in the cuts match those of the fills, while minimising the distance of movement. Simpson's rule is a staple of scientific data analysis and engineering; it is widely used, for example, by Naval architects to numerically determine hull offsets and cross-sectional areas to determine volumes and centroids of ships or lifeboats. There are therefore plenty of examples where irregular areas and volumes need to be determined by engineers.

At the end of this chapter, you should be able to:

- use the trapezoidal rule to determine irregular areas
- use the mid-ordinate rule to determine irregular areas
- use Simpson's rule to determine irregular areas
- estimate the volume of irregular solids
- determine the mean values of waveforms

28.1 Areas of irregular figures

Areas of irregular plane surfaces may be approximately determined by using

(a) a planimeter,

(b) the trapezoidal rule,

(c) the mid-ordinate rule, or

(d) Simpson's rule.

Such methods may be used by, for example, engineers estimating areas of indicator diagrams of steam engines, surveyors estimating areas of plots of land or naval architects estimating areas of water planes or transverse sections of ships.

(a) **A planimeter** is an instrument for directly measuring small areas bounded by an irregular curve. There are many different kinds of planimeters but all operate in a similar way. A pointer on the planimeter is used to trace around the boundary of the shape. This induces a movement in another part of the instrument and a reading of this is used to establish the area of the shape.

(b) **Trapezoidal rule**

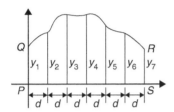

Figure 28.1

To determine the area *PQRS* in Fig. 28.1,

(i) Divide base *PS* into any number of equal intervals, each of width d (the greater the number of intervals, the greater the accuracy).
(ii) Accurately measure ordinates y_1, y_2, y_3, etc.
(iii) Area *PQRS*

$$= d\left[\frac{y_1 + y_7}{2} + y_2 + y_3 + y_4 + y_5 + y_6\right].$$

In general, the trapezoidal rule states

$$\textbf{Area} = \binom{\textbf{width of}}{\textbf{interval}}\left[\frac{1}{2}\binom{\textbf{first} + \textbf{last}}{\textbf{ordinate}}\right.$$
$$\left. + \binom{\textbf{sum of}}{\textbf{remaining}}{\textbf{ordinates}}\right]$$

(c) **Mid-ordinate rule**

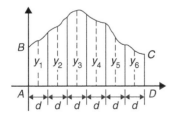

Figure 28.2

To determine the area *ABCD* of Fig. 28.2,

(i) Divide base *AD* into any number of equal intervals, each of width d (the greater the number of intervals, the greater the accuracy).
(ii) Erect ordinates in the middle of each interval (shown by broken lines in Fig. 28.2).
(iii) Accurately measure ordinates y_1, y_2, y_3, etc.
(iv) Area *ABCD*
$$= d(y_1 + y_2 + y_3 + y_4 + y_5 + y_6).$$

In general, the mid-ordinate rule states

Area = (width of interval)(sum of
mid-ordinates)

(d) **Simpson's rule**[*]
To determine the area *PQRS* of Fig. 28.1,

(i) Divide base *PS* into an **even** number of intervals, each of width d (the greater the number of intervals, the greater the accuracy).
(ii) Accurately measure ordinates y_1, y_2, y_3, etc.
(iii) Area $PQRS = \dfrac{d}{3}[(y_1 + y_7) + 4(y_2 + y_4 + y_6)$
$$+ 2(y_3 + y_5)]$$

In general, Simpson's rule states

$$\textbf{Area} = \frac{1}{3}\binom{\textbf{width of}}{\textbf{interval}}\left[\binom{\textbf{first} + \textbf{last}}{\textbf{ordinate}}\right.$$
$$\left. + 4\binom{\textbf{sum of even}}{\textbf{ordinates}} + 2\binom{\textbf{sum of remaining}}{\textbf{odd ordinates}}\right]$$

Problem 1. A car starts from rest and its speed is measured every second for 6 s.

Time t (s)	0	1	2	3	4	5	6
Speed v (m/s)	0	2.5	5.5	8.75	12.5	17.5	24.0

Determine the distance travelled in 6 seconds (i.e. the area under the v/t graph), using (a) the trapezoidal rule (b) the mid-ordinate rule (c) Simpson's rule

A graph of speed/time is shown in Fig. 28.3.

(a) **Trapezoidal rule** (see (b) above)
The time base is divided into 6 strips, each of width 1 s, and the length of the ordinates measured.

---

*Who was Simpson? – **Thomas Simpson** FRS (20 August 1710 – 14 May 1761) was the British mathematician who invented Simpson's rule to approximate definite integrals. To find out more go to www.routledge.com/cw/bird

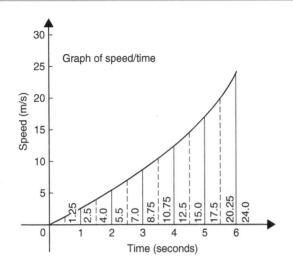

Figure 28.3

Thus,

$$\text{area} = (1)\left[\left(\frac{0+24.0}{2}\right) + 2.5 + 5.5\right.$$

$$\left. + 8.75 + 12.5 + 17.5\right] = \mathbf{58.75\,m}$$

(b) **Mid-ordinate rule** (see (c) above)
The time base is divided into 6 strips each of width 1 s. Mid-ordinates are erected as shown in Fig. 28.3 by the broken lines. The length of each mid-ordinate is measured. Thus,

$$\text{area} = (1)[1.25 + 4.0 + 7.0 + 10.75 + 15.0$$

$$+ 20.25] = \mathbf{58.25\,m}$$

(c) **Simpson's rule** (see (d) above)
The time base is divided into 6 strips each of width 1 s and the length of the ordinates measured. Thus,

$$\text{area} = \frac{1}{3}(1)[(0 + 24.0) + 4(2.5 + 8.75$$

$$+ 17.5) + 2(5.5 + 12.5)] = \mathbf{58.33\,m}$$

Problem 2. A river is 15 m wide. Soundings of the depth are made at equal intervals of 3 m across the river and are as shown below.

Depth (m)	0	2.2	3.3	4.5	4.2	2.4	0

Calculate the cross-sectional area of the flow of water at this point using Simpson's rule

From (d) above,

$$\text{Area} = \frac{1}{3}(3)[(0+0) + 4(2.2+4.5+2.4) + 2(3.3+4.2)]$$

$$= (1)[0 + 36.4 + 15] = \mathbf{51.4\,m^2}$$

Now try the following Practice Exercise

1. Plot a graph of $y = 3x - x^2$ by completing a table of values of y from $x = 0$ to $x = 3$. Determine the area enclosed by the curve, the x-axis and ordinates $x = 0$ and $x = 3$ by (a) the trapezoidal rule (b) the mid-ordinate rule (c) Simpson's rule.

2. Plot the graph of $y = 2x^2 + 3$ between $x = 0$ and $x = 4$. Estimate the area enclosed by the curve, the ordinates $x = 0$ and $x = 4$ and the x-axis by an approximate method.

3. The velocity of a car at 1 second intervals is given in the following table.

Time t (s)	0	1	2	3	4	5	6
Velocity v (m/s)	0	2.0	4.5	8.0	14.0	21.0	29.0

Determine the distance travelled in 6 seconds (i.e. the area under the v/t graph) using Simpson's rule.

4. The shape of a piece of land is shown in Fig. 28.4. To estimate the area of the land, a surveyor takes measurements at intervals of 50 m, perpendicular to the straight portion with the results shown (the dimensions being in metres). Estimate the area of the land in hectares (1 ha $= 10^4\,\text{m}^2$).

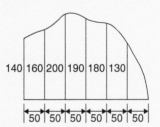

Figure 28.4

5. The deck of a ship is 35 m long. At equal intervals of 5 m the width is given by the following table.

Width (m)	0	2.8	5.2	6.5	5.8	4.1	3.0	2.3

Estimate the area of the deck.

28.2 Volumes of irregular solids

If the cross-sectional areas $A_1, A_2, A_3, \ldots$ of an irregular solid bounded by two parallel planes are known at equal intervals of width d (as shown in Fig. 28.5), by Simpson's rule

Volume, $V = \dfrac{d}{3}[(A_1 + A_7) + 4(A_2 + A_4 + A_6)$

$$+ 2(A_3 + A_5)]$$

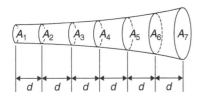

Figure 28.5

Problem 3. A tree trunk is 12 m in length and has a varying cross-section. The cross-sectional areas at intervals of 2 m measured from one end are 0.52, 0.55, 0.59, 0.63, 0.72, 0.84 and 0.97 m². Estimate the volume of the tree trunk

A sketch of the tree trunk is similar to that shown in Fig. 28.5 above, where $d = 2$ m, $A_1 = 0.52$ m², $A_2 = 0.55$ m², and so on.
Using Simpson's rule for volumes gives

Volume $= \dfrac{2}{3}[(0.52 + 0.97) + 4(0.55 + 0.63 + 0.84)$

$$+ 2(0.59 + 0.72)]$$

$$= \dfrac{2}{3}[1.49 + 8.08 + 2.62] = \mathbf{8.13\,m^3}$$

Problem 4. The areas of seven horizontal cross-sections of a water reservoir at intervals of 10 m are 210, 250, 320, 350, 290, 230 and 170 m². Calculate the capacity of the reservoir in litres

Using Simpson's rule for volumes gives

Volume $= \dfrac{10}{3}[(210 + 170) + 4(250 + 350 + 230)$

$$+ 2(320 + 290)]$$

$$= \dfrac{10}{3}[380 + 3320 + 1220] = 16\,400\,\text{m}^3$$

$16\,400\,\text{m}^3 = 16\,400 \times 10^6\,\text{cm}^3$.

Since 1 litre $= 1000\,\text{cm}^3$,

capacity of reservoir $= \dfrac{16400 \times 10^6}{1000}$ litres

$$= 16\,400\,000 = \mathbf{16.4 \times 10^6\ litres}$$

Now try the following Practice Exercise

Practice Exercise 111   Volumes of irregular solids (answers on page 434)

1. The areas of equidistantly spaced sections of the underwater form of a small boat are as follows:
   1.76, 2.78, 3.10, 3.12, 2.61, 1.24 and 0.85 m².
   Determine the underwater volume if the sections are 3 m apart.

2. To estimate the amount of earth to be removed when constructing a cutting, the cross-sectional area at intervals of 8 m were estimated as follows:
   0, 2.8, 3.7, 4.5, 4.1, 2.6 and 0 m³
   Estimate the volume of earth to be excavated.

3. The circumference of a 12 m long log of timber of varying circular cross-section is measured at intervals of 2 m along its length and the results are as follows. Estimate the volume of the timber in cubic metres.

Distance from one end (m)	0	2	4	6
Circumference (m)	2.80	3.25	3.94	4.32

Distance from one end (m)	8	10	12
Circumference (m)	5.16	5.82	6.36

28.3 Mean or average values of waveforms

The mean or average value, y, of the waveform shown in Fig. 28.6 is given by

$$y = \frac{\text{area under curve}}{\text{length of base, } b}$$

If the mid-ordinate rule is used to find the area under the curve, then

$$y = \frac{\text{sum of mid-ordinates}}{\text{number of mid-ordinates}}$$

$$\left(= \frac{y_1 + y_2 + y_3 + y_4 + y_5 + y_6 + y_7}{7} \text{ for Fig. 28.6} \right)$$

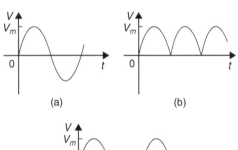

Figure 28.6

For a **sine wave**, the mean or average value

(a) over one complete cycle is **zero** (see Fig. 28.7(a)),

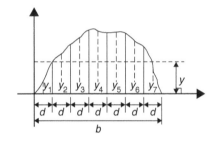

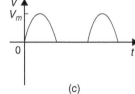

Figure 28.7

(b) over half a cycle is **0.637 × maximum value** or $\dfrac{2}{\pi}$ **× maximum value,**

(c) of a full-wave rectified waveform (see Fig. 28.7(b)) is **0.637 × maximum value,**

(d) of a half-wave rectified waveform (see Fig. 28.7(c)) is **0.318 × maximum value** or $\dfrac{1}{\pi}$ **× maximum value.**

Problem 5. Determine the average values over half a cycle of the periodic waveforms shown in Fig. 28.8

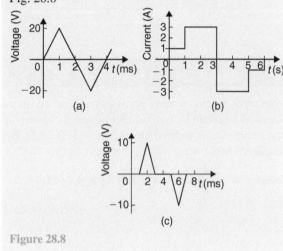

Figure 28.8

(a) Area under triangular waveform (a) for a half cycle is given by

$$\text{Area} = \frac{1}{2}(\text{base})(\text{perpendicular height})$$

$$= \frac{1}{2}(2 \times 10^{-3})(20) = 20 \times 10^{-3} \, \text{Vs}$$

Average value of waveform

$$= \frac{\text{area under curve}}{\text{length of base}}$$

$$= \frac{20 \times 10^{-3} \text{Vs}}{2 \times 10^{-3} \text{s}} = \mathbf{10 \, V}$$

(b) Area under waveform (b) for a half cycle
$$= (1 \times 1) + (3 \times 2) = 7 \, \text{As}$$

Average value of waveform $= \dfrac{\text{area under curve}}{\text{length of base}}$

$$= \frac{7 \, \text{As}}{3 \, \text{s}} = \mathbf{2.33 \, A}$$

(c) A half cycle of the voltage waveform (c) is completed in 4 ms.

$$\text{Area under curve} = \frac{1}{2}\{(3 - 1)10^{-3}\}(10)$$

$$= 10 \times 10^{-3} \, \text{Vs}$$

Average value of waveform = $\dfrac{\text{area under curve}}{\text{length of base}}$

$$= \frac{10 \times 10^{-3}\,\text{Vs}}{4 \times 10^{-3}\,\text{s}} = \mathbf{2.5\,V}$$

Mean value over one cycle = $\dfrac{\text{area under curve}}{\text{length of base}}$

$$= \frac{3 \times 10^{-3}\,\text{As}}{5 \times 10^{-3}\,\text{s}}$$

$$= \mathbf{0.6\,A}$$

Problem 6. Determine the mean value of current over one complete cycle of the periodic waveforms shown in Fig. 28.9

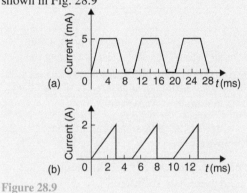

(a)

(b)

Figure 28.9

(a) One cycle of the trapezoidal waveform (a) is completed in 10 ms (i.e. the periodic time is 10 ms).

Area under curve = area of trapezium

$$= \frac{1}{2}(\text{sum of parallel sides})(\text{perpendicular distance between parallel sides})$$

$$= \frac{1}{2}\{(4+8) \times 10^{-3}\}(5 \times 10^{-3})$$

$$= 30 \times 10^{-6}\,\text{As}$$

Mean value over one cycle = $\dfrac{\text{area under curve}}{\text{length of base}}$

$$= \frac{30 \times 10^{-6}\,\text{As}}{10 \times 10^{-3}\,\text{s}}$$

$$= \mathbf{3\,mA}$$

(b) One cycle of the sawtooth waveform (b) is completed in 5 ms.

$$\text{Area under curve} = \frac{1}{2}(3 \times 10^{-3})(2)$$

$$= 3 \times 10^{-3}\,\text{As}$$

Problem 7. The power used in a manufacturing process during a 6 hour period is recorded at intervals of 1 hour as shown below.

Time (h)	0	1	2	3	4	5	6
Power (kW)	0	14	29	51	45	23	0

Plot a graph of power against time and, by using the mid-ordinate rule, determine (a) the area under the curve and (b) the average value of the power

The graph of power/time is shown in Fig. 28.10.

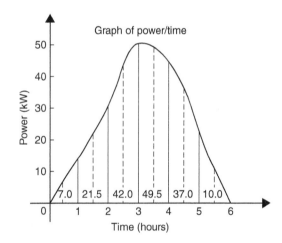

Figure 28.10

(a) The time base is divided into 6 equal intervals, each of width 1 hour. Mid-ordinates are erected (shown by broken lines in Fig. 28.10) and measured. The values are shown in Fig. 28.10.

Area under curve

$$= (\text{width of interval})(\text{sum of mid-ordinates})$$

$$= (1)[7.0 + 21.5 + 42.0 + 49.5 + 37.0 + 10.0]$$

$$= \mathbf{167\,kWh} \text{ (i.e. a measure of electrical energy)}$$

(b) **Average value of waveform** $= \dfrac{\text{area under curve}}{\text{length of base}}$

$$= \dfrac{167\,\text{kWh}}{6\,\text{h}}$$

$$= \textbf{27.83\,kW}$$

Alternatively, average value

$$= \dfrac{\text{sum of mid-ordinates}}{\text{number of mid-ordinates}}$$

Problem 8. Fig. 28.11 shows a sinusoidal output voltage of a full-wave rectifier. Determine, using the mid-ordinate rule with 6 intervals, the mean output voltage

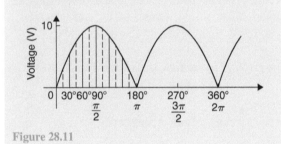

Figure 28.11

One cycle of the output voltage is completed in π radians or $180°$. The base is divided into 6 intervals, each of width $30°$. The mid-ordinate of each interval will lie at $15°, 45°, 75°$, etc.

At $15°$ the height of the mid-ordinate is $10\sin 15° = 2.588\,\text{V}$,

At $45°$ the height of the mid-ordinate is $10\sin 45° = 7.071\,\text{V}$, and so on.

The results are tabulated below.

Mid-ordinate	Height of mid-ordinate
$15°$	$10\sin 15° = 2.588\,\text{V}$
$45°$	$10\sin 45° = 7.071\,\text{V}$
$75°$	$10\sin 75° = 9.659\,\text{V}$
$105°$	$10\sin 105° = 9.659\,\text{V}$
$135°$	$10\sin 135° = 7.071\,\text{V}$
$165°$	$10\sin 165° = 2.588\,\text{V}$
Sum of mid-ordinates $= 38.636\,\text{V}$	

Mean or average value of output voltage

$$= \dfrac{\text{sum of mid-ordinates}}{\text{number of mid-ordinates}} = \dfrac{38.636}{6} = \textbf{6.439\,V}$$

(With a larger number of intervals a more accurate answer may be obtained.)

For a sine wave the actual mean value is $0.637 \times$ maximum value, which in this problem gives $6.37\,\text{V}$.

Problem 9. An indicator diagram for a steam engine is shown in Fig. 28.12. The base line has been divided into 6 equally spaced intervals and the lengths of the 7 ordinates measured with the results shown in centimetres. Determine

(a) the area of the indicator diagram using Simpson's rule

(b) the mean pressure in the cylinder given that 1 cm represents 100 kPa.

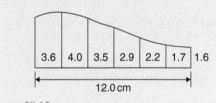

Figure 28.12

(a) The width of each interval is $\dfrac{12.0}{6}$ cm. Using Simpson's rule,

$$\textbf{area} = \frac{1}{3}(2.0)[(3.6+1.6)+4(4.0+2.9+1.7)$$
$$+\,2(3.5+2.2)]$$
$$= \frac{2}{3}[5.2+34.4+11.4] = \textbf{34\,cm}^2$$

(b) Mean height of ordinates $= \dfrac{\text{area of diagram}}{\text{length of base}}$

$$= \dfrac{34}{12} = 2.83\,\text{cm}$$

Since 1 cm represents 100 kPa,

mean pressure in the cylinder
$$= 2.83\,\text{cm} \times 100\,\text{kPa/cm} = \textbf{283\,kPa}$$

Now try the following Practice Exercise

1. Determine the mean value of the periodic
    waveforms shown in Fig. 28.13 over a half
    cycle.

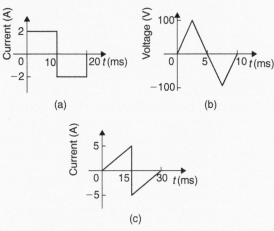

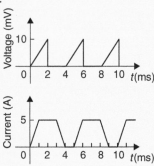

(c)

Figure 28.13

2. Find the average value of the periodic wave-
    forms shown in Fig. 28.14 over one complete
    cycle.

Figure 28.14

3. An alternating current has the following values
    at equal intervals of 5 ms:

Time (ms)	0	5	10	15	20	25	30
Current (A)	0	0.9	2.6	4.9	5.8	3.5	0

    Plot a graph of current against time and esti-
    mate the area under the curve over the 30 ms
    period, using the mid-ordinate rule, and deter-
    mine its mean value.

4. Determine, using an approximate method, the
    average value of a sine wave of maximum
    value 50 V for (a) a half cycle (b) a complete
    cycle.

5. An indicator diagram of a steam engine is
    12 cm long. Seven evenly spaced ordinates,
    including the end ordinates, are measured as
    follows:

    5.90, 5.52, 4.22, 3.63, 3.32, 3.24 and 3.16 cm.

    Determine the area of the diagram and the
    mean pressure in the cylinder if 1 cm repre-
    sents 90 kPa.

**For fully worked solutions to each of the problems in Practice Exercises 110 to 112 in this chapter,
go to the website:**
www.routledge.com/cw/bird

This assignment covers the material contained in Chapters 27 and 28. *The marks available are shown in brackets at the end of each question.*

1. A rectangular block of alloy has dimensions of 60 mm by 30 mm by 12 mm. Calculate the volume of the alloy in cubic centimetres. (3)

2. Determine how many cubic metres of concrete are required for a 120 m long path, 400 mm wide and 10 cm deep. (3)

3. Find the volume of a cylinder of radius 5.6 cm and height 15.5 cm. Give the answer correct to the nearest cubic centimetre. (3)

4. A garden roller is 0.35 m wide and has a diameter of 0.20 m. What area will it roll in making 40 revolutions? (4)

5. Find the volume of a cone of height 12.5 cm and base diameter 6.0 cm, correct to 1 decimal place. (3)

6. Find (a) the volume and (b) the total surface area of the right-angled triangular prism shown in Fig. RT11.1. (9)

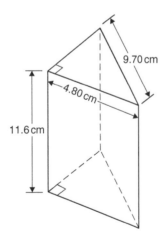

Figure RT11.1

7. A pyramid having a square base has a volume of 86.4 cm^3. If the perpendicular height is 20 cm, determine the length of each side of the base. (4)

8. A copper pipe is 80 m long. It has a bore of 80 mm and an outside diameter of 100 mm. Calculate, in cubic metres, the volume of copper in the pipe. (4)

9. Find (a) the volume and (b) the surface area of a sphere of diameter 25 mm. (4)

10. A piece of alloy with dimensions 25 mm by 60 mm by 1.60 m is melted down and recast into a cylinder whose diameter is 150 mm. Assuming no wastage, calculate the height of the cylinder in centimetres, correct to 1 decimal place. (4)

11. Determine the volume (in cubic metres) and the total surface area (in square metres) of a solid metal cone of base radius 0.5 m and perpendicular height 1.20 m. Give answers correct to 2 decimal places. (6)

12. A rectangular storage container has dimensions 3.2 m by 90 cm by 60 cm. Determine its volume in (a) m^3 (b) cm^3 (4)

13. Calculate (a) the volume and (b) the total surface area of a 10 cm by 15 cm rectangular pyramid of height 20 cm. (8)

14. A water container is of the form of a central cylindrical part 3.0 m long and diameter 1.0 m, with a hemispherical section surmounted at each end as shown in Fig. RT11.2. Determine the maximum capacity of the container, correct to the nearest litre. (1 litre = 1000 cm^3) (5)

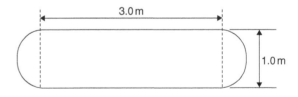

Figure RT11.2

15. Find the total surface area of a bucket consisting of an inverted frustum of a cone of slant height 35.0 cm and end diameters 60.0 cm and 40.0 cm. (4)

16. A boat has a mass of 20 000 kg. A model of the boat is made to a scale of 1 to 80. If the model is made of the same material as the boat, determine the mass of the model (in grams). (3)

17. Plot a graph of $y = 3x^2 + 5$ from $x = 1$ to $x = 4$. Estimate, correct to 2 decimal places, using 6 intervals, the area enclosed by the curve, the ordinates $x = 1$ and $x = 4$, and the x-axis by
    (a) the trapezoidal rule
    (b) the mid-ordinate rule
    (c) Simpson's rule.   (16)

18. A circular cooling tower is 20 m high. The inside diameter of the tower at different heights is given in the following table.

Height (m)	0	5.0	10.0	15.0	20.0
Diameter (m)	16.0	13.3	10.7	8.6	8.0

Determine the area corresponding to each diameter and hence estimate the capacity of the tower in cubic metres.   (7)

19. A vehicle starts from rest and its velocity is measured every second for 6 seconds, with the following results.

Time t (s)	0	1	2	3	4	5	6
Velocity v (m/s)	0	1.2	2.4	3.7	5.2	6.0	9.2

Using Simpson's rule, calculate
(a) the distance travelled in 6 s (i.e. the area under the v/t graph),
(b) the average speed over this period.   (6)

Chapter 29

Vectors

Why it is important to understand: Vectors

Vectors are an important part of the language of science, mathematics and engineering. They are used to discuss multivariable calculus, electrical circuits with oscillating currents, stress and strain in structures and materials, and flows of atmospheres and fluids, and they have many other applications. Resolving a vector into components is a precursor to computing things with or about a vector quantity. Because position, velocity, acceleration, force, momentum and angular momentum are all vector quantities, resolving vectors into components is a most important skill required in any engineering studies.

At the end of this chapter, you should be able to:

- distinguish between scalars and vectors
- recognise how vectors are represented
- add vectors using the nose-to-tail method
- add vectors using the parallelogram method
- resolve vectors into their horizontal and vertical components
- add vectors by calculation using horizontal and vertical components
- perform vector subtraction
- understand relative velocity
- understand i, j, k notation

29.1 Introduction

This chapter initially explains the difference between scalar and vector quantities and shows how a vector is drawn and represented.

Any object that is acted upon by an external force will respond to that force by moving in the line of the force. However, if two or more forces act simultaneously, the result is more difficult to predict; the ability to add two or more vectors then becomes important.

This chapter thus shows how vectors are added and subtracted, both by drawing and by calculation, and how finding the resultant of two or more vectors has many uses in engineering. (Resultant means the single vector which would have the same effect as the individual vectors.) Relative velocities and vector i, j, k notation are also briefly explained.

29.2 Scalars and vectors

The time taken to fill a water tank may be measured as, say, 50 s. Similarly, the temperature in a room may be measured as, say, 16°C or the mass of a bearing may be measured as, say, 3 kg. Quantities such as time, temperature and mass are entirely defined by a numerical value and are called **scalars** or **scalar quantities**.

Not all quantities are like this. Some are defined by more than just size; some also have direction. For example, the velocity of a car may be 90 km/h due west, a force of 20 N may act vertically downwards, or an acceleration of 10 m/s² may act at 50° to the horizontal.

Quantities such as velocity, force and acceleration, which **have both a magnitude and a direction**, are called **vectors**.

Now try the following Practice Exercise

Practice Exercise 113   Scalar and vector quantities (answers on page 434)

1. State the difference between scalar and vector quantities.

   In Problems 2 to 9, state whether the quantities given are scalar or vector.

2. A temperature of 70°C

3. 5 m³ volume

4. A downward force of 20 N

5. 500 J of work

6. 30 cm² area

7. A south-westerly wind of 10 knots

8. 50 m distance

9. An acceleration of 15 m/s² at 60° to the horizontal

29.3   Drawing a vector

A vector quantity can be represented graphically by a line, drawn so that

(a) the **length** of the line denotes the magnitude of the quantity, and

(b) the **direction** of the line denotes the direction in which the vector quantity acts.

An arrow is used to denote the sense, or direction, of the vector.

The arrow end of a vector is called the 'nose' and the other end the 'tail'. For example, a force of 9 N acting at 45° to the horizontal is shown in Fig. 29.1. Note that an angle of **+45°** is drawn from the horizontal and moves **anticlockwise**.

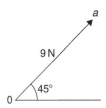

Figure 29.1

A velocity of 20 m/s at −60° is shown in Fig. 29.2. Note that an angle of **−60°** is drawn from the horizontal and moves **clockwise**.

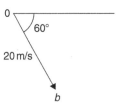

Figure 29.2

29.3.1   Representing a vector

There are a number of ways of representing vector quantities. These include

(a) Using bold print.

(b) $\overrightarrow{AB}$ where an arrow above two capital letters denotes the sense of direction, where A is the starting point and B the end point of the vector.

(c) $\overline{AB}$ or $\overline{a}$; i.e. a line over the top of letter.

(d) $\underline{a}$; i.e. underlined letter.

The force of 9 N at 45° shown in Fig. 29.1 may be represented as

$$\mathbf{0a} \quad \text{or} \quad \overrightarrow{0a} \quad \text{or} \quad \overline{0a}$$

The magnitude of the force is 0a

Similarly, the velocity of 20 m/s at −60° shown in Fig. 29.2 may be represented as

$$\mathbf{0b} \quad \text{or} \quad \overrightarrow{0b} \quad \text{or} \quad \overline{0b}$$

The magnitude of the velocity is 0b

In this chapter a vector quantity is denoted by **bold print**.

29.4 Addition of vectors by drawing

Adding two or more vectors by drawing assumes that a ruler, pencil and protractor are available. Results obtained by drawing are naturally not as accurate as those obtained by calculation.

(a) **Nose-to-tail method**
   Two force vectors, F_1 and F_2, are shown in Fig. 29.3. When an object is subjected to more than one force, the resultant of the forces is found by the addition of vectors.

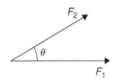

Figure 29.3

To add forces F_1 and F_2,
   (i) Force F_1 is drawn to scale horizontally, shown as **0a** in Fig. 29.4.
   (ii) From the nose of F_1, force F_2 is drawn at angle θ to the horizontal, shown as **ab**
   (iii) The resultant force is given by length **0b**, which may be measured.
This procedure is called the **'nose-to-tail'** or **'triangle' method**.

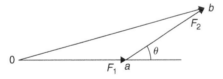

Figure 29.4

(b) **Parallelogram method**
   To add the two force vectors, F_1 and F_2 of Fig. 29.3,
   (i) A line *cb* is constructed which is parallel to and equal in length to **0a** (see Fig. 29.5).
   (ii) A line *ab* is constructed which is parallel to and equal in length to **0c**
   (iii) The resultant force is given by the diagonal of the parallelogram; i.e. length **0b**
This procedure is called the **'parallelogram' method**.

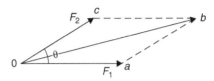

Figure 29.5

Problem 1. A force of 5 N is inclined at an angle of 45° to a second force of 8 N, both forces acting at a point. Find the magnitude of the resultant of these two forces and the direction of the resultant with respect to the 8 N force by (a) the nose-to-tail method and (b) the parallelogram method

The two forces are shown in Fig. 29.6. (Although the 8 N force is shown horizontal, it could have been drawn in any direction.)

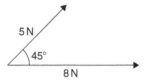

Figure 29.6

(a) **Nose-to-tail method**
   (i) The 8 N force is drawn horizontally 8 units long, shown as **0a** in Fig. 29.7.
   (ii) From the nose of the 8 N force, the 5 N force is drawn 5 units long at an angle of 45° to the horizontal, shown as **ab**
   (iii) The resultant force is given by length **0b** and is measured as **12 N** and angle θ is measured as **17°**

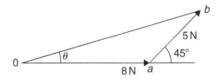

Figure 29.7

(b) **Parallelogram method**
   (i) In Fig. 29.8, a line is constructed which is parallel to and equal in length to the 8 N force.
   (ii) A line is constructed which is parallel to and equal in length to the 5 N force.

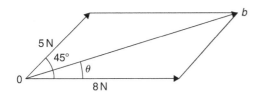

Figure 29.8

(iii) The resultant force is given by the diagonal of the parallelogram, i.e. length $0b$, and is measured as **12 N** and angle θ is measured as **17°**

Thus, **the resultant of the two force vectors in Fig. 29.6 is 12 N at 17° to the 8 N force**.

Problem 2. Forces of 15 and 10 N are at an angle of 90° to each other as shown in Fig. 29.9. Find, by drawing, the magnitude of the resultant of these two forces and the direction of the resultant with respect to the 15 N force

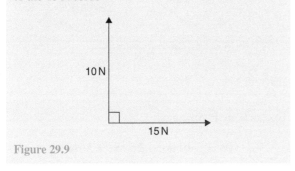

Figure 29.9

Using the nose-to-tail method,

(i) The 15 N force is drawn horizontally 15 units long, as shown in Fig. 29.10.

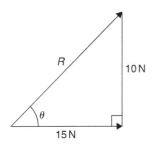

Figure 29.10

(ii) From the nose of the 15 N force, the 10 N force is drawn 10 units long at an angle of 90° to the horizontal as shown.

(iii) The resultant force is shown as R and is measured as **18 N** and angle θ is measured as **34°**

Thus, **the resultant of the two force vectors is 18 N at 34° to the 15 N force**.

Problem 3. Velocities of 10 m/s, 20 m/s and 15 m/s act as shown in Fig. 29.11. Determine, by drawing, the magnitude of the resultant velocity and its direction relative to the horizontal

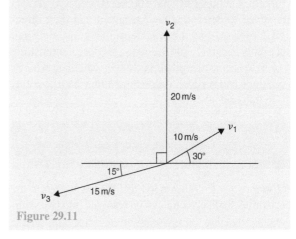

Figure 29.11

When more than 2 vectors are being added the nose-to-tail method is used. The order in which the vectors are added does not matter. In this case the order taken is v_1, then v_2, then v_3. However, if a different order is taken the same result will occur.

(i) v_1 is drawn 10 units long at an angle of 30° to the horizontal, shown as $\mathbf{0}a$ in Fig. 29.12.

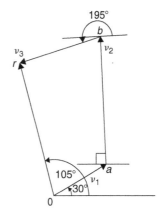

Figure 29.12

(ii) From the nose of v_1, v_2 is drawn 20 units long at an angle of 90° to the horizontal, shown as $\mathbf{ab}$

(iii) From the nose of v_2, v_3 is drawn 15 units long at an angle of 195° to the horizontal, shown as **br**

(iv) The resultant velocity is given by length **0r** and is measured as **22 m/s** and the angle measured to the horizontal is **105°**

Thus, **the resultant of the three velocities is 22 m/s at 105° to the horizontal**.

Worked examples 1 to 3 have demonstrated how vectors are added to determine their resultant and their direction. However, drawing to scale is time-consuming and not highly accurate. The following sections demonstrate how to determine resultant vectors by calculation using horizontal and vertical components and, where possible, by Pythagoras' theorem.

29.5 Resolving vectors into horizontal and vertical components

A force vector F is shown in Fig. 29.13 at angle θ to the horizontal. Such a vector can be resolved into two components such that the vector addition of the components is equal to the original vector.

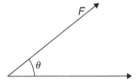

Figure 29.13

The two components usually taken are a **horizontal component** and a **vertical component**. If a right-angled triangle is constructed as shown in Fig. 29.14, $0a$ is called the horizontal component of F and ab is called the vertical component of F.

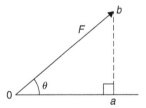

Figure 29.14

From trigonometry (see Chapter 21 and remember SOH CAH TOA),

$$\cos\theta = \frac{0a}{0b}, \text{ from which } 0a = 0b\cos\theta = F\cos\theta$$

i.e. **the horizontal component of $F = F\cos\theta$**, and

$$\sin\theta = \frac{ab}{0b} \text{ from which, } ab = 0b\sin\theta = F\sin\theta$$

i.e. **the vertical component of $F = F\sin\theta$**.

> **Problem 4.** Resolve the force vector of 50 N at an angle of 35° to the horizontal into its horizontal and vertical components

The **horizontal component** of the 50 N force, $0a = 50\cos 35° = \mathbf{40.96\,N}$
The **vertical component** of the 50 N force, $ab = 50\sin 35° = \mathbf{28.68\,N}$
The horizontal and vertical components are shown in Fig. 29.15.

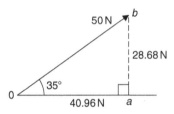

Figure 29.15

(To check: by Pythagoras,

$$0b = \sqrt{40.96^2 + 28.68^2} = 50\,\text{N}$$

and $\quad \theta = \tan^{-1}\left(\dfrac{28.68}{40.96}\right) = 35°$

Thus, the vector addition of components 40.96 N and 28.68 N is 50 N at 35°)

> **Problem 5.** Resolve the velocity vector of 20 m/s at an angle of −30° to the horizontal into horizontal and vertical components

The **horizontal component** of the 20 m/s velocity, $0a = 20\cos(-30°) = \mathbf{17.32\,m/s}$
The **vertical component** of the 20 m/s velocity, $ab = 20\sin(-30°) = \mathbf{-10\,m/s}$
The horizontal and vertical components are shown in Fig. 29.16.

> **Problem 6.** Resolve the displacement vector of 40 m at an angle of 120° into horizontal and vertical components

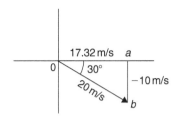

Figure 29.16

The **horizontal component** of the 40 m displacement,
$0a = 40\cos 120° = -20.0\text{m}$
The **vertical component** of the 40 m displacement,
$ab = 40\sin 120° = 34.64\text{m}$
The horizontal and vertical components are shown in
Fig. 29.17.

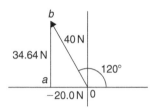

Figure 29.17

29.6 Addition of vectors by calculation

Two force vectors, F_1 and F_2, are shown in Fig. 29.18,
F_1 being at an angle of θ_1 and F_2 at an angle of θ_2

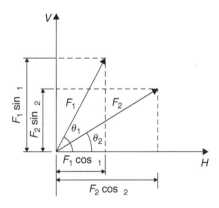

Figure 29.18

A method of adding two vectors together is to use
horizontal and vertical components.
The horizontal component of force F_1 is $F_1\cos\theta_1$ and
the horizontal component of force F_2 is $F_2\cos\theta_2$. The

total horizontal component of the two forces,

$$H = F_1\cos\theta_1 + F_2\cos\theta_2$$

The vertical component of force F_1 is $F_1\sin\theta_1$ and the
vertical component of force F_2 is $F_2\sin\theta_2$. The total
vertical component of the two forces,

$$V = F_1\sin\theta_1 + F_2\sin\theta_2$$

Since we have H and V, the resultant of F_1 and F_2
is obtained by using the theorem of Pythagoras. From
Fig. 29.19,

$$0b^2 = H^2 + V^2$$

i.e. $$\textbf{resultant} = \sqrt{H^2 + V^2} \text{ at an angle}$$
$$\text{given by } \theta = \tan^{-1}\left(\frac{V}{H}\right)$$

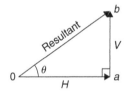

Figure 29.19

> **Problem 7.** A force of 5 N is inclined at an angle
> of 45° to a second force of 8 N, both forces acting at
> a point. Calculate the magnitude of the resultant of
> these two forces and the direction of the resultant
> with respect to the 8 N force

The two forces are shown in Fig. 29.20.

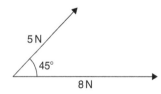

Figure 29.20

The horizontal component of the 8 N force is $8\cos 0°$
and the horizontal component of the 5 N force is
$5\cos 45°$. The total horizontal component of the two
forces,

$$H = 8\cos 0° + 5\cos 45° = 8 + 3.5355 = \textbf{11.5355}$$

The vertical component of the 8 N force is $8\sin 0°$ and the vertical component of the 5 N force is $5\sin 45°$. The total vertical component of the two forces,

$$V = 8\sin 0° + 5\sin 45° = 0 + 3.5355 = \mathbf{3.5355}$$

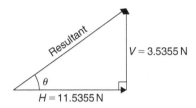

Figure 29.21

From Fig. 29.21, magnitude of resultant vector

$$= \sqrt{H^2 + V^2}$$
$$= \sqrt{11.5355^2 + 3.5355^2} = \mathbf{12.07\,N}$$

The direction of the resultant vector,

$$\theta = \tan^{-1}\left(\frac{V}{H}\right) = \tan^{-1}\left(\frac{3.5355}{11.5355}\right)$$
$$= \tan^{-1} 0.30648866\ldots = \mathbf{17.04°}$$

Thus, **the resultant of the two forces is a single vector of 12.07 N at 17.04° to the 8 N vector.**

Problem 8. Forces of 15 N and 10 N are at an angle of 90° to each other as shown in Fig. 29.22. Calculate the magnitude of the resultant of these two forces and its direction with respect to the 15 N force

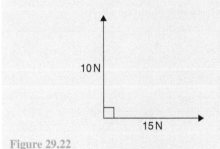

Figure 29.22

The horizontal component of the 15 N force is $15\cos 0°$ and the horizontal component of the 10 N force is $10\cos 90°$. The total horizontal component of the two velocities,

$$H = 15\cos 0° + 10\cos 90° = 15 + 0 = \mathbf{15}$$

The vertical component of the 15 N force is $15\sin 0°$ and the vertical component of the 10 N force is $10\sin 90°$. The total vertical component of the two velocities,

$$V = 15\sin 0° + 10\sin 90° = 0 + 10 = \mathbf{10}$$

Magnitude of resultant vector
$$= \sqrt{H^2 + V^2} = \sqrt{15^2 + 10^2} = \mathbf{18.03\,N}$$
The direction of the resultant vector,
$$\theta = \tan^{-1}\left(\frac{V}{H}\right) = \tan^{-1}\left(\frac{10}{15}\right) = \mathbf{33.69°}$$

Thus, **the resultant of the two forces is a single vector of 18.03 N at 33.69° to the 15 N vector.**

There is an alternative method of calculating the resultant vector in this case. If we used the triangle method, the diagram would be as shown in Fig. 29.23.

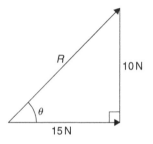

Figure 29.23

Since a right-angled triangle results, we could use Pythagoras' theorem without needing to go through the procedure for horizontal and vertical components. In fact, the horizontal and vertical components are 15 N and 10 N respectively.

This is, of course, a special case. **Pythagoras can only be used when there is an angle of 90° between vectors.** This is demonstrated in worked Problem 9.

Problem 9. Calculate the magnitude and direction of the resultant of the two acceleration vectors shown in Fig. 29.24.

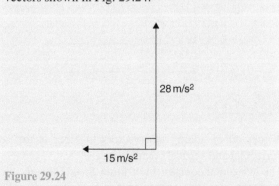

Figure 29.24

The 15 m/s² acceleration is drawn horizontally, shown as **0a** in Fig. 29.25.

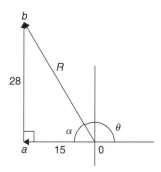

Figure 29.25

From the nose of the 15 m/s² acceleration, the 28 m/s² acceleration is drawn at an angle of 90° to the horizontal, shown as **ab**
The resultant acceleration, R, is given by length **0b**
Since a right-angled triangle results, the theorem of Pythagoras may be used.

$$0b = \sqrt{15^2 + 28^2} = \textbf{31.76 m/s}^2$$

and $\quad \alpha = \tan^{-1}\left(\dfrac{28}{15}\right) = \textbf{61.82}°$

Measuring from the horizontal,
$\theta = 180° - 61.82° = \textbf{118.18}°$
Thus, **the resultant of the two accelerations is a single vector of 31.76 m/s² at 118.18° to the horizontal**.

Problem 10. Velocities of 10 m/s, 20 m/s and 15 m/s act as shown in Fig. 29.26. Calculate the magnitude of the resultant velocity and its direction relative to the horizontal

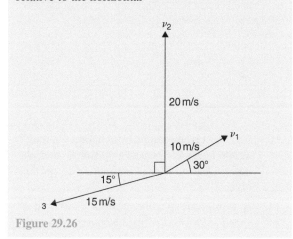

Figure 29.26

The horizontal component of the 10 m/s velocity
$= 10 \cos 30° = 8.660$ m/s,

the horizontal component of the 20 m/s velocity is
$20 \cos 90° = 0$ m/s
and the horizontal component of the 15 m/s velocity is
$15 \cos 195° = -14.489$ m/s
The total horizontal component of the three velocities,

$$H = 8.660 + 0 - 14.489 = -\textbf{5.829 m/s}$$

The vertical component of the 10 m/s velocity
$= 10 \sin 30° = 5$ m/s,
the vertical component of the 20 m/s velocity is
$20 \sin 90° = 20$ m/s
and the vertical component of the 15 m/s velocity is
$15 \sin 195° = -3.882$ m/s
The total vertical component of the three forces,
$$V = 5 + 20 - 3.882 = \textbf{21.118 m/s}$$

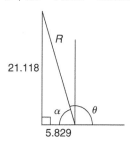

Figure 29.27

From Fig. 29.27, magnitude of resultant vector,

$$R = \sqrt{H^2 + V^2} = \sqrt{5.829^2 + 21.118^2} = \textbf{21.91 m/s}$$

The direction of the resultant vector,

$$\alpha = \tan^{-1}\left(\frac{V}{H}\right) = \tan^{-1}\left(\frac{21.118}{5.829}\right) = \textbf{74.57}°$$

Measuring from the horizontal,
$$\theta = 180° - 74.57° = \textbf{105.43}°$$
Thus, **the resultant of the three velocities is a single vector of 21.91 m/s at 105.43° to the horizontal**.

Now try the following Practice Exercise

Practice Exercise 114   Addition of vectors by calculation (answers on page 434)

1. A force of 7 N is inclined at an angle of 50° to a second force of 12 N, both forces acting at a point. Calculate the magnitude of the resultant of the two forces and the direction of the resultant with respect to the 12 N force.

2. Velocities of 5 m/s and 12 m/s act at a point at 90° to each other. Calculate the resultant velocity and its direction relative to the 12 m/s velocity.

3. Calculate the magnitude and direction of the resultant of the two force vectors shown in Fig. 29.28.

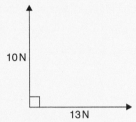

Figure 29.28

4. Calculate the magnitude and direction of the resultant of the two force vectors shown in Fig. 29.29.

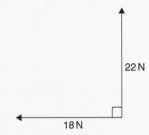

Figure 29.29

5. A displacement vector s_1 is 30 m at $0°$. A second displacement vector s_2 is 12 m at $90°$. Calculate the magnitude and direction of the resultant vector $s_1 + s_2$

6. Three forces of 5 N, 8 N and 13 N act as shown in Fig. 29.30. Calculate the magnitude and direction of the resultant force.

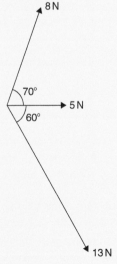

Figure 29.30

7. If velocity $v_1 = 25$ m/s at $60°$ and $v_2 = 15$ m/s at $-30°$, calculate the magnitude and direction of $v_1 + v_2$

8. Calculate the magnitude and direction of the resultant vector of the force system shown in Fig. 29.31.

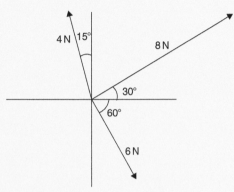

Figure 29.31

9. Calculate the magnitude and direction of the resultant vector of the system shown in Fig. 29.32.

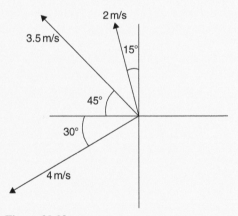

Figure 29.32

10. An object is acted upon by two forces of magnitude 10 N and 8 N at an angle of $60°$ to each other. Determine the resultant force on the object.

11. A ship heads in a direction of E $20°$S at a speed of 20 knots while the current is 4 knots in a direction of N $30°$E. Determine the speed and actual direction of the ship.

29.7   Vector subtraction

In Fig. 29.33, a force vector F is represented by oa. The vector $(-oa)$ can be obtained by drawing a vector from o in the opposite sense to oa but having the same magnitude, shown as ob in Fig. 29.33; i.e. $ob = (-oa)$

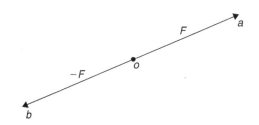

Figure 29.33

For two vectors acting at a point, as shown in Fig. 29.34(a), the resultant of vector addition is

$$os = oa + ob$$

Fig. 29.34(b) shows vectors $ob + (-oa)$ that is, $ob - oa$ and the vector equation is $ob - oa = od$. Comparing od in Fig. 29.34(b) with the broken line ab in Fig. 29.34(a) shows that the second diagonal of the parallelogram method of vector addition gives the magnitude and direction of vector subtraction of oa from ob

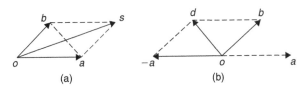

(a)   (b)

Figure 29.34

Problem 11. Accelerations of $a_1 = 1.5\,\text{m/s}^2$ at $90°$ and $a_2 = 2.6\,\text{m/s}^2$ at $145°$ act at a point. Find $a_1 + a_2$ and $a_1 - a_2$ by (a) drawing a scale vector diagram and (b) calculation

(a) The scale vector diagram is shown in Fig. 29.35. By measurement,

$$a_1 + a_2 = 3.7\ \text{m/s}^2\ \text{at}\ 126°$$

$$a_1 - a_2 = 2.1\ \text{m/s}^2\ \text{at}\ 0°$$

(b) Resolving horizontally and vertically gives
Horizontal component of $a_1 + a_2$,
$$H = 1.5\cos 90° + 2.6\cos 145° = -\textbf{2.13}$$

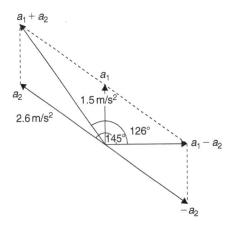

Figure 29.35

Vertical component of $a_1 + a_2$,
$$V = 1.5\sin 90° + 2.6\sin 145° = \textbf{2.99}$$
From Fig. 29.36, the magnitude of $a_1 + a_2$,
$$R = \sqrt{(-2.13)^2 + 2.99^2} = \textbf{3.67}\,\textbf{m/s}^2$$

In Fig. 29.36, $\alpha = \tan^{-1}\left(\dfrac{2.99}{2.13}\right) = 54.53°$ and
$\theta = 180° - 54.53° = 125.47°$
Thus, $a_1 + a_2 = \textbf{3.67}\,\textbf{m/s}^2$ **at** $\textbf{125.47°}$

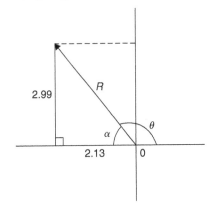

Figure 29.36

Horizontal component of $a_1 - a_2$
$$= 1.5\cos 90° - 2.6\cos 145° = \textbf{2.13}$$
Vertical component of $a_1 - a_2$
$$= 1.5\sin 90° - 2.6\sin 145° = 0$$
Magnitude of $a_1 - a_2 = \sqrt{2.13^2 + 0^2}$
$$= \textbf{2.13}\,\textbf{m/s}^2$$
Direction of $a_1 - a_2 = \tan^{-1}\left(\dfrac{0}{2.13}\right) = \textbf{0°}$
Thus, $a_1 - a_2 = \textbf{2.13}\,\textbf{m/s}^2$ **at** $\textbf{0°}$

Problem 12. Calculate the resultant of
(a) $v_1 - v_2 + v_3$ and (b) $v_2 - v_1 - v_3$ when
$v_1 = 22$ units at $140°$, $v_2 = 40$ units at $190°$ and
$v_3 = 15$ units at $290°$

(a) The vectors are shown in Fig. 29.37.

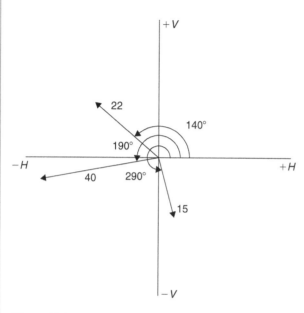

Figure 29.37

The horizontal component of $v_1 - v_2 + v_3$
$= (22\cos 140°) - (40\cos 190°) + (15\cos 290°)$
$= (-16.85) - (-39.39) + (5.13) = \textbf{27.67 units}$
The vertical component of $v_1 - v_2 + v_3$
$= (22\sin 140°) - (40\sin 190°) + (15\sin 290°)$
$= (14.14) - (-6.95) + (-14.10) = \textbf{6.99 units}$
The magnitude of the resultant,
$R = \sqrt{27.67^2 + 6.99^2} = \textbf{28.54 units}$
The direction of the resultant R
$= \tan^{-1}\left(\dfrac{6.99}{27.67}\right) = 14.18°$
Thus, $v_1 - v_2 + v_3 = \textbf{28.54 units at 14.18°}$

(b) The horizontal component of $v_2 - v_1 - v_3$
$= (40\cos 190°) - (22\cos 140°) - (15\cos 290°)$
$= (-39.39) - (-16.85) - (5.13) = \textbf{-27.67 units}$
The vertical component of $v_2 - v_1 - v_3$
$= (40\sin 190°) - (22\sin 140°) - (15\sin 290°)$
$= (-6.95) - (14.14) - (-14.10) = \textbf{-6.99 units}$
From Fig. 29.38, the magnitude of the resultant, $R = \sqrt{(-27.67)^2 + (-6.99)^2} = \textbf{28.54 units}$
and $\alpha = \tan^{-1}\left(\dfrac{6.99}{27.67}\right) = 14.18°$, from which,
$\theta = 180° + 14.18° = 194.18°$

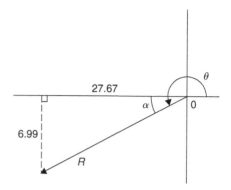

Figure 29.38

Thus, $v_2 - v_1 - v_3 = \textbf{28.54 units at 194.18°}$
This result is as expected, since $v_2 - v_1 - v_3 = -(v_1 - v_2 + v_3)$ and the vector 28.54 units at $194.18°$ is minus times (i.e. is $180°$ out of phase with) the vector 28.54 units at $14.18°$

Now try the following Practice Exercise

1. Forces of $F_1 = 40$N at $45°$ and $F_2 = 30$N at $125°$ act at a point. Determine by drawing and by calculation (a) $F_1 + F_2$ (b) $F_1 - F_2$

2. Calculate the resultant of (a) $v_1 + v_2 - v_3$ (b) $v_3 - v_2 + v_1$ when $v_1 = 15$ m/s at $85°$, $v_2 = 25$ m/s at $175°$ and $v_3 = 12$ m/s at $235°$

29.8 Relative velocity

For relative velocity problems, some fixed datum point needs to be selected. This is often a fixed point on the earth's surface. In any vector equation, only the start and finish points affect the resultant vector of a system. Two different systems are shown in Fig. 29.39, but, in each of the systems, the resultant vector is ad.

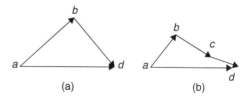

Figure 29.39

The vector equation of the system shown in Fig. 29.39(a) is

$$ad = ab + bd$$

and that for the system shown in Fig. 29.39(b) is

$$ad = ab + bc + cd$$

Thus, in vector equations of this form, only the first and last letters, a and d, respectively, fix the magnitude and direction of the resultant vector. This principle is used in relative velocity problems.

Problem 13. Two cars, P and Q, are travelling towards the junction of two roads which are at right angles to one another. Car P has a velocity of 45 km/h due east and car Q a velocity of 55 km/h due south. Calculate (a) the velocity of car P relative to car Q and (b) the velocity of car Q relative to car P

(a) The directions of the cars are shown in Fig. 29.40(a), which is called a **space diagram**. The velocity diagram is shown in Fig. 29.40(b), in which pe is taken as the velocity of car P relative to point e on the earth's surface. The velocity of P relative to Q is vector pq and the vector equation is $pq = pe + eq$. Hence, the vector directions are as shown, eq being in the opposite direction to qe

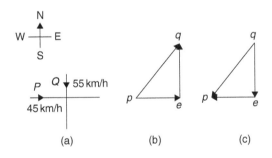

(a)   (b)   (c)

Figure 29.40

From the geometry of the vector triangle, the magnitude of $pq = \sqrt{45^2 + 55^2} = 71.06$ km/h and the direction of $pq = \tan^{-1}\left(\dfrac{55}{45}\right) = 50.71°$

That is, **the velocity of car P relative to car Q is 71.06 km/h at 50.71°**

(b) The velocity of car Q relative to car P is given by the vector equation $qp = qe + ep$ and the vector diagram is as shown in Fig. 29.40(c), having ep opposite in direction to pe

From the geometry of this vector triangle, the magnitude of $qp = \sqrt{45^2 + 55^2} = 71.06$ m/s and the direction of $qp = \tan^{-1}\left(\dfrac{55}{45}\right) = 50.71°$ but must lie in the third quadrant; i.e. the required angle is $180° + 50.71° = 230.71°$

That is, **the velocity of car Q relative to car P is 71.06 m/s at 230.71°**

Now try the following Practice Exercise

Practice Exercise 116   Relative velocity (answers on page 434)

1. A car is moving along a straight horizontal road at 79.2 km/h and rain is falling vertically downwards at 26.4 km/h. Find the velocity of the rain relative to the driver of the car.

2. Calculate the time needed to swim across a river 142 m wide when the swimmer can swim at 2 km/h in still water and the river is flowing at 1 km/h. At what angle to the bank should the swimmer swim?

3. A ship is heading in a direction N 60°E at a speed which in still water would be 20 km/h. It is carried off course by a current of 8 km/h in a direction of E 50°S. Calculate the ship's actual speed and direction.

29.9   *i, j* and *k* notation

A method of completely specifying the direction of a vector in space relative to some reference point is to use three unit vectors, i, j and k, mutually at right angles to each other, as shown in Fig. 29.41.

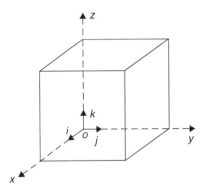

Figure 29.41

Calculations involving vectors given in i, j, k notation are carried out in exactly the same way as standard algebraic calculations, as shown in the worked examples below.

Problem 14. Determine
$(3i + 2j + 2k) - (4i - 3j + 2k)$

$(3i + 2j + 2k) - (4i - 3j + 2k)$

$$= 3i + 2j + 2k - 4i + 3j - 2k$$

$$= -i + 5j$$

Problem 15. Given $p = 3i + 2k$,
$q = 4i - 2j + 3k$ and $r = -3i + 5j - 4k$,
determine
(a) $-r$ (b) $3p$ (c) $2p + 3q$ (d) $-p + 2r$
(e) $0.2p + 0.6q - 3.2r$

(a) $\quad -r = -(-3i + 5j - 4k) = +3i - 5j + 4k$

(b) $\quad 3p = 3(3i + 2k) = 9i + 6k$

(c) $\quad 2p + 3q = 2(3i + 2k) + 3(4i - 2j + 3k)$

$$= 6i + 4k + 12i - 6j + 9k$$

$$= 18i - 6j + 13k$$

(d) $\quad -p + 2r = -(3i + 2k) + 2(-3i + 5j - 4k)$

$$= -3i - 2k + (-6i + 10j - 8k)$$

$$= -3i - 2k - 6i + 10j - 8k$$

$$= -9i + 10j - 10k$$

(e) $\quad 0.2p + 0.6q - 3.2r$

$$= 0.2(3i + 2k) + 0.6(4i - 2j + 3k)$$
$$-3.2(-3i + 5j - 4k)$$

$$= 0.6i + 0.4k + 2.4i - 1.2j + 1.8k + 9.6i$$
$$-16j + 12.8k$$

$$= 12.6i - 17.2j + 15k$$

Now try the following Practice Exercise

Practice Exercise 117 $\quad i, j, k$ notation
(answers on page 434)

Given that $p = 2i + 0.5j - 3k, q = -i + j + 4k$ and $r = 6j - 5k$, evaluate and simplify the following vectors in i, j, k form.

1. $-q$	2. $2p$
3. $q + r$	4. $-q + 2p$
5. $3q + 4r$	6. $q - 2p$
7. $p + q + r$	8. $p + 2q + 3r$
9. $2p + 0.4q + 0.5r$	10. $7r - 2q$

For fully worked solutions to each of the problems in Practice Exercises 113 to 117 in this chapter,
go to the website:
www.routledge.com/cw/bird

Methods of adding alternating waveforms

Why it is important to understand: Methods of adding alternating waveforms

In electrical engineering, a phasor is a rotating vector representing a quantity such as an alternating current or voltage that varies sinusoidally. Sometimes it is necessary when studying sinusoidal quantities to add together two alternating waveforms, for example in an a.c. series circuit, that are not in phase with each other. Electrical engineers, electronics engineers, electronic engineering technicians and aircraft engineers all use phasor diagrams to visualise complex constants and variables. So, given oscillations to add and subtract, the required rotating vectors are constructed, called a phasor diagram, and graphically the resulting sum and/or difference oscillations are added or calculated. Phasors may be used to analyse the behaviour of electrical and mechanical systems that have reached a kind of equilibrium called sinusoidal steady state. Hence, discovering different methods of combining sinusoidal waveforms is of some importance in certain areas of engineering.

At the end of this chapter, you should be able to:

- determine the resultant of two phasors by graph plotting
- determine the resultant of two or more phasors by drawing
- determine the resultant of two phasors by the sine and cosine rules
- determine the resultant of two or more phasors by horizontal and vertical components

30.1 Combining two periodic functions

There are a number of instances in engineering and science where waveforms have to be combined and where it is required to determine the single phasor (called the resultant) that could replace two or more separate phasors. Uses are found in electrical alternating current theory, in mechanical vibrations, in the addition of forces and with sound waves.

There are a number of methods of determining the resultant waveform. These include

(a) drawing the waveforms and adding graphically

(b) drawing the phasors and measuring the resultant

(c) using the cosine and sine rules

(d) using horizontal and vertical components.

30.2 Plotting periodic functions

This may be achieved by sketching the separate functions on the same axes and then adding (or subtracting) ordinates at regular intervals. This is demonstrated in the following worked problems.

Problem 1. Plot the graph of $y_1 = 3\sin A$ from $A = 0°$ to $A = 360°$. On the same axes plot $y_2 = 2\cos A$. By adding ordinates, plot $y_R = 3\sin A + 2\cos A$ and obtain a sinusoidal expression for this resultant waveform

$y_1 = 3\sin A$ and $y_2 = 2\cos A$ are shown plotted in Fig. 30.1. Ordinates may be added at, say, $15°$ intervals. For example,

at $0°$, $y_1 + y_2 = 0 + 2 = 2$

at $15°$, $y_1 + y_2 = 0.78 + 1.93 = 2.71$

at $120°$, $y_1 + y_2 = 2.60 + -1 = 1.6$

at $210°$, $y_1 + y_2 = -1.50 - 1.73 = -3.23$, and so on.

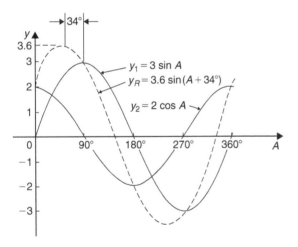

Figure 30.1

The resultant waveform, shown by the broken line, has the same period, i.e. $360°$, and thus the same frequency as the single phasors. The maximum value, or amplitude, of the resultant is 3.6. The resultant waveform **leads** $y_1 = 3\sin A$ by $34°$ or $34 \times \dfrac{\pi}{180}$rad $= 0.593$rad. The sinusoidal expression for the resultant waveform is

$$y_R = 3.6\sin(A + 34°) \text{ or } y_R = 3.6\sin(A + 0.593)$$

Problem 2. Plot the graphs of $y_1 = 4\sin\omega t$ and $y_2 = 3\sin(\omega t - \pi/3)$ on the same axes, over one cycle. By adding ordinates at intervals plot $y_R = y_1 + y_2$ and obtain a sinusoidal expression for the resultant waveform

$y_1 = 4\sin\omega t$ and $y_2 = 3\sin(\omega t - \pi/3)$ are shown plotted in Fig. 30.2.

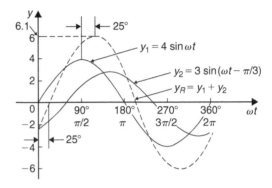

Figure 30.2

Ordinates are added at $15°$ intervals and the resultant is shown by the broken line. The amplitude of the resultant is 6.1 and it **lags** y_1 by $25°$ or 0.436rad.
Hence, the sinusoidal expression for the resultant waveform is

$$y_R = 6.1\sin(\omega t - 0.436)$$

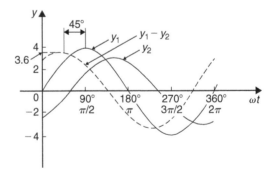

Figure 30.3

Problem 3. Determine a sinusoidal expression for $y_1 - y_2$ when $y_1 = 4\sin\omega t$ and $y_2 = 3\sin(\omega t - \pi/3)$

y_1 and y_2 are shown plotted in Fig. 30.3. At $15°$ intervals y_2 is subtracted from y_1. For example, at $0°$, $y_1 - y_2 = 0 - (-2.6) = +2.6$

at $30°$, $y_1 - y_2 = 2 - (-1.5) = +3.5$

at $150°$, $y_1 - y_2 = 2 - 3 = -1$, and so on.

The amplitude, or peak, value of the resultant (shown by the broken line) is 3.6 and it leads y_1 by $45°$ or 0.79 rad. Hence,

$$y_1 - y_2 = 3.6\sin(\omega t + 0.79)$$

Problem 4. Two alternating currents are given by $i_1 = 20\sin\omega t$ amperes and $i_2 = 10\sin\left(\omega t + \dfrac{\pi}{3}\right)$ amperes. By drawing the waveforms on the same axes and adding, determine the sinusoidal expression for the resultant $i_1 + i_2$

i_1 and i_2 are shown plotted in Fig. 30.4. The resultant waveform for $i_1 + i_2$ is shown by the broken line. It has the same period, and hence frequency, as i_1 and i_2

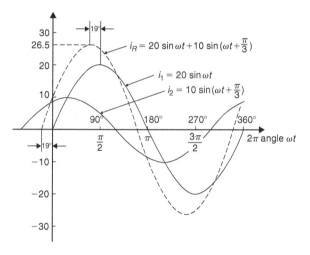

Figure 30.4

The amplitude or peak value is 26.5 A. The resultant waveform leads the waveform of $i_1 = 20\sin\omega t$ by $19°$ or 0.33 rad.

Hence, the sinusoidal expression for the resultant $i_1 + i_2$ is given by

$$i_R = i_1 + i_2 = 26.5\sin(\omega t + 0.33)\,\text{A}$$

Now try the following Practice Exercise

1. Plot the graph of $y = 2\sin A$ from $A = 0°$ to $A = 360°$. On the same axes plot $y = 4\cos A$. By adding ordinates at intervals plot $y = 2\sin A + 4\cos A$ and obtain a sinusoidal expression for the waveform.

2. Two alternating voltages are given by $v_1 = 10\sin\omega t$ volts and $v_2 = 14\sin(\omega t + \pi/3)$ volts. By plotting v_1 and v_2 on the same axes over one cycle obtain a sinusoidal expression for (a) $v_1 + v_2$ (b) $v_1 - v_2$

3. Express $12\sin\omega t + 5\cos\omega t$ in the form $A\sin(\omega t \pm \alpha)$ by drawing and measurement.

30.3 Determining resultant phasors by drawing

The resultant of two periodic functions may be found from their relative positions when the time is zero. For example, if $y_1 = 4\sin\omega t$ and $y_2 = 3\sin(\omega t - \pi/3)$ then each may be represented as **phasors** as shown in Fig. 30.5, y_1 being 4 units long and drawn horizontally and y_2 being 3 units long, lagging y_1 by $\pi/3$ radians or $60°$. To determine the resultant of $y_1 + y_2$, y_1 is drawn horizontally as shown in Fig. 30.6 and y_2 is joined to the end of y_1 at $60°$ to the horizontal. The resultant is given by y_R. This is the same as the diagonal of a parallelogram that is shown completed in Fig. 30.7.

Figure 30.5

Resultant y_R, in Figs 30.6 and 30.7, may be determined by drawing the phasors and their directions to scale and measuring using a ruler and protractor. In this example, y_R is measured as 6 units long and angle ϕ is measured as $25°$

$$25° = 25 \times \frac{\pi}{180}\text{radians} = 0.44\text{ rad}$$

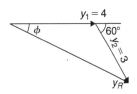

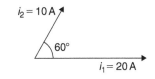

Figure 30.9

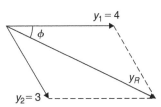

Figure 30.6

Figure 30.7

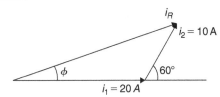

Figure 30.10

Hence, summarising, by drawing,

$$y_R = y_1 + y_2 = 4\sin\omega t + 3\sin(\omega t - \pi/3)$$
$$= 6\sin(\omega t - 0.44)$$

If the resultant phasor, $y_R = y_1 - y_2$ is required then y_2 is still 3 units long but is drawn in the opposite direction, as shown in Fig. 30.8.

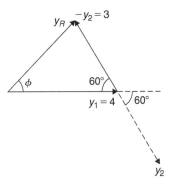

Figure 30.8

Problem 5. Two alternating currents are given by $i_1 = 20\sin\omega t$ amperes and $i_2 = 10\sin\left(\omega t + \dfrac{\pi}{3}\right)$ amperes. Determine $i_1 + i_2$ by drawing phasors

The relative positions of i_1 and i_2 at time $t = 0$ are shown as phasors in Fig. 30.9, where $\dfrac{\pi}{3}$ rad $= 60°$. The phasor diagram in Fig. 30.10 is drawn to scale with a ruler and protractor.

The resultant i_R is shown and is measured as 26 A and angle ϕ as 19° or 0.33 rad leading i_1. Hence, by drawing and measuring,

$$i_R = i_1 + i_2 = 26\sin(\omega t + 0.33)\,\text{A}$$

Problem 6. For the currents in Problem 5, determine $i_1 - i_2$ by drawing phasors

At time $t = 0$, current i_1 is drawn 20 units long horizontally as shown by $0a$ in Fig. 30.11. Current i_2 is shown, drawn 10 units long in a broken line and leading by 60°. The current $-i_2$ is drawn in the opposite direction to the broken line of i_2, shown as ab in Fig. 30.11. The resultant i_R is given by $0b$ lagging by angle ϕ

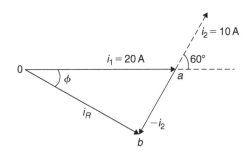

Figure 30.11

By measurement, $i_R = 17\,\text{A}$ and $\phi = 30°$ or 0.52 rad. Hence, by drawing phasors,

$$i_R = i_1 - i_2 = 17\sin(\omega t - 0.52)\,\text{A}$$

Now try the following Practice Exercise

Practice Exercise 119 Determining resultant phasors by drawing (answers on page 434)

1. Determine a sinusoidal expression for $2\sin\theta + 4\cos\theta$ by drawing phasors.

2. If $v_1 = 10\sin\omega t$ volts and $v_2 = 14\sin(\omega t + \pi/3)$ volts, determine by drawing phasors sinusoidal expressions for (a) $v_1 + v_2$ (b) $v_1 - v_2$

3. Express $12\sin\omega t + 5\cos\omega t$ in the form $A\sin(\omega t \pm \alpha)$ by drawing phasors.

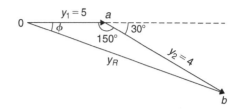

Figure 30.13

30.4 Determining resultant phasors by the sine and cosine rules

As stated earlier, the resultant of two periodic functions may be found from their relative positions when the time is zero. For example, if $y_1 = 5\sin\omega t$ and $y_2 = 4\sin(\omega t - \pi/6)$ then each may be represented by phasors as shown in Fig. 30.12, y_1 being 5 units long and drawn horizontally and y_2 being 4 units long, lagging y_1 by $\pi/6$ radians or 30°. To determine the resultant of $y_1 + y_2$, y_1 is drawn horizontally as shown in Fig. 30.13 and y_2 is joined to the end of y_1 at $\pi/6$ radians; i.e. 30° to the horizontal. The resultant is given by y_R

Using the cosine rule on triangle $0ab$ of Fig. 30.13 gives

$$y_R^2 = 5^2 + 4^2 - [2(5)(4)\cos 150°]$$
$$= 25 + 16 - (-34.641) = 75.641$$

from which $\qquad y_R = \sqrt{75.641} = 8.697$

Using the sine rule, $\qquad \dfrac{8.697}{\sin 150°} = \dfrac{4}{\sin\phi}$

from which $\qquad \sin\phi = \dfrac{4\sin 150°}{8.697} = 0.22996$

and $\qquad \phi = \sin^{-1} 0.22996$
$$= 13.29° \text{ or } 0.232\,\text{rad}$$

Hence, $y_R = y_1 + y_2 = 5\sin\omega t + 4\sin(\omega t - \pi/6)$
$$= \mathbf{8.697\sin(\omega t - 0.232)}$$

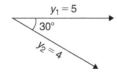

Figure 30.12

Problem 7. Given $y_1 = 2\sin\omega t$ and $y_2 = 3\sin(\omega t + \pi/4)$, obtain an expression, by calculation, for the resultant, $y_R = y_1 + y_2$

When time $t = 0$, the position of phasors y_1 and y_2 are as shown in Fig. 30.14(a). To obtain the resultant, y_1 is drawn horizontally, 2 units long, and y_2 is drawn 3 units long at an angle of $\pi/4$ rad or 45° and joined to the end of y_1 as shown in Fig. 30.14(b).

From Fig. 30.14(b), and using the cosine rule,

$$y_R^2 = 2^2 + 3^2 - [2(2)(3)\cos 135°]$$
$$= 4 + 9 - [-8.485] = 21.485$$

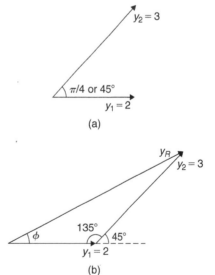

Figure 30.14

Hence, $\qquad y_R = \sqrt{21.485} = 4.6352$

Using the sine rule $\qquad \dfrac{3}{\sin\phi} = \dfrac{4.6352}{\sin 135°}$

from which $\sin\phi = \dfrac{3\sin 135°}{4.6352} = 0.45765$

Hence, $\phi = \sin^{-1} 0.45765$

$= 27.24°$ or $0.475\,\text{rad}$

Thus, by calculation, $y_R = 4.635\sin(\omega t + 0.475)$

Problem 8. Determine

$$20\sin\omega t + 10\sin\left(\omega t + \frac{\pi}{3}\right)$$

using the cosine and sine rules

From the phasor diagram of Fig. 30.15 and using the cosine rule,

$$i_R^2 = 20^2 + 10^2 - [2(20)(10)\cos 120°] = 700$$

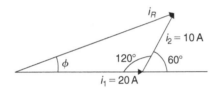

Figure 30.15

Hence, $i_R = \sqrt{700} = \textbf{26.46\,A}$

Using the sine rule gives $\dfrac{10}{\sin\phi} = \dfrac{26.46}{\sin 120°}$

from which $\sin\phi = \dfrac{10\sin 120°}{26.46}$

$= 0.327296$

and $\phi = \sin^{-1} 0.327296 = \textbf{19.10°}$

$= 19.10 \times \dfrac{\pi}{180} = \textbf{0.333\,rad}$

Hence, by cosine and sine rules,

$$i_R = i_1 + i_2 = \textbf{26.46}\sin(\omega t + \textbf{0.333})\,\textbf{A}$$

Now try the following Practice Exercise

Practice Exercise 120   Resultant phasors by the sine and cosine rules (answers on page 435)

1. Determine, using the cosine and sine rules, a sinusoidal expression for
   $$y = 2\sin A + 4\cos A$$

2. Given $v_1 = 10\sin\omega t$ volts and $v_2 = 14\sin(\omega t + \pi/3)$ volts, use the cosine

and sine rules to determine sinusoidal expressions for (a) $v_1 + v_2$ (b) $v_1 - v_2$

In Problems 3 to 5, express the given expressions in the form $A\sin(\omega t \pm \alpha)$ by using the cosine and sine rules.

3. $12\sin\omega t + 5\cos\omega t$

4. $7\sin\omega t + 5\sin\left(\omega t + \dfrac{\pi}{4}\right)$

5. $6\sin\omega t + 3\sin\left(\omega t - \dfrac{\pi}{6}\right)$

6. The sinusoidal currents in two parallel branches of an electrical network are $400\sin\omega t$ and $750\sin(\omega t - \pi/3)$, both measured in milliamperes. Determine the total current flowing into the parallel arrangement. Give the answer in sinusoidal form and in amperes.

30.5 Determining resultant phasors by horizontal and vertical components

If a right-angled triangle is constructed as shown in Fig. 30.16, $0a$ is called the horizontal component of F and ab is called the vertical component of F

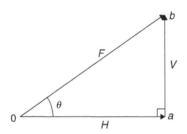

Figure 30.16

From trigonometry (see Chapter 21 and remember SOH CAH TOA),

$$\cos\theta = \frac{0a}{0b}, \text{ from which } 0a = 0b\cos\theta = F\cos\theta$$

i.e. **the horizontal component of F, $H = F\cos\theta$,** and

$$\sin\theta = \frac{ab}{0b}, \text{ from which } ab = 0b\sin\theta = F\sin\theta$$

i.e. **the vertical component of F, $V = F\sin\theta$.**

Determining resultant phasors by horizontal and vertical components is demonstrated in the following worked problems.

The relative positions of v_1 and v_2 at time $t = 0$ are shown in Fig. 30.17(a) and the phasor diagram is shown in Fig. 30.17(b).

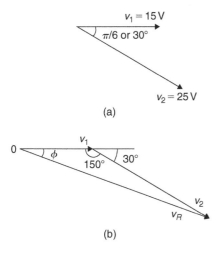

(a)

(b)

Figure 30.17

The horizontal component of v_R,
$$H = 15 \cos 0° + 25 \cos(-30°) = \mathbf{36.65\,V}$$
The vertical component of v_R,
$$V = 15 \sin 0° + 25 \sin(-30°) = \mathbf{-12.50\,V}$$

Hence, $v_R = \sqrt{36.65^2 + (-12.50)^2}$
                        by Pythagoras' theorem
                = **38.72 volts**

$$\tan \phi = \frac{V}{H} = \frac{-12.50}{36.65} = -0.3411$$

from which $\phi = \tan^{-1}(-0.3411)$
                = **−18.83°** or **−0.329 radians**.

Hence, $v_R = v_1 + v_2 = \mathbf{38.72 \sin(\omega t - 0.329)\,V}$

The horizontal component of v_R,
$$H = 15 \cos 0° - 25 \cos(-30°) = \mathbf{-6.65\,V}$$

The vertical component of v_R,
$$V = 15 \sin 0° - 25 \sin(-30°) = \mathbf{12.50\,V}$$

Hence, $v_R = \sqrt{(-6.65)^2 + (12.50)^2}$
                        by Pythagoras' theorem
                = **14.16 volts**

$$\tan \phi = \frac{V}{H} = \frac{12.50}{-6.65} = -1.8797$$

from which $\phi = \tan^{-1}(-1.8797)$
                = **118.01°** or **2.06 radians**.

Hence, $v_R = v_1 - v_2 = \mathbf{14.16 \sin(\omega t + 2.06)\,V}$

The phasor diagram is shown in Fig. 30.18.

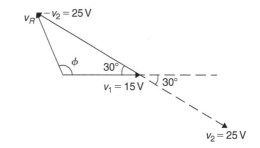

Figure 30.18

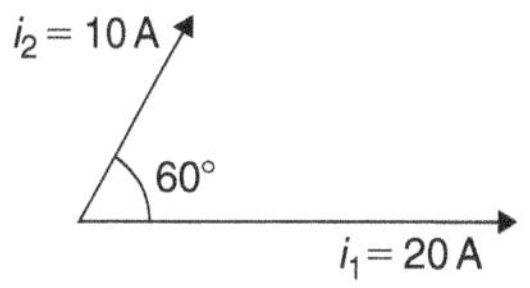

Figure 30.19

From the phasors shown in Fig. 30.19,
Total horizontal component,
$$H = 20 \cos 0° + 10 \cos 60° = \mathbf{25.0}$$
Total vertical component,
$$V = 20 \sin 0° + 10 \sin 60° = \mathbf{8.66}$$
By Pythagoras, the resultant,
$$i_R = \sqrt{[25.0^2 + 8.66^2]} = \mathbf{26.46\,A}$$
Phase angle, $\phi = \tan^{-1} \left(\dfrac{8.66}{25.0} \right)$
                = **19.11°** or **0.333 rad**

Hence, by using horizontal and vertical components,

$$20\sin \omega t + 10\sin\left(\omega t + \frac{\pi}{3}\right) = 26.46\sin(\omega t + 0.333)$$

Now try the following Practice Exercise

Practice Exercise 121   Resultant phasors by horizontal and vertical components (answers on page 435)

In Problems 1 to 5, express the combination of periodic functions in the form $A\sin(\omega t \pm \alpha)$ by horizontal and vertical components.

1. $7\sin \omega t + 5\sin\left(\omega t + \dfrac{\pi}{4}\right)$

2. $6\sin \omega t + 3\sin\left(\omega t - \dfrac{\pi}{6}\right)$

3. $i = 25\sin \omega t - 15\sin\left(\omega t + \dfrac{\pi}{3}\right)$

4. $v = 8\sin \omega t - 5\sin\left(\omega t - \dfrac{\pi}{4}\right)$

5. $x = 9\sin\left(\omega t + \dfrac{\pi}{3}\right) - 7\sin\left(\omega t - \dfrac{3\pi}{8}\right)$

6. The voltage drops across two components when connected in series across an a.c. supply are $v_1 = 200\sin 314.2t$ and $v_2 = 120\sin(314.2t - \pi/5)$ volts respectively. Determine

    (a) the voltage of the supply (given by $v_1 + v_2$) in the form $A\sin(\omega t \pm \alpha)$

    (b) the frequency of the supply

7. If the supply to a circuit is $v = 20\sin 628.3t$ volts and the voltage drop across one of the components is $v_1 = 15\sin(628.3t - 0.52)$ volts, calculate

    (a) the voltage drop across the remainder of the circuit, given by $v - v_1$, in the form $A\sin(\omega t \pm \alpha)$

    (b) the supply frequency

    (c) the periodic time of the supply.

8. The voltages across three components in a series circuit when connected across an a.c. supply are $v_1 = 25\sin\left(300\pi t + \dfrac{\pi}{6}\right)$ volts, $v_2 = 40\sin\left(300\pi t - \dfrac{\pi}{4}\right)$ volts and $v_3 = 50\sin\left(300\pi t + \dfrac{\pi}{3}\right)$ volts. Calculate

    (a) the supply voltage, in sinusoidal form, in the form $A\sin(\omega t \pm \alpha)$

    (b) the frequency of the supply

    (c) the periodic time.

9. In an electrical circuit, two components are connected in series. The voltage across the first component is given by $80\sin(\omega t + \pi/3)$ volts, and the voltage across the second component is given by $150\sin(\omega t - \pi/4)$ volts. Determine the total supply voltage to the two components. Give the answer in sinusoidal form.

For fully worked solutions to each of the problems in Practice Exercises 118 to 121 in this chapter, go to the website:
www.routledge.com/cw/bird

This assignment covers the material contained in Chapters 29 and 30. *The marks available are shown in brackets at the end of each question.*

1. State the difference between scalar and vector quantities. (2)

2. State whether the following are scalar or vector quantities.
   (a) A temperature of 50°C.
   (b) $2\,m^3$ volume.
   (c) A downward force of 10 N.
   (d) 400 J of work.
   (e) $20\,cm^2$ area.
   (f) A south-easterly wind of 20 knots.
   (g) 40 m distance.
   (h) An acceleration of $25\,m/s^2$ at 30° to the horizontal. (8)

3. A velocity vector of 16 m/s acts at an angle of −40° to the horizontal. Calculate its horizontal and vertical components, correct to 3 significant figures. (4)

4. Calculate the resultant and direction of the displacement vectors shown in Fig. RT12.1, correct to 2 decimal places. (6)

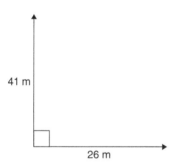

Figure RT12.1

5. Calculate the resultant and direction of the force vectors shown in Fig. RT12.2, correct to 2 decimal places. (6)

6. If acceleration $a_1 = 11\,m/s^2$ at 70° and $a_2 = 19\,m/s^2$ at −50°, calculate the magnitude and direction of $a_1 + a_2$, correct to 2 decimal places. (8)

7. If velocity $v_1 = 36$ m/s at 52° and $v_2 = 17$ m/s at −15°, calculate the magnitude and direction of $v_1 - v_2$, correct to 2 decimal places. (8)

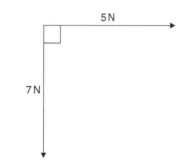

Figure RT12.2

8. Forces of 10 N, 16 N and 20 N act as shown in Fig. RT12.3. Determine the magnitude of the resultant force and its direction relative to the 16 N force
   (a) by scaled drawing.
   (b) by calculation. (13)

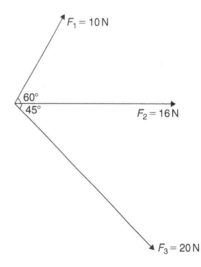

Figure RT12.3

9. For the three forces shown in Fig. RT12.3, calculate the resultant of $F_1 - F_2 - F_3$ and its direction relative to force F_2 (9)

10. Two cars, A and B, are travelling towards cross-roads. A has a velocity of 60 km/h due south and B a velocity of 75 km/h due west. Calculate the velocity of A relative to B. (6)

11. Given $a = -3i + 3j + 5k$, $b = 2i - 5j + 7k$ and $c = 3i + 6j - 4k$, determine the following: (i) $-4b$ (ii) $a + b - c$ (iii) $5b - 3a$ (6)

12. Calculate the magnitude and direction of the resultant vector of the displacement system shown in Fig. RT12.4. (9)

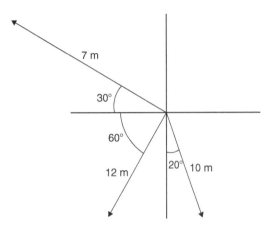

Fig. RT12.4

13. The instantaneous values of two alternating voltages are given by

$$v_1 = 150\sin\left(\omega t + \frac{\pi}{3}\right) \text{ volts}$$

and

$$v_2 = 90\sin\left(\omega t - \frac{\pi}{6}\right) \text{ volts}$$

Plot the two voltages on the same axes to scales of $1\,\text{cm} = 50$ volts and $1\,\text{cm} = \pi/6$. Obtain a sinusoidal expression for the resultant of v_1 and v_2 in the form $R\sin(\omega t + \alpha)$ (a) by adding ordinates at intervals and (b) by calculation. (15)

Multiple choice questions Test 4
Areas, volumes and vectors
This test covers the material in Chapters 25 to 30

All questions have only one correct answer (Answers on page 440).

1. A hollow shaft has an outside diameter of 6.0 cm and an inside diameter of 4.0 cm. The cross-sectional area of the shaft is:

   (a) 6283 mm^2 (b) 1257 mm^2
   (c) 1571 mm^2 (d) 628 mm^2

2. The speed of a car at 1 second intervals is given in the following table:

Time t (s)	0	1	2	3	4	5	6
Speed v (m/s)	0	2.5	5.0	9.0	15.0	22.0	30.0

   The distance travelled in 6 s (i.e. the area under the v/t graph) using the trapezoidal rule is:

   (a) 83.5 m   (b) 68 m   (c) 68.5 m   (d) 204 m

3. An arc of a circle of length 5.0 cm subtends an angle of 2 radians. The circumference of the circle is:

   (a) 2.5 cm   (b) 10.0 cm   (c) 5.0 cm   (d) 15.7 cm

4. A force of 4 N is inclined at an angle of 45° to a second force of 7 N, both forces acting at a point, as shown in Fig. M4.1. The magnitude of the resultant of these two forces and the direction of the resultant with respect to the 7 N force is:

   (a) 10.2 N at 16°   (b) 5 N at 146°
   (c) 11 N at 135°   (d) 3 N at 45°

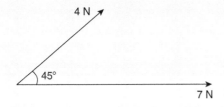

Figure M4.1

5. The mean value of a sine wave over half a cycle is:

   (a)   0.318 × maximum value

   (b)   0.707 × maximum value

   (c)   the peak value

   (d)   0.637 × maximum value

6. The area of the path shown shaded in Fig. M4.2 is:

   (a) 300 m^2 (b) 234 m^2

   (c) 124 m^2 (d) 66 m^2

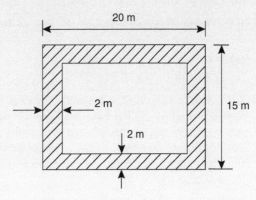

Figure M4.2

Questions 7 and 8 relate to the following information.

Two voltage phasors V_1 and V_2 are shown in Fig. M4.3.

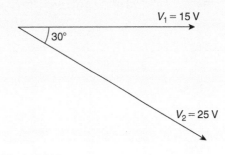

Figure M4.3

7. The resultant $V_1 + V_2$ is given by:

   (a)   14.16 V at 62° to V_1

   (b)   38.72 V at −19° to V_1

   (c)   38.72 V at 161° to V_1

   (d)   14.16 V at 118° to V_1

8. The resultant $V_1 - V_2$ is given by:

   (a) 38.72 V at $-19°$ to V_1

   (b) 14.16 V at $62°$ to V_1

   (c) 38.72 V at $161°$ to V_1

   (d) 14.16 V at $118°$ to V_1

9. A vehicle has a mass of 2000 kg. A model of the vehicle is made to a scale of 1 to 100. If the vehicle and model are made of the same material, the mass of the model is:

   (a) 2 g   (b) 20 kg

   (c) 200 g   (d) 20 g

10. If the circumference of a circle is 100 mm, its area is:

    (a) 314.2 cm^2 (b) 7.96 cm^2

    (c) 31.83 mm^2 (d) 78.54 cm^2

11. An indicator diagram for a steam engine is as shown in Fig. M4.4. The base has been divided into 6 equally spaced intervals and the lengths of the 7 ordinates measured, with the results shown in centimetres. Using Simpson's rule the area of the indicator diagram is:

    (a) 32 cm^2 (b) 17.9 cm^2

    (c) 16 cm^2 (d) 96 cm^2

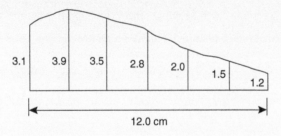

3.1   3.9   3.5   2.8   2.0   1.5   1.2

12.0 cm

Figure M4.4

12. The magnitude of the resultant of velocities of 3 m/s at $20°$ and 7 m/s at $120°$ when acting simultaneously at a point is:

    (a) 21 m/s   (b) 10 m/s

    (c) 7.12 m/s   (d) 4 m/s

13. The equation of a circle is $x^2 + y^2 - 2x + 4y - 4 = 0$. Which of the following statements is correct?

    (a) The circle has centre $(1, -2)$ and radius 4

    (b) The circle has centre $(-1, 2)$ and radius 2

    (c) The circle has centre $(-1, -2)$ and radius 4

    (d) The circle has centre $(1, -2)$ and radius 3

14. The surface area of a sphere of diameter 40 mm is:

    (a) 50.27 cm^2 (b) 33.51 cm^2

    (c) 268.08 cm^2 (d) 201.06 cm^2

15. A water tank is in the shape of a rectangular prism having length 1.5 m, breadth 60 cm and height 300 mm. If 1 litre = 1000 cm^3, the capacity of the tank is:

    (a) 27 litre   (b) 2.7 litre

    (c) 2700 litre   (d) 270 litre

16. Three forces of 2 N, 3 N and 4 N act as shown in Fig. M4.5. The magnitude of the resultant force is:

    (a) 8.08 N   (b) 9 N

    (c) 7.17 N   (d) 1 N

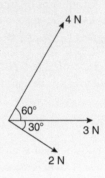

Figure M4.5

17. A pendulum of length 1.2 m swings through an angle of $12°$ in a single swing. The length of arc traced by the pendulum bob is:

    (a) 14.40 cm   (b) 25.13 cm

    (c) 10.00 cm   (d) 45.24 cm

18. A wheel on a car has a diameter of 800 mm. If the car travels 5 miles, the number of complete revolutions the wheel makes (given $1 \text{ km} = \frac{5}{8}$ mile) is:

    (a) 1989   (b) 1591

    (c) 3183   (d) 10000

19. A rectangular building is shown on a building plan having dimensions 20 mm by 10 mm. If the plan is drawn to a scale of 1 to 300, the true area of the building in m^2 is:

    (a) 60000 m^2 (b) 18 m^2

    (c) 0.06 m^2 (d) 1800 m^2

20. The total surface area of a cylinder of length 20 cm and diameter 6 cm is:

    (a) 56.55 cm^2 (b) 433.54 cm^2

    (c) 980.18 cm^2 (d) 226.19 cm^2

21. The total surface area of a solid hemisphere of diameter 6.0 cm is:

    (a) 84.82 cm^2 (b) 339.3 cm^2

    (c) 226.2 cm^2 (d) 56.55 cm^2

22. The outside measurements of a picture frame are 80 cm by 30 cm. If the frame is 3 cm wide, the area of the metal used to make the frame is:

    (a) 624 cm^2 (b) 2079 cm^2

    (c) 660 mm^2 (d) 588 cm^2

Questions 23 and 24 relate to the following information.

Two alternating voltages are given by:

$v_1 = 2 \sin \omega t$ and $v_2 = 3 \sin \left(\omega t + \dfrac{\pi}{4} \right)$ volts.

23. Which of the phasor diagrams shown in Fig. M4.6 represents $v_R = v_1 + v_2$?

    (a) (i)   (b) (ii)   (c) (iii)   (d) (iv)

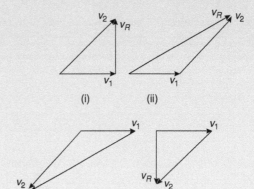

(i)   (ii)

(iii)   (iv)

Figure M4.6

24. Which of the phasor diagrams shown in Fig. M4.6 represents $v_R = v_1 - v_2$?

    (a) (i)   (b) (ii)   (c) (iii)   (d) (iv)

25. A cylindrical, copper pipe, 1.8 m long, has an outside diameter of 300 mm and an inside diameter of 180 mm. The volume of copper in the pipe, in cubic metres is:

    (a) 0.3257 m^2 (b) 0.0814 m^2

    (c) 8.143 m^2 (d) 814.3 m^2

The companion website for this book contains the above multiple-choice test. If you prefer to attempt the test online then visit:

www.routledge.com/cw/bird

For a copy of this multiple choice test, go to:

www.routledge.com/cw/bird

COMPANION @ WEBSITE

Chapter 31

Presentation of statistical data

Why it is important to understand: Presentation of statistical data

Statistics is the study of the collection, organisation, analysis, and interpretation of data. It deals with all aspects of this, including the planning of data collection in terms of the design of surveys and experiments. Statistics is applicable to a wide variety of academic disciplines, including natural and social sciences, engineering, government and business. Statistical methods can be used for summarising or describing a collection of data. Engineering statistics combines engineering and statistics. Design of experiments is a methodology for formulating scientific and engineering problems using statistical models. Quality control and process control use statistics as a tool to manage conformance to specifications of manufacturing processes and their products. Time and methods engineering use statistics to study repetitive operations in manufacturing in order to set standards and find optimum manufacturing procedures. Reliability engineering measures the ability of a system to perform for its intended function (and time) and has tools for improving performance. Probabilistic design involves the use of probability in product and system design. System identification uses statistical methods to build mathematical models of dynamical systems from measured data. System identification also includes the optimal design of experiments for efficiently generating informative data for fitting such models. This chapter introduces the presentation of statistical data.

At the end of this chapter, you should be able to:

- distinguish between discrete and continuous data
- present data diagrammatically – pictograms, horizontal and vertical bar charts, percentage component bar charts, pie diagrams
- produce a tally diagram for a set of data
- form a frequency distribution from a tally diagram
- construct a histogram from a frequency distribution
- construct a frequency polygon from a frequency distribution
- produce a cumulative frequency distribution from a set of grouped data
- construct an ogive from a cumulative frequency distribution

31.1   Some statistical terminology

31.1.1   Discrete and continuous data

Data are obtained largely by two methods:

(a) By counting – for example, the number of stamps sold by a post office in equal periods of time.

(b) By measurement – for example, the heights of a group of people.

When data are obtained by counting and only whole numbers are possible, the data are called **discrete**. Measured data can have any value within certain limits and are called **continuous**.

Problem 1. Data are obtained on the topics given below. State whether they are discrete or continuous data.

(a) The number of days on which rain falls in a month for each month of the year.

(b) The mileage travelled by each of a number of salesmen.

(c) The time that each of a batch of similar batteries lasts.

(d) The amount of money spent by each of several families on food.

(a) The number of days on which rain falls in a given month must be an integer value and is obtained by **counting** the number of days. Hence, these data are **discrete**.

(b) A salesman can travel any number of miles (and parts of a mile) between certain limits and these data are **measured**. Hence, the data are **continuous**.

(c) The time that a battery lasts is **measured** and can have any value between certain limits. Hence, these data are **continuous**.

(d) The amount of money spent on food can only be expressed correct to the nearest pence, the amount being **counted**. Hence, these data are **discrete**.

Now try the following Practice Exercise

Practice Exercise 122   Discrete and continuous data (answers on page 435)

In the following problems, state whether data relating to the topics given are discrete or continuous.

1. (a) The amount of petrol produced daily, for each of 31 days, by a refinery.
   (b) The amount of coal produced daily by each of 15 miners.
   (c) The number of bottles of milk delivered daily by each of 20 milkmen.
   (d) The size of 10 samples of rivets produced by a machine.

2. (a) The number of people visiting an exhibition on each of 5 days.
   (b) The time taken by each of 12 athletes to run 100 metres.
   (c) The value of stamps sold in a day by each of 20 post offices.
   (d) The number of defective items produced in each of 10 one-hour periods by a machine.

31.1.2   Further statistical terminology

A **set** is a group of data and an individual value within the set is called a **member** of the set. Thus, if the masses of five people are measured correct to the nearest 0.1 kilogram and are found to be 53.1 kg, 59.4 kg, 62.1 kg, 77.8 kg and 64.4 kg then the set of masses in kilograms for these five people is

$$\{53.1, 59.4, 62.1, 77.8, 64.4\}$$

and one of the members of the set is 59.4

A set containing all the members is called a **population**. Some members selected at random from a population are called a **sample**. Thus, all car registration numbers form a population but the registration numbers of, say, 20 cars taken at random throughout the country are a sample drawn from that population.

The number of times that the value of a member occurs in a set is called the **frequency** of that member. Thus, in the set $\{2, 3, 4, 5, 4, 2, 4, 7, 9\}$, member 4 has a frequency of three, member 2 has a frequency of 2 and the other members have a frequency of one.

The **relative frequency** with which any member of a set occurs is given by the ratio

$$\frac{\text{frequency of member}}{\text{total frequency of all members}}$$

For the set $\{2, 3, 5, 4, 7, 5, 6, 2, 8\}$, the relative frequency of member 5 is $\frac{2}{9}$. Often, relative frequency is expressed as a percentage and the **percentage relative frequency** is

$$(\text{relative frequency} \times 100)\%$$

31.2 Presentation of ungrouped data

Ungrouped data can be presented diagrammatically in several ways and these include

(a) **pictograms**, in which pictorial symbols are used to represent quantities (see Problem 2),

(b) **horizontal bar charts**, having data represented by equally spaced horizontal rectangles (see Problem 3), and

(c) **vertical bar charts**, in which data are represented by equally spaced vertical rectangles (see Problem 4).

Trends in ungrouped data over equal periods of time can be presented diagrammatically by a **percentage component bar chart**. In such a chart, equally spaced rectangles of any width, but whose height corresponds to 100%, are constructed. The rectangles are then subdivided into values corresponding to the percentage relative frequencies of the members (see Problem 5).

A **pie diagram** is used to show diagrammatically the parts making up the whole. In a pie diagram, the area of a circle represents the whole and the areas of the sectors of the circle are made proportional to the parts which make up the whole (see Problem 6).

Problem 2. The number of television sets repaired in a workshop by a technician in 6 one-month periods is as shown below. Present these data as a pictogram

Month	January	February	March
Number repaired	11	6	15

Month	April	May	June
Number repaired	9	13	8

Each symbol shown in Fig. 31.1 represents two television sets repaired. Thus, in January, $5\frac{1}{2}$ symbols are used to represent the 11 sets repaired; in February, 3 symbols are used to represent the 6 sets repaired, and so on.

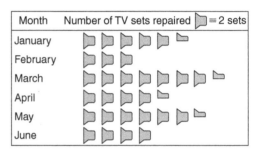

Figure 31.1

Problem 3. The distance in miles travelled by four salesmen in a week are as shown below.

Salesman	P	Q	R	S
Distance travelled (miles)	413	264	597	143

Use a horizontal bar chart to represent these data diagrammatically

Equally spaced horizontal rectangles of any width, but whose length is proportional to the distance travelled, are used. Thus, the length of the rectangle for salesman P is proportional to 413 miles, and so on. The horizontal bar chart depicting these data is shown in Fig. 31.2.

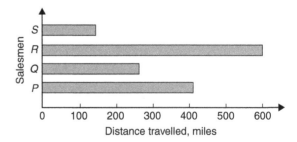

Figure 31.2

Problem 4. The number of issues of tools or materials from a store in a factory is observed for 7 one-hour periods in a day and the results of the survey are as follows.

Period	1	2	3	4	5	6	7
Number of issues	34	17	9	5	27	13	6

Present these data on a vertical bar chart

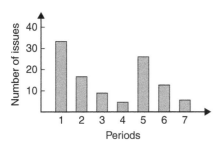

Figure 31.3

In a vertical bar chart, equally spaced vertical rectangles of any width, but whose height is proportional to the quantity being represented, are used. Thus, the height of the rectangle for period 1 is proportional to 34 units, and so on. The vertical bar chart depicting these data is shown in Fig. 31.3.

Problem 5. The numbers of various types of dwellings sold by a company annually over a three-year period are as shown below. Draw percentage component bar charts to present these data

	Year 1	Year 2	Year 3
4-roomed bungalows	24	17	7
5-roomed bungalows	38	71	118
4-roomed houses	44	50	53
5-roomed houses	64	82	147
6-roomed houses	30	30	25

A table of percentage relative frequency values, correct to the nearest 1%, is the first requirement. Since

$$\text{percentage relative frequency}$$
$$= \frac{\text{frequency of member} \times 100}{\text{total frequency}}$$

then for 4-roomed bungalows in year 1

$$\text{percentage relative frequency}$$
$$= \frac{24 \times 100}{24 + 38 + 44 + 64 + 30} = 12\%$$

The percentage relative frequencies of the other types of dwellings for each of the three years are similarly calculated and the results are as shown in the table below.

	Year 1	Year 2	Year 3
4-roomed bungalows	12%	7%	2%
5-roomed bungalows	19%	28%	34%
4-roomed houses	22%	20%	15%
5-roomed houses	32%	33%	42%
6-roomed houses	15%	12%	7%

The percentage component bar chart is produced by constructing three equally spaced rectangles of any width, corresponding to the three years. The heights of the rectangles correspond to 100% relative frequency and are subdivided into the values in the table of percentages shown above. A key is used (different types of shading or different colour schemes) to indicate corresponding percentage values in the rows of the table of percentages. The percentage component bar chart is shown in Fig. 31.4.

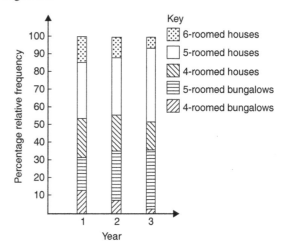

Figure 31.4

Problem 6. The retail price of a product costing £2 is made up as follows: materials 10 p, labour 20 p, research and development 40 p, overheads 70 p, profit 60 p. Present these data on a pie diagram

A circle of any radius is drawn. The area of the circle represents the whole, which in this case is £2. The circle is subdivided into sectors so that the areas of the sectors are proportional to the parts; i.e. the parts which make up the total retail price. For the area of a sector to be proportional to a part, the angle at the centre of the circle must be proportional to that part. The whole, £2 or 200p, corresponds to 360°. Therefore,

$$10\text{p corresponds to } 360 \times \frac{10}{200} \text{ degrees, i.e. } 18°$$

$$20\text{p corresponds to } 360 \times \frac{20}{200} \text{ degrees, i.e. } 36°$$

and so on, giving the angles at the centre of the circle for the parts of the retail price as 18°, 36°, 72°, 126° and 108°, respectively.
The pie diagram is shown in Fig. 31.5.

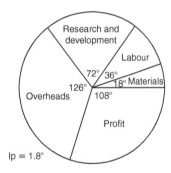

Figure 31.5

Problem 7.

(a) Using the data given in Fig. 31.2 only, calculate the amount of money paid to each salesman for travelling expenses if they are paid an allowance of 37p per mile.

(b) Using the data presented in Fig. 31.4, comment on the housing trends over the three-year period.

(c) Determine the profit made by selling 700 units of the product shown in Fig. 31.5

(a) By measuring the length of rectangle P, the mileage covered by salesman P is equivalent to

413 miles. Hence **salesman P** receives a travelling allowance of

$$\frac{£413 \times 37}{100} \text{ i.e. } \textbf{£152.81}$$

Similarly, for **salesman Q**, the miles travelled are 264 and his allowance is

$$\frac{£264 \times 37}{100} \text{ i.e. } \textbf{£97.68}$$

Salesman R travels 597 miles and he receives

$$\frac{£597 \times 37}{100} \text{ i.e. } \textbf{£220.89}$$

Finally, **salesman S** receives

$$\frac{£143 \times 37}{100} \text{ i.e. } \textbf{£52.91}$$

(b) An analysis of Fig. 31.4 shows that 5-roomed bungalows and 5-roomed houses are becoming more popular, the greatest change in the three years being a 15% increase in the sales of 5-roomed bungalows.

(c) Since 1.8° corresponds to 1p and the profit occupies 108° of the pie diagram, the profit per unit is

$$\frac{108 \times 1}{1.8} \text{ i.e. } 60\text{p}$$

The profit when selling 700 units of the product is

$$£\frac{700 \times 60}{100} \text{ i.e. } \textbf{£420}$$

Now try the following Practice Exercise

Practice Exercise 123   Presentation of ungrouped data (answers on page 435)

1. The number of vehicles passing a stationary observer on a road in 6 ten-minute intervals is as shown. Draw a pictogram to represent these data.

Period of time	1	2	3	4	5	6
Number of vehicles	35	44	62	68	49	41

2. The number of components produced by a factory in a week is as shown below.

Day	Mon	Tues	Wed	Thur	Fri
Number of components	1580	2190	1840	2385	1280

Show these data on a pictogram.

3. For the data given in Problem 1 above, draw a horizontal bar chart.

4. Present the data given in Problem 2 above on a horizontal bar chart.

5. For the data given in Problem 1 above, construct a vertical bar chart.

6. Depict the data given in Problem 2 above on a vertical bar chart.

7. A factory produces three different types of components. The percentages of each of these components produced for 3 one-month periods are as shown below. Show this information on percentage component bar charts and comment on the changing trend in the percentages of the types of component produced.

Month	1	2	3
Component P	20	35	40
Component Q	45	40	35
Component R	35	25	25

8. A company has five distribution centres and the mass of goods in tonnes sent to each centre during 4 one-week periods is as shown.

Week	1	2	3	4
Centre A	147	160	174	158
Centre B	54	63	77	69
Centre C	283	251	237	211
Centre D	97	104	117	144
Centre E	224	218	203	194

Use a percentage component bar chart to present these data and comment on any trends.

9. The employees in a company can be split into the following categories: managerial 3, supervisory 9, craftsmen 21, semi-skilled 67, others 44. Show these data on a pie diagram.

10. The way in which an apprentice spent his time over a one-month period is as follows: drawing office 44 hours, production 64 hours, training 12 hours, at college 28 hours. Use a pie diagram to depict this information.

11. (a) With reference to Fig. 31.5, determine the amount spent on labour and materials to produce 1650 units of the product.
    (b) If in year 2 of Fig. 31.4 1% corresponds to 2.5 dwellings, how many bungalows are sold in that year?

12. (a) If the company sells 23 500 units per annum of the product depicted in Fig. 31.5, determine the cost of their overheads per annum.
    (b) If 1% of the dwellings represented in year 1 of Fig. 31.4 corresponds to 2 dwellings, find the total number of houses sold in that year.

31.3 Presentation of grouped data

When the number of members in a set is small, say ten or less, the data can be represented diagrammatically without further analysis, by means of pictograms, bar charts, percentage components bar charts or pie diagrams (as shown in Section 31.2).

For sets having more than ten members, those members having similar values are grouped together in **classes** to form a **frequency distribution**. To assist in accurately counting members in the various classes, a **tally diagram** is used (see Problems 8 and 12).

A **frequency distribution** is merely a table showing classes and their corresponding frequencies (see Problems 8 and 12). The new set of values obtained by forming a frequency distribution is called **grouped data**. The terms used in connection with grouped data are shown in Fig. 31.6(a). The size or range of a class is given by the **upper class boundary value** minus the **lower class boundary value** and in Fig. 31.6(b) is $7.65 - 7.35$; i.e. 0.30. The **class interval** for the class shown in Fig. 31.6(b) is 7.4 to 7.6 and the class mid-point value is given by

$$\frac{\text{(upper class boundary value)} + \text{(lower class boundary value)}}{2}$$

and in Fig. 31.6(b) is

$$\frac{7.65 + 7.35}{2} \quad \text{i.e. 7.5}$$

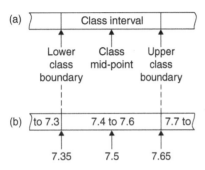

Figure 31.6

One of the principal ways of presenting grouped data diagrammatically is to use a **histogram**, in which the **areas** of vertical, adjacent rectangles are made proportional to frequencies of the classes (see Problem 9). When class intervals are equal, the heights of the rectangles of a histogram are equal to the frequencies of the classes. For histograms having unequal class intervals, the area must be proportional to the frequency. Hence, if the class interval of class A is twice the class interval of class B, then for equal frequencies the height of the rectangle representing A is half that of B (see Problem 11).

Another method of presenting grouped data diagrammatically is to use a **frequency polygon**, which is the graph produced by plotting frequency against class mid-point values and joining the co-ordinates with straight lines (see Problem 12).

A **cumulative frequency distribution** is a table showing the cumulative frequency for each value of upper class boundary. The cumulative frequency for a particular value of upper class boundary is obtained by adding the frequency of the class to the sum of the previous frequencies. A cumulative frequency distribution is formed in Problem 13.

The curve obtained by joining the co-ordinates of cumulative frequency (vertically) against upper class boundary (horizontally) is called an **ogive** or a **cumulative frequency distribution curve** (see Problem 13).

Problem 8. The data given below refer to the gain of each of a batch of 40 transistors, expressed correct to the nearest whole number. Form a frequency distribution for these data having seven classes

81	83	87	74	76	89	82	84
86	76	77	71	86	85	87	88
84	81	80	81	73	89	82	79
81	79	78	80	85	77	84	78
83	79	80	83	82	79	80	77

The **range** of the data is the value obtained by taking the value of the smallest member from that of the largest member. Inspection of the set of data shows that range $= 89 - 71 = 18$. The size of each class is given approximately by the range divided by the number of classes. Since seven classes are required, the size of each class is $18 \div 7$; that is, approximately 3. To achieve seven equal classes spanning a range of values from 71 to 89, the class intervals are selected as 70–72, 73–75, and so on. To assist with accurately determining the number in each class, a **tally diagram** is produced, as shown in Table 31.1(a). This is obtained by listing the classes in the left-hand column and then inspecting each of the 40 members of the set in turn and allocating them to the appropriate classes by putting '1's in the appropriate rows. Every fifth '1' allocated to a particular row is shown as an oblique line crossing the four previous '1's, to help with final counting.

Table 31.1(a)

Class	Tally
70–72	1
73–75	11
76–78	⧸⧸⧸⧸ 11
79–81	⧸⧸⧸⧸ ⧸⧸⧸⧸ 11
82–84	⧸⧸⧸⧸ 1111
85–87	⧸⧸⧸⧸ 1
88–90	111

A **frequency distribution** for the data is shown in Table 31.1(b) and lists classes and their corresponding frequencies, obtained from the tally diagram. (Class mid-point values are also shown in the table, since they are used for constructing the histogram for these data (see Problem 9).)

Table 31.1 (b)

Class	Class mid-point	Frequency
70–72	71	1
73–75	74	2
76–78	77	7
79–81	80	12
82–84	83	9
85–87	86	6
88–90	89	3

Problem 9. Construct a histogram for the data given in Table 31.1(b)

The histogram is shown in Fig. 31.7. The width of the rectangles corresponds to the upper class boundary values minus the lower class boundary values and the heights of the rectangles correspond to the class frequencies. The easiest way to draw a histogram is to mark the class mid-point values on the horizontal scale and draw the rectangles symmetrically about the appropriate class mid-point values and touching one another.

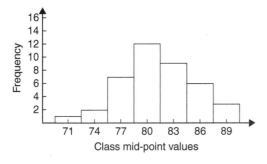

Figure 31.7

Problem 10. The amount of money earned weekly by 40 people working part-time in a factory, correct to the nearest £10, is shown below. Form a frequency distribution having 6 classes for these data

80	90	70	110	90	160	110	80
140	30	90	50	100	110	60	100
80	90	110	80	100	90	120	70
130	170	80	120	100	110	40	110
50	100	110	90	100	70	110	80

Inspection of the set given shows that the majority of the members of the set lie between £80 and £110 and that there is a much smaller number of extreme values ranging from £30 to £170. If equal class intervals are selected, the frequency distribution obtained does not give as much information as one with unequal class intervals. Since the majority of the members lie between £80 and £100, the class intervals in this range are selected to be smaller than those outside of this range. There is no unique solution and one possible solution is shown in Table 31.2.

Table 31.2

Class	Frequency
20–40	2
50–70	6
80–90	12
100–110	14
120–140	4
150–170	2

Problem 11. Draw a histogram for the data given in Table 31.2

When dealing with unequal class intervals, the histogram must be drawn so that the areas (and not the heights) of the rectangles are proportional to the frequencies of the classes. The data given are shown in columns 1 and 2 of Table 31.3. Columns 3 and 4 give the upper and lower class boundaries, respectively. In column 5, the class ranges (i.e. upper class boundary minus lower class boundary values) are listed. The heights of the rectangles are proportional to the ratio $\dfrac{\text{frequency}}{\text{class range}}$, as shown in column 6. The histogram is shown in Fig. 31.8.

Table 31.3

1 Class	2 Frequency	3 Upper class boundary	4 Lower class boundary	5 Class range	6 Height of rectangle
20–40	2	45	15	30	$\dfrac{2}{30} = \dfrac{1}{15}$
50–70	6	75	45	30	$\dfrac{6}{30} = \dfrac{3}{15}$
80–90	12	95	75	20	$\dfrac{12}{20} = \dfrac{9}{15}$
100–110	14	115	95	20	$\dfrac{14}{20} = \dfrac{10\frac{1}{2}}{15}$
120–140	4	145	115	30	$\dfrac{4}{30} = \dfrac{2}{15}$
150–170	2	175	145	30	$\dfrac{2}{30} = \dfrac{1}{15}$

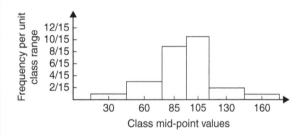

Figure 31.8

Problem 12. The masses of 50 ingots in kilograms are measured correct to the nearest 0.1 kg and the results are as shown below. Produce a frequency distribution having about seven classes for these data and then present the grouped data as a frequency polygon and a histogram

8.0	8.6	8.2	7.5	8.0	9.1	8.5	7.6	8.2	7.8
8.3	7.1	8.1	8.3	8.7	7.8	8.7	8.5	8.4	8.5
7.7	8.4	7.9	8.8	7.2	8.1	7.8	8.2	7.7	7.5
8.1	7.4	8.8	8.0	8.4	8.5	8.1	7.3	9.0	8.6
7.4	8.2	8.4	7.7	8.3	8.2	7.9	8.5	7.9	8.0

The **range** of the data is the member having the largest value minus the member having the smallest value. Inspection of the set of data shows that
range $= 9.1 - 7.1 = \mathbf{2.0}$

The size of each class is given approximately by

$$\frac{\text{range}}{\text{number of classes}}$$

Since about seven classes are required, the size of each class is $2.0 \div 7$, i.e. approximately 0.3, and thus the **class limits** are selected as 7.1 to 7.3, 7.4 to 7.6, 7.7 to 7.9, and so on. The **class mid-point** for the 7.1 to 7.3 class is

$$\frac{7.35 + 7.05}{2} \text{ i.e. } 7.2$$

the class mid-point for the 7.4 to 7.6 class is

$$\frac{7.65 + 7.35}{2} \text{ i.e. } 7.5$$

and so on.

To assist with accurately determining the number in each class, a **tally diagram** is produced as shown in Table 31.4. This is obtained by listing the classes in the left-hand column and then inspecting each of the 50 members of the set of data in turn and allocating it to the appropriate class by putting a '1' in the appropriate row. Each fifth '1' allocated to a particular row is marked as an oblique line to help with final counting.

A **frequency distribution** for the data is shown in Table 31.5 and lists classes and their corresponding frequencies. Class mid-points are also shown in this table since they are used when constructing the frequency polygon and histogram.

Table 31.4

Class	Tally
7.1 to 7.3	111
7.4 to 7.6	̶I̶I̶I̶I̶
7.7 to 7.9	̶I̶I̶I̶I̶ 1111
8.0 to 8.2	̶I̶I̶I̶I̶ ̶I̶I̶I̶I̶ 1111
8.3 to 8.5	̶I̶I̶I̶I̶ ̶I̶I̶I̶I̶ 1
8.6 to 8.8	̶I̶I̶I̶I̶ 1
8.9 to 9.1	11

Table 31.5

Class	Class mid-point	Frequency
7.1 to 7.3	7.2	3
7.4 to 7.6	7.5	5
7.7 to 7.9	7.8	9
8.0 to 8.2	8.1	14
8.3 to 8.5	8.4	11
8.6 to 8.8	8.7	6
8.9 to 9.1	9.0	2

A **frequency polygon** is shown in Fig. 31.9, the co-ordinates corresponding to the class mid-point/frequency values given in Table 31.5. The co-ordinates are joined by straight lines and the polygon is 'anchored-down' at each end by joining to the next class mid-point value and zero frequency.

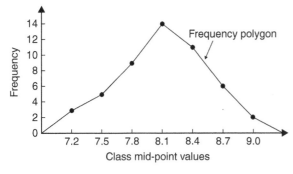

Figure 31.9

A **histogram** is shown in Fig. 31.10, the width of a rectangle corresponding to (upper class boundary value – lower class boundary value) and height corresponding

to the class frequency. The easiest way to draw a histogram is to mark class mid-point values on the horizontal scale and to draw the rectangles symmetrically about the appropriate class mid-point values and touching one another. A histogram for the data given in Table 31.5 is shown in Fig. 31.10.

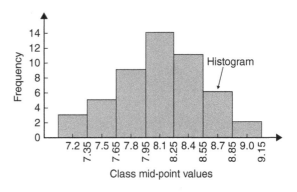

Figure 31.10

Problem 13. The frequency distribution for the masses in kilograms of 50 ingots is

7.1 to 7.3	3
7.4 to 7.6	5
7.7 to 7.9	9
8.0 to 8.2	14
8.3 to 8.5	11
8.6 to 8.8	6
8.9 to 9.1	2

Form a cumulative frequency distribution for these data and draw the corresponding ogive

A **cumulative frequency distribution** is a table giving values of cumulative frequency for the values of upper class boundaries and is shown in Table 31.6. Columns 1 and 2 show the classes and their frequencies. Column 3 lists the upper class boundary values for the classes given in column 1. Column 4 gives the cumulative frequency values for all frequencies less than the upper class boundary values given in column 3. Thus, for example, for the 7.7 to 7.9 class shown in row 3, the cumulative frequency value is the sum of all frequencies having values of less than 7.95, i.e. $3 + 5 + 9 = 17$, and so on.

Table 31.6

1	2	3	4
Class	Frequency	Upper class boundary less than	Cumulative frequency
7.1–7.3	3	7.35	3
7.4–7.6	5	7.65	8
7.7–7.9	9	7.95	17
8.0–8.2	14	8.25	31
8.3–8.5	11	8.55	42
8.6–8.8	6	8.85	48
8.9–9.1	2	9.15	50

The **ogive** for the cumulative frequency distribution given in Table 31.6 is shown in Fig. 31.11. The co-ordinates corresponding to each upper class boundary/cumulative frequency value are plotted and the co-ordinates are joined by straight lines (not the best curve drawn through the co-ordinates as in experimental work). The ogive is 'anchored' at its start by adding the co-ordinate (7.05, 0)

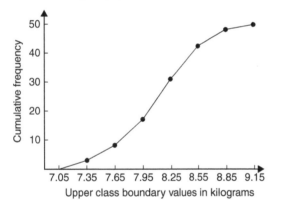

Figure 31.11

Now try the following Practice Exercise

Practice Exercise 124   Presentation of grouped data (answers on page 435)

1. The mass in kilograms, correct to the nearest one-tenth of a kilogram, of 60 bars of metal are as shown. Form a frequency distribution of about eight classes for these data.

39.8	40.1	40.3	40.0	40.6	39.7	40.0	40.4	39.6	39.3
39.6	40.7	40.2	39.9	40.3	40.2	40.4	39.9	39.8	40.0
40.2	40.1	40.3	39.7	39.9	40.5	39.9	40.5	40.0	39.9
40.1	40.8	40.0	40.0	40.1	40.2	40.1	40.0	40.2	39.9
39.7	39.8	40.4	39.7	39.9	39.5	40.1	40.1	39.9	40.2
39.5	40.6	40.0	40.1	39.8	39.7	39.5	40.2	39.9	40.3

2. Draw a histogram for the frequency distribution given in the solution of Problem 1.

3. The information given below refers to the value of resistance in ohms of a batch of 48 resistors of similar value. Form a frequency distribution for the data, having about six classes, and draw a frequency polygon and histogram to represent these data diagrammatically.

21.0	22.4	22.8	21.5	22.6	21.1	21.6	22.3
22.9	20.5	21.8	22.2	21.0	21.7	22.5	20.7
23.2	22.9	21.7	21.4	22.1	22.2	22.3	21.3
22.1	21.8	22.0	22.7	21.7	21.9	21.1	22.6
21.4	22.4	22.3	20.9	22.8	21.2	22.7	21.6
22.2	21.6	21.3	22.1	21.5	22.0	23.4	21.2

4. The time taken in hours to the failure of 50 specimens of a metal subjected to fatigue failure tests are as shown. Form a frequency distribution, having about eight classes and unequal class intervals, for these data.

28	22	23	20	12	24	37	28	21	25
21	14	30	23	27	13	23	7	26	19
24	22	26	3	21	24	28	40	27	24
20	25	23	26	47	21	29	26	22	33
27	9	13	35	20	16	20	25	18	22

5. Form a cumulative frequency distribution and hence draw the ogive for the frequency distribution given in the solution to Problem 3.

6. Draw a histogram for the frequency distribution given in the solution to Problem 4.

7. The frequency distribution for a batch of 50 capacitors of similar value, measured in microfarads, is

10.5–10.9	2
11.0–11.4	7
11.5–11.9	10
12.0–12.4	12
12.5–12.9	11
13.0–13.4	8

Form a cumulative frequency distribution for these data.

8. Draw an ogive for the data given in the solution of Problem 7.

9. The diameter in millimetres of a reel of wire is measured in 48 places and the results are as shown.

2.10	2.29	2.32	2.21	2.14	2.22
2.28	2.18	2.17	2.20	2.23	2.13
2.26	2.10	2.21	2.17	2.28	2.15
2.16	2.25	2.23	2.11	2.27	2.34
2.24	2.05	2.29	2.18	2.24	2.16
2.15	2.22	2.14	2.27	2.09	2.21
2.11	2.17	2.22	2.19	2.12	2.20
2.23	2.07	2.13	2.26	2.16	2.12

(a) Form a frequency distribution of diameters having about 6 classes.

(b) Draw a histogram depicting the data.

(c) Form a cumulative frequency distribution.

(d) Draw an ogive for the data.

For fully worked solutions to each of the problems in Practice Exercises 122 to 124 in this chapter, go to the website:
www.routledge.com/cw/bird

COMPANION
@
WEBSITE

Chapter 32

Mean, median, mode and standard deviation

Why it is important to understand: Mean, median, mode and standard deviation

Statistics is a field of mathematics that pertains to data analysis. In many real-life situations, it is helpful to describe data by a single number that is most representative of the entire collection of numbers. Such a number is called a measure of central tendency; the most commonly used measures are mean, median, mode and standard deviation, the latter being the average distance between the actual data and the mean. Statistics is important in the field of engineering since it provides tools to analyse collected data. For example, a chemical engineer may wish to analyse temperature measurements from a mixing tank. Statistical methods can be used to determine how reliable and reproducible the temperature measurements are, how much the temperature varies within the data set, what future temperatures of the tank may be, and how confident the engineer can be in the temperature measurements made. When performing statistical analysis on a set of data, the mean, median, mode and standard deviation are all helpful values to calculate; this chapter explains how to determine these measures of central tendency.

At the end of this chapter, you should be able to:

- determine the mean, median and mode for a set of ungrouped data
- determine the mean, median and mode for a set of grouped data
- draw a histogram from a set of grouped data
- determine the mean, median and mode from a histogram
- calculate the standard deviation from a set of ungrouped data
- calculate the standard deviation from a set of grouped data
- determine the quartile values from an ogive
- determine quartile, decile and percentile values from a set of data

32.1 Measures of central tendency

A single value, which is representative of a set of values, may be used to give an indication of the general size of the members in a set, the word 'average' often being used to indicate the single value. The statistical term used for 'average' is the 'arithmetic mean' or just the 'mean'.

Other measures of central tendency may be used and these include the **median** and the **modal** values.

32.2 Mean, median and mode for discrete data

32.2.1 Mean

The **arithmetic mean value** is found by adding together the values of the members of a set and dividing by the number of members in the set. Thus, the mean of the set of numbers $\{4, 5, 6, 9\}$ is

$$\frac{4+5+6+9}{4} \text{ i.e. } 6$$

In general, the mean of the set $\{x_1, x_2, x_3, \ldots x_n\}$ is

$$\bar{x} = \frac{x_1 + x_2 + x_3 + \cdots + x_n}{n} \text{ written as } \frac{\sum x}{n}$$

where $\sum$ is the Greek letter 'sigma' and means 'the sum of' and $\bar{x}$ (called x-bar) is used to signify a mean value.

32.2.2 Median

The **median value** often gives a better indication of the general size of a set containing extreme values. The set $\{7, 5, 74, 10\}$ has a mean value of 24, which is not really representative of any of the values of the members of the set. The median value is obtained by

(a) **ranking** the set in ascending order of magnitude, and

(b) selecting the value of the **middle member** for sets containing an odd number of members or finding the value of the mean of the two middle members for sets containing an even number of members.

For example, the set $\{7, 5, 74, 10\}$ is ranked as $\{5, 7, 10, 74\}$ and, since it contains an even number of members (four in this case), the mean of 7 and 10 is taken, giving a median value of 8.5. Similarly, the set $\{3, 81, 15, 7, 14\}$ is ranked as $\{3, 7, 14, 15, 81\}$ and the median value is the value of the middle member, i.e. 14

32.2.3 Mode

The **modal value**, or **mode**, is the most commonly occurring value in a set. If two values occur with the same frequency, the set is '**bi-modal**'. The set $\{5, 6, 8, 2, 5, 4, 6, 5, 3\}$ has a modal value of 5, since the member having a value of 5 occurs the most, i.e. three times.

Problem 1. Determine the mean, median and mode for the set $\{2, 3, 7, 5, 5, 13, 1, 7, 4, 8, 3, 4, 3\}$

The mean value is obtained by adding together the values of the members of the set and dividing by the number of members in the set. Thus,

mean value, $\bar{x}$

$$= \frac{2+3+7+5+5+13+1+7+4+8+3+4+3}{13}$$

$$= \frac{65}{13} = \mathbf{5}$$

To obtain the median value the set is ranked, that is, placed in ascending order of magnitude, and since the set contains an odd number of members the value of the middle member is the median value. Ranking the set gives $\{1, 2, 3, 3, 3, 4, 4, 5, 5, 7, 7, 8, 13\}$. The middle term is the seventh member; i.e. 4. Thus, the **median value is 4**

The **modal value** is the value of the most commonly occurring member and is **3**, which occurs three times, all other members only occurring once or twice.

Problem 2. The following set of data refers to the amount of money in £s taken by a news vendor for 6 days. Determine the mean, median and modal values of the set
$$\{27.90, 34.70, 54.40, 18.92, 47.60, 39.68\}$$

Mean value

$$= \frac{27.90+34.70+54.40+18.92+47.60+39.68}{6}$$

$$= \mathbf{£37.20}$$

The ranked set is
$$\{18.92, 27.90, 34.70, 39.68, 47.60, 54.40\}$$
Since the set has an even number of members, the mean of the middle two members is taken to give the median value; i.e.

$$\text{median value} = \frac{34.70+39.68}{2} = \mathbf{£37.19}$$

Since no two members have the same value, this set has **no mode**.

Now try the following Practice Exercise

Practice Exercise 125   Mean, median and mode for discrete data (answers on page 436)

In Problems 1 to 4, determine the mean, median and modal values for the sets given.

1. {3, 8, 10, 7, 5, 14, 2, 9, 8}

2. {26, 31, 21, 29, 32, 26, 25, 28}

3. {4.72, 4.71, 4.74, 4.73, 4.72, 4.71, 4.73, 4.72}

4. {73.8, 126.4, 40.7, 141.7, 28.5, 237.4, 157.9}

32.3 Mean, median and mode for grouped data

The mean value for a set of grouped data is found by determining the sum of the (frequency × class mid-point values) and dividing by the sum of the frequencies; i.e.

$$\textbf{mean value}, \bar{x} = \frac{f_1 x_1 + f_2 x_2 + \cdots f_n x_n}{f_1 + f_2 + \cdots + f_n} = \frac{\sum(fx)}{\sum f}$$

where f is the frequency of the class having a mid-point value of x, and so on.

Problem 3.   The frequency distribution for the value of resistance in ohms of 48 resistors is as shown. Determine the mean value of resistance

20.5–20.9	3
21.0–21.4	10
21.5–21.9	11
22.0–22.4	13
22.5–22.9	9
23.0–23.4	2

The class mid-point/frequency values are 20.7 3, 21.2 10, 21.7 11, 22.2 13, 22.7 9 and 23.2 2
For grouped data, the mean value is given by

$$\bar{x} = \frac{\sum(fx)}{\sum f}$$

where f is the class frequency and x is the class mid-point value. Hence mean value,

$$\bar{x} = \frac{\begin{array}{c}(3 \times 20.7) + (10 \times 21.2) + (11 \times 21.7)\\ + (13 \times 22.2) + (9 \times 22.7) + (2 \times 23.2)\end{array}}{48}$$

$$= \frac{1052.1}{48} = 21.919\ldots$$

i.e. **the mean value is 21.9 ohms**, correct to 3 significant figures.

32.3.1 Histograms

The mean, median and modal values for grouped data may be determined from a **histogram**. In a histogram, frequency values are represented vertically and variable values horizontally. The mean value is given by the value of the variable corresponding to a vertical line drawn through the centroid of the histogram. The median value is obtained by selecting a variable value such that the area of the histogram to the left of a vertical line drawn through the selected variable value is equal to the area of the histogram on the right of the line. The modal value is the variable value obtained by dividing the width of the highest rectangle in the histogram in proportion to the heights of the adjacent rectangles. The method of determining the mean, median and modal values from a histogram is shown in Problem 4.

Problem 4.   The time taken in minutes to assemble a device is measured 50 times and the results are as shown. Draw a histogram depicting the data and hence determine the mean, median and modal values of the distribution

14.5–15.5	5
16.5–17.5	8
18.5–19.5	16
20.5–21.5	12
22.5–23.5	6
24.5–25.5	3

The histogram is shown in Fig. 32.1. The mean value lies at the centroid of the histogram. With reference to

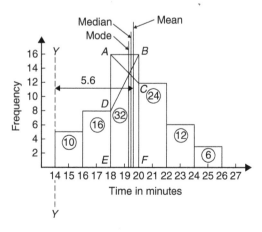

Figure 32.1

any arbitrary axis, say YY shown at a time of 14 minutes, the position of the horizontal value of the centroid can be obtained from the relationship $AM = \sum(am)$, where A is the area of the histogram, M is the horizontal distance of the centroid from the axis YY, a is the area of a rectangle of the histogram and m is the distance of the centroid of the rectangle from YY. The areas of the individual rectangles are shown circled on the histogram giving a total area of 100 square units. The positions, m, of the centroids of the individual rectangles are $1, 3, 5, \ldots$ units from YY. Thus

$$100M = (10 \times 1) + (16 \times 3) + (32 \times 5) + (24 \times 7)$$
$$+ (12 \times 9) + (6 \times 11)$$

i.e. $M = \dfrac{560}{100} = 5.6$ units from YY

Thus, the position of the **mean** with reference to the time scale is $14 + 5.6$, i.e. **19.6 minutes**.

The median is the value of time corresponding to a vertical line dividing the total area of the histogram into two equal parts. The total area is 100 square units, hence the vertical line must be drawn to give 50 units of area on each side. To achieve this with reference to Fig. 32.1, rectangle $ABFE$ must be split so that $50 - (10 + 16)$ units of area lie on one side and $50 - (24 + 12 + 6)$ units of area lie on the other. This shows that the area of $ABFE$ is split so that 24 units of area lie to the left of the line and 8 units of area lie to the right; i.e. the vertical line must pass through 19.5 minutes. Thus, the **median value** of the distribution is **19.5 minutes**.

The mode is obtained by dividing the line AB, which is the height of the highest rectangle, proportionally to the heights of the adjacent rectangles. With reference to Fig. 32.1, this is achieved by joining AC and BD and drawing a vertical line through the point of intersection of these two lines. This gives the **mode** of the distribution, which is **19.3 minutes**.

Now try the following Practice Exercise

Practice Exercise 126   Mean, median and mode for grouped data (answers on page 436)

1. 21 bricks have a mean mass of 24.2 kg and 29 similar bricks have a mass of 23.6 kg. Determine the mean mass of the 50 bricks.

2. The frequency distribution given below refers to the heights in centimetres of 100 people.

Determine the mean value of the distribution, correct to the nearest millimetre.

150–156	5
157–163	18
164–170	20
171–177	27
178–184	22
185–191	8

3. The gain of 90 similar transistors is measured and the results are as shown. By drawing a histogram of this frequency distribution, determine the mean, median and modal values of the distribution.

83.5–85.5	6
86.5–88.5	39
89.5–91.5	27
92.5–94.5	15
95.5–97.5	3

4. The diameters, in centimetres, of 60 holes bored in engine castings are measured and the results are as shown. Draw a histogram depicting these results and hence determine the mean, median and modal values of the distribution.

2.011–2.014	7
2.016–2.019	16
2.021–2.024	23
2.026–2.029	9
2.031–2.034	5

32.4   Standard deviation

32.4.1   Discrete data

The standard deviation of a set of data gives an indication of the amount of dispersion, or the scatter, of members of the set from the measure of central tendency. Its value is the root-mean-square value of the members of the set and for discrete data is obtained as follows.

(i) Determine the measure of central tendency, usually the mean value, (occasionally the median or modal values are specified).

(ii) Calculate the deviation of each member of the set from the mean, giving

$$(x_1 - \overline{x}), (x_2 - \overline{x}), (x_3 - \overline{x}), \ldots$$

(iii) Determine the squares of these deviations; i.e.
$(x_1 - \overline{x})^2, (x_2 - \overline{x})^2, (x_3 - \overline{x})^2, \ldots$

(iv) Find the sum of the squares of the deviations, i.e.
$(x_1 - \overline{x})^2 + (x_2 - \overline{x})^2 + (x_3 - \overline{x})^2, \ldots$

(v) Divide by the number of members in the set, n, giving
$$\frac{(x_1 - \overline{x})^2 + (x_2 - \overline{x})^2 + (x^3 - \overline{x})^2 + \cdots}{n}$$

(vi) Determine the square root of (v)

The standard deviation is indicated by σ (the Greek letter small 'sigma') and is written mathematically as

$$\text{standard deviation}, \sigma = \sqrt{\left\{\frac{\sum(x - \overline{x})^2}{n}\right\}}$$

where x is a member of the set, $\overline{x}$ is the mean value of the set and n is the number of members in the set. The value of standard deviation gives an indication of the distance of the members of a set from the mean value. The set $\{1, 4, 7, 10, 13\}$ has a mean value of 7 and a standard deviation of about 4.2. The set $\{5, 6, 7, 8, 9\}$ also has a mean value of 7 but the standard deviation is about 1.4. This shows that the members of the second set are mainly much closer to the mean value than the members of the first set. The method of determining the standard deviation for a set of discrete data is shown in Problem 5.

Problem 5. Determine the standard deviation from the mean of the set of numbers $\{5, 6, 8, 4, 10, 3\}$, correct to 4 significant figures

The arithmetic mean, $\overline{x} = \dfrac{\sum x}{n}$

$$= \frac{5 + 6 + 8 + 4 + 10 + 3}{6} = 6$$

Standard deviation, $\sigma = \sqrt{\left\{\dfrac{\sum(x - \overline{x})^2}{n}\right\}}$

The $(x - \overline{x})^2$ values are $(5 - 6)^2, (6 - 6)^2, (8 - 6)^2,$ $(4 - 6)^2, (10 - 6)^2$ and $(3 - 6)^2$

The sum of the $(x - \overline{x})^2$ values,

i.e. $\sum(x - \overline{x})^2$, is $1 + 0 + 4 + 4 + 16 + 9 = 34$

and $\dfrac{\sum(x - \overline{x})^2}{n} = \dfrac{34}{6} = 5.\dot{6}$

since there are 6 members in the set.
Hence, **standard deviation,**

$$\sigma = \sqrt{\left\{\frac{\sum(x - \overline{x})^2}{n}\right\}} = \sqrt{5.\dot{6}} = \mathbf{2.380,}$$

correct to 4 significant figures.

32.4.2 Grouped data

For grouped data,

$$\text{standard deviation}, \sigma = \sqrt{\left\{\frac{\sum\{f(x - \overline{x})^2\}}{\sum f}\right\}}$$

where f is the class frequency value, x is the class mid-point value and $\overline{x}$ is the mean value of the grouped data. The method of determining the standard deviation for a set of grouped data is shown in Problem 6.

Problem 6. The frequency distribution for the values of resistance in ohms of 48 resistors is as shown. Calculate the standard deviation from the mean of the resistors, correct to 3 significant figures

20.5–20.9	3
21.0–21.4	10
21.5–21.9	11
22.0–22.4	13
22.5–22.9	9
23.0–23.4	2

The standard deviation for grouped data is given by

$$\sigma = \sqrt{\left\{\frac{\sum\{f(x - \overline{x})^2\}}{\sum f}\right\}}$$

From Problem 3, the distribution mean value is $\overline{x} = 21.92$, correct to 2 significant figures.

The 'x-values' are the class mid-point values, i.e. $20.7, 21.2, 21.7, \ldots$

Thus, the $(x - \overline{x})^2$ values are $(20.7 - 21.92)^2,$ $(21.2 - 21.92)^2, (21.7 - 21.92)^2, \ldots$

and the $f(x - \overline{x})^2$ values are $3(20.7 - 21.92)^2,$ $10(21.2 - 21.92)^2, 11(21.7 - 21.92)^2, \ldots$

The $\sum f(x - \overline{x})^2$ values are

$$4.4652 + 5.1840 + 0.5324 + 1.0192$$
$$+ 5.4756 + 3.2768 = 19.9532$$

$$\frac{\sum\{f(x-\bar{x})^2\}}{\sum f} = \frac{19.9532}{48} = 0.41569$$

and **standard deviation,**

$$\sigma = \sqrt{\left\{\frac{\sum\{f(x-\bar{x})^2\}}{\sum f}\right\}} = \sqrt{0.41569}$$

$$= \mathbf{0.645},\text{ correct to 3 significant figures.}$$

Now try the following Practice Exercise

1. Determine the standard deviation from the mean of the set of numbers

   $\{35, 22, 25, 23, 28, 33, 30\}$

   correct to 3 significant figures.

2. The values of capacitances, in microfarads, of ten capacitors selected at random from a large batch of similar capacitors are

   34.3, 25.0, 30.4, 34.6, 29.6, 28.7,

   33.4, 32.7, 29.0 and 31.3

   Determine the standard deviation from the mean for these capacitors, correct to 3 significant figures.

3. The tensile strength in megapascals for 15 samples of tin were determined and found to be

   34.61, 34.57, 34.40, 34.63, 34.63, 34.51,

   34.49, 34.61, 34.52, 34.55, 34.58, 34.53,

   34.44, 34.48 and 34.40

   Calculate the mean and standard deviation from the mean for these 15 values, correct to 4 significant figures.

4. Calculate the standard deviation from the mean for the mass of the 50 bricks given in Problem 1 of Practice Exercise 126, page 341, correct to 3 significant figures.

5. Determine the standard deviation from the mean, correct to 4 significant figures, for the heights of the 100 people given in Problem 2 of Practice Exercise 126, page 341.

6. Calculate the standard deviation from the mean for the data given in Problem 4 of Practice Exercise 126, page 341, correct to 3 decimal places.

32.5 Quartiles, deciles and percentiles

Other measures of dispersion which are sometimes used are the quartile, decile and percentile values. The **quartile values** of a set of discrete data are obtained by selecting the values of members which divide the set into four equal parts. Thus, for the set {2, 3, 4, 5, 5, 7, 9, 11, 13, 14, 17} there are 11 members and the values of the members dividing the set into four equal parts are 4, 7 and 13. These values are signified by Q_1, Q_2 and Q_3 and called the first, second and third quartile values, respectively. It can be seen that the second quartile value, Q_2, is the value of the middle member and hence is the median value of the set.

For grouped data the ogive may be used to determine the quartile values. In this case, points are selected on the vertical cumulative frequency values of the ogive, such that they divide the total value of cumulative frequency into four equal parts. Horizontal lines are drawn from these values to cut the ogive. The values of the variable corresponding to these cutting points on the ogive give the quartile values (see Problem 7).

When a set contains a large number of members, the set can be split into ten parts, each containing an equal number of members. These ten parts are then called **deciles**. For sets containing a very large number of members, the set may be split into one hundred parts, each containing an equal number of members. One of these parts is called a **percentile**.

Problem 7. The frequency distribution given below refers to the overtime worked by a group of craftsmen during each of 48 working weeks in a year. Draw an ogive for these data and hence determine the quartile values.

25–29	5
30–34	4
35–39	7
40–44	11
45–49	12
50–54	8
55–59	1

The cumulative frequency distribution (i.e. upper class boundary/cumulative frequency values) is

29.5 5, 34.5 9, 39.5 16, 44.5 27,
49.5 39, 54.5 47, 59.5 48

The ogive is formed by plotting these values on a graph, as shown in Fig. 32.2. The total frequency is divided into four equal parts, each having a range of $48 \div 4$, i.e. 12. This gives cumulative frequency values of 0 to

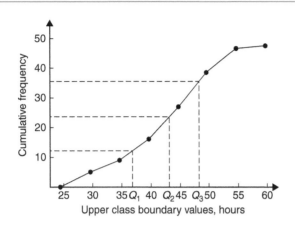

Figure 32.2

12 corresponding to the first quartile, 12 to 24 corresponding to the second quartile, 24 to 36 corresponding to the third quartile and 36 to 48 corresponding to the fourth quartile of the distribution; i.e. the distribution is divided into four equal parts. The quartile values are those of the variable corresponding to cumulative frequency values of 12, 24 and 36, marked Q_1, Q_2 and Q_3 in Fig. 32.2. These values, correct to the nearest hour, are **37 hours, 43 hours and 48 hours**, respectively. The Q_2 value is also equal to the median value of the distribution. One measure of the dispersion of a distribution is called the **semi-interquartile range** and is given by $(Q_3 - Q_1) \div 2$ and is $(48 - 37) \div 2$ in this case; i.e. **$5\frac{1}{2}$ hours**.

Problem 8. Determine the numbers contained in the (a) 41st to 50th percentile group and (b) 8th decile group of the following set of numbers.

14 22 17 21 30 28 37 7 23 32
24 17 20 22 27 19 26 21 15 29

The set is ranked, giving

7 14 15 17 17 19 20 21 21 22
22 23 24 26 27 28 29 30 32 37

(a) There are 20 numbers in the set, hence the first 10% will be the two numbers 7 and 14, the second

---

10% will be 15 and 17, and so on. Thus, the 41st to 50th percentile group will be the numbers **21 and 22**

(b) The first decile group is obtained by splitting the ranked set into 10 equal groups and selecting the first group; i.e. the numbers 7 and 14. The second decile group is the numbers 15 and 17, and so on. Thus, the 8th decile group contains the numbers **27 and 28**

Now try the following Practice Exercise

Practice Exercise 128   Quartiles, deciles and percentiles (answers on page 436)

1. The number of working days lost due to accidents for each of 12 one-month periods are as shown. Determine the median and first and third quartile values for this data.

27 37 40 28 23 30 35 24 30 32 31 28

2. The number of faults occurring on a production line in a nine-week period are as shown below. Determine the median and quartile values for the data.

30 27 25 24 27 37 31 27 35

3. Determine the quartile values and semi-interquartile range for the frequency distribution given in Problem 2 of Practice Exercise 126, page 341.

4. Determine the numbers contained in the 5th decile group and in the 61st to 70th percentile groups for the following set of numbers.

40 46 28 32 37 42 50 31 48 45
32 38 27 33 40 35 25 42 38 41

5. Determine the numbers in the 6th decile group and in the 81st to 90th percentile group for the following set of numbers.

43 47 30 25 15 51 17 21 37 33 44 56 40 49 22
36 44 33 17 35 58 51 35 44 40 31 41 55 50 16

For fully worked solutions to each of the problems in Practice Exercises 125 to 128 in this chapter, go to the website:
www.routledge.com/cw/bird

Chapter 33

Probability

Why it is important to understand: Probability

Engineers deal with uncertainty in their work, often with precision and analysis, and probability theory is widely used to model systems in engineering and scientific applications. There are a number of examples of where probability is used in engineering. For example, with electronic circuits, scaling down the power and energy of such circuits reduces the reliability and predictability of many individual elements, but the circuits must nevertheless be engineered so that the overall circuit is reliable. Centres for disease control need to decide whether to institute massive vaccination or other preventative measures in the face of globally threatening, possibly mutating diseases in humans and animals. System designers must weigh the costs and benefits of measures for reliability and security, such as levels of backups and firewalls, in the face of uncertainty about threats from equipment failures or malicious attackers. Models incorporating probability theory have been developed and are continuously being improved for understanding the brain, gene pools within populations, weather and climate forecasts, microelectronic devices, and imaging systems such as computer aided tomography (CAT) scan and radar. The electric power grid, including power generating stations, transmission lines, and consumers, is a complex system with many redundancies; however, breakdowns occur, and guidance for investment comes from modelling the most likely sequences of events that could cause outage. Similar planning and analysis is done for communication networks, transportation networks, water and other infrastructure. Probabilities, permutations and combinations are used daily in many different fields that range from gambling and games, to mechanical or structural failure rates, to rates of detection in medical screening. Uncertainty is clearly all around us, in our daily lives and in many professions. Standard deviation is widely used when results of opinion polls are described. The language of probability theory lets people break down complex problems, and argue about pieces of them with each other, and then aggregate information about subsystems to analyse a whole system. This chapter briefly introduces the important subject of probability.

At the end of this chapter, you should be able to:

- define probability
- define expectation, dependent event, independent event and conditional probability
- state the addition and multiplication laws of probability
- use the laws of probability in simple calculations
- use the laws of probability in practical situations

33.1 Introduction to probability

33.1.1 Probability

The **probability** of something happening is the likelihood or chance of it happening. Values of probability lie between 0 and 1, where 0 represents an absolute impossibility and 1 represents an absolute certainty. The probability of an event happening usually lies somewhere between these two extreme values and is expressed as either a proper or decimal fraction. Examples of probability are

that a length of copper wire has zero resistance at $100°C$	0
that a fair, six-sided dice will stop with a 3 upwards	$\frac{1}{6}$ or 0.1667
that a fair coin will land with a head upwards	$\frac{1}{2}$ or 0.5
that a length of copper wire has some resistance at $100°C$	1

If p is the probability of an event happening and q is the probability of the same event not happening, then the total probability is $p + q$ and is equal to unity, since it is an absolute certainty that the event either will or will not occur; i.e. $p + q = 1$

> **Problem 1.** Determine the probabilities of selecting at random (a) a man and (b) a woman from a crowd containing 20 men and 33 women

(a) The probability of selecting at random a man, p, is given by the ratio

$$\frac{\text{number of men}}{\text{number in crowd}}$$

i.e. $p = \frac{20}{20 + 33} = \frac{\mathbf{20}}{\mathbf{53}}$ or **0.3774**

(b) The probability of selecting at random a woman, q, is given by the ratio

$$\frac{\text{number of women}}{\text{number in crowd}}$$

i.e. $q = \frac{33}{20 + 33} = \frac{\mathbf{33}}{\mathbf{53}}$ or **0.6226**

(Check: the total probability should be equal to 1:

$p = \frac{20}{53}$ and $q = \frac{33}{53}$, thus the total probability,

$p + q = \frac{20}{53} + \frac{33}{53} = 1$

hence no obvious error has been made.)

33.1.2 Expectation

The **expectation**, E, of an event happening is defined in general terms as the product of the probability p of an event happening and the number of attempts made, n; i.e. $E = pn$

Thus, since the probability of obtaining a 3 upwards when rolling a fair dice is 1/6, the expectation of getting a 3 upwards on four throws of the dice is

$$\frac{1}{6} \times 4, \text{i.e.} \frac{2}{3}$$

Thus **expectation is the average occurrence of an event**.

> **Problem 2.** Find the expectation of obtaining a 4 upwards with 3 throws of a fair dice

Expectation is the average occurrence of an event and is defined as the probability times the number of attempts. The probability, p, of obtaining a 4 upwards for one throw of the dice is 1/6.

If 3 attempts are made, $n = 3$ and the expectation, E, is pn, i.e.

$$E = \frac{1}{6} \times 3 = \frac{1}{2} \quad \text{or} \quad \mathbf{0.50}$$

33.1.3 Dependent events

A **dependent event** is one in which the probability of an event happening affects the probability of another event happening. Let 5 transistors be taken at random from a batch of 100 transistors for test purposes and the probability of there being a defective transistor, p_1, be determined. At some later time, let another 5 transistors be taken at random from the 95 remaining transistors in the batch and the probability of there being a defective transistor, p_2, be determined. The value of p_2 is different from p_1 since the batch size has effectively altered from 100 to 95; i.e. probability p_2 is dependent on probability p_1. Since 5 transistors are drawn and then another 5 transistors are drawn without replacing the first 5, the second random selection is said to be **without replacement**.

33.1.4 Independent events

An independent event is one in which the probability of an event happening does not affect the probability of another event happening. If 5 transistors are taken at random from a batch of transistors and the probability of a defective transistor, p_1, is determined and the process is repeated after the original 5 have been replaced in the batch to give p_2, then p_1 is equal to p_2. Since the 5 transistors are replaced between draws, the second selection is said to be **with replacement**.

33.2 Laws of probability

33.2.1 The addition law of probability

The addition law of probability is recognised by the word 'or' joining the probabilities.
If p_A is the probability of event A happening and p_B is the probability of event B happening, the probability of **event A or event B** happening is given by $p_A + p_B$
Similarly, the probability of events **A or B or C or ...N** happening is given by

$$p_A + p_B + p_C + \cdots + p_N$$

33.2.2 The multiplication law of probability

The multiplication law of probability is recognised by the word '**and**' joining the probabilities.
If p_A is the probability of event A happening and p_B is the probability of event B happening, the probability of **event A and event B** happening is given by $p_A \times p_B$
Similarly, the probability of events **A and B and C and ...N** happening is given by

$$p_A \times p_B \times p_C \times \cdots \times p_N$$

Here are some worked problems to demonstrate probability.

Problem 3. Calculate the probabilities of selecting at random

(a) the winning horse in a race in which 10 horses are running and

(b) the winning horses in both the first and second races if there are 10 horses in each race

(a) Since only one of the ten horses can win, the probability of selecting at random the winning horse is

$$\frac{\text{number of winners}}{\text{number of horses}} \text{ i.e.} \frac{1}{10} \quad \text{or} \quad \mathbf{0.10}$$

(b) The probability of selecting the winning horse in the first race is $\dfrac{1}{10}$
The probability of selecting the winning horse in the second race is $\dfrac{1}{10}$
The probability of selecting the winning horses in the first **and** second race is given by the multiplication law of probability; i.e.

$$\textbf{probability} = \frac{1}{10} \times \frac{1}{10} = \frac{1}{100} \quad \text{or} \quad \mathbf{0.01}$$

Problem 4. The probability of a component failing in one year due to excessive temperature is 1/20, that due to excessive vibration is 1/25 and that due to excessive humidity is 1/50. Determine the probabilities that during a one-year period a component

(a) fails due to excessive temperature and excessive vibration,

(b) fails due to excessive vibration or excessive humidity,

(c) will not fail because of both excessive temperature and excessive humidity

Let p_A be the probability of failure due to excessive temperature, then

$$p_A = \frac{1}{20} \quad \text{and} \quad \overline{p_A} = \frac{19}{20}$$

(where $\overline{p_A}$ is the probability of not failing)

Let p_B be the probability of failure due to excessive vibration, then

$$p_B = \frac{1}{25} \quad \text{and} \quad \overline{p_B} = \frac{24}{25}$$

Let p_C be the probability of failure due to excessive humidity, then

$$p_C = \frac{1}{50} \quad \text{and} \quad \overline{p_C} = \frac{49}{50}$$

(a) The probability of a component failing due to excessive temperature **and** excessive vibration is given by

$$p_A \times p_B = \frac{1}{20} \times \frac{1}{25} = \frac{1}{500} \quad \text{or} \quad \mathbf{0.002}$$

(b) The probability of a component failing due to excessive vibration **or** excessive humidity is

$$p_B + p_C = \frac{1}{25} + \frac{1}{50} = \frac{3}{50} \quad \text{or} \quad \mathbf{0.06}$$

(c) The probability that a component will not fail due to excessive temperature **and** will not fail due to excess humidity is

$$\overline{p_A} \times \overline{p_C} = \frac{19}{20} \times \frac{49}{50} = \frac{931}{1000} \quad \text{or} \quad \mathbf{0.931}$$

Problem 5. A batch of 100 capacitors contains 73 which are within the required tolerance values and 17 which are below the required tolerance values, the remainder being above the required tolerance values. Determine the probabilities that, when randomly selecting a capacitor and then a second capacitor,

(a) both are within the required tolerance values when selecting with replacement,

(b) the first one drawn is below and the second one drawn is above the required tolerance value, when selection is without replacement

(a) The probability of selecting a capacitor within the required tolerance values is 73/100. The first capacitor drawn is now replaced and a second one is drawn from the batch of 100. The probability of this capacitor being within the required tolerance values is also 73/100. Thus, the probability of selecting a capacitor within the required tolerance values for both the first **and** the second draw is

$$\frac{73}{100} \times \frac{73}{100} = \frac{5329}{10000} \quad \text{or} \quad \mathbf{0.5329}$$

(b) The probability of obtaining a capacitor below the required tolerance values on the first draw is 17/100. There are now only 99 capacitors left in the batch, since the first capacitor is not replaced. The probability of drawing a capacitor above the required tolerance values on the second draw is 10/99, since there are $(100 - 73 - 17)$, i.e. 10, capacitors above the required tolerance value. Thus, the probability of randomly selecting a capacitor below the required tolerance values and subsequently randomly selecting a capacitor above the tolerance values is

$$\frac{17}{100} \times \frac{10}{99} = \frac{170}{9900} = \frac{17}{990} \quad \text{or} \quad \mathbf{0.0172}$$

Now try the following Practice Exercise

Practice Exercise 129   Laws of probability (answers on page 436)

1. In a batch of 45 lamps 10 are faulty. If one lamp is drawn at random, find the probability of it being (a) faulty (b) satisfactory.

2. A box of fuses are all of the same shape and size and comprises 23 2 A fuses, 47 5 A fuses and 69 13 A fuses. Determine the probability of selecting at random (a) a 2 A fuse (b) a 5 A fuse (c) a 13 A fuse.

3. (a) Find the probability of having a 2 upwards when throwing a fair 6-sided dice.

   (b) Find the probability of having a 5 upwards when throwing a fair 6-sided dice.

   (c) Determine the probability of having a 2 and then a 5 on two successive throws of a fair 6-sided dice.

4. Determine the probability that the total score is 8 when two like dice are thrown.

5. The probability of event A happening is $\dfrac{3}{5}$ and the probability of event B happening is $\dfrac{2}{3}$. Calculate the probabilities of

   (a) both A and B happening

   (b) only event A happening, i.e. event A happening and event B not happening

   (c) only event B happening

   (d) either A, or B, or A and B happening

6. When testing 1000 soldered joints, 4 failed during a vibration test and 5 failed due to having a high resistance. Determine the probability of a joint failing due to

   (a) vibration

   (b) high resistance

   (c) vibration or high resistance

   (d) vibration and high resistance

Here are some further worked problems on probability.

Problem 6. A batch of 40 components contains 5 which are defective. A component is drawn at random from the batch and tested and then a second component is drawn. Determine the probability that neither of the components is defective when drawn (a) with replacement and (b) without replacement

(a) **With replacement**

The probability that the component selected on the first draw is satisfactory is 35/40 i.e. 7/8. The component is now replaced and a second draw is made. The probability that this component is also satisfactory is 7/8. Hence, the probability that both the first component drawn and the second component drawn are satisfactory is

$$\frac{7}{8} \times \frac{7}{8} = \frac{49}{64} \quad \text{or} \quad \textbf{0.7656}$$

(b) **Without replacement**

The probability that the first component drawn is satisfactory is 7/8. There are now only 34 satisfactory components left in the batch and the batch number is 39. Hence, the probability of drawing a satisfactory component on the second draw is 34/39. Thus, the probability that the first component drawn and the second component drawn are satisfactory i.e. neither is defective is

$$\frac{7}{8} \times \frac{34}{39} = \frac{238}{312} \quad \text{or} \quad \textbf{0.7628}$$

Problem 7. A batch of 40 components contains 5 which are defective. If a component is drawn at random from the batch and tested and then a second component is drawn at random, calculate the probability of having one defective component, both (a) with replacement and (b) without replacement

The probability of having one defective component can be achieved in two ways. If p is the probability of drawing a defective component and q is the probability of drawing a satisfactory component, then the probability of having one defective component is given by drawing a satisfactory component and then a defective component **or** by drawing a defective component and then a satisfactory one; i.e. by $q \times p + p \times q$

(a) **With replacement**

$$p = \frac{5}{40} = \frac{1}{8} \quad \text{and} \quad q = \frac{35}{40} = \frac{7}{8}$$

Hence, the probability of having one defective component is

$$\frac{1}{8} \times \frac{7}{8} + \frac{7}{8} \times \frac{1}{8}$$

i.e. $$\frac{7}{64} + \frac{7}{64} = \frac{7}{32} \quad \text{or} \quad \textbf{0.2188}$$

(b) **Without replacement**

$$p_1 = \frac{1}{8} \text{ and } q_1 = \frac{7}{8} \text{ on the first of the two draws}$$

The batch number is now 39 for the second draw, thus,

$$p_2 = \frac{5}{39} \quad \text{and} \quad q_2 = \frac{35}{39}$$

$$p_1 q_2 + q_1 p_2 = \frac{1}{8} \times \frac{35}{39} + \frac{7}{8} \times \frac{5}{39}$$

$$= \frac{35 + 35}{312}$$

$$= \frac{70}{312} \quad \text{or} \quad \textbf{0.2244}$$

Problem 8. A box contains 74 brass washers, 86 steel washers and 40 aluminium washers. Three washers are drawn at random from the box without replacement. Determine the probability that all three are steel washers

Assume, for clarity of explanation, that a washer is drawn at random, then a second, then a third (although this assumption does not affect the results obtained). The total number of washers is

$$74 + 86 + 40, \quad \text{i.e.} \quad 200$$

The probability of randomly selecting a steel washer on the first draw is 86/200. There are now 85 steel washers in a batch of 199. The probability of randomly selecting a steel washer on the second draw is 85/199. There are now 84 steel washers in a batch of 198. The probability of randomly selecting a steel washer on the third draw is 84/198. Hence, the probability of selecting a steel washer on the first draw **and** the second draw **and** the third draw is

$$\frac{86}{200} \times \frac{85}{199} \times \frac{84}{198} = \frac{614040}{7880400} = \textbf{0.0779}$$

Problem 9. For the box of washers given in Problem 8 above, determine the probability that there are no aluminium washers drawn when three washers are drawn at random from the box without replacement

The probability of not drawing an aluminium washer on the first draw is $1 - \left(\dfrac{40}{200}\right)$ i.e. 160/200. There are now 199 washers in the batch of which 159 are not made of aluminium. Hence, the probability of not drawing an aluminium washer on the second draw is 159/199. Similarly, the probability of not drawing an aluminium washer on the third draw is 158/198. Hence the probability of not drawing an aluminium washer on the first **and** second **and** third draws is

$$\frac{160}{200} \times \frac{159}{199} \times \frac{158}{198} = \frac{4019520}{7880400} = \mathbf{0.5101}$$

Problem 10. For the box of washers in Problem 8 above, find the probability that there are two brass washers and either a steel or an aluminium washer when three are drawn at random, without replacement

Two brass washers (A) and one steel washer (B) can be obtained in any of the following ways.

1st draw	2nd draw	3rd draw
A	A	B
A	B	A
B	A	A

Two brass washers and one aluminium washer (C) can also be obtained in any of the following ways.

1st draw	2nd draw	3rd draw
A	A	C
A	C	A
C	A	A

Thus, there are six possible ways of achieving the combinations specified. If A represents a brass washer, B a steel washer and C an aluminium washer, the combinations and their probabilities are as shown.

Draw			Probability			
First	Second	Third				
A	A	B	$\dfrac{74}{200} \times \dfrac{73}{199} \times \dfrac{86}{198} = 0.0590$			
A	B	A	$\dfrac{74}{200} \times \dfrac{86}{199} \times \dfrac{73}{198} = 0.0590$			
B	A	A	$\dfrac{86}{200} \times \dfrac{74}{199} \times \dfrac{73}{198} = 0.0590$			
A	A	C	$\dfrac{74}{200} \times \dfrac{73}{199} \times \dfrac{40}{198} = 0.0274$			
A	C	A	$\dfrac{74}{200} \times \dfrac{40}{199} \times \dfrac{73}{198} = 0.0274$			
C	A	A	$\dfrac{40}{200} \times \dfrac{74}{199} \times \dfrac{73}{198} = 0.0274$			

The probability of having the first combination **or** the second **or** the third, and so on, is given by the sum of the probabilities; i.e. by $3 \times 0.0590 + 3 \times 0.0274$, i.e. **0.2592**

Now try the following Practice Exercise

Practice Exercise 130   Laws of probability (answers on page 436)

1. The probability that component A will operate satisfactorily for 5 years is 0.8 and that B will operate satisfactorily over that same period of time is 0.75. Find the probabilities that in a 5 year period
   (a) both components will operate satisfactorily
   (b) only component A will operate satisfactorily
   (c) only component B will operate satisfactorily

2. In a particular street, 80% of the houses have landline telephones. If two houses selected at random are visited, calculate the probabilities that

(a) they both have a telephone

(b) one has a telephone but the other does not

3. Veroboard pins are packed in packets of 20 by a machine. In a thousand packets, 40 have less than 20 pins. Find the probability that if 2 packets are chosen at random, one will contain less than 20 pins and the other will contain 20 pins or more.

4. A batch of 1 kW fire elements contains 16 which are within a power tolerance and 4 which are not. If 3 elements are selected at random from the batch, calculate the probabilities that

(a) all three are within the power tolerance

(b) two are within but one is not within the power tolerance

5. An amplifier is made up of three transistors, A, B and C. The probabilities of A, B or C being defective are 1/20, 1/25 and 1/50, respectively. Calculate the percentage of amplifiers produced

(a) which work satisfactorily

(b) which have just one defective transistor

6. A box contains 14 40 W lamps, 28 60 W lamps and 58 25 W lamps, all the lamps being of the same shape and size. Three lamps are drawn at random from the box, first one, then a second, then a third. Determine the probabilities of

(a) getting one 25 W, one 40 W and one 60 W lamp with replacement

(b) getting one 25 W, one 40 W and one 60 W lamp without replacement

(c) getting either one 25 W and two 40 W or one 60 W and two 40 W lamps with replacement

For fully worked solutions to each of the problems in Practice Exercises 129 and 130 in this chapter, go to the website:
www.routledge.com/cw/bird

Revision Test 13: Presentation of statistical data, mean, median, mode, standard deviation and probability

This assignment covers the material contained in Chapters 31–33. *The marks available are shown in brackets at the end of each question.*

1. A company produces five products in the following proportions:

Product A	24
Product B	6
Product C	15
Product D	9
Product E	18

   Draw (a) a horizontal bar chart and (b) a pie diagram to represent these data visually. (9)

2. State whether the data obtained on the following topics are likely to be discrete or continuous.
   (a) the number of books in a library
   (b) the speed of a car
   (c) the time to failure of a light bulb (3)

3. Draw a histogram, frequency polygon and ogive for the data given below which refer to the diameter of 50 components produced by a machine.

Class intervals	Frequency
1.30–1.32 mm	4
1.33–1.35 mm	7
1.36–1.38 mm	10
1.39–1.41 mm	12
1.42–1.44 mm	8
1.45–1.47 mm	5
1.48–1.50 mm	4

   (16)

4. Determine the mean, median and modal values for the following lengths given in metres:
   28, 20, 44, 30, 32, 30, 28, 34, 26, 28 (6)

5. The length in millimetres of 100 bolts is as shown below.

50–56	6
57–63	16
64–70	22
71–77	30
78–84	19
85–91	7

   Determine for the sample
   (a) the mean value
   (b) the standard deviation, correct to 4 significant figures (10)

6. The number of faulty components in a factory in a 12 week period is

   14 12 16 15 10 13 15 11 16 19 17 19

   Determine the median and the first and third quartile values. (7)

7. Determine the probability of winning a prize in a lottery by buying 10 tickets when there are 10 prizes and a total of 5000 tickets sold. (4)

8. A sample of 50 resistors contains 44 which are within the required tolerance value, 4 which are below and the remainder which are above. Determine the probability of selecting from the sample a resistor which is

   (a) below the required tolerance
   (b) above the required tolerance

   Now two resistors are selected at random from the sample. Determine the probability, correct to 3 decimal places, that neither resistor is defective when drawn

   (c) with replacement
   (d) without replacement
   (e) If a resistor is drawn at random from the batch and tested and then a second resistor is drawn from those left, calculate the probability of having one defective component when selection is without replacement. (15)

For lecturers/instructors/teachers, fully worked solutions to each of the problems in Revision Test 13, together with a full marking scheme, are available at the website:

www.routledge.com/cw/bird

Multiple choice questions Test 5
Statistics
This test covers the material in Chapters 31 to 33

All questions have only one correct answer (Answers on page 440).

1. A pie diagram is shown in Fig. M5.1 where P, Q, R and S represent the salaries of four employees of a firm. P earns £24000 p.a. Employee S earns:
   - (a) £40000
   - (b) £36000
   - (c) £20000
   - (d) £24000

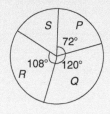

Figure M5.1

Questions 2 to 5 relate to the following information:

The capacitance (in pF) of 6 capacitors is as follows:
$\{5, 6, 8, 5, 10, 2\}$

2. The median value is:
   - (a) 36 pF
   - (b) 6 pF
   - (c) 5.5 pF
   - (d) 5 pF

3. The modal value is:
   - (a) 36 pF
   - (b) 6 pF
   - (c) 5.5 pF
   - (d) 5 pF

4. The mean value is:
   - (a) 36 pF
   - (b) 6 pF
   - (c) 5.5 pF
   - (d) 5 pF

5. The standard deviation is:
   - (a) 2.66 pF
   - (b) 2.52 pF
   - (c) 2.45 pF
   - (d) 6.33 pF

6. The curve obtained by joining the co-ordinates of cumulative frequency against upper class boundary values is called;
   - (a) a histogram
   - (b) a frequency polygon
   - (c) a tally diagram
   - (d) an ogive

Questions 7 to 9 relate to the following information:

A box contains 35 brass washers, 40 steel washers and 25 aluminium washers. 3 washers are drawn at random from the box without replacement.

7. The probability that all three are steel washers is:
   - (a) 0.0611
   - (b) 1.200
   - (c) 0.0640
   - (d) 1.182

8. The probability that there are no aluminium washers is:
   - (a) 2.250 (b) 0.418 (c) 0.014 (d) 0.422

9. The probability that there are two brass washers and either a steel or an aluminium washer is:
   - (a) 0.071 (b) 0.687 (c) 0.239 (d) 0.343

10. The number of faults occurring on a production line in a 9 week period are as shown:

    32 29 27 26 29 39 33 29 37

    The third quartile value is:
    - (a) 29 (b) 35 (c) 31 (d) 28

Questions 11 and 12 relate to the following information:

The frequency distribution for the values of resistance in ohms of 40 transistors is as follows:
15.5 – 15.9 3, 16.0 – 16.4 10, 16.5 – 16.9 13,
17.0 – 17.4 8, 17.5 – 17.9 6

11. The mean value of the resistance is:
    - (a) 16.75 Ω (b) 1.0 Ω (c) 15.85 Ω (d) 16.95 Ω

12. The standard deviation is:
    - (a) 0.335 Ω (b) 0.251 Ω (c) 0.682 Ω (d) 0.579 Ω

Questions 13 to 15 relate to the following information.

The probability of a component failing in 1 year due to excessive temperature is $\dfrac{1}{16}$, due to excessive vibration is $\dfrac{1}{20}$ and due to excessive humidity is $\dfrac{1}{40}$

13. The probability that a component fails due to excessive temperature and excessive vibration is:

(a) $\dfrac{285}{320}$ (b) $\dfrac{1}{320}$ (c) $\dfrac{9}{80}$ (d) $\dfrac{1}{800}$

14. The probability that a component fails due to excessive vibration or excessive humidity is:

(a) 0.00125 (b) 0.00257 (c) 0.0750 (d) 0.1125

15. The probability that a component will not fail because of both excessive temperature and excessive humidity is:

(a) 0.914 (b) 1.913 (c) 0.00156 (d) 0.0875

The companion website for this book contains the above multiple-choice test. If you prefer to attempt the test online then visit:

www.routledge.com/cw/bird

For a copy of this multiple choice test, go to:

www.routledge.com/cw/bird

Introduction to differentiation

At the end of this chapter, you should be able to:

- state that calculus comprises two parts – differential and integral calculus
- understand functional notation
- describe the gradient of a curve and limiting value
- differentiate $y = ax^n$ by the general rule
- differentiate sine and cosine functions
- differentiate exponential and logarithmic functions

34.1 Introduction to calculus

Calculus is a branch of mathematics involving or leading to calculations dealing with continuously varying functions such as velocity and acceleration, rates of change and maximum and minimum values of curves. Calculus has widespread applications in science and engineering and is used to solve complicated problems for which algebra alone is insufficient.

Calculus is a subject that falls into two parts:

(a) **differential calculus** (or **differentiation**),

(b) **integral calculus** (or **integration**).

This chapter provides an introduction to differentiation and applies differentiation to rates of change. Chapter 35 introduces integration and applies it to determine areas under curves.

Further applications of differentiation and integration are explored in *Engineering Mathematics* (Bird, 2014).

34.2 Functional notation

In an equation such as $y = 3x^2 + 2x - 5$, y is said to be a function of x and may be written as $y = f(x)$

An equation written in the form $f(x) = 3x^2 + 2x - 5$ is termed **functional notation**. The value of $f(x)$ when $x = 0$ is denoted by $f(0)$, and the value of $f(x)$ when $x = 2$ is denoted by $f(2)$, and so on. Thus, when $f(x) = 3x^2 + 2x - 5$,

$$f(0) = 3(0)^2 + 2(0) - 5 = -5$$

and $f(2) = 3(2)^2 + 2(2) - 5 = 11$, and so on.

Problem 1. If $f(x) = 4x^2 - 3x + 2$, find $f(0), f(3), f(-1)$ and $f(3) - f(-1)$

$$f(x) = 4x^2 - 3x + 2$$

$$f(0) = 4(0)^2 - 3(0) + 2 = \mathbf{2}$$

$$f(3) = 4(3)^2 - 3(3) + 2 = 36 - 9 + 2 = \mathbf{29}$$

$$f(-1) = 4(-1)^2 - 3(-1) + 2 = 4 + 3 + 2 = \mathbf{9}$$

$$f(3) - f(-1) = 29 - 9 = \mathbf{20}$$

Problem 2. Given that $f(x) = 5x^2 + x - 7$, determine (a) $f(-2)$ (b) $f(2) \div f(1)$

(a) $f(-2) = 5(-2)^2 + (-2) - 7 = 20 - 2 - 7 = \mathbf{11}$

(b) $f(2) = 5(2)^2 + 2 - 7 = 15$

$\quad f(1) = 5(1)^2 + 1 - 7 = -1$

$$f(2) \div f(1) = \frac{15}{-1} = \mathbf{-15}$$

Now try the following Practice Exercise

Practice Exercise 131   Functional notation (answers on page 436)

1. If $f(x) = 6x^2 - 2x + 1$, find $f(0), f(1), f(2), f(-1)$ and $f(-3)$

2. If $f(x) = 2x^2 + 5x - 7$, find $f(1), f(2), f(-1), f(2) - f(-1)$

3. Given $f(x) = 3x^3 + 2x^2 - 3x + 2$, prove that $f(1) = \frac{1}{7}f(2)$

34.3 The gradient of a curve

If a tangent is drawn at a point P on a curve, the gradient of this tangent is said to be the **gradient of the curve** at P. In Fig. 34.1, the gradient of the curve at P is equal to the gradient of the tangent PQ.

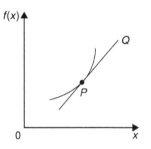

Figure 34.1

For the curve shown in Fig. 34.2, let the points A and B have co-ordinates (x_1, y_1) and (x_2, y_2), respectively. In functional notation, $y_1 = f(x_1)$ and $y_2 = f(x_2)$, as shown.

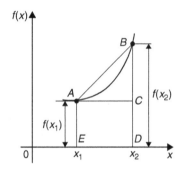

Figure 34.2

The gradient of the chord AB

$$= \frac{BC}{AC} = \frac{BD - CD}{ED} = \frac{f(x_2) - f(x_1)}{(x_2 - x_1)}$$

For the curve $f(x) = x^2$ shown in Fig. 34.3,

(a) the gradient of chord AB

$$= \frac{f(3) - f(1)}{3 - 1} = \frac{9 - 1}{2} = 4$$

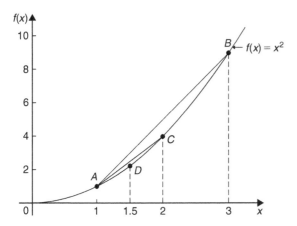

Figure 34.3

(b) the gradient of chord AC

$$= \frac{f(2) - f(1)}{2 - 1} = \frac{4 - 1}{1} = 3$$

(c) the gradient of chord AD

$$= \frac{f(1.5) - f(1)}{1.5 - 1} = \frac{2.25 - 1}{0.5} = 2.5$$

(d) if E is the point on the curve $(1.1, f(1.1))$ then the gradient of chord AE

$$= \frac{f(1.1) - f(1)}{1.1 - 1} = \frac{1.21 - 1}{0.1} = 2.1$$

(e) if F is the point on the curve $(1.01, f(1.01))$ then the gradient of chord AF

$$= \frac{f(1.01) - f(1)}{1.01 - 1} = \frac{1.0201 - 1}{0.01} = 2.01$$

Thus, as point B moves closer and closer to point A, the gradient of the chord approaches nearer and nearer to the value 2. This is called the **limiting value** of the gradient of the chord AB and when B coincides with A the chord becomes the tangent to the curve.

Now try the following Practice Exercise

Practice Exercise 132   The gradient of a curve (answers on page 436)

1. Plot the curve $f(x) = 4x^2 - 1$ for values of x from $x = -1$ to $x = +4$. Label the

co-ordinates $(3, f(3))$ and $(1, f(1))$ as J and K, respectively. Join points J and K to form the chord JK. Determine the gradient of chord JK. By moving J nearer and nearer to K, determine the gradient of the tangent of the curve at K.

34.4 Differentiation from first principles

In Fig. 34.4, A and B are two points very close together on a curve, δx (delta x) and δy (delta y) representing small increments in the x and y directions, respectively.

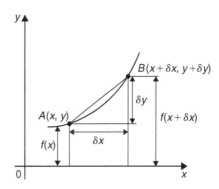

Figure 34.4

Gradient of chord $AB = \dfrac{\delta y}{\delta x}$

however, $\delta y = f(x + \delta x) - f(x)$

Hence, $\dfrac{\delta y}{\delta x} = \dfrac{f(x + \delta x) - f(x)}{\delta x}$

As δx approaches zero, $\dfrac{\delta y}{\delta x}$ approaches a limiting value and the gradient of the chord approaches the gradient of the tangent at A.

When determining the gradient of a tangent to a curve there **are two notations** used. The gradient of the curve at A in Fig. 34.4 can either be written as

$$\lim_{\delta x \to 0} \frac{\delta y}{\delta x} \quad \text{or} \quad \lim_{\delta x \to 0} \left\{ \frac{f(x + \delta x) - f(x)}{\delta x} \right\}$$

In **Leibniz**[*] notation, $\dfrac{dy}{dx} = \underset{\delta x \to 0}{\text{limit}} \dfrac{\delta y}{\delta x}$

In **functional notation,**

$$f'(x) = \underset{\delta x \to 0}{\text{limit}} \left\{ \dfrac{f(x + \delta x) - f(x)}{\delta x} \right\}$$

$\dfrac{dy}{dx}$ is the same as $f'(x)$ and is called the **differential coefficient** or the **derivative**. The process of finding the differential coefficient is called **differentiation**. Summarising, the differential coefficient,

$$\dfrac{dy}{dx} = f'(x) = \underset{\delta x \to 0}{\text{limit}} \dfrac{\delta y}{\delta x} = \underset{\delta x \to 0}{\text{limit}} \left\{ \dfrac{f(x + \delta x) - f(x)}{\delta x} \right\}$$

> **Problem 3.** Differentiate from first principles
> $f(x) = x^2$

To 'differentiate from first principles' means 'to find $f'(x)$' using the expression

$$f'(x) = \underset{\delta x \to 0}{\text{limit}} \left\{ \dfrac{f(x + \delta x) - f(x)}{\delta x} \right\}$$

[*]Who was Leibniz? – **Gottfried Wilhelm Leibniz** (sometimes von Leibniz) (1 July 1646 – 14 November 1716) was a German mathematician and philosopher. Leibniz developed infinitesimal calculus and invented the Leibniz wheel. To find out more go to www.routledge.com/cw/bird

$f(x) = x^2$ and substituting $(x + \delta x)$ for x gives $f(x + \delta x) = (x + \delta x)^2 = x^2 + 2x\delta x + \delta x^2$, hence,

$$f'(x) = \underset{\delta x \to 0}{\text{limit}} \left\{ \dfrac{(x^2 + 2x\delta x + \delta x^2) - (x^2)}{\delta x} \right\}$$

$$= \underset{\delta x \to 0}{\text{limit}} \left\{ \dfrac{2x\delta x + \delta x^2}{\delta x} \right\} = \underset{\delta x \to 0}{\text{limit}} \{2x + \delta x\}$$

As $\delta x \to 0$, $\{2x + \delta x\} \to \{2x + 0\}$.

Thus, $f'(x) = 2x$ i.e. **the differential coefficient of x^2 is $2x$**

This means that the general equation for the gradient of the curve $f(x) = x^2$ is $2x$. If the gradient is required at, say, $x = 3$, then gradient $= 2(3) = \mathbf{6}$

Differentiation from first principles can be a lengthy process and we do not want to have to go through this procedure every time we want to differentiate a function. In reality we do not have to because from the above procedure has evolved a set **general rule**, which we consider in the following section.

34.5 Differentiation of $y = ax^n$ by the general rule

From differentiation by first principles, a general rule for differentiating ax^n emerges where a and n are any constants. This rule is

$$\text{if } \boldsymbol{y = ax^n} \text{ then } \dfrac{dy}{dx} = anx^{n-1}$$

or $\quad$ if $\boldsymbol{f(x) = ax^n}$ then $\boldsymbol{f'(x) = anx^{n-1}}$

When differentiating, results can be expressed in a number of ways. For example,

(a) if $y = 3x^2$ then $\dfrac{dy}{dx} = 6x$

(b) if $f(x) = 3x^2$ then $f'(x) = 6x$

(c) the differential coefficient of $3x^2$ is $6x$

(d) the derivative of $3x^2$ is $6x$

(e) $\dfrac{d}{dx}(3x^2) = 6x$

34.5.1 Revision of some laws of indices

$\dfrac{1}{x^a} = x^{-a}$ $\quad$ For example, $\dfrac{1}{x^2} = x^{-2}$ and $x^{-5} = \dfrac{1}{x^5}$

$\sqrt{x} = x^{\frac{1}{2}}$ $\quad$ For example, $\sqrt{5} = 5^{\frac{1}{2}}$ and

$\qquad 16^{\frac{1}{2}} = \sqrt{16} = \pm 4$ and $\dfrac{1}{\sqrt{x}} = \dfrac{1}{x^{\frac{1}{2}}} = x^{-\frac{1}{2}}$

$\sqrt[a]{x^b} = x^{\frac{b}{a}}$ For example, $\sqrt[3]{x^5} = x^{\frac{5}{3}}$ and $x^{\frac{4}{3}} = \sqrt[3]{x^4}$

and $\dfrac{1}{\sqrt[3]{x^7}} = \dfrac{1}{x^{\frac{7}{3}}} = x^{-\frac{7}{3}}$

$x^0 = 1$ For example, $7^0 = 1$ and $43.5^0 = 1$

Here are some worked problems to demonstrate the general rule for differentiating $y = ax^n$

Problem 4.   Differentiate the following with respect to x: $y = 4x^7$

Comparing $y = 4x^7$ with $y = ax^n$ shows that $a = 4$ and $n = 7$. Using the general rule,

$$\frac{dy}{dx} = anx^{n-1} = (4)(7)x^{7-1} = 28x^6$$

Problem 5.   Differentiate the following with respect to x: $y = \dfrac{3}{x^2}$

$y = \dfrac{3}{x^2} = 3x^{-2}$, hence $a = 3$ and $n = -2$ in the general rule.

$$\frac{dy}{dx} = anx^{n-1} = (3)(-2)x^{-2-1} = -6x^{-3} = -\frac{6}{x^3}$$

Problem 6.   Differentiate the following with respect to x: $y = 5\sqrt{x}$

$y = 5\sqrt{x} = 5x^{\frac{1}{2}}$, hence $a = 5$ and $n = \dfrac{1}{2}$ in the general rule.

$$\frac{dy}{dx} = anx^{n-1} = (5)\left(\frac{1}{2}\right)x^{\frac{1}{2}-1}$$

$$= \frac{5}{2}x^{-\frac{1}{2}} = \frac{5}{2x^{\frac{1}{2}}} = \frac{5}{2\sqrt{x}}$$

Problem 7.   Differentiate $y = 4$

$y = 4$ may be written as $y = 4x^0$; i.e. in the general rule $a = 4$ and $n = 0$. Hence,

$$\frac{dy}{dx} = (4)(0)x^{0-1} = 0$$

The equation $y = 4$ represents a **straight horizontal line** and the gradient of a horizontal line is zero, hence the result could have been determined on inspection. In general, **the differential coefficient of a constant is always zero**.

Problem 8.   Differentiate $y = 7x$

Since $y = 7x$, i.e. $y = 7x^1$, in the general rule $a = 7$ and $n = 1$. Hence,

$$\frac{dy}{dx} = (7)(1)x^{1-1} = 7x^0 = 7 \qquad \text{since } x^0 = 1$$

The gradient of the line $y = 7x$ is 7 (from $y = mx + c$), hence the result could have been obtained by inspection. In general, **the differential coefficient of kx, where k is a constant, is always k.**

Problem 9.   Find the differential coefficient of

$$y = \frac{2}{3}x^4 - \frac{4}{x^3} + 9$$

$$y = \frac{2}{3}x^4 - \frac{4}{x^3} + 9$$

i.e. $$y = \frac{2}{3}x^4 - 4x^{-3} + 9$$

$$\frac{dy}{dx} = \left(\frac{2}{3}\right)(4)x^{4-1} - (4)(-3)x^{-3-1} + 0$$

$$= \frac{8}{3}x^3 + 12x^{-4}$$

i.e. $$\frac{dy}{dx} = \frac{8}{3}x^3 + \frac{12}{x^4}$$

Problem 10.   If $f(t) = 4t + \dfrac{1}{\sqrt{t^3}}$ find $f'(t)$

$$f(t) = 4t + \frac{1}{\sqrt{t^3}} = 4t + \frac{1}{t^{\frac{3}{2}}} = 4t^1 + t^{-\frac{3}{2}}$$

Hence, $$f'(t) = (4)(1)t^{1-1} + \left(-\frac{3}{2}\right)t^{-\frac{3}{2}-1}$$

$$= 4t^0 - \frac{3}{2}t^{-\frac{5}{2}}$$

i.e. $$f'(t) = 4 - \frac{3}{2t^{\frac{5}{2}}} = 4 - \frac{3}{2\sqrt{t^5}}$$

Problem 11.   Determine $\dfrac{dy}{dx}$ given $y = \dfrac{3x^2 - 5x}{2x}$

$$y = \frac{3x^2 - 5x}{2x} = \frac{3x^2}{2x} - \frac{5x}{2x} = \frac{3}{2}x - \frac{5}{2}$$

Hence, $$\frac{dy}{dx} = \frac{3}{2} \text{ or } 1.5$$

Problem 12. Find the differential coefficient of
$$y = \frac{2}{5}x^3 - \frac{4}{x^3} + 4\sqrt{x^5} + 7$$

$$y = \frac{2}{5}x^3 - \frac{4}{x^3} + 4\sqrt{x^5} + 7$$

i.e. $$y = \frac{2}{5}x^3 - 4x^{-3} + 4x^{\frac{5}{2}} + 7$$

$$\frac{dy}{dx} = \left(\frac{2}{5}\right)(3)x^{3-1} - (4)(-3)x^{-3-1}$$

$$+ (4)\left(\frac{5}{2}\right)x^{\frac{5}{2}-1} + 0$$

$$= \frac{6}{5}x^2 + 12x^{-4} + 10x^{\frac{3}{2}}$$

i.e. $$\frac{dy}{dx} = \frac{6}{5}x^2 + \frac{12}{x^4} + 10\sqrt{x^3}$$

Problem 13. Differentiate $y = \dfrac{(x+2)^2}{x}$ with respect to x

$$y = \frac{(x+2)^2}{x} = \frac{x^2 + 4x + 4}{x} = \frac{x^2}{x} + \frac{4x}{x} + \frac{4}{x}$$

i.e. $$y = x^1 + 4 + 4x^{-1}$$

Hence, $$\frac{dy}{dx} = 1x^{1-1} + 0 + (4)(-1)x^{-1-1}$$

$$= x^0 - 4x^{-2} = 1 - \frac{4}{x^2} \qquad \text{(since } x^0 = 1\text{)}$$

Problem 14. Find the gradient of the curve
$y = 2x^2 - \dfrac{3}{x}$ at $x = 2$

$$y = 2x^2 - \frac{3}{x} = 2x^3 - 3x^{-1}$$

$$\text{Gradient} = \frac{dy}{dx} = (2)(2)x^{2-1} - (3)(-1)x^{-1-1}$$

$$= 4x + 3x^{-2}$$

$$= 4x + \frac{3}{x^2}$$

When $x = 2$, gradient $= 4x + \dfrac{3}{x^2} = 4(2) + \dfrac{3}{(2)^2}$

$$= 8 + \frac{3}{4} = \mathbf{8.75}$$

Problem 15. Find the gradient of the curve
$y = 3x^4 - 2x^2 + 5x - 2$ at the points $(0, -2)$
and $(1, 4)$

The gradient of a curve at a given point is given by the corresponding value of the derivative.
Thus, since $y = 3x^4 - 2x^2 + 5x - 2$,

the **gradient** $= \dfrac{dy}{dx} = \mathbf{12x^3 - 4x + 5}$

At the point $(0, -2)$, $x = 0$, thus

the **gradient** $= 12(0)^3 - 4(0) + 5 = \mathbf{5}$

At the point $(1, 4)$, $x = 1$, thus

the **gradient** $= 12(1)^3 - 4(1) + 5 = \mathbf{13}$

Now try the following Practice Exercise

Practice Exercise 133   Differentiation of
$y = ax^n$ by the general rule (answers on
page 436)

In Problems 1 to 20, determine the differential coefficients with respect to the variable.

1. $y = 7x^4$

2. $y = 2x + 1$

3. $y = x^2 - x$

4. $y = 2x^3 - 5x + 6$

5. $y = \dfrac{1}{x}$

6. $y = 12$

7. $y = x - \dfrac{1}{x^2}$

8. $y = 3x^5 - 2x^4 + 5x^3 + x^2 - 1$

9. $y = \dfrac{2}{x^3}$

10. $y = 4x(1 - x)$

11. $y = \sqrt{x}$

12. $y = \sqrt{t^3}$

13. $y = 6 + \dfrac{1}{x^3}$

14. $y = 3x - \dfrac{1}{\sqrt{x}} + \dfrac{1}{x}$

15. $y = (x + 1)^2$

16. $y = x + 3\sqrt{x}$

17. $y = (1 - x)^2$

18. $y = \dfrac{5}{x^2} - \dfrac{1}{\sqrt{x^7}} + 2$

19. $y = 3(t - 2)^2$

20. $y = \dfrac{(x + 2)^2}{x}$

21. Find the gradient of the following curves at the given points.

(a) $y = 3x^2$ at $x = 1$

(b) $y = \sqrt{x}$ at $x = 9$

(c) $y = x^3 + 3x - 7$ at $x = 0$

(d) $y = \dfrac{1}{\sqrt{x}}$ at $x = 4$

(e) $y = \dfrac{1}{x}$ at $x = 2$

(f) $y = (2x + 3)(x - 1)$ at $x = -2$

22.   Differentiate $f(x) = 6x^2 - 3x + 5$ and find the gradient of the curve at

    (a) $x = -1$ (b) $x = 2$

23.   Find the differential coefficient of $y = 2x^3 + 3x^2 - 4x - 1$ and determine the gradient of the curve at $x = 2$

24.   Determine the derivative of $y = -2x^3 + 4x + 7$ and determine the gradient of the curve at $x = -1.5$

34.6   Differentiation of sine and cosine functions

Fig. 34.5(a) shows a graph of $y = \sin x$. The gradient is continually changing as the curve moves from 0 to A to B to C to D. The gradient, given by $\dfrac{dy}{dx}$, may be plotted in a corresponding position below $y = \sin x$, as shown in Fig. 34.5(b).

At 0, the gradient is positive and is at its steepest. Hence, $0'$ is a maximum positive value. Between 0 and A the gradient is positive but is decreasing in value until at A the gradient is zero, shown as A'. Between A and B the gradient is negative but is increasing in value until at B the gradient is at its steepest. Hence B' is a maximum negative value.

If the gradient of $y = \sin x$ is further investigated between B and C and C and D then the resulting graph of $\dfrac{dy}{dx}$ is seen to be a **cosine wave**.

Hence the rate of change of $\sin x$ is $\cos x$, i.e.

$$\textbf{if } y = \sin x \textbf{ then } \frac{dy}{dx} = \cos x$$

It may also be shown that

$$\textbf{if } y = \sin ax, \frac{dy}{dx} = a \cos ax \qquad (1)$$

$$\text{(where } a \text{ is a constant)}$$

and $\textbf{if } y = \sin(ax + \alpha), \dfrac{dy}{dx} = a \cos(ax + \alpha)$ (2)

$$\text{(where } a \text{ and } \alpha \text{ are constants).}$$

If a similar exercise is followed for $y = \cos x$ then the graphs of Fig. 34.6 result, showing $\dfrac{dy}{dx}$ to be a graph of $\sin x$ but displaced by π radians.

If each point on the curve $y = \sin x$ (as shown in Fig. 34.5(a)) were to be made negative (i.e. $+\dfrac{\pi}{2}$ made $-\dfrac{\pi}{2}$, $-\dfrac{3\pi}{2}$ made $+\dfrac{3\pi}{2}$, and so on) then the graph shown in Fig. 34.6(b) would result. This latter graph therefore represents the curve of $-\sin x$

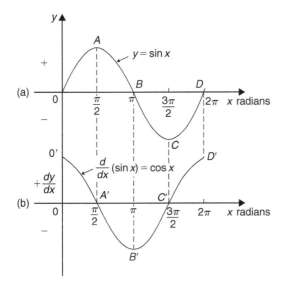

Figure 34.5

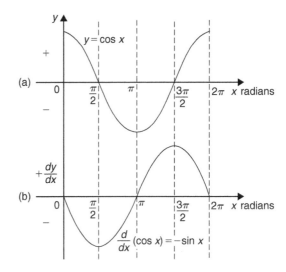

Figure 34.6

Thus,

$$\text{if } y = \cos x, \frac{dy}{dx} = -\sin x$$

It may also be shown that

$$\text{if } y = \cos ax, \frac{dy}{dx} = -a \sin ax \qquad (3)$$

(where a is a constant)

and if $y = \cos(ax + \alpha), \dfrac{dy}{dx} = -a \sin(ax + \alpha)$ (4)

(where a and α are constants).

Problem 16. Find the differential coefficient of $y = 7 \sin 2x - 3 \cos 4x$

$$\frac{dy}{dx} = (7)(2 \cos 2x) - (3)(-4 \sin 4x)$$

from equations (1) and (3)

$$= 14 \cos 2x + 12 \sin 4x$$

Problem 17. Differentiate the following with respect to the variable (a) $y = 2 \sin 5\theta$
(b) $f(t) = 3 \cos 2t$

(a) $y = 2 \sin 5\theta$

$$\frac{dy}{d\theta} = (2)(5 \cos 5\theta) = \mathbf{10 \cos 5\theta}$$

from equation (1)

(b) $f(t) = 3 \cos 2t$

$$f'(t) = (3)(-2 \sin 2t) = \mathbf{-6 \sin 2t}$$

from equation (3)

Problem 18. Differentiate the following with respect to the variable
(a) $f(\theta) = 5 \sin(100\pi\theta - 0.40)$
(b) $f(t) = 2 \cos(5t + 0.20)$

(a) If $f(\theta) = 5 \sin(100\pi\theta - 0.40)$

$$f'(\theta) = 5[100\pi \cos(100\pi\theta - 0.40)]$$

from equation (2), where $a = 100\pi$

$$= \mathbf{500\pi \cos(100\pi\theta - 0.40)}$$

(b) If $f(t) = 2 \cos(5t + 0.20)$

$$f'(t) = 2[-5 \sin(5t + 0.20)]$$

from equation (4), where $a = 5$

$$= \mathbf{-10 \sin(5t + 0.20)}$$

Problem 19. An alternating voltage is given by $v = 100 \sin 200t$ volts, where t is the time in seconds. Calculate the rate of change of voltage when (a) $t = 0.005$ s and (b) $t = 0.01$ s

$v = 100 \sin 200t$ volts. The rate of change of v is given by $\dfrac{dv}{dt}$

$$\frac{dv}{dt} = (100)(200 \cos 200t) = 20\,000 \cos 200t$$

(a) When $t = 0.005$ s,

$$\frac{dv}{dt} = 20\,000 \cos(200)(0.005) = 20\,000 \cos 1$$

$\cos 1$ means 'the cosine of 1 radian' (make sure your calculator is on radians, not degrees). Hence,

$$\frac{dv}{dt} = \mathbf{10\,806 \text{ volts per second}}$$

(b) When $t = 0.01$ s,

$$\frac{dv}{dt} = 20\,000 \cos(200)(0.01) = 20\,000 \cos 2$$

Hence,

$$\frac{dv}{dt} = \mathbf{-8323 \text{ volts per second}}$$

Now try the following Practice Exercise

Practice Exercise 134   Differentiation of sine and cosine functions (answers on page 436)

1. Differentiate with respect to x: (a) $y = 4 \sin 3x$
    (b) $y = 2 \cos 6x$

2. Given $f(\theta) = 2 \sin 3\theta - 5 \cos 2\theta$, find $f'(\theta)$

3. Find the gradient of the curve $y = 2 \cos \dfrac{1}{2}x$ at $x = \dfrac{\pi}{2}$

4. Determine the gradient of the curve $y = 3 \sin 2x$ at $x = \dfrac{\pi}{3}$

5. An alternating current is given by $i = 5 \sin 100t$ amperes, where t is the time in seconds. Determine the rate of change of current $\left(\text{i.e. } \dfrac{di}{dt}\right)$ when $t = 0.01$ seconds.

6. $v = 50 \sin 40t$ volts represents an alternating voltage, v, where t is the time in seconds. At

a time of 20×10^{-3} seconds, find the rate of change of voltage $\left(\text{i.e. } \dfrac{dv}{dt}\right)$

7. If $f(t) = 3\sin(4t + 0.12) - 2\cos(3t - 0.72)$, determine $f'(t)$

34.7 Differentiation of e^{ax} and $\ln ax$

A graph of $y = e^x$ is shown in Fig. 34.7(a). The gradient of the curve at any point is given by $\dfrac{dy}{dx}$ and is continually changing. By drawing tangents to the curve at many points on the curve and measuring the gradient of the tangents, values of $\dfrac{dy}{dx}$ for corresponding values of x may be obtained. These values are shown graphically in Fig. 34.7(b).

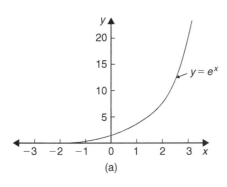

(a)

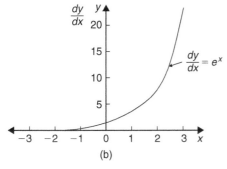

(b)

Figure 34.7

The graph of $\dfrac{dy}{dx}$ against x is identical to the original graph of $y = e^x$. It follows that

$$\text{if } y = e^x, \text{ then } \frac{dy}{dx} = e^x$$

It may also be shown that

$$\text{if } y = e^{ax}, \text{ then } \frac{dy}{dx} = ae^{ax}$$

Therefore,

$$\text{if } y = 2e^{6x}, \text{ then } \frac{dy}{dx} = (2)(6e^{6x}) = 12e^{6x}$$

A graph of $y = \ln x$ is shown in Fig. 34.8(a). The gradient of the curve at any point is given by $\dfrac{dy}{dx}$ and is continually changing. By drawing tangents to the curve at many points on the curve and measuring the gradient of the tangents, values of $\dfrac{dy}{dx}$ for corresponding values of x may be obtained. These values are shown graphically in Fig. 34.8(b).

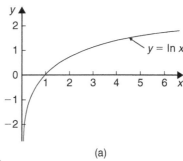

(a)

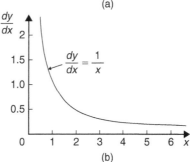

(b)

Figure 34.8

The graph of $\dfrac{dy}{dx}$ against x is the graph of $\dfrac{dy}{dx} = \dfrac{1}{x}$. It follows that

$$\text{if } y = \ln x, \text{ then } \frac{dy}{dx} = \frac{1}{x}$$

It may also be shown that

$$\text{if } y = \ln ax, \text{ then } \frac{dy}{dx} = \frac{1}{x}$$

(Note that, in the latter expression, the constant a does not appear in the $\dfrac{dy}{dx}$ term.) Thus, if $y = \ln 4x$, then $\dfrac{dy}{dx} = \dfrac{1}{x}$

Problem 20. Differentiate the following with respect to the variable (a) $y = 3e^{2x}$ (b) $f(t) = \dfrac{4}{3e^{5t}}$

(a) If $y = 3e^{2x}$ then $\dfrac{dy}{dx} = (3)(2e^{2x}) = \mathbf{6e^{2x}}$

(b) If $f(t) = \dfrac{4}{3e^{5t}} = \dfrac{4}{3}e^{-5t}$, then

$$f'(t) = \dfrac{4}{3}(-5e^{-5t}) = -\dfrac{20}{3}e^{-5t} = -\dfrac{\mathbf{20}}{\mathbf{3e^{5t}}}$$

Problem 21. Differentiate $y = 5\ln 3x$

If $y = 5\ln 3x$, then $\dfrac{dy}{dx} = (5)\left(\dfrac{1}{x}\right) = \dfrac{\mathbf{5}}{\mathbf{x}}$

Now try the following Practice Exercise

Practice Exercise 135   Differentiation of e^{ax} and $\ln ax$ (answers on page 437)

1. Differentiate with respect to x: (a) $y = 5e^{3x}$
   (b) $y = \dfrac{2}{7e^{2x}}$

2. Given $f(\theta) = 5\ln 2\theta - 4\ln 3\theta$, determine $f'(\theta)$

3. If $f(t) = 4\ln t + 2$, evaluate $f'(t)$ when $t = 0.25$

4. Find the gradient of the curve
   $y = 2e^x - \dfrac{1}{4}\ln 2x$ at $x = \dfrac{1}{2}$ correct to 2 decimal places.

5. Evaluate $\dfrac{dy}{dx}$ when $x = 1$, given
   $y = 3e^{4x} - \dfrac{5}{2e^{3x}} + 8\ln 5x$. Give the answer correct to 3 significant figures.

34.8 Summary of standard derivatives

The standard derivatives used in this chapter are summarised in Table 34.1 and are true for all real values of x.

Table 34.1

y or $f(x)$	$\dfrac{dy}{dx}$ or $f'(x)$
ax^n	anx^{n-1}
$\sin ax$	$a\cos ax$
$\cos ax$	$-a\sin ax$
e^{ax}	ae^{ax}
$\ln ax$	$\dfrac{1}{x}$

Problem 22. Find the gradient of the curve $y = 3x^2 - 7x + 2$ at the point $(1, -2)$

If $y = 3x^2 - 7x + 2$, then gradient $= \dfrac{dy}{dx} = 6x - 7$
At the point $(1, -2), x = 1$,
hence **gradient** $= 6(1) - 7 = \mathbf{-1}$

Problem 23. If $y = \dfrac{3}{x^2} - 2\sin 4x + \dfrac{2}{e^x} + \ln 5x$, determine $\dfrac{dy}{dx}$

$$y = \dfrac{3}{x^2} - 2\sin 4x + \dfrac{2}{e^x} + \ln 5x$$
$$= 3x^{-2} - 2\sin 4x + 2e^{-x} + \ln 5x$$

$$\dfrac{dy}{dx} = 3(-2x^{-3}) - 2(4\cos 4x) + 2(-e^{-x}) + \dfrac{1}{x}$$
$$= -\dfrac{6}{x^3} - 8\cos 4x - \dfrac{2}{e^x} + \dfrac{1}{x}$$

Now try the following Practice Exercise

Practice Exercise 136   Standard derivatives (answers on page 437)

1. Find the gradient of the curve
   $y = 2x^4 + 3x^3 - x + 4$ at the points
   (a) $(0, 4)$ (b) $(1, 8)$

2. Differentiate with respect to x:
   $y = \dfrac{2}{x^2} + 2\ln 2x - 2(\cos 5x + 3\sin 2x) - \dfrac{2}{e^{3x}}$

34.9   Successive differentiation

When a function $y = f(x)$ is differentiated with respect to x, the differential coefficient is written as $\dfrac{dy}{dx}$ or $f'(x)$. If the expression is differentiated again, the second differential coefficient is obtained and is written as $\dfrac{d^2 y}{dx^2}$ (pronounced dee two y by dee x squared) or $f''(x)$ (pronounced f double-dash x). By successive differentiation further higher derivatives such as $\dfrac{d^3 y}{dx^3}$ and $\dfrac{d^4 y}{dx^4}$ may be obtained. Thus,

if $y = 5x^4$, $\dfrac{dy}{dx} = 20x^3$, $\dfrac{d^2 y}{dx^2} = 60x^2$, $\dfrac{d^3 y}{dx^3} = 120x$,

$\dfrac{d^4 y}{dx^4} = 120$ and $\dfrac{d^5 y}{dx^5} = 0$

Problem 24.   If $f(x) = 4x^5 - 2x^3 + x - 3$, find $f''(x)$

$$f(x) = 4x^5 - 2x^3 + x - 3$$
$$f'(x) = 20x^4 - 6x^2 + 1$$
$$\mathbf{f''(x) = 80x^3 - 12x} \quad \text{or} \quad \mathbf{4x(20x^2 - 3)}$$

Problem 25.   Given $y = \dfrac{2}{3}x^3 - \dfrac{4}{x^2} + \dfrac{1}{2x} - \sqrt{x}$, determine $\dfrac{d^2 y}{dx^2}$

$$y = \frac{2}{3}x^3 - \frac{4}{x^2} + \frac{1}{2x} - \sqrt{x}$$
$$= \frac{2}{3}x^3 - 4x^{-2} + \frac{1}{2}x^{-1} - x^{\frac{1}{2}}$$
$$\frac{dy}{dx} = \left(\frac{2}{3}\right)(3x^2) - 4(-2x^{-3})$$
$$+ \left(\frac{1}{2}\right)(-1x^{-2}) - \frac{1}{2}x^{-\frac{1}{2}}$$

i.e. $\dfrac{dy}{dx} = 2x^2 + 8x^{-3} - \dfrac{1}{2}x^{-2} - \dfrac{1}{2}x^{-\frac{1}{2}}$

$$\frac{d^2 y}{dx^2} = 4x + (8)(-3x^{-4}) - \left(\frac{1}{2}\right)(-2x^{-3})$$
$$- \left(\frac{1}{2}\right)\left(-\frac{1}{2}x^{-\frac{3}{2}}\right)$$

$$= 4x - 24x^{-4} + 1x^{-3} + \frac{1}{4}x^{-\frac{3}{2}}$$

i.e. $\dfrac{d^2 y}{dx^2} = 4x - \dfrac{24}{x^4} + \dfrac{1}{x^3} + \dfrac{1}{4\sqrt{x^3}}$

Now try the following Practice Exercise

Practice Exercise 137   Successive differentiation (answers on page 437)

1.   If $y = 3x^4 + 2x^3 - 3x + 2$, find   (a) $\dfrac{d^2 y}{dx^2}$
     (b) $\dfrac{d^3 y}{dx^3}$

2.   If $y = 4x^2 + \dfrac{1}{x}$ find $\dfrac{d^2 y}{dx^2}$

3.   (a)   Given $f(t) = \dfrac{2}{5}t^2 - \dfrac{1}{t^3} + \dfrac{3}{t} - \sqrt{t} + 1$, determine $f''(t)$
     (b)   Evaluate $f''(t)$ in part (a) when $t = 1$

4.   If $y = 3\sin 2t + \cos t$, find $\dfrac{d^2 y}{dx^2}$

5.   If $f(\theta) = 2\ln 4\theta$, show that $f''(\theta) = -\dfrac{2}{\theta^2}$

34.10   Rates of change

If a quantity y depends on and varies with a quantity x then the rate of change of y with respect to x is $\dfrac{dy}{dx}$.

Thus, for example, the rate of change of pressure p with height h is $\dfrac{dp}{dh}$

A rate of change with respect to time is usually just called 'the rate of change', the 'with respect to time' being assumed. Thus, for example, a rate of change of current, i, is $\dfrac{di}{dt}$ and a rate of change of temperature, θ, is $\dfrac{d\theta}{dt}$, and so on.

Here are some worked problems to demonstrate practical examples of rates of change.

Problem 26.   The length L metres of a certain metal rod at temperature $t°C$ is given by $L = 1 + 0.00003t + 0.0000004t^2$. Determine the rate of change of length, in mm/°C, when the temperature is (a) 100°C   (b) 250°C

The rate of change of length means $\dfrac{dL}{dt}$

Since length $L = 1 + 0.00003t + 0.0000004t^2$, then

$$\frac{dL}{dt} = 0.00003 + 0.0000008t.$$

(a) **When $t = 100°C$,**

$$\frac{dL}{dt} = 0.00003 + (0.0000008)(100)$$

$$= 0.00011 \, \text{m/}°\text{C} = \mathbf{0.11\,mm/°C}.$$

(b) **When $t = 250°C$,**

$$\frac{dL}{dt} = 0.00003 + (0.0000008)(250)$$

$$= 0.00023 \, \text{m/}°\text{C} = \mathbf{0.23\,mm/°C}.$$

Problem 27. The luminous intensity I candelas of a lamp at varying voltage V is given by $I = 5 \times 10^{-4} V^2$. Determine the voltage at which the light is increasing at a rate of 0.4 candelas per volt

The rate of change of light with respect to voltage is given by $\dfrac{dI}{dV}$

Since $I = 5 \times 10^{-4} V^2$

$$\frac{dI}{dV} = (5 \times 10^{-4})(2V) = 10 \times 10^{-4} V = 10^{-3} V$$

When the light is increasing at 0.4 candelas per volt then

$$+0.4 = 10^{-3} V, \text{ from which}$$

$$\textbf{voltage}, V = \frac{0.4}{10^{-3}} = 0.4 \times 10^{+3} = \mathbf{400\,volts}$$

Problem 28. Newton's law of cooling is given by $\theta = \theta_0 e^{-kt}$, where the excess of temperature at zero time is $\theta_0 \,°\text{C}$ and at time t seconds is $\theta\,°\text{C}$. Determine the rate of change of temperature after 50 s, given that $\theta_0 = 15°C$ and $k = -0.02$

The rate of change of temperature is $\dfrac{d\theta}{dt}$

Since $\theta = \theta_0 e^{-kt}$, then $\dfrac{d\theta}{dt} = (\theta_0)(-ke^{-kt})$

$$= -k\theta_0 e^{-kt}$$

When $\theta_0 = 15, k = -0.02$ and $t = 50$, then

$$\frac{d\theta}{dt} = -(-0.02)(15)e^{-(-0.02)(50)}$$

$$= 0.30 \, e^1 = \mathbf{0.815°C/s}$$

Problem 29. The pressure p of the atmosphere at height h above ground level is given by $p = p_0 e^{-h/c}$, where p_0 is the pressure at ground level and c is a constant. Determine the rate of change of pressure with height when $p_0 = 10^5$ pascals and $c = 6.2 \times 10^4$ at 1550 metres

The rate of change of pressure with height is $\dfrac{dp}{dh}$

Since $p = p_0 e^{-h/c}$, then

$$\frac{dp}{dh} = (p_0)\left(-\frac{1}{c}e^{-h/c}\right) = -\frac{p_0}{c}e^{-h/c}$$

When $p_0 = 10^5, c = 6.2 \times 10^4$ and $h = 1550$, then

rate of change of pressure,

$$\frac{dp}{dh} = -\frac{10^5}{6.2 \times 10^4}e^{-(1550/6.2 \times 10^4)}$$

$$= -\frac{10}{6.2}e^{-0.025} = \mathbf{-1.573\,Pa/m}$$

Now try the following Practice Exercise

Practice Exercise 138   Rates of change
(answers on page 437)

1. An alternating current, i amperes, is given by $i = 10\sin 2\pi ft$, where f is the frequency in hertz and t is the time in seconds. Determine the rate of change of current when $t = 12\,\text{ms}$, given that $f = 50\,\text{Hz}$.

2. The luminous intensity, I candelas, of a lamp is given by $I = 8 \times 10^{-4} V^2$, where V is the voltage. Find

   (a)   the rate of change of luminous intensity with voltage when $V = 100$ volts

(b) the voltage at which the light is increasing at a rate of 0.5 candelas per volt.

3. The voltage across the plates of a capacitor at any time t seconds is given by $v = Ve^{-t/CR}$, where V, C and R are constants. Given $V = 200\,\text{V}, C = 0.10\,\mu\text{F}$ and $R = 2\,\text{M}\Omega$, find

(a) the initial rate of change of voltage

(b) the rate of change of voltage after 0.2 s

4. The pressure p of the atmosphere at height h above ground level is given by $p = p_0 e^{-h/c}$, where p_0 is the pressure at ground level and c is a constant. Determine the rate of change of pressure with height when $p_0 = 1.013 \times 10^5$ pascals and $c = 6.05 \times 10^4$ at 1450 metres.

For fully worked solutions to each of the problems in Practice Exercises 131 to 138 in this chapter, go to the website:
www.routledge.com/cw/bird

Chapter 35

Introduction to integration

Why it is important to understand: Introduction to integration

Engineering is all about problem solving and many problems in engineering can be solved using calculus. Physicists, chemists, engineers and many other scientific and technical specialists use calculus in their everyday work; it is a technique of fundamental importance. Both integration and differentiation have numerous applications in engineering and science and some typical examples include determining areas, mean and rms values, volumes of solids of revolution, centroids, second moments of area, differential equations and Fourier series. Standard integrals are introduced in this chapter, together with one application – finding the area under a curve. For any further studies in engineering, differential and integral calculus are unavoidable.

At the end of this chapter, you should be able to:

- understand that integration is the reverse process of differentiation
- determine integrals of the form ax^n where n is fractional, zero, or a positive or negative integer
- integrate standard functions – $\cos ax$, $\sin ax$, e^{ax}, $\dfrac{1}{x}$
- evaluate definite integrals
- determine the area under a curve

35.1 The process of integration

The process of integration reverses the process of differentiation. In differentiation, if $f(x) = 2x^2$ then $f'(x) = 4x$. Thus, the integral of $4x$ is $2x^2$; i.e. integration is the process of moving from $f'(x)$ to $f(x)$. By similar reasoning, the integral of $2t$ is t^2

Integration is a process of summation or adding parts together and an elongated S, shown as $\int$, is used to replace the words 'the integral of'. Hence, from above, $\int 4x = 2x^2$ and $\int 2t$ is t^2

In differentiation, the differential coefficient $\dfrac{dy}{dx}$ indicates that a function of x is being differentiated with respect to x, the dx indicating that it is 'with respect to x'

In integration the variable of integration is shown by adding d (the variable) after the function to be integrated. Thus,

$$\int 4x\,dx \text{ means 'the integral of } 4x \text{ with respect to } x\text{'},$$

and $\int 2t\,dt$ means 'the integral of $2t$ with respect to t'

As stated above, the differential coefficient of $2x^2$ is $4x$, hence; $\int 4x\,dx = 2x^2$. However, the differential coefficient of $2x^2 + 7$ is also $4x$. Hence, $\int 4x\,dx$ could also be equal to $2x^2 + 7$. To allow for the possible presence of a constant, whenever the process of integration is performed a constant c is added to the result. Thus,

$$\int 4x\,dx = 2x^2 + c \quad \text{and} \quad \int 2t\,dt = t^2 + c$$

c is called the **arbitrary constant of integration**.

35.2   The general solution of integrals of the form ax^n

The general solution of integrals of the form $\int ax^n\,dx$, where a and n are constants and $n \neq -1$ is given by

$$\int ax^n\,dx = \frac{ax^{n+1}}{n+1} + c$$

Using this rule gives

(i) $\displaystyle\int 3x^4\,dx = \frac{3x^{4+1}}{4+1} + c = \frac{3}{5}x^5 + c$

(ii) $\displaystyle\int \frac{4}{9}t^3\,dt\,dx = \frac{4}{9}\left(\frac{t^{3+1}}{3+1}\right) + c = \frac{4}{9}\left(\frac{t^4}{4}\right) + c$

$$= \frac{1}{9}t^4 + c$$

(iii) $\displaystyle\int \frac{2}{x^2}\,dx = \int 2x^{-2}\,dx = \frac{2x^{-2+1}}{-2+1} + c$

$$= \frac{2x^{-1}}{-1} + c = -\frac{2}{x} + c$$

(iv) $\displaystyle\int \sqrt{x}\,dx = \int x^{\frac{1}{2}}\,dx = \frac{x^{\frac{1}{2}+1}}{\frac{1}{2}+1} + c = \frac{x^{\frac{3}{2}}}{\frac{3}{2}} + c$

$$= \frac{2}{3}\sqrt{x^3} + c$$

Each of these results may be checked by differentiation.

(a) The integral of a constant k is $kx + c$. For example,

$$\int 8\,dx = 8x + c \quad \text{and} \quad \int 5\,dt = 5t + c$$

(b) When a sum of several terms is integrated the result is the sum of the integrals of the separate terms. For example,

$$\int (3x + 2x^2 - 5)\,dx$$

$$= \int 3x\,dx + \int 2x^2\,dx - \int 5\,dx$$

$$= \frac{3x^2}{2} + \frac{2x^3}{3} - 5x + c$$

35.3   Standard integrals

From Chapter 34, $\dfrac{d}{dx}(\sin ax) = a\cos ax$. Since integration is the reverse process of differentiation, it follows that

$$\int a\cos ax\,dx = \sin ax + c$$

or $\displaystyle\int \cos ax\,dx = \frac{1}{a}\sin ax + c$

By similar reasoning

$$\int \sin ax\,dx = -\frac{1}{a}\cos ax + c$$

$$\int e^{ax}\,dx = \frac{1}{a}e^{ax} + c$$

and $\displaystyle\int \frac{1}{x}\,dx = \ln x + c$

From above, $\displaystyle\int ax^n\,dx = \frac{ax^{n+1}}{n+1} + c$ except when

$$n = -1$$

When $n = -1$, $\displaystyle\int x^{-1}\,dx = \int \frac{1}{x}\,dx = \ln x + c$

A list of **standard integrals** is summarised in Table 35.1.

Table 35.1 Standard integrals

y	$\int y\,dx$
1. $\int ax^n$	$\dfrac{ax^{n+1}}{n+1} + c$ (except when $n = -1$)
2. $\int \cos ax\,dx$	$\dfrac{1}{a}\sin ax + c$
3. $\int \sin ax\,dx$	$-\dfrac{1}{a}\cos ax + c$
4. $\int e^{ax}\,dx$	$\dfrac{1}{a}e^{ax} + c$
5. $\int \dfrac{1}{x}\,dx$	$\ln x + c$

Problem 1.   Determine $\int 7x^2\,dx$

The standard integral, $\displaystyle\int ax^n\,dx = \frac{ax^{n+1}}{n+1} + c$

When $a = 7$ and $n = 2$,

$$\int 7x^2\,dx = \frac{7x^{2+1}}{2+1} + c = \frac{7x^3}{3} + c \quad \text{or} \quad \frac{7}{3}x^3 + c$$

Problem 2. Determine $\int 2t^3\,dt$

When $a = 2$ and $n = 3$,

$$\int 2t^3\,dt = \frac{2t^{3+1}}{3+1} + c = \frac{2t^4}{4} + c = \frac{1}{2}t^4 + c$$

Note that each of the results in worked examples 1 and 2 may be checked by differentiating them.

Problem 3. Determine $\int 8\,dx$

$\int 8\,dx$ is the same as $\int 8x^0\,dx$ and, using the general rule when $a = 8$ and $n = 0$, gives

$$\int 8x^0\,dx = \frac{8x^{0+1}}{0+1} + c = 8x + c$$

In general, if k is a constant then $\int k\,dx = kx + c$.

Problem 4. Determine $\int 2x\,dx$

When $a = 2$ and $n = 1$,

$$\int 2x\,dx = \int 2x^1\,dx = \frac{2x^{1+1}}{1+1} + c = \frac{2x^2}{2} + c$$
$$= x^2 + c$$

Problem 5. Determine $\int \left(3 + \frac{2}{5}x - 6x^2\right) dx$

$\int \left(3 + \frac{2}{5}x - 6x^2\right) dx$ may be written as

$\int 3\,dx + \int \frac{2}{5}x\,dx - \int 6x^2\,dx$

i.e. each term is integrated separately. (This splitting up of terms only applies, however, for addition and subtraction.) Hence,

$$\int \left(3 + \frac{2}{5}x - 6x^2\right) dx$$
$$= 3x + \left(\frac{2}{5}\right)\frac{x^{1+1}}{1+1} - (6)\frac{x^{2+1}}{2+1} + c$$
$$= 3x + \left(\frac{2}{5}\right)\frac{x^2}{2} - (6)\frac{x^3}{3} + c = 3x + \frac{1}{5}x^2 - 2x^3 + c$$

Note that when an integral contains more than one term there is no need to have an arbitrary constant for each; just a single constant c at the end is sufficient.

Problem 6. Determine $\int \left(\frac{2x^3 - 3x}{4x}\right) dx$

Rearranging into standard integral form gives

$$\int \left(\frac{2x^3 - 3x}{4x}\right) dx = \int \left(\frac{2x^3}{4x} - \frac{3x}{4x}\right) dx$$
$$= \int \left(\frac{1}{2}x^2 - \frac{3}{4}\right) dx = \left(\frac{1}{2}\right)\frac{x^{2+1}}{2+1} - \frac{3}{4}x + c$$
$$= \left(\frac{1}{2}\right)\frac{x^3}{3} - \frac{3}{4}x + c = \frac{1}{6}x^3 - \frac{3}{4}x + c$$

Problem 7. Determine $\int (1-t)^2\,dt$

Rearranging $\int (1-t)^2\,dt$ gives

$$\int (1 - 2t + t^2)\,dt = t - \frac{2t^{1+1}}{1+1} + \frac{t^{2+1}}{2+1} + c$$
$$= t - \frac{2t^2}{2} + \frac{t^3}{3} + c$$
$$= t - t^2 + \frac{1}{3}t^3 + c$$

This example shows that functions often have to be rearranged into the standard form of $\int ax^n\,dx$ before it is possible to integrate them.

Problem 8. Determine $\int \frac{5}{x^2}\,dx$

$$\int \frac{5}{x^2}\,dx = \int 5x^{-2}\,dx$$

Using the standard integral, $\int ax^n\,dx$, when $a = 5$ and $n = -2$, gives

$$\int 5x^{-2}\,dx = \frac{5x^{-2+1}}{-2+1} + c = \frac{5x^{-1}}{-1} + c$$
$$= -5x^{-1} + c = -\frac{5}{x} + c$$

Problem 9. Determine $\int 3\sqrt{x}\,dx$

For fractional powers it is necessary to appreciate $\sqrt[n]{a^m} = a^{\frac{m}{n}}$

$$\int 3\sqrt{x}\,dx = \int 3x^{\frac{1}{2}}\,dx = \frac{3x^{\frac{1}{2}+1}}{\frac{1}{2}+1} + c = \frac{3x^{\frac{3}{2}}}{\frac{3}{2}} + c$$

$$= 2x^{\frac{3}{2}} + c = \mathbf{2\sqrt{x^3} + c}$$

Problem 10. Determine $\displaystyle\int \frac{-5}{9\sqrt[4]{t^3}}\,dt$

$$\int \frac{-5}{9\sqrt[4]{t^3}}\,dt = \int \frac{-5}{9t^{\frac{3}{4}}}\,dt = \int \left(-\frac{5}{9}\right)t^{-\frac{3}{4}}\,dt$$

$$= \left(-\frac{5}{9}\right)\frac{t^{-\frac{3}{4}+1}}{-\frac{3}{4}+1} + c$$

$$= \left(-\frac{5}{9}\right)\frac{t^{\frac{1}{4}}}{\frac{1}{4}} + c = \left(-\frac{5}{9}\right)\left(\frac{4}{1}\right)t^{\frac{1}{4}} + c$$

$$= -\frac{20}{9}\sqrt[4]{t} + c$$

Problem 11. Determine $\int 4\cos 3x\,dx$

From 2 of Table 35.1,

$$\int 4\cos 3x\,dx = (4)\left(\frac{1}{3}\right)\sin 3x + c$$

$$= \frac{4}{3}\sin 3x + c$$

Problem 12. Determine $\int 5\sin 2\theta\,d\theta$

From 3 of Table 35.1,

$$\int 5\sin 2\theta\,d\theta = (5)\left(-\frac{1}{2}\right)\cos 2\theta + c$$

$$= -\frac{5}{2}\cos 2\theta + c$$

Problem 13. Determine $\int 5e^{3x}\,dx$

From 4 of Table 35.1,

$$\int 5e^{3x}\,dx = (5)\left(\frac{1}{3}\right)e^{3x} + c$$

$$= \frac{5}{3}e^{3x} + c$$

Problem 14. Determine $\displaystyle\int \frac{2}{3e^{4t}}\,dt$

$$\int \frac{2}{3e^{4t}}\,dt = \int \frac{2}{3}e^{-4t}\,dt$$

$$= \left(\frac{2}{3}\right)\left(-\frac{1}{4}\right)e^{-4t} + c$$

$$= -\frac{1}{6}e^{-4t} + c = -\frac{1}{6e^{4t}} + c$$

Problem 15. Determine $\displaystyle\int \frac{3}{5x}\,dx$

From 5 of Table 35.1,

$$\int \frac{3}{5x}\,dx = \int \left(\frac{3}{5}\right)\left(\frac{1}{x}\right)\,dx$$

$$= \frac{3}{5}\ln x + c$$

Problem 16. Determine $\displaystyle\int \left(\frac{2x^2 + 1}{x}\right)\,dx$

$$\int \left(\frac{2x^2 + 1}{x}\right)\,dx = \int \left(\frac{2x^2}{x} + \frac{1}{x}\right)\,dx$$

$$= \int \left(2x + \frac{1}{x}\right)\,dx = \frac{2x^2}{2} + \ln x + c$$

$$= x^2 + \ln x + c$$

Now try the following Practice Exercise

Practice Exercise 139   Standard integrals (answers on page 437)

Determine the following integrals.

1. (a) $\int 4\,dx$ (b) $\int 7x\,dx$

2. (a) $\int 5x^3\,dx$ (b) $\int 3t^7\,dt$

3. (a) $\int \frac{2}{5}x^2\,dx$ (b) $\int \frac{5}{6}x^3\,dx$

4. (a) $\int (2x^4 - 3x)\,dx$ (b) $\int (2 - 3t^3)\,dt$

5. (a) $\int \left(\frac{3x^2 - 5x}{x}\right)\,dx$ (b) $\int (2+\theta)^2\,d\theta]$

6. (a) $\int (2+\theta)(3\theta - 1)\, d\theta$

    (b) $\int (3x - 2)(x^2 + 1)\, dx$

7. (a) $\int \dfrac{4}{3x^2}\, dx$ (b) $\int \dfrac{3}{4x^4}\, dx$

8. (a) $2\int \sqrt{x^3}\, dx$ (b) $\int \dfrac{1}{4} \sqrt[4]{x^5}\, dx$

9. (a) $\int \dfrac{-5}{\sqrt{t^3}}\, dt$ (b) $\int \dfrac{3}{7\sqrt[5]{x^4}}\, dx$

10. (a) $\int 3\cos 2x\, dx$ (b) $\int 7\sin 3\theta\, d\theta$

11. (a) $\int 3\sin \dfrac{1}{2}x\, dx$ (b) $\int 6\cos \dfrac{1}{3}x\, dx$

12. (a) $\int \dfrac{3}{4} e^{2x}\, dx$ (b) $\dfrac{2}{3}\int \dfrac{dx}{e^{5x}}$

13. (a) $\int \dfrac{2}{3x}\, dx$ (b) $\int \left(\dfrac{u^2 - 1}{u}\right) du$

14. (a) $\int \dfrac{(2+3x)^2}{\sqrt{x}}\, dx$ (b) $\int \left(\dfrac{1}{t} + 2t\right)^2 dt$

35.4   Definite integrals

Integrals containing an arbitrary constant c in their results are called **indefinite integrals** since their precise value cannot be determined without further information. **Definite integrals** are those in which limits are applied. If an expression is written as $[x]_a^b$, b is called the **upper limit** and a the **lower limit**. The operation of applying the limits is defined as $[x]_a^b = (b) - (a)$

For example, the increase in the value of the integral x^2 as x increases from 1 to 3 is written as $\int_1^3 x^2 dx$. Applying the limits gives

$$\int_1^3 x^2 dx = \left[\dfrac{x^3}{3} + c\right]_1^3 = \left(\dfrac{3^3}{3} + c\right) - \left(\dfrac{1^3}{3} + c\right)$$

$$= (9 + c) - \left(\dfrac{1}{3} + c\right) = 8\dfrac{2}{3}$$

Note that the c term always cancels out when limits are applied and it need not be shown with definite integrals.

Problem 17.   Evaluate $\displaystyle\int_1^2 3x\, dx$

$$\int_1^2 3x\, dx = \left[\dfrac{3x^2}{2}\right]_1^2 = \left\{\dfrac{3}{2}(2)^2\right\} - \left\{\dfrac{3}{2}(1)^2\right\}$$

$$= 6 - 1\dfrac{1}{2} = \mathbf{4\dfrac{1}{2}}$$

Problem 18.   Evaluate $\displaystyle\int_{-2}^3 (4 - x^2)\, dx$

$$\int_{-2}^3 (4 - x^2)\, dx = \left[4x - \dfrac{x^3}{3}\right]_{-2}^3$$

$$= \left\{4(3) - \dfrac{(3)^3}{3}\right\} - \left\{4(-2) - \dfrac{(-2)^3}{3}\right\}$$

$$= \{12 - 9\} - \left\{-8 - \dfrac{-8}{3}\right\}$$

$$= \{3\} - \left\{-5\dfrac{1}{3}\right\} = \mathbf{8\dfrac{1}{3}}$$

Problem 19.   Evaluate $\displaystyle\int_0^2 x(3 + 2x)\, dx$

$$\int_0^2 x(3 + 2x)\, dx = \int_0^2 (3x + 2x^2)\, dx = \left[\dfrac{3x^2}{2} + \dfrac{2x^3}{3}\right]_0^2$$

$$= \left\{\dfrac{3(2)^2}{2} + \dfrac{2(2)^3}{3}\right\} - \{0 + 0\}$$

$$= 6 + \dfrac{16}{3} = 11\dfrac{1}{3} \text{ or } \mathbf{11.33}$$

Problem 20.   Evaluate $\displaystyle\int_{-1}^1 \left(\dfrac{x^4 - 5x^2 + x}{x}\right) dx$

$$\int_{-1}^1 \left(\dfrac{x^4 - 5x^2 + x}{x}\right) dx$$

$$= \int_{-1}^1 \left(\dfrac{x^4}{x} - \dfrac{5x^2}{x} + \dfrac{x}{x}\right) dx$$

$$= \int_{-1}^1 \left(x^3 - 5x + 1\right) dx = \left[\dfrac{x^4}{4} - \dfrac{5x^2}{2} + x\right]_{-1}^1$$

$$= \left\{\frac{1}{4} - \frac{5}{2} + 1\right\} - \left\{\frac{(-1)^4}{4} - \frac{5(-1)^2}{2} + (-1)\right\}$$

$$= \left\{\frac{1}{4} - \frac{5}{2} + 1\right\} - \left\{\frac{1}{4} - \frac{5}{2} - 1\right\} = \mathbf{2}$$

Problem 21. Evaluate $\int_1^2 \left(\frac{1}{x^2} + \frac{2}{x}\right) dx$ correct to 3 decimal places

$$\int_1^2 \left(\frac{1}{x^2} + \frac{2}{x}\right) dx$$

$$= \int_1^2 \left\{x^{-2} + 2\left(\frac{1}{x}\right)\right\} dx = \left[\frac{x^{-2+1}}{-2+1} + 2\ln x\right]_1^2$$

$$= \left[\frac{x^{-1}}{-1} + 2\ln x\right]_1^2 = \left[-\frac{1}{x} + 2\ln x\right]_1^2$$

$$= \left(-\frac{1}{2} + 2\ln 2\right) - \left(-\frac{1}{1} + 2\ln 1\right) = \mathbf{1.886}$$

Problem 22. Evaluate $\int_0^{\pi/2} 3\sin 2x\, dx$

$$\int_0^{\pi/2} 3\sin 2x\, dx$$

$$= \left[(3)\left(-\frac{1}{2}\right)\cos 2x\right]_0^{\pi/2} = \left[-\frac{3}{2}\cos 2x\right]_0^{\pi/2}$$

$$= \left\{-\frac{3}{2}\cos 2\left(\frac{\pi}{2}\right)\right\} - \left\{-\frac{3}{2}\cos 2(0)\right\}$$

$$= \left\{-\frac{3}{2}\cos \pi\right\} - \left\{-\frac{3}{2}\cos 0\right\}$$

$$= \left\{-\frac{3}{2}(-1)\right\} - \left\{-\frac{3}{2}(1)\right\}$$

$$= \frac{3}{2} + \frac{3}{2} = \mathbf{3}$$

Problem 23. Evaluate $\int_1^2 4\cos 3t\, dt$

$$\int_1^2 4\cos 3t\, dt = \left[(4)\left(\frac{1}{3}\right)\sin 3t\right]_1^2 = \left[\frac{4}{3}\sin 3t\right]_1^2$$

$$= \left\{\frac{4}{3}\sin 6\right\} - \left\{\frac{4}{3}\sin 3\right\}$$

Note that limits of trigonometric functions are always expressed in **radians** – thus, for example, sin 6 means the sine of 6 radians $= -0.279415\ldots$ Hence,

$$\int_1^2 4\cos 3t\, dt = \left\{\frac{4}{3}(-0.279415\ldots)\right\} - \left\{\frac{4}{3}(0.141120\ldots)\right\}$$

$$= (-0.37255) - (0.18816)$$

$$= \mathbf{-0.5607}$$

Problem 24. Evaluate $\int_1^2 4e^{2x}\, dx$ correct to 4 significant figures

$$\int_1^2 4e^{2x}\, dx = \left[\frac{4}{2}e^{2x}\right]_1^2 = 2\left[e^{2x}\right]_1^2 = 2[e^4 - e^2]$$

$$= 2[54.5982 - 7.3891]$$

$$= \mathbf{94.42}$$

Problem 25. Evaluate $\int_1^4 \frac{3}{4u}\, du$ correct to 4 significant figures

$$\int_1^4 \frac{3}{4u}\, du = \left[\frac{3}{4}\ln u\right]_1^4 = \frac{3}{4}[\ln 4 - \ln 1]$$

$$= \frac{3}{4}[1.3863 - 0] = \mathbf{1.040}$$

Now try the following Practice Exercise

Practice Exercise 140   Definite integrals (answers on page 437)

In Problems 1 to 10, evaluate the definite integrals (where necessary, correct to 4 significant figures).

1. (a) $\int_1^2 x\, dx$ (b) $\int_1^2 (x-1)\, dx$

2. (a) $\int_1^4 5x^2 dx$ (b) $\int_{-1}^1 -\frac{3}{4}t^2\, dt$

3. (a) $\int_{-1}^2 (3-x^2)\, dx$ (b) $\int_1^3 (x^2-4x+3)\, dx$

4. (a) $\int_1^2 (x^3-3x)\, dx$ (b) $\int_1^2 (x^2-3x+3)\, dx$

5. (a) $\int_0^4 2\sqrt{x}\, dx$ (b) $\int_2^3 \frac{1}{x^2}\, dx$

6. (a) $\displaystyle\int_0^\pi \frac{3}{2}\cos\theta\,d\theta$ (b) $\displaystyle\int_0^{\pi/2} 4\cos\theta\,d\theta$

7. (a) $\displaystyle\int_{\pi/6}^{\pi/3} 2\sin 2\theta\,d\theta$ (b) $\displaystyle\int_0^2 3\sin t\,dt$

8. (a) $\displaystyle\int_0^1 5\cos 3x\,dx$

    (b) $\displaystyle\int_{\pi/4}^{\pi/2}(3\sin 2x - 2\cos 3x)\,dx$

9. (a) $\displaystyle\int_0^1 3e^{3t}\,dt$ (b) $\displaystyle\int_{-1}^2 \frac{2}{3e^{2x}}\,dx$

10. (a) $\displaystyle\int_2^3 \frac{2}{3x}\,dx$ (b) $\displaystyle\int_1^3 \frac{2x^2+1}{x}\,dx$

11. The volume of liquid in a tank is given by:
$v = \int_{t_1}^{t_2} q\,dt$.
Determine the volume of a chemical, given $q = (5 - 0.05t + 0.003t^2)$ m^3/s, $t_1 = 0$ and $t_2 = 16$ s

35.5 The area under a curve

The area shown shaded in Fig. 35.1 may be determined using approximate methods such as the trapezoidal rule, the mid-ordinate rule or Simpson's rule (see Chapter 28) or, more precisely, by using integration.

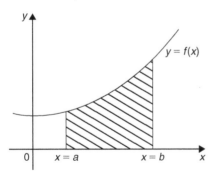

Figure 35.1

The shaded area in Fig. 35.1 is given by

$$\text{shaded area} = \int_a^b y\,dx = \int_a^b f(x)\,dx$$

Thus, determining the area under a curve by integration merely involves evaluating a definite integral, as shown in Section 35.4.

There are several instances in engineering and science where the area beneath a curve needs to be accurately determined. For example, the areas between the limits of a

     velocity/time graph gives distance travelled,

     force/distance graph gives work done,

     voltage/current graph gives power, and so on.

Should a curve drop below the x-axis then $y(=f(x))$ becomes negative and $\int f(x)\,dx$ is negative. When determining such areas by integration, a negative sign is placed before the integral. For the curve shown in Fig. 35.2, the total shaded area is given by (area $E +$ area $F +$ area G).

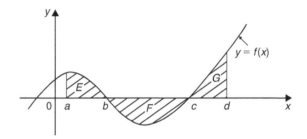

Figure 35.2

By integration,

$$\text{total shaded area} = \int_a^b f(x)\,dx - \int_b^c f(x)\,dx$$
$$+ \int_c^d f(x)\,dx$$

(Note that this is **not** the same as $\displaystyle\int_a^d f(x)\,dx$)

It is usually necessary to sketch a curve in order to check whether it crosses the x-axis.

> **Problem 26.** Determine the area enclosed by $y = 2x + 3$, the x-axis and ordinates $x = 1$ and $x = 4$

$y = 2x + 3$ is a straight line graph as shown in Fig. 35.3, in which the required area is shown shaded.
By integration,

$$\text{shaded area} = \int_1^4 y\,dx = \int_1^4 (2x+3)\,dx = \left[\frac{2x^2}{2} + 3x\right]_1^4$$

$$= [(16 + 12) - (1 + 3)] = \mathbf{24\ square\ units}$$

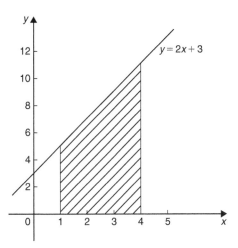

Figure 35.3

(This answer may be checked since the shaded area is a trapezium: area of trapezium $= \dfrac{1}{2}$ (sum of parallel sides)(perpendicular distance between parallel sides) $= \dfrac{1}{2}(5+11)(3) = $ **24 square units**.)

Problem 27. The velocity v of a body t seconds after a certain instant is given by $v = (2t^2 + 5)\,\text{m/s}$. Find by integration how far it moves in the interval from $t = 0$ to $t = 4\,\text{s}$

Since $2t^2 + 5$ is a quadratic expression, the curve $v = 2t^2 + 5$ is a parabola cutting the v-axis at $v = 5$, as shown in Fig. 35.4.

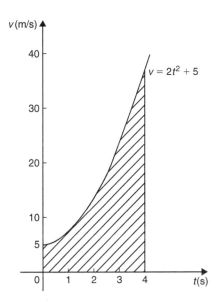

Figure 35.4

The distance travelled is given by the area under the v/t curve (shown shaded in Fig. 35.4). By integration,

$$\text{shaded area} = \int_0^4 v\,dt = \int_0^4 (2t^2 + 5)\,dt = \left[\frac{2t^3}{3} + 5t \right]_0^4$$

$$= \left(\frac{2(4^3)}{3} + 5(4) \right) - (0)$$

i.e. **distance travelled = 62.67 m**

Problem 28. Sketch the graph $y = x^3 + 2x^2 - 5x - 6$ between $x = -3$ and $x = 2$ and determine the area enclosed by the curve and the x-axis

A table of values is produced and the graph sketched as shown in Fig. 35.5, in which the area enclosed by the curve and the x-axis is shown shaded.

x	−3	−2	−1	0	1	2
y	0	4	0	−6	−8	0

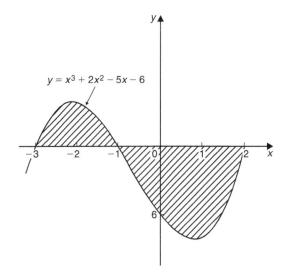

Figure 35.5

Shaded area $= \displaystyle\int_{-3}^{-1} y\,dx - \int_{-1}^{2} y\,dx$, the minus sign before the second integral being necessary since the enclosed area is below the x-axis. Hence,

shaded area $= \int_{-3}^{-1} (x^3 + 2x^2 - 5x - 6)\, dx$

$$- \int_{-1}^{2} (x^3 + 2x^2 - 5x - 6)\, dx$$

$$= \left[\frac{x^4}{4} + \frac{2x^3}{3} - \frac{5x^2}{2} - 6x \right]_{-3}^{-1}$$

$$- \left[\frac{x^4}{4} + \frac{2x^3}{3} - \frac{5x^2}{2} - 6x \right]_{-1}^{2}$$

$$= \left[\left\{ \frac{1}{4} - \frac{2}{3} - \frac{5}{2} + 6 \right\} - \left\{ \frac{81}{4} - 18 - \frac{45}{2} + 18 \right\} \right]$$

$$- \left[\left\{ 4 + \frac{16}{3} - 10 - 12 \right\} - \left\{ \frac{1}{4} - \frac{2}{3} - \frac{5}{2} + 6 \right\} \right]$$

$$= \left[\left\{ 3\frac{1}{12} \right\} - \left\{ -2\frac{1}{4} \right\} \right] - \left[\left\{ -12\frac{2}{3} \right\} - \left\{ 3\frac{1}{12} \right\} \right]$$

$$= \left[5\frac{1}{3} \right] - \left[-15\frac{3}{4} \right]$$

$$= 21\frac{1}{12} \quad \text{or} \quad \textbf{21.08 square units}$$

Problem 29. Determine the area enclosed by the curve $y = 3x^2 + 4$, the x-axis and ordinates $x = 1$ and $x = 4$ by (a) the trapezoidal rule, (b) the mid-ordinate rule, (c) Simpson's rule and (d) integration.

The curve $y = 3x^2 + 4$ is shown plotted in Fig. 35.6. The trapezoidal rule, the mid-ordinate rule and Simpson's rule are discussed in Chapter 28, page 291.

(a) **By the trapezoidal rule**

$$\text{area} = \left(\begin{array}{c} \text{width of} \\ \text{interval} \end{array} \right) \left[\frac{1}{2} \left(\begin{array}{c} \text{first} + \text{last} \\ \text{ordinate} \end{array} \right) \right.$$

$$\left. + \begin{array}{c} \text{sum of} \\ \text{remaining} \\ \text{ordinates} \end{array} \right]$$

Selecting six intervals each of width 0.5 gives

$$\text{area} = (0.5) \left[\frac{1}{2}(7 + 52) + 10.75 + 16 \right.$$

$$\left. + 22.75 + 31 + 40.75 \right]$$

$$= \textbf{75.375 square units}$$

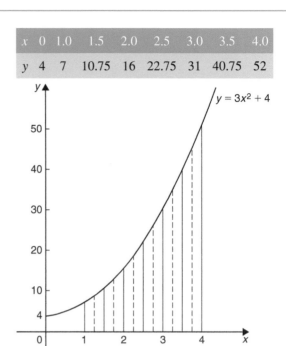

x	0	1.0	1.5	2.0	2.5	3.0	3.5	4.0
y	4	7	10.75	16	22.75	31	40.75	52

Figure 35.6

(b) **By the mid-ordinate rule**

$$\text{area} = (\text{width of interval})(\text{sum of mid-ordinates})$$

Selecting six intervals, each of width 0.5, gives the mid-ordinates as shown by the broken lines in Fig. 35.6. Thus,

$$\text{area} = (0.5)(8.7 + 13.2 + 19.2 + 26.7$$

$$+ 35.7 + 46.2)$$

$$= \textbf{74.85 square units}$$

(c) **By Simpson's rule***

$$\text{area} = \frac{1}{3} \left(\begin{array}{c} \text{width of} \\ \text{interval} \end{array} \right) \left[\left(\begin{array}{c} \text{first} + \text{last} \\ \text{ordinates} \end{array} \right) \right.$$

$$+ 4 \left(\begin{array}{c} \text{sum of even} \\ \text{ordinates} \end{array} \right)$$

$$\left. + 2 \left(\begin{array}{c} \text{sum of remaining} \\ \text{odd ordinates} \end{array} \right) \right]$$

*Who was Simpson? To find out more go to www.routledge.com/cw/bird

Selecting six intervals, each of width 0.5, gives

$$\text{area} = \frac{1}{3}(0.5)[(7+52)+4(10.75+22.75$$

$$+40.75)+2(16+31)]$$

$$= \textbf{75 square units}$$

(d) **By integration**

$$\text{shaded area} = \int_1^4 y\,dx$$

$$= \int_1^4 (3x^2+4)\,dx = \left[x^3+4x\right]_1^4$$

$$= (64+16)-(1+4)$$

$$= \textbf{75 square units}$$

Integration gives the precise value for the area under a curve. In this case, Simpson's rule is seen to be the most accurate of the three approximate methods.

> **Problem 30.** Find the area enclosed by the curve $y = \sin 2x$, the x-axis and the ordinates $x = 0$ and $x = \dfrac{\pi}{3}$

A sketch of $y = \sin 2x$ is shown in Fig. 35.7. (Note that $y = \sin 2x$ has a period of $\dfrac{2\pi}{2}$ i.e. π radians.)

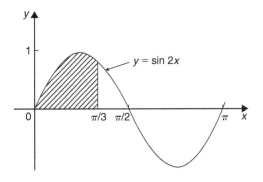

Figure 35.7

$$\text{Shaded area} = \int_0^{\pi/3} y\,dx$$

$$= \int_0^{\pi/3} \sin 2x\,dx = \left[-\frac{1}{2}\cos 2x\right]_0^{\pi/3}$$

$$= \left\{-\frac{1}{2}\cos\frac{2\pi}{3}\right\} - \left\{-\frac{1}{2}\cos 0\right\}$$

$$= \left\{-\frac{1}{2}\left(-\frac{1}{2}\right)\right\} - \left\{-\frac{1}{2}(1)\right\}$$

$$= \frac{1}{4} + \frac{1}{2} = \frac{3}{4} \text{ or } \textbf{0.75 square units}$$

Now try the following Practice Exercise

Practice Exercise 141   Area under curves (answers on page 437)

Unless otherwise stated all answers are in square units.

1. Show by integration that the area of a rectangle formed by the line $y = 4$, the ordinates $x = 1$ and $x = 6$ and the x-axis is 20 square units.

2. Show by integration that the area of the triangle formed by the line $y = 2x$, the ordinates $x = 0$ and $x = 4$ and the x-axis is 16 square units.

3. Sketch the curve $y = 3x^2 + 1$ between $x = -2$ and $x = 4$. Determine by integration the area enclosed by the curve, the x-axis and ordinates $x = -1$ and $x = 3$. Use an approximate method to find the area and compare your result with that obtained by integration.

4. The force F newtons acting on a body at a distance x metres from a fixed point is given by $F = 3x + 2x^2$. If work done $= \displaystyle\int_{x_1}^{x_2} F\,dx$, determine the work done when the body moves from the position where $x_1 = 1$ m to that when $x_2 = 3$m.

In Problems 5 to 9, sketch graphs of the given equations and then find the area enclosed between the curves, the horizontal axis and the given ordinates.

5. $y = 5x$; $x = 1, x = 4$

6. $y = 2x^2 - x + 1$; $x = -1, x = 2$

7. $y = 2\sin 2x$; $x = 0, x = \dfrac{\pi}{4}$

8. $y = 5\cos 3t$; $t = 0, t = \dfrac{\pi}{6}$

9. $y = (x - 1)(x - 3)$; $x = 0, x = 3$

10. The velocity v of a vehicle t seconds after a certain instant is given by $v = \left(3t^2 + 4\right)$ m/s. Determine how far it moves in the interval from $t = 1$ s to $t = 5$ s.

For fully worked solutions to each of the problems in Practice Exercises 139 to 141 in this chapter, go to the website:
www.routledge.com/cw/bird

This assignment covers the material contained in Chapters 34 and 35. *The marks available are shown in brackets at the end of each question.*

1. Differentiate the following functions with respect to x.
   (a) $y = 5x^2 - 4x + 9$ (b) $y = x^4 - 3x^2 - 2$
   (4)

2. Given $y = 2(x-1)^2$, find $\dfrac{dy}{dx}$ (3)

3. If $y = \dfrac{3}{x}$ determine $\dfrac{dy}{dx}$ (2)

4. Given $f(t) = \sqrt{t^5}$, find $f'(t)$ (2)

5. Determine the derivative of $y = 5 - 3x + \dfrac{4}{x^2}$ (3)

6. Calculate the gradient of the curve $y = 3\cos\dfrac{x}{3}$ at $x = \dfrac{\pi}{4}$, correct to 3 decimal places. (4)

7. Find the gradient of the curve $f(x) = 7x^2 - 4x + 2$ at the point $(1, 5)$ (3)

8. If $y = 5\sin 3x - 2\cos 4x$ find $\dfrac{dy}{dx}$ (2)

9. Determine the value of the differential coefficient of $y = 5\ln 2x - \dfrac{3}{e^{2x}}$ when $x = 0.8$, correct to 3 significant figures. (4)

10. If $y = 5x^4 - 3x^3 + 2x^2 - 6x + 5$, find (a) $\dfrac{dy}{dx}$
    (b) $\dfrac{d^2y}{dx^2}$ (4)

11. Newton's law of cooling is given by $\theta = \theta_0 e^{-kt}$, where the excess of temperature at zero time is $\theta_0\,^\circ$C and at time t seconds is $\theta\,^\circ$C. Determine the rate of change of temperature after 40 s, correct to 3 decimal places, given that $\theta_0 = 16\,^\circ$C and $k = -0.01$ (4)

In Problems 12 to 15, determine the indefinite integrals.

12. (a) $\displaystyle\int (x^2 + 4)dx$ (b) $\displaystyle\int \dfrac{1}{x^3}dx$ (4)

13. (a) $\displaystyle\int \left(\dfrac{2}{\sqrt{x}} + 3\sqrt{x}\right)dx$ (b) $\displaystyle\int 3\sqrt{t^5}dt$ (4)

14. (a) $\displaystyle\int \dfrac{2}{\sqrt[3]{x^2}}dx$ (b) $\displaystyle\int \left(e^{0.5x} + \dfrac{1}{3x} - 2\right)dx$ (6)

15. (a) $\displaystyle\int (2+\theta)^2\,d\theta$

    (b) $\displaystyle\int \left(\cos\dfrac{1}{2}x + \dfrac{3}{x} - e^{2x}\right)dx$ (6)

Evaluate the integrals in Problems 16 to 19, each, where necessary, correct to 4 significant figures.

16. (a) $\displaystyle\int_1^3 (t^2 - 2t)dt$ (b) $\displaystyle\int_{-1}^2 \left(2x^3 - 3x^2 + 2\right)dx$ (6)

17. (a) $\displaystyle\int_0^{\pi/3} 3\sin 2t\,dt$ (b) $\displaystyle\int_{\pi/4}^{3\pi/4} \cos\dfrac{1}{3}x\,dx$ (7)

18. (a) $\displaystyle\int_1^2 \left(\dfrac{2}{x^2} + \dfrac{1}{x} + \dfrac{3}{4}\right)dx$

    (b) $\displaystyle\int_1^2 \left(\dfrac{3}{x} - \dfrac{1}{x^3}\right)dx$ (8)

19. (a) $\displaystyle\int_0^1 \left(\sqrt{x} + 2e^x\right)dx$ (b) $\displaystyle\int_1^2 \left(r^3 - \dfrac{1}{r}\right)dr$ (6)

In Problems 20 to 22, find the area bounded by the curve, the x-axis and the given ordinates. Assume answers are in square units. Give answers correct to 2 decimal places where necessary.

20. $y = x^2;\quad x = 0, x = 2$ (3)

21. $y = 3x - x^2;\quad x = 0, x = 3$ (3)

22. $y = (x-2)^2;\quad x = 1, x = 2$ (4)

23. Find the area enclosed between the curve $y = \sqrt{x} + \dfrac{1}{\sqrt{x}}$, the horizontal axis and the ordinates $x = 1$ and $x = 4$. Give the answer correct to 2 decimal places. (5)

24. The force F newtons acting on a body at a distance x metres from a fixed point is given by $F = 2x + 3x^2$. If work done $= \displaystyle\int_{x_1}^{x_2} F\,dx$, determine the work done when the body moves from the position when $x = 1$ m to that when $x = 4$ m. (3)

For lecturers/instructors/teachers, fully worked solutions to each of the problems in Revision Test 14, together with a full marking scheme, are available at the website:
www.routledge.com/cw/bird

Multiple choice questions Test 6
Differentiation and integration
This test covers the material in Chapters 34 and 35

All questions have only one correct answer (Answers on page 440).

1. Differentiating $y = 4x^5$ gives:

   (a) $\dfrac{dy}{dx} = \dfrac{2}{3}x^6$ (b) $\dfrac{dy}{dx} = 20x^4$

   (c) $\dfrac{dy}{dx} = 4x^6$ (d) $\dfrac{dy}{dx} = 5x^4$

2. $\int (5 - 3t^2)\, dt$ is equal to:

   (a) $5 - t^3 + c$ (b) $-3t^3 + c$

   (c) $-6t + c$ (d) $5t - t^3 + c$

3. The gradient of the curve $y = -2x^3 + 3x + 5$ at $x = 2$ is:

   (a) -21 (b) 27   (c) -16 (d) -5

4. $\int \left(\dfrac{5x - 1}{x} \right) dx$ is equal to:

   (a) $5x - \ln x + c$ (b) $\dfrac{5x^2 - x}{\dfrac{x^2}{2}}$

   (c) $\dfrac{5x^2}{2} + \dfrac{1}{x^2} + c$ (d) $5x + \dfrac{1}{x^2} + c$

5. If $y = 5\sqrt{x^3} - 2$, $\dfrac{dy}{dx}$ is equal to:

   (a) $\dfrac{15}{2}\sqrt{x}$ (b) $2\sqrt{x^5} - 2x + c$

   (c) $\dfrac{5}{2}\sqrt{x} - 2$ (d) $5\sqrt{x} - 2x$

6. The value of $\int_0^1 (3\sin 2\theta - 4\cos\theta)\, d\theta$, correct to 4 significant figures, is:

   (a) -0.06890 (b) -1.242

   (c) -2.742 (d) -1.569

7. An alternating current is given by $i = 4\sin 150t$ amperes, where t is the time in seconds. The rate of change of current at $t = 0.025$ s is:

   (a) -492.3 A/s   (b) 3.99 A/s

   (c) -3.28 A/s   (d) 598.7 A/s

8. A vehicle has a velocity $v = (2 + 3t)$ m/s after t seconds. The distance travelled is equal to the area

under the v/t graph. In the first 3 seconds the vehicle has travelled:

   (a) 11 m   (b) 33 m   (c) 13.5 m   (d) 19.5 m

9. Differentiating $i = 3\sin 2t - 2\cos 3t$ with respect to t gives:

   (a) $3\cos 2t + 2\sin 3t$ (b) $6(\sin 2t - \cos 3t)$

   (c) $\dfrac{3}{2}\cos 2t + \dfrac{2}{3}\sin 3t$ (d) $6(\cos 2t + \sin 3t)$

10. The area, in square units, enclosed by the curve $y = 2x + 3$, the x-axis and ordinates $x = 1$ and $x = 4$ is:

    (a) 28   (b) 2   (c) 24   (d) 39

11. Given $y = 3e^x + 2\ln 3x$, $\dfrac{dy}{dx}$ is equal to:

    (a) $6e^x + \dfrac{2}{3x}$ (b) $3e^x + \dfrac{2}{x}$

    (c) $6e^x + \dfrac{2}{x}$ (d) $3e^x + \dfrac{2}{3}$

12. $\int \dfrac{2}{9}t^3\, dt$ is equal to:

    (a) $\dfrac{t^4}{18} + c$ (b) $\dfrac{2}{3}t^2 + c$

    (c) $\dfrac{2}{9}t^4 + c$ (d) $\dfrac{2}{9}t^3 + c$

13. Given $f(t) = 3t^4 - 2$, $f'(t)$ is equal to:

    (a) $12t^3 - 2$ (b) $\dfrac{3}{4}t^5 - 2t + c$

    (c) $12t^3$ (d) $3t^5 - 2$

14. Evaluating $\int_0^{\pi/3} 3\sin 3x\, dx$ gives:

    (a) 1.503 (b) 2   (c) -18 (d) 6

15. The current i in a circuit at time t seconds is given by $i = 0.20(1 - e^{-20t})$ A. When time $t = 0.1$ s, the rate of change of current is:

    (a) -1.022 A/s   (b) 0.173 A/s

    (c) 0.541 A/s   (d) 0.373 A/s

16. $\int (5\sin 3t - 3\cos 5t)\,dt$ is equal to:

(a) $-5\cos 3t + 3\sin 5t + c$

(b) $15(\cos 3t + \sin 3t) + c$

(c) $-\dfrac{5}{3}\cos 3t - \dfrac{3}{5}\sin 5t + c$

(d) $\dfrac{3}{5}\cos 3t - \dfrac{5}{3}\sin 5t + c$

17. The gradient of the curve $y = 4x^2 - 7x + 3$ at the point $(1, 0)$ is

(a) 3   (b) 1   (c) 0   (d) -7

18. $\int (\sqrt{x} - 3)\,dx$ is equal to:

(a) $\dfrac{3}{2}\sqrt{x^3} - 3x + c$ (b) $\dfrac{2}{3}\sqrt{x^3} + c$

(c) $\dfrac{1}{2\sqrt{x}} + c$ (d) $\dfrac{2}{3}\sqrt{x^3} - 3x + c$

19. An alternating voltage is given by $v = 10\sin 300t$ volts, where t is the time in seconds. The rate of change of voltage when $t = 0.01$ s is:

(a) -2996 V/s   (b) 157 V/s

(c) -2970 V/s   (d) 0.523 V/s

20. The area enclosed by the curve $y = 3\cos 2\theta$, the ordinates $\theta = 0$ and $\theta = \dfrac{\pi}{4}$ and the θ axis is:

(a) -3 (b) 6   (c) 1.5   (d) 3

21. If $f(t) = 5t - \dfrac{1}{\sqrt{t}}$, $f'(t)$ is equal to:

(a) $5 + \dfrac{1}{2\sqrt{t^3}}$ (b) $5 - 2\sqrt{t}$

(c) $\dfrac{5t^2}{2} - 2\sqrt{t} + c$ (d) $5 + \dfrac{1}{\sqrt{t^3}}$

22. $\int \left(1 + \dfrac{4}{e^{2x}}\right)\,dx$ is equal to:

(a) $\dfrac{8}{e^{2x}} + c$ (b) $x - \dfrac{2}{e^{2x}} + c$

(c) $x + \dfrac{4}{e^{2x}}$ (d) $x - \dfrac{8}{e^{2x}} + c$

23. If $y = 3x^2 - \ln 5x$ then $\dfrac{d^2 y}{dx^2}$ is equal to:

(a) $6 + \dfrac{1}{5x^2}$ (b) $6x - \dfrac{1}{x}$

(c) $6 - \dfrac{1}{5x}$ (d) $6 + \dfrac{1}{x^2}$

24. Evaluating $\displaystyle\int_1^2 2e^{3t}\,dt$, correct to 4 significant figures, gives:

(a) 2300   (b) 255.6   (c) 766.7   (d) 282.3

25. An alternating current, i amperes, is given by $i = 100\sin 2\pi f t$ amperes, where f is the frequency in hertz and t is the time in seconds. The rate of change of current when $t = 12$ ms and $f = 50$ Hz is:

(a) 31348 A/s   (b) -58.78 A/s

(c) 627.0 A/s   (d) -25416 A/s

Number sequences

At the end of this chapter, you should be able to:

- define and determine simple number sequences
- calculate the nth term of a series
- calculate the nth term of an AP
- calculate the sum of n terms of an AP
- calculate the nth term of a GP
- calculate the sum of n terms of a GP
- calculate the sum to infinity of a GP

36.1 Simple sequences

A set of numbers which are connected by a definite law is called a **series** or a **sequence of numbers**. Each of the numbers in the series is called a **term** of the series.

For example, $1, 3, 5, 7, \ldots$ is a series obtained by adding 2 to the previous term, and $2, 8, 32, 128, \ldots$ is a sequence obtained by multiplying the previous term by 4

Problem 1. Determine the next two terms in the series: $3, 6, 9, 12, \ldots$

We notice that the sequence $3, 6, 9, 12, \ldots$ progressively increases by 3, thus the next two terms will be **15 and 18**

Problem 2. Find the next two terms in the series: $9, 5, 1, \ldots$

We notice that each term in the series $9, 5, 1, \ldots$ progressively decreases by 4, thus the next two terms will be $1 - 4$, i.e. **−3** and $-3 - 4$, i.e. **−7**

Problem 3. Determine the next two terms in the series: $2, 6, 18, 54, \ldots$

We notice that the second term, 6, is three times the first term, the third term, 18, is three times the second term, and that the fourth term, 54, is three times the third term. Hence the fifth term will be $3 \times 54 = $ **162**, and the sixth term will be $3 \times 162 = $ **486**

Now try the following Practice Exercise

Practice Exercise 142   Simple sequences (answers on page 437)

Determine the next two terms in each of the following series:

1. $5, 9, 13, 17, \ldots$ 2. $3, 6, 12, 24, \ldots$

3. $112, 56, 28, \ldots$ 4. $12, 7, 2, \ldots$

5. $2, 5, 10, 17, 26, 37, \ldots$ 6. $1, 0.1, 0.01, \ldots$

7. $4, 9, 19, 34, \ldots$

36.2   The nth term of a series

If a series is represented by a general expression, say, $2n + 1$, where n is an integer (i.e. a whole number), then by substituting $n = 1, 2, 3, \ldots$ the terms of the series can be determined; in this example, the first three terms will be:

$2(1) + 1, \quad 2(2) + 1, \quad 2(3) + 1, \ldots, \quad$ i.e. $3, 5, 7, \ldots$

What is the nth term of the sequence $1, 3, 5, 7, \ldots$? First, we notice that the gap between each term is 2, hence the law relating the numbers is: '$2n +$ something'.
The second term, $3 = 2n +$ something, hence when $n = 2$ (i.e. the second term of the series), then $3 = 4 +$ something, and the 'something' must be -1.
Thus, **the nth term of $1, 3, 5, 7, \ldots$ is $2n - 1$**

Hence the fifth term is given by $2(5) - 1 = 9$, and the twentieth term is $2(20) - 1 = 39$, and so on.

Problem 4. The nth term of a sequence is given by $3n + 1$. Write down the first four terms.

The first four terms of the series $3n + 1$ will be:

$3(1) + 1, \quad 3(2) + 1, \quad 3(3) + 1 \quad$ and $\quad 3(4) + 1$

i.e. **4, 7, 10 and 13**

Problem 5. The nth term of a series is given by $4n - 1$. Write down the first four terms.

The first four terms of the series $4n - 1$ will be:

$4(1) - 1, \quad 4(2) - 1, \quad 4(3) - 1 \quad$ and $\quad 4(4) - 1$

i.e. **3, 7, 11 and 15**

Problem 6. Find the nth term of the series: $1, 4, 7, \ldots$

We notice that the gap between each of the given three terms is 3, hence the law relating the numbers is: '$3n +$ something'.
The second term, $\quad 4 = 3n +$ something,
so when $n = 2$, then $\quad 4 = 6 +$ something,
so the 'something' must be -2 (from simple equations)
Thus, **the nth term of the series $1, 4, 7, \ldots$ is: $3n - 2$**

Problem 7. Find the nth term of the sequence: $3, 9, 15, 21, \ldots$ Hence determine the 15th term of the series.

We notice that the gap between each of the given four terms is 6, hence the law relating the numbers is: '$6n +$ something'.
The second term, $\quad 9 = 6n +$ something,
so when $n = 2$, then $\quad 9 = 12 +$ something,
so the 'something' must be -3
Thus, **the nth term of the series $3, 9, 15, 21, \ldots$ is: $6n - 3$**
The 15th term of the series is given by $6n - 3$ when $n = 15$
Hence, **the 15th term of the series $3, 9, 15, 21, \ldots$ is: $6(15) - 3 = $ 87**

Problem 8. Find the nth term of the series: $1, 4, 9, 16, 25, \ldots$

This is a special series and does not follow the pattern of the previous examples. Each of the terms in the given series are square numbers,

i.e. $1, 4, 9, 16, 25, \ldots \equiv 1^2, 2^2, 3^2, 4^2, 5^2, \ldots$

Hence the nth term is: n^2

Now try the following Practice Exercise

Practice Exercise 143   The nth term of a series (answers on page 437)

1. The nth term of a sequence is given by $2n - 1$. Write down the first four terms.

2. The nth term of a sequence is given by $3n + 4$. Write down the first five terms.

3. Write down the first four terms of the sequence given by $5n + 1$.

In Problems 4 to 8, find the nth term in the series:

4. $5, 10, 15, 20, \ldots$ 5. $4, 10, 16, 22, \ldots$

6. $3, 5, 7, 9, \ldots$ 7. $2, 6, 10, 14, \ldots$

8. $9, 12, 15, 18, \ldots$

9. Write down the next two terms of the series: $1, 8, 27, 64, 125, \ldots$

36.3   Arithmetic progressions

When a sequence has a constant difference between successive terms it is called an **arithmetic progression** (often abbreviated to AP).

Examples include:

(i) $1, 4, 7, 10, 13, \ldots$ where the **common difference** is 3

and   (ii) $a, a + d, a + 2d, a + 3d, \ldots$ where the common difference is d.

36.3.1   General expression for the nth term of an AP

If the first term of an AP is 'a' and the common difference is 'd' then:

the nth term is: $a + (n - 1)d$

In example (i) above, the 7th term is given by $1 + (7 - 1)3 = \mathbf{19}$, which may be readily checked.

36.3.2   Sum of n terms of an AP

The sum S of an AP can be obtained by multiplying the average of all the terms by the number of terms.

The average of all the terms $= \dfrac{a + l}{2}$, where 'a' is the first term and 'l' is the last term, i.e. $l = a + (n - 1)d$, for n terms.

Hence, the sum of n terms, $S_n = n\left(\dfrac{a + l}{2}\right)$

$$= \frac{n}{2}\{a + [a + (n - 1)d]\}$$

i.e. $$S_n = \frac{n}{2}[2a + (n - 1)d]$$

For example, the sum of the first 7 terms of the series $1, 4, 7, 10, 13, \ldots$ is given by:

$$S_7 = \frac{7}{2}[2(1) + (7 - 1)3] \quad \text{since } a = 1 \quad \text{and} \quad d = 3$$

$$= \frac{7}{2}[2 + 18] = \frac{7}{2}[20] = \mathbf{70}$$

Here are some worked problems to help understanding of arithmetic progressions.

Problem 9. Determine (a) the ninth, and (b) the sixteenth term of the series $2, 7, 12, 17, \ldots$

$2, 7, 12, 17, \ldots$ is an arithmetic progression with a common difference, d, of 5

(a)   The nth term of an AP is given by $a + (n - 1)d$
Since the first term $a = 2, d = 5$ and $n = 9$ then the 9th term is: $2 + (9 - 1)5 = 2 + (8)(5)$
$$= 2 + 40 = \mathbf{42}$$

(b)   The 16th term is: $2 + (16 - 1)5 = 2 + (15)(5)$
$$= 2 + 75 = \mathbf{77}$$

Problem 10. The 6th term of an AP is 17 and the 13th term is 38. Determine the 19th term

The nth term of an AP is: $a + (n - 1)d$

The 6th term is: $a + 5d = 17$ (1)

The 13th term is: $a + 12d = 38$ (2)

Equation (2) $-$ equation (1) gives: $7d = 21$ from which,
$$d = \frac{21}{7} = 3$$

Substituting in equation (1) gives: $a + 15 = 17$ from which, $a = 2$

Hence, the 19th term is: $a+(n-1)d = 2+(19-1)3$

$$= 2+(18)(3)$$
$$= 2+54 = \mathbf{56}$$

Problem 11. Determine the number of the term whose value is 22 in the series $2.5, 4, 5.5, 7, \ldots$

$2.5, 4, 5.5, 7, \ldots$ is an AP where $a=2.5$ and $d=1.5$

Hence, if the nth term is 22 then:

$$a+(n-1)d = 22$$

i.e. $\qquad 2.5+(n-1)(1.5) = 22$

i.e. $\qquad (n-1)(1.5) = 22-2.5 = 19.5$

and $\qquad n-1 = \dfrac{19.5}{1.5} = 13$

from which, $\qquad \boldsymbol{n = 13+1 = 14}$

i.e. **the 14th term of the AP is 22**

Problem 12. Find the sum of the first 12 terms of the series $5, 9, 13, 17, \ldots$

$5, 9, 13, 17, \ldots$ is an AP where $a=5$ and $d=4$
The sum of n terms of an AP, $S_n = \dfrac{n}{2}[2a+(n-1)d]$

Hence, the sum of the first 12 terms,

$$S_{12} = \dfrac{12}{2}[2(5)+(12-1)4]$$
$$= 6[10+44] = 6(54) = \mathbf{324}$$

Problem 13. Find the sum of the first 21 terms of the series $3.5, 4.1, 4.7, 5.3, \ldots$

$3.5, 4.1, 4.7, 5.3, \ldots$ is an AP where $a=3.5$ and $d=0.6$

The sum of the first 21 terms,

$$S_{21} = \dfrac{21}{2}[2a+(n-1)d]$$
$$= \dfrac{21}{2}[2(3.5)+(21-1)0.6]$$
$$= \dfrac{21}{2}[7+12]$$
$$= \dfrac{21}{2}(19) = \dfrac{399}{2} = \mathbf{199.5}$$

Now try the following Practice Exercise

Practice Exercise 144   Arithmetic progressions (answers on page 438)

1. Find the 11th term of the series $8, 14, 20, 26, \ldots$

2. Find the 17th term of the series $11, 10.7, 10.4, 10.1, \ldots$

3. The 7th term of a series is 29 and the 11th term is 54. Determine the 16th term.

4. Find the 15th term of an arithmetic progression of which the first term is 2.5 and the 10th term is 16.

5. Determine the number of the term which is 29 in the series $7, 9.2, 11.4, 13.6, \ldots$

6. Find the sum of the first 11 terms of the series $4, 7, 10, 13, \ldots$

7. Determine the sum of the series $6.5, 8.0, 9.5, 11.0, \ldots, 32$

Here are some further worked problems on arithmetic progressions.

Problem 14. The sum of 7 terms of an AP is 35 and the common difference is 1.2. Determine the first term of the series.

$n=7, d=1.2$ and $S_7 = 35$. Since the sum of n terms of an AP is given by:

$$S_n = \dfrac{n}{2}[2a+(n-1)d]$$

Then $\quad 35 = \dfrac{7}{2}[2a+(7-1)1.2] = \dfrac{7}{2}[2a+7.2]$

Hence, $\qquad \dfrac{35\times2}{7} = 2a+7.2$

i.e. $\qquad 10 = 2a+7.2$

Thus, $\qquad 2a = 10-7.2 = 2.8$

from which, $\qquad a = \dfrac{2.8}{2} = 1.4$

i.e. **the first term, $a = 1.4$**

Problem 15. Three numbers are in arithmetic progression. Their sum is 15 and their product is 80. Determine the three numbers.

Let the three numbers be $(a-d), a$ and $(a+d)$

Then, $(a-d)+a+(a+d) = 15$

i.e. $3a = 15$

from which, $a = 5$

Also, $a(a-d)(a+d) = 80$

i.e. $a(a^2 - d^2) = 80$

Since $a = 5$, $5(5^2 - d^2) = 80$

i.e. $125 - 5d^2 = 80$

and $125 - 80 = 5d^2$

i.e. $45 = 5d^2$

from which, $d^2 = \dfrac{45}{5} = 9$

Hence, $d = \sqrt{9} = \pm 3$

The three numbers are thus: $(5-3), 5$ and $(5+3)$, i.e. **2, 5 and 8**

Problem 16. Find the sum of all the numbers between 0 and 207 which are exactly divisible by 3.

The series $3, 6, 9, 12, \ldots 207$ is an AP whose first term, $a = 3$ and common difference, $d = 3$

The last term is: $a + (n-1)d = 207$

i.e. $3 + (n-1)3 = 207$

from which, $(n-1) = \dfrac{207 - 3}{3} = 68$

Hence, $n = 68 + 1 = 69$

The sum of all 69 terms is given by:

$$S_{69} = \frac{n}{2}[2a + (n-1)d]$$

$$= \frac{69}{2}[2(3) + (69-1)3]$$

$$= \frac{69}{2}[6 + 204] = \frac{69}{2}(210) = \mathbf{7245}$$

Problem 17. The first, 12th and last term of an arithmetic progression are: $4, 31.5$, and 376.5, respectively. Determine (a) the number of terms in the series, (b) the sum of all the terms, and (c) the 80th term.

(a) Let the AP be $a, a+d, a+2d, \ldots, a+(n-1)d$, where $a = 4$

The 12th term is: $a + (12-1)d = 31.5$

i.e. $4 + 11d = 31.5$

from which, $11d = 31.5 - 4 = 27.5$

Hence, $d = \dfrac{27.5}{11} = 2.5$

The last term is: $a + (n-1)d$

i.e. $4 + (n-1)(2.5) = 376.5$

i.e. $(n-1) = \dfrac{376.5 - 4}{2.5} = \dfrac{372.5}{2.5} = 149$

Hence, the number of terms in the series, $n = 149 + 1 = 150$

(b) **Sum of all the terms,**

$$S_{150} = \frac{n}{2}[2a + (n-1)d]$$

$$= \frac{150}{2}[2(4) + (150-1)(2.5)]$$

$$= 75[8 + (149)(2.5)] = 75[8 + 372.5]$$

$$= 75(380.5) = \mathbf{28537.5}$$

(c) **The 80th term is:** $a + (n-1)d = 4 + (80-1)(2.5)$

$$= 4 + (79)(2.5)$$

$$= 4 + 197.5$$

$$= \mathbf{201.5}$$

Now try the following Practice Exercise

Practice Exercise 145   Arithmetic progressions (answers on page 438)

1. The sum of 15 terms of an arithmetic progression is 202.5 and the common difference is 2. Find the first term of the series.

2. Three numbers are in arithmetic progression. Their sum is 9 and their product is 20.25. Determine the three numbers.

3. Find the sum of all the numbers between 5 and 250 which are exactly divisible by 4.

4. Find the number of terms of the series $5, 8, 11, \ldots$ of which the sum is 1025.

5. Insert four terms between 5 and 22.5 to form an arithmetic progression.

6. The first, tenth and last terms of an arithmetic progression are 9, 40.5 and 425.5, respectively. Find (a) the number of terms, (b) the sum of all the terms, and (c) the 70th term.

7. On commencing employment a man is paid a salary of £16 000 per annum and receives annual increments of £480. Determine his salary in the 9th year and calculate the total he will have received in the first 12 years.

8. An oil company bores a hole 80 m deep. Estimate the cost of boring if the cost is £30 for drilling the first metre with an increase in cost of £2 per metre for each succeeding metre.

36.4 Geometric progressions

When a sequence has a constant ratio between successive terms it is called a **geometric progression** (often abbreviated to GP). The constant is called the **common ratio, r**.

Examples include

(i) $1, 2, 4, 8, \ldots$ where the common ratio is 2

and (ii) $a, ar, ar^2, ar^3, \ldots$ where the common ratio is r

36.4.1 General expression for the nth term of a GP

If the first term of a GP is 'a' and the common ratio is 'r', then

the nth term is: ar^{n-1}

which can be readily checked from the above examples. For example, the 8th term of the GP $1, 2, 4, 8, \ldots$ is $(1)(2)^7 = 128$, since 'a' $= 1$ and 'r' $= 2$

36.4.2 Sum of n terms of a GP

Let a GP be $a, ar, ar^2, ar^3, \ldots, ar^{n-1}$

then the sum of n terms,

$$S_n = a + ar + ar^2 + ar^3 + \cdots + ar^{n-1} \cdots \quad (1)$$

Multiplying throughout by r gives:

$$rS_n = ar + ar^2 + ar^3 + ar^4 + \cdots ar^{n-1} + ar^n \cdots \quad (2)$$

Subtracting equation (2) from equation (1) gives:

$$S_n - rS_n = a - ar^n$$

i.e. $\qquad S_n(1 - r) = a(1 - r^n)$

Thus, **the sum of n terms, $S_n = \dfrac{a(1 - r^n)}{(1 - r)}$** which is valid when $r < 1$

Subtracting equation (1) from equation (2) gives:

$$S_n = \frac{a(r^n - 1)}{(r - 1)} \quad \text{which is valid when } r > 1$$

For example, the sum of the first 8 terms of the GP $1, 2, 4, 8, 16, \ldots$ is given by:

$$S_8 = \frac{1(2^8 - 1)}{(2 - 1)} \quad \text{since '}a\text{'} = 1 \text{ and } r = 2$$

i.e. $\quad S_8 = \dfrac{1(256 - 1)}{1} = 255$

36.4.3 Sum to infinity of a GP

When the common ratio r of a GP is less than unity, the sum of n terms,

$$S_n = \frac{a(1 - r^n)}{(1 - r)}, \text{ which may be written as:}$$

$$S_n = \frac{a}{(1 - r)} - \frac{ar^n}{(1 - r)}$$

Since $r < 1, r^n$ becomes less as n increases, i.e. $r^n \to 0$ as $n \to \infty$

Hence, $\qquad \dfrac{ar^n}{(1 - r)} \to 0 \text{ as } n \to \infty$

Thus, $\qquad S_n \to \dfrac{a}{(1 - r)} \text{ as } n \to \infty$

The quantity $\dfrac{a}{(1 - r)}$ is called the **sum to infinity, S_∞**, and is the limiting value of the sum of an infinite number of terms,

i.e. $\quad S_\infty = \dfrac{a}{(1 - r)}$ which is valid when $-1 < r < 1$

For example, the sum to infinity of the GP: $1, \dfrac{1}{2}, \dfrac{1}{4}, \ldots$ is:

$$S_\infty = \frac{1}{1 - \frac{1}{2}}, \quad \text{since } a = 1 \text{ and } r = \frac{1}{2}$$

i.e. $\qquad S_\infty = 2$

Here are some worked problems to help understanding of geometric progressions.

Problem 18. Determine the tenth term of the series $3, 6, 12, 24, \ldots$

$3, 6, 12, 24, \ldots$ is a geometric progression with a common ratio r of 2.

The nth term of a GP is ar^{n-1}, where 'a' is the first term.

Hence, **the 10th term is**:
$(3)(2)^{10-1} = (3)(2)^9 = 3(512) = \mathbf{1536}$

Problem 19. Find the sum of the first 7 terms of the series, $0.5, 1.5, 4.5, 13.5, \ldots$

$0.5, 1.5, 4.5, 13.5, \ldots$ is a GP with a common ratio $r = 3$

The sum of n terms, $S_n = \dfrac{a(r^n - 1)}{(r - 1)}$

Hence, **the sum of the first 7 terms**,

$$S_7 = \frac{0.5(3^7 - 1)}{(3 - 1)} = \frac{0.5(2187 - 1)}{2} = \mathbf{546.5}$$

Problem 20. The first term of a geometric progression is 12 and the fifth term is 55. Determine the 8th term and the 11th term.

The 5th term is given by: $ar^4 = 55$, where the first term, $a = 12$

Hence, $r^4 = \dfrac{55}{a} = \dfrac{55}{12}$ and $r = \sqrt[4]{\left(\dfrac{55}{12}\right)}$

$$= 1.4631719\ldots$$

The 8th term is: $ar^7 = (12)(1.4631719\ldots)^7$

$$= \mathbf{172.3}$$

The 11th term is: $ar^{10} = (12)(1.4631719\ldots)^{10}$

$$= \mathbf{539.7}$$

Problem 21. Which term of the series $2187, 729, 243, \ldots$ is $\dfrac{1}{9}$?

$2187, 729, 243, \ldots$ is a GP with a common ratio, $r = \dfrac{1}{3}$ and first term, $a = 2187$

The nth term of a GP is given by: ar^{n-1}

Hence, $\dfrac{1}{9} = (2187)\left(\dfrac{1}{3}\right)^{n-1}$ from which,

$$\left(\frac{1}{3}\right)^{n-1} = \frac{1}{(9)(2187)} = \frac{1}{3^2 3^7} = \frac{1}{3^9} = \left(\frac{1}{3}\right)^9$$

Thus, $(n - 1) = 9$ from which, $\mathbf{n = 9 + 1 = 10}$

i.e. $\dfrac{1}{9}$ **is the 10th term of the GP**

Problem 22. Find the sum of the first 9 terms of the series $72.0, 57.6, 46.08, \ldots$

The common ratio, $r = \dfrac{ar}{a} = \dfrac{57.6}{72.0} = 0.8$

$\left(\text{also } \dfrac{ar^2}{ar} = \dfrac{46.08}{57.6} = 0.8\right)$

The sum of 9 terms, $S_9 = \dfrac{a(1 - r^n)}{(1 - r)}$

$$= \frac{72.0(1 - 0.8^9)}{(1 - 0.8)}$$

$$= \frac{72.0(1 - 0.1342)}{0.2}$$

$$= \mathbf{311.7}$$

Problem 23. Find the sum to infinity of the series $3, 1, \dfrac{1}{3}, \ldots$

$3, 1, \dfrac{1}{3}, \ldots$ is a GP of common ratio, $r = \dfrac{1}{3}$

The sum to infinity, $S_\infty = \dfrac{a}{1 - r} = \dfrac{3}{1 - \frac{1}{3}} = \dfrac{3}{\frac{2}{3}}$

$$= \frac{9}{2} = \mathbf{4.5}$$

Now try the following Practice Exercise

Practice Exercise 146   Geometric progressions (answers on page 438)

1. Find the 10th term of the series $5, 10, 20, 40, \ldots$

2. Determine the sum of the first 7 terms of the series $0.25, 0.75, 2.25, 6.75, \ldots$

3. The 1st term of a geometric progression is 4 and the 6th term is 128. Determine the 8th and 11th terms.

4. Find the sum of the first 7 terms of the series $2, 5, 12.5, \ldots$ (correct to 4 significant figures).

5. Determine the sum to infinity of the series $4, 2, 1, \ldots$

6. Find the sum to infinity of the series $2\frac{1}{2}, -1\frac{1}{4}, \frac{5}{8}, \ldots$

Here are some further worked problems on geometric progressions.

Problem 24. In a geometric progression the sixth term is 8 times the third term and the sum of the seventh and eighth terms is 192. Determine (a) the common ratio, (b) the first term, and (c) the sum of the fifth to eleventh terms, inclusive.

(a) Let the GP be $a, ar, ar^2, ar^3, \ldots, ar^{n-1}$

   The 3rd term $= ar^2$ and the sixth term $= ar^5$

   The 6th term is 8 times the 3rd

   Hence, $ar^5 = 8ar^2$ from which,

   $r^3 = 8$ and $r = \sqrt[3]{8}$

   i.e.   **the common ratio $r = 2$**

(b) The sum of the 7th and 8th terms is 192. Hence $ar^6 + ar^7 = 192$

   Since $r = 2$,   then $64a + 128a = 192$

   $192a = 192$ from which, **the first term, $a = 1$**

(c) **The sum of the 5th to 11th terms** (inclusive) **is given by**:

$$S_{11} - S_4 = \frac{a(r^{11}-1)}{(r-1)} - \frac{a(r^4-1)}{(r-1)}$$

$$= \frac{1(2^{11}-1)}{(2-1)} - \frac{1(2^4-1)}{(2-1)}$$

$$= (2^{11}-1) - (2^4-1) = 2^{11} - 2^4$$

$$= 2048 - 16 = \mathbf{2032}$$

Problem 25. A tool hire firm finds that their net return from hiring tools is decreasing by 10% per annum. If their net gain on a certain tool this year is £400, find the possible total of all future profits from this tool (assuming the tool lasts for ever).

The net gain forms a series: £400 + £400 × 0.9 + £400 × 0.9² + ⋯, which is a GP with $a = 400$ and $r = 0.9$

The sum to infinity, $S_\infty = \dfrac{a}{(1-r)} = \dfrac{400}{(1-0.9)}$

$$= \mathbf{£4000}$$

$$= \textbf{total future profits}$$

Problem 26. If £100 is invested at compound interest of 8% per annum, determine (a) the value after 10 years, (b) the time, correct to the nearest year, it takes to reach more than £300.

(a) Let the GP be $a, ar, ar^2, \ldots ar^n$

   The first term, $a = £100$ and the common ratio, $r = 1.08$

   Hence, the second term is: $ar = (100)(1.08) = £108$, which is the value after 1 year, the third term is: $ar^2 = (100)(1.08)^2 = £116.64$, which is the value after 2 years, and so on.

   Thus, **the value after 10 years**

   $= ar^{10} = (100)(1.08)^{10} = \mathbf{£215.89}$

(b) When £300 has been reached, $300 = ar^n$

   i.e. $300 = 100(1.08)^n$

   and $3 = (1.08)^n$

   Taking logarithms to base 10 of both sides gives:

   $$\lg 3 = \lg(1.08)^n = n \lg(1.08)$$

   by the laws of logarithms

   from which, $n = \dfrac{\lg 3}{\lg 1.08} = 14.3$

   Hence, it will take 15 years to reach more than £300

Problem 27. A drilling machine is to have six speeds ranging from 50 rev/min to 750 rev/min. If the speeds form a geometric progression determine their values, each correct to the nearest whole number.

Let the GP of n terms be given by: $a, ar, ar^2, \ldots, ar^{n-1}$

The first term $a = 50$ rev/min

The 6th term is given by ar^{6-1}, which is 750 rev/min,

i.e. $ar^5 = 750$

from which, $r^5 = \dfrac{750}{a} = \dfrac{750}{50} = 15$

Thus, the common ratio, $r = \sqrt[5]{15} = 1.7188$

The first term is: $a = 50\,\text{rev/min}$

The second term is: $ar = (50)(1.7188) = 85.94,$

The third term is: $ar^2 = (50)(1.7188)^2 = 147.71,$

The fourth term is: $ar^3 = (50)(1.7188)^3 = 253.89,$

The fifth term is: $ar^4 = (50)(1.7188)^4 = 436.39,$

The sixth term is: $ar^5 = (50)(1.7188)^5 = 750.06$

Hence, correct to the nearest whole number, **the 6 speeds of the drilling machine are**:

50, 86, 148, 254, 436 and 750 rev/min

Now try the following Practice Exercise

Practice Exercise 147   Geometric progressions (answers on page 438)

1. In a geometric progression the 5th term is 9 times the 3rd term and the sum of the 6th and 7th terms is 1944. Determine (a) the common ratio, (b) the first term and (c) the sum of the 4th to 10th terms inclusive.

2. Which term of the series $3, 9, 27, \ldots$ is 59049?

3. The value of a lathe originally valued at £3000 depreciates 15% per annum. Calculate its value after 4 years. The machine is sold when its value is less than £550. After how many years is the lathe sold?

4. If the population of Great Britain is 64 million and is decreasing at 2.4% per annum, what will be the population in 5 years' time?

5. 100 g of a radioactive substance disintegrates at a rate of 3% per annum. How much of the substance is left after 11 years?

6. If £250 is invested at compound interest of 6% per annum, determine (a) the value after 15 years, (b) the time, correct to the nearest year, it takes to reach £750.

7. A drilling machine is to have 8 speeds ranging from 100 rev/min to 1000 rev/min. If the speeds form a geometric progression determine their values, each correct to the nearest whole number.

For fully worked solutions to each of the problems in Exercises 142 to 147 in this chapter, go to the website:
www.routledge.com/cw/bird

Chapter 37

Binary, octal and hexadecimal numbers

Why it is important to understand: Binary, octal and hexadecimal number

There are infinite ways to represent a number. The four commonly associated with modern computers and digital electronics are decimal, binary, octal and hexadecimal. All four number systems are equally capable of representing any number. Furthermore, a number can be perfectly converted between the various number systems without any loss of numeric value. At a first look, it seems like using any number system other than decimal is complicated and unnecessary. However, since the job of electrical and software engineers is to work with digital circuits, engineers require number systems that can best transfer information between the human world and the digital circuit world. Thus the way in which a number is represented can make it easier for the engineer to perceive the meaning of the number as it applies to a digital circuit, i.e. the appropriate number system can actually make things less complicated.

At the end of this chapter, you should be able to:

- recognise a binary number
- convert binary to decimal and vice-versa
- add binary numbers
- recognise an octal number
- convert decimal to binary via octal and vice-versa
- recognise a hexadecimal number
- convert from hexadecimal to decimal and vice-versa
- convert from binary to hexadecimal and vice-versa

37.1 Introduction

All data in modern computers is stored as series of **bits**, a bit being a **bi**nary digi**t**, and can have one of two values, the numbers 0 and 1. The most basic form of representing computer data is to represent a piece of data as a string of '1's and '0's, one for each bit. This is called a **binary** or base-2 number.

Because binary notation requires so many bits to represent relatively small numbers, two further compact notations are often used, called **octal** and **hexadecimal**. Computer programmers who design sequences of number codes instructing a computer what to do, would have a very difficult task if they were forced to work with nothing but long strings of '1's and '0's, the 'native language' of any digital circuit.

Octal notation represents data as base-8 numbers with each digit in an octal number representing three bits. Similarly, hexadecimal notation uses base-16 numbers, representing four bits with each digit. Octal numbers use only the digits 0–7, while hexadecimal numbers use all ten base-10 digits (0–9) and the letters A–F (representing the numbers 10–15).

This chapter explains how to convert between the decimal, binary, octal and hexadecimal systems.

37.2 Binary numbers

The system of numbers in everyday use is the **denary** or **decimal** system of numbers, using the digits 0 to 9. It has ten different digits (0, 1, 2, 3, 4, 5, 6, 7, 8 and 9) and is said to have a **radix** or **base** of 10.

The **binary** system of numbers has a radix of 2 and uses only the digits 0 and 1.

(a) Conversion of binary to decimal

The decimal number 234.5 is equivalent to

$$2 \times 10^2 + 3 \times 10^1 + 4 \times 10^0 + 5 \times 10^{-1}$$

In the binary system of numbers, the base is 2, so 1101.1 is equivalent to:

$$1 \times 2^3 + 1 \times 2^2 + 0 \times 2^1 + 1 \times 2^0 + 1 \times 2^{-1}$$

Thus, the decimal number equivalent to the binary number 1101.1 is:

$$8 + 4 + 0 + 1 + \frac{1}{2} \text{ that is } 13.5$$

i.e. $\mathbf{1101.1_2 = 13.5_{10}}$, the suffixes 2 and 10 denoting binary and decimal systems of numbers, respectively.

Problem 1. Convert 1010_2 to a decimal number.

From above:

$$1010_2 = 1 \times 2^3 + 0 \times 2^2 + 1 \times 2^1 + 0 \times 2^0$$
$$= 8 + 0 + 2 + 0$$
$$= \mathbf{10_{10}}$$

Problem 2. Convert 11011_2 to a decimal number.

$$11011_2 = 1 \times 2^4 + 1 \times 2^3 + 0 \times 2^2 + 1 \times 2^1 + 1 \times 2^0$$
$$= 16 + 8 + 0 + 2 + 1$$
$$= \mathbf{27_{10}}$$

Problem 3. Convert 0.1011_2 to a decimal fraction.

$$0.1011_2 = 1 \times 2^{-1} + 0 \times 2^{-2} + 1 \times 2^{-3} + 1 \times 2^{-4}$$
$$= 1 \times \frac{1}{2} + 0 \times \frac{1}{2^2} + 1 \times \frac{1}{2^3} + 1 \times \frac{1}{2^4}$$
$$= \frac{1}{2} + \frac{1}{8} + \frac{1}{16}$$
$$= 0.5 + 0.125 + 0.0625$$
$$= \mathbf{0.6875_{10}}$$

Problem 4. Convert 101.0101_2 to a decimal number.

$$101.0101_2 = 1 \times 2^2 + 0 \times 2^1 + 1 \times 2^0 + 0 \times 2^{-1}$$
$$+ 1 \times 2^{-2} + 0 \times 2^{-3} + 1 \times 2^{-4}$$
$$= 4 + 0 + 1 + 0 + 0.25 + 0 + 0.0625$$
$$= \mathbf{5.3125_{10}}$$

Now try the following Practice Exercise

Practice Exercise 148 Conversion of binary to decimal numbers (answers on page 438)

In Problems 1 to 4, convert the binary numbers given to decimal numbers.

1. (a) 110 (b) 1011 (c) 1110 (d) 1001

2. (a) 10101 (b) 11001 (c) 101101 (d) 110011

3. (a) 101010 (b) 111000
   (c) 1000001 (d) 10111000

4. (a) 0.1101 (b) 0.11001
   (c) 0.00111 (d) 0.01011

5. (a) 11010.11 (b) 10111.011
   (c) 110101.0111 (d) 11010101.10111

(b) Conversion of decimal to binary

An integer decimal number can be converted to a corresponding binary number by repeatedly dividing by 2 and noting the remainder at each stage, as shown below for 39_{10}

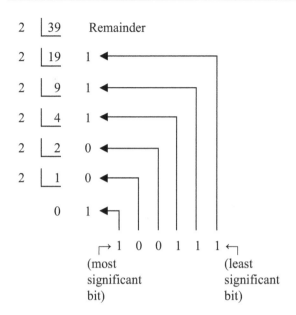

The result is obtained by writing the top digit of the remainder as the least significant bit, (the least significant bit is the one on the right). The bottom bit of the remainder is the most significant bit, i.e. the bit on the left.

Thus, $39_{10} = 100111_2$

The fractional part of a decimal number can be converted to a binary number by repeatedly multiplying by 2, as shown below for the fraction 0.625

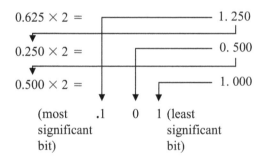

For fractions, the most significant bit of the result is the top bit obtained from the integer part of multiplication by 2. The least significant bit of the result is the bottom bit obtained from the integer part of multiplication by 2.

Thus, $0.625_{10} = 0.101_2$

Problem 5.   Convert 49_{10} to a binary number.

From above, repeatedly dividing by 2 and noting the remainder gives:

```
2  | 49      Remainder

2  | 24      1  ◄─────────────┐
                              │
2  | 12      0  ◄───────────┐ │
                            │ │
2  |  6      0  ◄─────────┐ │ │
                          │ │ │
2  |  3      0  ◄───────┐ │ │ │
                        │ │ │ │
2  |  1      1  ◄─────┐ │ │ │ │
                      │ │ │ │ │
   0         1  ◄───┐ │ │ │ │ │
```

┌→ 1   1   0   0   0   1 ←┐
(most   (least
significant   significant
bit)   bit)

Thus, $49_{10} = 110001_2$

Problem 6.   Convert 56_{10} to a binary number.

The integer part is repeatedly divided by 2, giving:

```
2  | 56      Remainder

2  | 28      0  ◄─────────────────┐

2  | 14      0  ◄───────────────┐

2  |  7      0  ◄─────────────┐

2  |  3      1  ◄───────────┐

2  |  1      1  ◄─────────┐

   0         1  ◄───────┐
```

                1   1   1   0   0   0

Thus, $56_{10} = 111000_2$

Problem 7.   Convert 0.40625_{10} to a binary number.

From above, repeatedly multiplying by 2 gives:

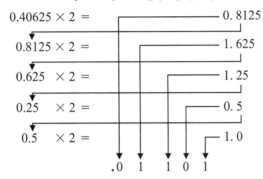

$$0.40625 \times 2 = \quad 0.8125$$
$$0.8125 \times 2 = \quad 1.625$$
$$0.625 \times 2 = \quad 1.25$$
$$0.25 \times 2 = \quad 0.5$$
$$0.5 \times 2 = \quad 1.0$$

.0 1 1 0 1

i.e. $\mathbf{0.40625_{10} = 0.01101_2}$

Problem 8. Convert 58.3125_{10} to a binary number.

The integer part is repeatedly divided by 2, giving:

2	$\underline{58}$	Remainder
2	$\underline{29}$	0
2	$\underline{14}$	1
2	$\underline{7}$	0
2	$\underline{3}$	1
2	$\underline{1}$	1
	0	1

1 1 1 0 1 0

The fractional part is repeatedly multiplied by 2, giving:

$$0.3125 \times 2 = \quad 0.625$$
$$0.625 \times 2 = \quad 1.25$$
$$0.25 \times 2 = \quad 0.5$$
$$0.5 \times 2 = \quad 1.0$$

.0 1 0 1

Thus, $\mathbf{58.3125_{10} = 111010.0101_2}$

Now try the following Practice Exercise

Practice Exercise 149   Conversion of decimal to binary numbers (answers on page 438)

In Problems 1 to 5, convert the decimal numbers given to binary numbers.

1. (a) 5   (b) 15   (c) 19   (d) 29
2. (a) 31   (b) 42   (c) 57   (d) 63
3. (a) 47   (b) 60   (c) 73   (d) 84
4. (a) 0.25   (b) 0.21875
   (c) 0.28125   (d) 0.59375
5. (a) 47.40625   (b) 30.8125
   (c) 53.90625   (d) 61.65625

(c) Binary addition

Binary addition of two/three bits is achieved according to the following rules:

	sum	carry			sum	carry
$0+0=0$		0		$0+0+0=0$		0
$0+1=1$		0		$0+0+1=1$		0
$1+0=1$		0		$0+1+0=1$		0
$1+1=0$		1		$0+1+1=0$		1
				$1+0+0=1$		0
				$1+0+1=0$		1
				$1+1+0=0$		1
				$1+1+1=1$		1

These rules are demonstrated in the following worked problems.

Problem 9. Perform the binary addition:
$1001 + 10110$

$$
\begin{array}{r}
1001 \\
+ \underline{10110} \\
\mathbf{\underline{11111}}
\end{array}
$$

Problem 10. Perform the binary addition:
$11111 + 10101$

$$
\begin{array}{r}
11111 \\
+ \underline{10101} \\
\text{sum } \mathbf{\underline{110100}} \\
\text{carry } 11111
\end{array}
$$

Problem 11. Perform the binary addition:
$1101001 + 1110101$

$$
\begin{array}{r}
1101001 \\
+\ \underline{1110101} \\
\text{sum } \mathbf{11011110} \\
\text{carry } 11 \qquad 1
\end{array}
$$

Problem 12. Perform the binary addition:
$1011101 + 1100001 + 110101$

$$
\begin{array}{r}
1011101 \\
1100001 \\
+\ \underline{110101} \\
\text{sum } \mathbf{11110011} \\
\text{carry } 11111\ 1
\end{array}
$$

Now try the following Practice Exercise

Practice Exercise 150   Binary addition (answers on page 438)

Perform the following binary additions:

1. $10 + 11$
2. $101 + 110$
3. $1101 + 111$
4. $1111 + 11101$
5. $110111 + 10001$
6. $10000101 + 10000101$
7. $11101100 + 111001011$
8. $110011010 + 11100011$
9. $10110 + 1011 + 11011$
10. $111 + 10101 + 11011$
11. $1101 + 1001 + 11101$
12. $100011 + 11101 + 101110$

37.3   Octal numbers

For decimal integers containing several digits, repeatedly dividing by 2 can be a lengthy process. In this case, it is usually easier to convert a decimal number to a binary number via the octal system of numbers. This system has a radix of 8, using the digits 0, 1, 2, 3, 4, 5, 6 and 7. The decimal number equivalent to the octal number 4317_8 is:

$$
4 \times 8^3 + 3 \times 8^2 + 1 \times 8^1 + 7 \times 8^0
$$

i.e. $4 \times 512 + 3 \times 64 + 1 \times 8 + 7 \times 1 = \mathbf{2255_{10}}$

An integer decimal number can be converted to a corresponding octal number by repeatedly dividing by 8 and noting the remainder at each stage, as shown below for 493_{10}

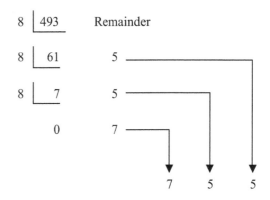

Thus, $\mathbf{493_{10} = 755_8}$

The fractional part of a decimal number can be converted to an octal number by repeatedly multiplying by 8, as shown below for the fraction 0.4375_{10}

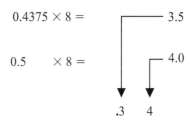

For fractions, the most significant bit is the top integer obtained by multiplication of the decimal fraction by 8, thus,

$$
\mathbf{0.4375_{10} = 0.34_8}
$$

The natural binary code for digits 0 to 7 is shown in Table 37.1, and an octal number can be converted to a binary number by writing down the three bits corresponding to the octal digit.

Table 37.1

Octal digit	Natural binary number
0	000
1	001
2	010
3	011
4	100
5	101
6	110
7	111

Thus, $\qquad$ $437_8 = 100\ 011\ 111_2$

and $\qquad$ $26.35_8 = 010\ 110.011\ 101_2$

The '0' on the extreme left does not signify anything, thus

$$26.35_8 = 10\ 110.011\ 101_2$$

Conversion of decimal to binary via octal is demonstrated in the following worked problems.

Problem 13. Convert 3714_{10} to a binary number via octal.

Dividing repeatedly by 8, and noting the remainder gives:

```
8 | 3714   Remainder
8 |  464      2 ───────────────┐
8 |   58      0 ──────────┐     │
8 |    7      2 ─────┐     │     │
     0        7 ┐    │     │     │
                ▼    ▼     ▼     ▼
                7    2     0     2
```

From Table 37.1, $\qquad 7202_8 = 111\ 010\ 000\ 010_2$

i.e. $\qquad$ $\mathbf{3714_{10} = 111\ 010\ 000\ 010_2}$

Problem 14. Convert 0.59375_{10} to a binary number via octal.

Multiplying repeatedly by 8, and noting the integer values, gives:

$$0.59375 \times 8 = \quad\longrightarrow\quad 4.75$$
$$0.75 \quad \times 8 = \qquad\qquad\longrightarrow 6.00$$
$$\qquad\qquad\qquad\qquad .4 \quad 6$$

Thus, $\qquad\qquad 0.59375_{10} = 0.46_8$

From Table 37.1, $\qquad 0.46_8 = 0.100\ 110_2$

i.e. $\qquad$ $\mathbf{0.59375_{10} = 0.100\ 11_2}$

Problem 15. Convert 5613.90625_{10} to a binary number via octal.

The integer part is repeatedly divided by 8, noting the remainder, giving:

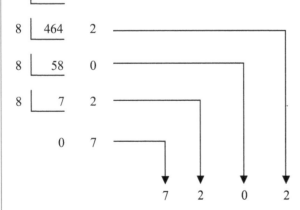

```
8 | 5613   Remainder
8 |  701      5 ──────────────────────┐
8 |   87      5 ─────────────────┐     │
8 |   10      7 ────────────┐     │     │
8 |    1      2 ───────┐     │     │     │
     0        1 ┐      │     │     │     │
                ▼      ▼     ▼     ▼     ▼
                1      2     7     5     5
```

This octal number is converted to a binary number, (see Table 37.1)

$$12755_8 = 001\ 010\ 111\ 101\ 101_2$$

i.e. $5613_{10} = 1\ 010\ 111\ 101\ 101_2$

The fractional part is repeatedly multiplied by 8, and noting the integer part, giving:

$$0.90625 \times 8 = \rule{2cm}{0pt} 7.25$$

$$0.25 \quad \times 8 = \rule{2cm}{0pt} 2.00$$

$$.7 \quad 2$$

This octal fraction is converted to a binary number, (see Table 37.1)

$$0.72_8 = 0.111\ 010_2$$

i.e. $\quad 0.90625_{10} = 0.111\ 01_2$

Thus, $\mathbf{5613.90625_{10} = 1\ 010\ 111\ 101\ 101.111\ 01_2}$

Problem 16. Convert $11\ 110\ 011.100\ 01_2$ to a decimal number via octal.

Grouping the binary number in three's from the binary point gives:

$$011\ 110\ 011.100\ 010_2$$

Using Table 37.1 to convert this binary number to an octal number gives:

$$363.42_8$$

and $363.42_8 = 3 \times 8^2 + 6 \times 8^1 + 3 \times 8^0$
$$+ 4 \times 8^{-1} + 2 \times 8^{-2}$$
$$= 192 + 48 + 3 + 0.5 + 0.03125_{10}$$
$$= 243.53125_{10}$$

i.e. $\mathbf{11\ 110\ 011.100\ 01_2 = 363.42_8 = 243.53125_{10}}$

Now try the following Practice Exercise

Practice Exercise 151   Conversion between decimal and binary numbers via octal (answers on page 438)

In Problems 1 to 3, convert the decimal numbers given to binary numbers, via octal.

1. (a) 343   (b) 572   (c) 1265

2. (a) 0.46875   (b) 0.6875   (c) 0.71875

3. (a) 247.09375 (b) 514.4375   (c) 1716.78125

4. Convert the following binary numbers to decimal numbers via octal:

(a) 111.011 1   (b) 101 001.01
(c) 1 110 011 011 010.001 1

37.4   Hexadecimal numbers

The hexadecimal system is particularly important in computer programming, since four bits (each consisting of a one or zero) can be succinctly expressed using a single hexadecimal digit. Two hexadecimal digits represent numbers from 0 to 255, a common range used, for example, to specify colours. Thus, in the HTML language of the web, colours are specified using three pairs of hexadecimal digits RRGGBB, where RR is the amount of red, GG the amount of green, and BB the amount of blue.

A **hexadecimal numbering system** has a radix of 16 and uses the following 16 distinct digits:

0, 1, 2, 3, 4, 5, 6, 7, 8, 9, A, B, C, D, E and F

'A' corresponds to 10 in the denary system, B to 11, C to 12, and so on.

(a) Converting from hexadecimal to decimal

For example, $1A_{16} = 1 \times 16^1 + A \times 16^0$
$$= 1 \times 16^1 + 10 \times 1$$
$$= 16 + 10 = 26$$

i.e. $\quad \mathbf{1A_{16} = 26_{10}}$

Similarly, $\quad \mathbf{2E_{16}} = 2 \times 16^1 + E \times 16^0$
$$= 2 \times 16^1 + 14 \times 16^0$$
$$= 32 + 14 = \mathbf{46_{10}}$$

and $\quad \mathbf{1BF_{16}} = 1 \times 16^2 + B \times 16^1 + F \times 16^0$
$$= 1 \times 16^2 + 11 \times 16^1 + 15 \times 16^0$$
$$= 256 + 176 + 15 = \mathbf{447_{10}}$$

Table 37.2 compares decimal, binary, octal and hexadecimal numbers and shows, for example, that

$$23_{10} = 10111_2 = 27_8 = 17_{16}$$

Problem 17. Convert the following hexadecimal numbers into their decimal equivalents:
(a) $7A_{16}$ (b) $3F_{16}$

(a) $7A_{16} = 7 \times 16^1 + A \times 16^0 = 7 \times 16 + 10 \times 1$
$$= 112 + 10 = 122$$
Thus, $\quad \mathbf{7A_{16} = 122_{10}}$

(b) $3F_{16} = 3 \times 16^1 + F \times 16^0 = 3 \times 16 + 15 \times 1$
$$= 48 + 15 = 63$$
Thus, $\quad \mathbf{3F_{16} = 63_{10}}$

Problem 18. Convert the following hexadecimal numbers into their decimal equivalents:
(a) $C9_{16}$ (b) BD_{16}

Table 37.2

Decimal	Binary	Octal	Hexadecimal
0	0000	0	0
1	0001	1	1
2	0010	2	2
3	0011	3	3
4	0100	4	4
5	0101	5	5
6	0110	6	6
7	0111	7	7
8	1000	10	8
9	1001	11	9
10	1010	12	A
11	1011	13	B
12	1100	14	C
13	1101	15	D
14	1110	16	E
15	1111	17	F
16	10000	20	10
17	10001	21	11
18	10010	22	12
19	10011	23	13
20	10100	24	14
21	10101	25	15
22	10110	26	16
23	10111	27	17
24	11000	30	18
25	11001	31	19
26	11010	32	1A
27	11011	33	1B
28	11100	34	1C
29	11101	35	1D
30	11110	36	1E
31	11111	37	1F
32	100000	40	20

(a) $C9_{16} = C \times 16^1 + 9 \times 16^0 = 12 \times 16 + 9 \times 1$
$$= 192 + 9 = 201$$

Thus, $\qquad C9_{16} = 201_{10}$

(b) $BD_{16} = B \times 16^1 + D \times 16^0 = 11 \times 16 + 13 \times 1$
$$= 176 + 13 = 189$$

Thus, $\qquad BD_{16} = 189_{10}$

Problem 19. Convert $1A4E_{16}$ into a decimal number.

$1A4E_{16} = 1 \times 16^3 + A \times 16^2 + 4 \times 16^1 + E \times 16^0$

$$= 1 \times 16^3 + 10 \times 16^2 + 4 \times 16^1 + 14 \times 16^0$$

$$= 1 \times 4096 + 10 \times 256 + 4 \times 16 + 14 \times 1$$

$$= 4096 + 2560 + 64 + 14 = 6734$$

Thus, $\mathbf{1A4E_{16} = 6734_{10}}$

(b) Converting from decimal to hexadecimal

This is achieved by repeatedly dividing by 16 and noting the remainder at each stage, as shown below for 26_{10}

Hence, $\qquad \mathbf{26_{10} = 1A_{16}}$

Similarly, for 446_{10}

Thus, $\qquad \mathbf{446_{10} = 1BE_{16}}$

Problem 20. Convert the following decimal numbers into their hexadecimal equivalents:
(a) 39_{10} (b) 107_{10}

(a)   16 | 39   Remainder

16 | 2 $7 \equiv 7_{16}$

0        $2 \equiv 2_{16}$

2   7

(most significant bit)   (least significant bit)

Hence, $\mathbf{39_{10} = 27_{16}}$

(b)   16 | 107   Remainder

16 | 6 $11 \equiv B_{16}$

0        $6 \equiv 6_{16}$

6   B

Hence, $\mathbf{107_{10} = 6B_{16}}$

Problem 21. Convert the following decimal numbers into their hexadecimal equivalents:
(a) 164_{10} (b) 239_{10}

(a)   16 | 164   Remainder

16 | 10 $4 \equiv 4_{16}$

0        $10 \equiv A_{16}$

A   4

Hence, $\mathbf{164_{10} = A4_{16}}$

(b)   16 | 239   Remainder

16 | 14 $15 \equiv F_{16}$

0        $14 \equiv E_{16}$

E   F

Hence, $\mathbf{239_{10} = EF_{16}}$

Now try the following Practice Exercise

Practice Exercise 152   Hexadecimal numbers (answers on page 438)

In Problems 1 to 4, convert the given hexadecimal numbers into their decimal equivalents.

1. $E7_{16}$ 2. $2C_{16}$ 3. 98_{16} 4. $2F1_{16}$

In Problems 5 to 8, convert the given decimal numbers into their hexadecimal equivalents.

5. 54_{10} 6. 200_{10} 7. 91_{10} 8. 238_{10}

(c) Converting from binary to hexadecimal

The binary bits are arranged in groups of four, starting from right to left, and a hexadecimal symbol is assigned to each group. For example, the binary number 1110011110101001

is initially grouped in

fours as: 1110   0111   1010   1001

and a hexadecimal symbol assigned to each group as: E   7   A   9

from Table 37.2

Hence, $\mathbf{1110011110101001_2 = E7A9_{16}}$

Problem 22. Convert the following binary numbers into their hexadecimal equivalents:
(a) 11010110_2 (b) 1100111_2

(a)   Grouping bits in fours from the

right gives: 1101   0110

and assigning hexadecimal symbols

to each group gives: D   6

from Table 37.2

Thus, $\mathbf{11010110_2 = D6_{16}}$

(b) Grouping bits in fours from the
right gives: 0110   0111
and assigning hexadecimal symbols
to each group gives: 6   7
from Table 37.2

Thus, $1100111_2 = 67_{16}$

> **Problem 23.** Convert the following binary numbers into their hexadecimal equivalents:
> (a) 11001111_2 (b) 110011110_2

(a) Grouping bits in fours from the
right gives: 1100   1111
and assigning hexadecimal symbols
to each group gives: C   F
from Table 37.2

Thus, $11001111_2 = CF_{16}$

(b) Grouping bits in fours
from the right gives: 0001   1001   1110
and assigning hexadecimal
symbols to each group gives: 1   9   E
from Table 37.2

Thus, $110011110_2 = 19E_{16}$

(d) Converting from hexadecimal to binary

The above procedure is reversed, thus, for example,

$6CF3_{16} = 0110\ 1100\ 1111\ 0011$ from Table 37.2

i.e. $6CF3_{16} = 110110011110011_2$

> **Problem 24.** Convert the following hexadecimal numbers into their binary equivalents:
> (a) $3F_{16}$ (b) $A6_{16}$

(a) Spacing out hexadecimal
digits gives: 3   F
and converting each into
binary gives: 0011   1111
from Table 37.2

Thus, $3F_{16} = 111111_2$

(b) Spacing out hexadecimal
digits gives: A   6
and converting each into
binary gives: 1010   0110
from Table 37.2

Thus, $A6_{16} = 10100110_2$

> **Problem 25.** Convert the following hexadecimal numbers into their binary equivalents:
> (a) $7B_{16}$ (b) $17D_{16}$

(a) Spacing out hexadecimal
digits gives: 7   B
and converting each into
binary gives: 0111   1011
from Table 37.2

Thus, $7B_{16} = 1111011_2$

(b) Spacing out hexadecimal
digits gives: 1   7   D
and converting each into
binary gives: 0001   0111   1101
from Table 37.2

Thus, $17D_{16} = 101111101_2$

Now try the following Practice Exercise

>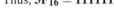
> **Practice Exercise 153**   **Hexadecimal numbers (answers on page 438)**
>
> In Problems 1 to 4, convert the given binary numbers into their hexadecimal equivalents.
>
> 1. 11010111_2 2. 11101010_2
> 3. 10001011_2 4. 10100101_2
>
> In Problems 5 to 8, convert the given hexadecimal numbers into their binary equivalents.
>
> 5. 37_{16} 6. ED_{16}
> 7. $9F_{16}$ 8. $A21_{16}$

For fully worked solutions to each of the problems in Practice Exercises 148 to 153 in this chapter, go to the website:
www.routledge.com/cw/bird

Chapter 38

Inequalities

Why it is important to understand: **Inequalities**

In mathematics, an inequality is a relation that holds between two values when they are different. A working knowledge of inequalities can be beneficial to the practising engineer, and inequalities are central to the definitions of all limiting processes, including differentiation and integration. When exact solutions are unavailable, inconvenient, or unnecessary, inequalities can be used to obtain error bounds for numerical approximation. Understanding and using inequalities is important in many branches of engineering.

At the end of this chapter, you should be able to:

- define an inequality
- state simple rules for inequalities
- solve simple inequalities
- solve inequalities involving a modulus
- solve inequalities involving quotients
- solve inequalities involving square functions
- solve quadratic inequalities

38.1 Introduction to inequalities

An **inequality** is any expression involving one of the symbols $<$, $>$, $\leq$ or $\geq$

$p < q$ means p is less than q

$p > q$ means p is greater than q

$p \leq q$ means p is less than or equal to q

$p \geq q$ means p is greater than or equal to q

38.1.1 Some simple rules

(i) When a quantity is **added or subtracted** to both sides of an inequality, the inequality still remains.

For example, if $p < 3$

then $\quad p + 2 < 3 + 2$ (adding 2 to both sides)

and $\quad p - 2 < 3 - 2$ (subtracting 2 from both sides)

(ii) When **multiplying or dividing** both sides of an inequality by a **positive** quantity, say 5, the inequality **remains the same**. For example,

$$\text{if } p > 4 \quad \text{then } 5p > 20 \quad \text{and} \quad \frac{p}{5} > \frac{4}{5}$$

(iii) When **multiplying or dividing** both sides of an inequality by a **negative** quantity, say -3, **the inequality is reversed**. For example,

$$\text{if } p > 1 \quad \text{then } -3p < -3 \quad \text{and} \quad \frac{p}{-3} < \frac{1}{-3}$$

(Note $>$ has changed to $<$ in each example)

To **solve an inequality** means finding all the values of the variable for which the inequality is true. Knowledge of simple equations (Chapter 11) and quadratic equations (Chapter 14) are needed in this chapter.

38.2 Simple inequalities

The solution of some simple inequalities, using only the rules given in Section 38.1, is demonstrated in the following worked problems.

> **Problem 1.** Solve the following inequalities:
> (a) $3 + x > 7$ (b) $3t < 6$
> (c) $z - 2 \geq 5$ (d) $\dfrac{p}{3} \leq 2$

(a) Subtracting 3 from both sides of the inequality: $3 + x > 7$ gives:

$$3 + x - 3 > 7 - 3, \text{ i.e. } \boldsymbol{x > 4}$$

Hence, all values of x greater than 4 satisfy the inequality.

(b) Dividing both sides of the inequality: $3t < 6$ by 3 gives:

$$\frac{3t}{3} < \frac{6}{3}, \text{ i.e. } \boldsymbol{t < 2}$$

Hence, all values of t less than 2 satisfy the inequality.

(c) Adding 2 to both sides of the inequality: $z - 2 \geq 5$ gives:

$$z - 2 + 2 \geq 5 + 2, \text{ i.e. } \boldsymbol{z \geq 7}$$

Hence, all values of z equal to or greater than 7 satisfy the inequality.

(d) Multiplying both sides of the inequality: $\dfrac{p}{3} \leq 2$ by 3 gives:

$$(3)\frac{p}{3} \leq (3)2, \text{ i.e. } \boldsymbol{p \leq 6}$$

Hence, all values of p equal to or less than 6 satisfy the inequality.

> **Problem 2.** Solve the inequality: $4x + 1 > x + 5$

Subtracting 1 from both sides of the inequality: $4x + 1 > x + 5$ gives:

$$4x > x + 4$$

Subtracting x from both sides of the inequality: $4x > x + 4$ gives:

$$3x > 4$$

Dividing both sides of the inequality: $3x > 4$ by 3 gives:

$$x > \frac{4}{3}$$

Hence all values of x greater than $\dfrac{4}{3}$ satisfy the inequality:

$$4x + 1 > x + 5$$

> **Problem 3.** Solve the inequality: $3 - 4t \leq 8 + t$

Subtracting 3 from both sides of the inequality: $3 - 4t \leq 8 + t$ gives:

$$-4t \leq 5 + t$$

Subtracting t from both sides of the inequality: $-4t \leq 5 + t$ gives:

$$-5t \leq 5$$

Dividing both sides of the inequality: $-5t \leq 5$ by -5 gives:

$$\boldsymbol{t \geq -1} \text{ (remembering to reverse the inequality)}$$

Hence, all values of t greater than or equal to -1 satisfy the inequality.

Now try the following Practice Exercise

> **Practice Exercise 154   Simple inequalities (answers on page 438)**
>
> Solve the following inequalities:
>
> 1. (a) $3t > 6$ (b) $2x < 10$
> 2. (a) $\dfrac{x}{2} > 1.5$ (b) $x + 2 \geq 5$
> 3. (a) $4t - 1 \leq 3$ (b) $5 - x \geq -1$
> 4. (a) $\dfrac{7 - 2k}{4} \leq 1$ (b) $3z + 2 > z + 3$
> 5. (a) $5 - 2y \leq 9 + y$ (b) $1 - 6x \leq 5 + 2x$

38.3 Inequalities involving a modulus

The **modulus** of a number is the size of the number, regardless of sign. Vertical lines enclosing the number denote a modulus.

For example, $|4|=4$ and $|-4|=4$ (the modulus of a number is never negative)

The inequality: $|t|<1$ means that all numbers whose actual size, regardless of sign, is less than 1, i.e. any value between -1 and $+1$.

Thus $|t|<1$ means $-1<t<1$

Similarly, $|x|>3$ means all numbers whose actual size, regardless of sign, is greater than 3, i.e. any value greater than 3 and any value less than -3.

Thus $|x|>3$ means $x>3$ and $x<-3$

Inequalities involving a modulus are demonstrated in the following worked problems.

Problem 4. Solve the following inequality:

$$|3x+1|<4$$

Since $|3x+1|<4$ then $-4<3x+1<4$

Now $-4<3x+1$ becomes $-5<3x$, i.e. $-\dfrac{5}{3}<x$

and $3x+1<4$ becomes $3x<3$, i.e. $x<1$

Hence, these two results together become $-\dfrac{5}{3}<x<1$ and mean that the inequality $|3x+1|<4$ is satisfied for any value of x greater than $-\dfrac{5}{3}$ but less than 1.

Problem 5. Solve the inequality: $|1+2t|\leq5$

Since $|1+2t|\leq5$ then $-5\leq1+2t\leq5$

Now $-5\leq1+2t$ becomes $-6\leq2t$ i.e. $-3\leq t$

and $1+2t\leq5$ becomes $2t\leq4$ i.e. $t\leq2$

Hence, these two results together become: $-3\leq t\leq2$

Problem 6. Solve the inequality: $|3z-4|>2$

$|3z-4|>2$ means $3z-4>2$ and $3z-4<-2$

i.e. $3z>6$ and $3z<2$

i.e. the inequality: $|3z-4|>2$ is satisfied when

$$z>2 \text{ and } z<\frac{2}{3}$$

Now try the following Practice Exercise

Practice Exercise 155   Inequalities involving a modulus (answers on page 438)

Solve the following inequalities:

1. $|t+1|<4$

2. $|y+3|\leq2$

3. $|2x-1|<4$

4. $|3t-5|>4$

5. $|1-k|\geq3$

38.4   Inequalities involving quotients

If $\dfrac{p}{q}>0$ then $\dfrac{p}{q}$ must be a **positive** value.

For $\dfrac{p}{q}$ to be positive, **either** p is positive **and** q is positive **or** p is negative **and** q is negative.

i.e. $\dfrac{+}{+}=+$ and $\dfrac{-}{-}=+$

If $\dfrac{p}{q}<0$ then $\dfrac{p}{q}$ must be a **negative** value.

For $\dfrac{p}{q}$ to be negative, **either** p is positive **and** q is negative **or** p is negative **and** q is positive.

i.e. $\dfrac{+}{-}=-$ and $\dfrac{-}{+}=-$

This reasoning is used when solving inequalities involving quotients, as demonstrated in the following worked problems.

Problem 7. Solve the inequality: $\dfrac{t+1}{3t-6}>0$

Since $\dfrac{t+1}{3t-6}>0$ then $\dfrac{t+1}{3t-6}$ must be **positive**.

For $\dfrac{t+1}{3t-6}$ to be positive,

either (i) $t+1>0$ **and** $3t-6>0$

or (ii) $t+1<0$ **and** $3t-6<0$

(i) If $t+1>0$ then $t>-1$ and if $3t-6>0$ then $3t>6$ and $t>2$

Both of the inequalities $t>-1$ and $t>2$ are only true when $t>2$,

i.e. the fraction $\dfrac{t+1}{3t-6}$ is positive when $t>2$

(ii) If $t+1<0$ then $t<-1$ and if $3t-6<0$ then $3t<6$ and $t<2$

Both of the inequalities $t<-1$ and $t<2$ are only true when $t<-1$,

i.e. the fraction $\dfrac{t+1}{3t-6}$ is positive when $t<-1$

Summarising, $\dfrac{t+1}{3t-6}>0$ when $t>2$ or $t<-1$

Problem 8. Solve the inequality: $\dfrac{2x+3}{x+2}\leq 1$

Since $\dfrac{2x+3}{x+2}\leq 1$ then $\dfrac{2x+3}{x+2}-1\leq 0$

i.e. $\dfrac{2x+3}{x+2}-\dfrac{x+2}{x+2}\leq 0$

i.e. $\dfrac{2x+3-(x+2)}{x+2}\leq 0$ or $\dfrac{x+1}{x+2}\leq 0$

For $\dfrac{x+1}{x+2}$ to be negative or zero,

   either (i) $x+1\leq 0$ **and** $x+2>0$

   or (ii) $x+1\geq 0$ **and** $x+2<0$

(i) If $x+1\leq 0$ then $x\leq -1$ and if $x+2>0$ then $x>-2$

(Note that $>$ is used for the denominator, not $\geq$; a zero denominator gives a value for the fraction which is impossible to evaluate.)

Hence, the inequality $\dfrac{x+1}{x+2}\leq 0$ is true when x is greater than -2 and less than or equal to -1, which may be written as $-2<x\leq -1$

(ii) If $x+1\geq 0$ then $x\geq -1$ and if $x+2<0$ then $x<-2$

It is not possible to satisfy both $x\geq -1$ and $x<-2$ thus no value of x satisfies (ii).

Summarising, $\dfrac{2x+3}{x+2}\leq 1$ when $-2<x\leq -1$

Now try the following Practice Exercise

Practice Exercise 156 Inequalities involving quotients (answers on page 439)

Solve the following inequalitites:

1. $\dfrac{x+4}{6-2x}\geq 0$ 2. $\dfrac{2t+4}{t-5}>1$

3. $\dfrac{3z-4}{z+5}\leq 2$ 4. $\dfrac{2-x}{x+3}\geq 4$

38.5 Inequalities involving square functions

The following two general rules apply when inequalities involve square functions:

(i) **if $x^2>k$ then $x>\sqrt{k}$ or $x<-\sqrt{k}$**   (1)

(ii) **if $x^2<k$ then $-\sqrt{k}<x<\sqrt{k}$**   (2)

These rules are demonstrated in the following worked problems.

Problem 9. Solve the inequality: $t^2>9$

Since $t^2>9$ then $t^2-9>0$, i.e. $(t+3)(t-3)>0$ by factorising.
For $(t+3)(t-3)$ to be positive,

   either (i) $(t+3)>0$ **and** $(t-3)>0$

   or (ii) $(t+3)<0$ **and** $(t-3)<0$

(i) If $(t+3)>0$ then $t>-3$ and if $(t-3)>0$ then $t>3$
Both of these are true only when $t>3$

(ii) If $(t+3)<0$ then $t<-3$ and if $(t-3)<0$ then $t<3$
Both of these are true only when $t<-3$

Summarising, $t^2>9$ when $t>3$ or $t<-3$

This demonstrates the general rule:

   if $x^2>k$ then $x>\sqrt{k}$ or $x<-\sqrt{k}$ (1)

Problem 10. Solve the inequality: $x^2>4$

From the general rule stated above in equation (1):

   if $x^2>4$ then $x>\sqrt{4}$ or $x<-\sqrt{4}$

i.e. the inequality: $x^2 > 4$ is satisfied when $x > 2$ or $x < -2$

Problem 11. Solve the inequality: $(2z+1)^2 > 9$

From equation (1), if $(2z+1)^2 > 9$ then

$$2z+1 > \sqrt{9} \quad \text{or} \quad 2z+1 < -\sqrt{9}$$

i.e. $2z+1 > 3 \quad \text{or} \quad 2z+1 < -3$

i.e. $2z > 2 \quad \text{or} \quad 2z < -4$

i.e. $z > 1 \quad \text{or} \quad z < -2$

Problem 12. Solve the inequality: $t^2 < 9$

Since $t^2 < 9$ then $t^2 - 9 < 0$, i.e. $(t+3)(t-3) < 0$ by factorising.
For $(t+3)(t-3)$ to be negative,

either (i) $(t+3) > 0$ **and** $(t-3) < 0$

or (ii) $(t+3) < 0$ **and** $(t-3) > 0$

(i) If $(t+3) > 0$ then $t > -3$ and if $(t-3) < 0$ then $t < 3$

Hence (i) is satisfied when $t > -3$ and $t < 3$ which may be written as: $-3 < t < 3$

(ii) If $(t+3) < 0$ then $t < -3$ and if $(t-3) > 0$ then $t > 3$

It is not possible to satisfy both $t < -3$ and $t > 3$, thus no value of t satisfies (ii).

Summarising, $t^2 < 9$ when $-3 < t < 3$ which means that all values of t between -3 and $+3$ will satisfy the inequality.

This demonstrates the general rule:

$$\text{if } x^2 < k \text{ then} -\sqrt{k} < x < \sqrt{k} \tag{2}$$

Problem 13. Solve the inequality: $x^2 < 4$

From the general rule stated above in equation (2):

$$\text{if } x^2 < 4 \text{ then } -\sqrt{4} < x < \sqrt{4}$$

i.e. the inequality: $x^2 < 4$ is satisfied when: $-2 < x < 2$

Problem 14. Solve the inequality: $(y-3)^2 \le 16$

From equation (2), $-\sqrt{16} \le (y-3) \le \sqrt{16}$

i.e. $-4 \le (y-3) \le 4$

from which, $3-4 \le y \le 4+3$

i.e. $-1 \le y \le 7$

Now try the following Practice Exercise

Practice Exercise 157 Inequalities involving square functions (answers on page 439)

Solve the following inequalities:

1. $z^2 > 16$ 2. $z^2 < 16$

3. $2x^2 \ge 6$ 4. $3k^2 - 2 \le 10$

5. $(t-1)^2 \le 36$ 6. $(t-1)^2 \ge 36$

7. $7 - 3y^2 \le -5$ 8. $(4k+5)^2 > 9$

38.6 Quadratic inequalities

Inequalities involving quadratic expressions are solved using either **factorisation** or **'completing the square'**. For example,

$$x^2 - 2x - 3 \text{ is factorised as } (x+1)(x-3)$$

and $6x^2 + 7x - 5$ is factorised as $(2x-1)(3x+5)$

If a quadratic expression does not factorise, then the technique of 'completing the square' is used. In general, the procedure for $x^2 + bx + c$ is:

$$x^2 + bx + c \equiv \left(x + \frac{b}{2}\right)^2 + c - \left(\frac{b}{2}\right)^2$$

For example, $x^2 + 4x - 7$ does not factorise; completing the square gives:

$$x^2 + 4x - 7 \equiv (x+2)^2 - 7 - 2^2 \equiv (x+2)^2 - 11$$

Similarly,

$$x^2 + 6x - 5 \equiv (x+3)^2 - 5 - 3^2 \equiv (x-3)^2 - 14$$

Solving quadratic inequalities is demonstrated in the following worked problems.

Problem 15. Solve the inequality:
$$x^2 + 2x - 3 > 0$$

Since $x^2 + 2x - 3 > 0$ then $(x-1)(x+3) > 0$ by factorising.
For the product $(x-1)(x+3)$ to be positive,

either   (i) $(x-1) > 0$ and $(x+3) > 0$

or   (ii) $(x-1) < 0$ and $(x+3) < 0$

(i)   Since $(x-1) > 0$ then $x > 1$ and since $(x+3) > 0$ then $x > -3$

     Both of these inequalities are satisfied only when
$x > 1$

(ii)   Since $(x-1) < 0$ then $x < 1$ and since $(x+3) < 0$ then $x < -3$

     Both of these inequalities are satisfied only when
$x < -3$

Summarising, $x^2 + 2x - 3 > 0$ is satisfied when either
$x > 1$ or $x < -3$

Problem 16. Solve the inequality: $t^2 - 2t - 8 < 0$

Since $t^2 - 2t - 8 < 0$ then $(t-4)(t+2) < 0$ by factorising.
For the product $(t-4)(t+2)$ to be negative,

either   (i) $(t-4) > 0$ and $(t+2) < 0$

or   (ii) $(t-4) < 0$ and $(t+2) > 0$

(i)   Since $(t-4) > 0$ then $t > 4$ and since $(t+2) < 0$ then $t < -2$
     It is not possible to satisfy both $t > 4$ and $t < -2$, thus no value of t satisfies the inequality (i)

(ii)   Since $(t-4) < 0$ then $t < 4$ and since $(t+2) > 0$ then $t > -2$

Hence, (ii) is satisfied when $-2 < t < 4$

Summarising, $t^2 - 2t - 8 < 0$ is satisfied when
$-2 < t < 4$

Problem 17. Solve the inequality:
$$x^2 + 6x + 3 < 0$$

$x^2 + 6x + 3$ does not factorise; completing the square gives:
$$x^2 + 6x + 3 \equiv (x+3)^2 + 3 - 3^2$$
$$\equiv (x+3)^2 - 6$$
The inequality thus becomes: $(x+3)^2 - 6 < 0$ or
$$(x+3)^2 < 6$$

From equation (2), $-\sqrt{6} < (x+3) < \sqrt{6}$

from which, $(-\sqrt{6} - 3) < x < (\sqrt{6} - 3)$

Hence, $x^2 + 6x + 3 < 0$ is satisfied when
$-5.45 < x < -0.55$, correct to 2 decimal places.

Problem 18. Solve the inequality:
$$y^2 - 8y - 10 \geq 0$$

$y^2 - 8y - 10$ does not factorise; completing the square gives:
$$y^2 - 8y - 10 \equiv (y-4)^2 - 10 - 4^2$$
$$\equiv (y-4)^2 - 26$$
The inequality thus becomes: $(y-4)^2 - 26 \geq 0$ or
$$(y-4)^2 \geq 26$$

From equation (1), $(y-4) \geq \sqrt{26}$ or $(y-4) \leq -\sqrt{26}$

from which,   **$y \geq 4 + \sqrt{26}$ or $y \leq 4 - \sqrt{26}$**

Hence, $y^2 - 8y - 10 \geq 0$ is satisfied when **$y \geq 9.10$ or $y \leq -1.10$**, correct to 2 decimal places.

Now try the following Practice Exercise

Practice Exercise 158   Quadratic inequalities (answers on page 439)

Solve the following inequalities:

1. $x^2 - x - 6 > 0$

2. $t^2 + 2t - 8 \leq 0$

3. $2x^2 + 3x - 2 < 0$

4. $y^2 - y - 20 \geq 0$

5. $z^2 + 4z + 4 \leq 4$

6. $x^2 + 6x + 6 \leq 0$

7. $t^2 - 4t - 7 \geq 0$

8. $k^2 + k - 3 \geq 0$

For fully worked solutions to each of the problems in Practice Exercises 154 to 158 in this chapter, go to the website:
www.routledge.com/cw/bird

Chapter 39

Graphs with logarithmic scales

Why it is important to understand: Graphs with logarithmic scales

As mentioned in previous chapters, graphs are important tools for analysing and displaying data between two experimental quantities and that many times situations occur in which the relationship between the variables is not linear. By manipulation, a straight line graph may be plotted to produce a law relating the two variables. Knowledge of logarithms may be used to simplify plotting the relation between one variable and another. In particular, we consider those situations in which one of the variables requires scaling because the range of its data values is very large in comparison to the range of the other variable. Log-log and log-linear graph paper is available to make the plotting process easier.

At the end of this chapter, you should be able to:

- understand logarithmic scales
- understand log-log and log-linear graph paper
- plot a graph of the form $y = ax^n$ using log-log graph paper and determine constants 'a' and 'n'
- plot a graph of the form $y = ab^x$ using log-linear graph paper and determine constants 'a' and 'b'
- plot a graph of the form $y = ae^{kx}$ using log-linear graph paper and determine constants 'a' and 'k'

39.1 Logarithmic scales and logarithmic graph paper

Graph paper is available where the scale markings along the horizontal and vertical axes are proportional to the logarithms of the numbers. Such graph paper is called **log-log graph paper**.

A **logarithmic scale** is shown in Fig. 39.1 where the distance between, say 1 and 2, is proportional to lg 2 − lg 1, i.e. 0.3010 of the total distance from 1 to 10. Similarly, the distance between 7 and 8 is proportional to lg 8 − lg 7, i.e. 0.05799 of the total distance from 1 to 10. Thus the distance between markings progressively decreases as the numbers increase from 1 to 10.

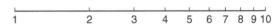

Figure 39.1

With log-log graph paper the scale markings are from 1 to 9, and this pattern can be repeated several times. The number of times the pattern of markings is repeated on an axis signifies the number of **cycles**. When the vertical axis has, say, 3 sets of values from 1 to 9, and the horizontal axis has, say, 2 sets of values from 1 to 9, then this log-log graph paper is called 'log 3 cycle × 2 cycle' (see Fig. 39.2). Many different arrangements are available ranging from 'log 1 cycle × 1 cycle' through to 'log 5 cycle × 5 cycle'.

To depict a set of values, say, from 0.4 to 161, on an axis of log-log graph paper, 4 cycles are required, from 0.1 to 1, 1 to 10, 10 to 100 and 100 to 1000.

39.2 Graphs of the form $y = ax^n$

Taking logarithms to a base of 10 of both sides of $y = ax^n$

gives: $\lg y = \lg(ax^n) = \lg a + \lg x^n$

i.e. $\lg y = n\lg x + \lg a$

which compares with $Y = mX + c$

Thus, by plotting $\lg y$ vertically against $\lg x$ horizontally, a straight line results, i.e. the equation $y = ax^n$ is reduced to linear form. With log-log graph paper available x and y may be plotted directly, without having first to determine their logarithms, as shown in Chapter 18.

Problem 1. Experimental values of two related quantities x and y are shown below:

x	0.41	0.63	0.92	1.36	2.17	3.95
y	0.45	1.21	2.89	7.10	20.79	82.46

The law relating x and y is believed to be $y = ax^b$, where 'a' and 'b' are constants. Verify that this law is true and determine the approximate values of 'a' and 'b'

If $y = ax^b$ then $\lg y = b \lg x + \lg a$, from above, which is of the form $Y = mX + c$, showing that to produce a straight line graph $\lg y$ is plotted vertically against $\lg x$ horizontally. x and y may be plotted directly on to log-log graph paper as shown in Fig. 39.2. The values of y range from 0.45 to 82.46 and 3 cycles are needed (i.e. 0.1 to 1, 1 to 10 and 10 to 100). The values of x range from 0.41 to 3.95 and 2 cycles are needed (i.e. 0.1 to 1 and 1 to 10). Hence 'log 3 cycle × 2 cycle' is used as shown in Fig. 39.2 where the axes are marked and the points plotted. **Since the points lie on a straight line the law $y = ax^b$ is verified**.

To evaluate constants 'a' and 'b':

Method 1. Any two points on the straight line, say points A and C, are selected, and AB and BC are

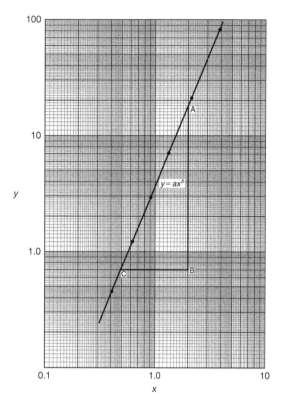

Figure 39.2

measured (say in centimetres). Then,

$$\text{gradient, } b = \frac{AB}{BC} = \frac{11.5\,\text{units}}{5\,\text{units}} = \mathbf{2.3}$$

Since $\lg y = b \lg x + \lg a$, when $x = 1$, $\lg x = 0$ and $\lg y = \lg a$.

The straight line crosses the ordinate $x = 1.0$ at $y = 3.5$. Hence, $\lg a = \lg 3.5$, i.e. $\mathbf{a = 3.5}$

Method 2. Any two points on the straight line, say points A and C, are selected. A has co-ordinates (2, 17.25) and C has co-ordinates (0.5, 0.7).

Since $y = ax^b$ then $17.25 = a(2)^b$ (1)

and $0.7 = a(0.5)^b$ (2)

i.e. two simultaneous equations are produced and may be solved for 'a' and 'b'.

Dividing equation (1) by equation (2) to eliminate 'a' gives:

$$\frac{17.25}{0.7} = \frac{(2)^b}{(0.5)^b} = \left(\frac{2}{0.5}\right)^b$$

i.e. $24.643 = (4)^b$

Taking logarithms of both sides gives:

$\lg 24.643 = b \lg 4,$

i.e. $b = \dfrac{\lg 24.643}{\lg 4}$

$= \textbf{2.3},$ correct to 2 significant figures.

Substituting $b = 2.3$ in equation (1) gives:
$17.25 = a(2)^{2.3}$

i.e. $a = \dfrac{17.25}{(2)^{2.3}} = \dfrac{17.25}{4.925}$

$= \textbf{3.5},$ correct to 2 significant figures

Hence, the law of the graph is: $\boldsymbol{y = 3.5x^{2.3}}$

Problem 2. The power dissipated by a resistor was measured for varying values of current flowing in the resistor and the results are as shown:

Current, I amperes	1.4	4.7	6.8	9.1	11.2	13.1
Power, P watts	49	552	1156	2070	3136	4290

Prove that the law relating current and power is of the form $P = RI^n$, where R and n are constants, and determine the law. Hence calculate the power when the current is 12 A and the current when the power is 1000 W.

Since $P = RI^n$ then $\lg P = n \lg I + \lg R$, which is of the form $Y = mX + c$, showing that to produce a straight line graph $\lg P$ is plotted vertically against $\lg I$ horizontally. Power values range from 49 to 4290, hence 3 cycles of log-log graph paper are needed (10 to 100, 100 to 1000 and 1000 to 10 000). Current values range from 1.4 to 13.1, hence 2 cycles of log-log graph paper are needed (1 to 10 and 10 to 100). Thus 'log 3 cycles × 2 cycles' is used as shown in Fig. 39.3 (or, if not available, graph paper having a larger number of cycles per axis can be used).
The co-ordinates are plotted and **a straight line results which proves that the law relating current and power is of the form** $\boldsymbol{P = RI^n}$

Gradient of straight line, $n = \dfrac{AB}{BC} = \dfrac{14 \text{ units}}{7 \text{ units}} = 2$

At point $C, I = 2$ and $P = 100$. Substituting these values into $P = RI^n$ gives:

$$100 = R(2)^2$$

Hence, $R = \dfrac{100}{2^2} = 25$

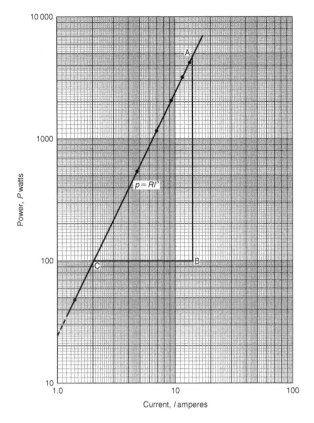

Figure 39.3

which may have been found from the intercept on the $I = 1.0$ axis in Fig. 39.3.
Hence, the law of the graph is: $\boldsymbol{P = 25I^2}$
When current $I = 12$,
power $P = 25(12)^2 = \textbf{3600 watts}$ (which may be read from the graph).

When power $P = 1000, 1000 = 25I^2$

Hence $I^2 = \dfrac{1000}{25} = 40$

from which,   **current** $I = \sqrt{40} = \textbf{6.32A}$

Problem 3. The pressure p and volume v of a gas are believed to be related by a law of the form $p = cv^n$, where c and n are constants. Experimental values of p and corresponding values of v obtained in a laboratory are:

p pascals	2.28×10^5	8.04×10^5	2.03×10^6
v m^3	3.2×10^{-2}	1.3×10^{-2}	6.7×10^{-3}

p pascals	5.05×10^6	1.82×10^7
v m^3	3.5×10^{-3}	1.4×10^{-3}

Verify that the law is true and determine approximate values of c and n.

Since $p = cv^n$, then $\lg p = n \lg v + \lg c$, which is of the form $Y = mX + c$, showing that to produce a straight line graph $\lg p$ is plotted vertically against $\lg v$ horizontally. The co-ordinates are plotted on 'log 3 cycle × 2 cycle' graph paper as shown in Fig. 39.4. With the data expressed in standard form, the axes are marked in standard form also. Since a straight line results the law $p = cv^n$ is verified.

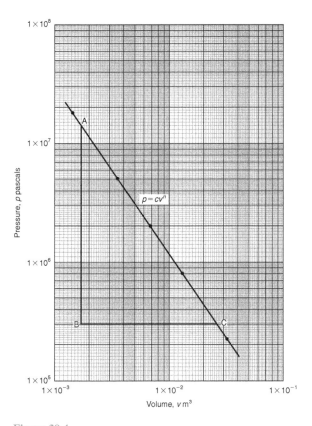

Figure 39.4

The straight line has a negative gradient and the value of the gradient is given by:

$$\frac{AB}{BC} = \frac{14\,\text{units}}{10\,\text{units}} = 1.4, \text{ hence } n = \mathbf{-1.4}$$

Selecting any point on the straight line, say point C, having co-ordinates $(2.63 \times 10^{-2}, 3 \times 10^5)$, and substituting these values in $p = cv^n$

gives: $3 \times 10^5 = c(2.63 \times 10^{-2})^{-1.4}$

Hence, $c = \dfrac{3 \times 10^5}{(2.63 \times 10^{-2})^{-1.4}} = \dfrac{3 \times 10^5}{(0.0263)^{-1.4}}$

$$= \frac{3 \times 10^5}{1.63 \times 10^2}$$

$$= \mathbf{1840}, \text{ correct to 3 significant figures.}$$

Hence, the law of the graph is: $p = 1840\,v^{-1.4}$ or $pv^{1.4} = 1840$

Now try the following Practice Exercise

Practice Exercise 159   Logarithmic graphs of the form $y = ax^n$ (answers on page 439)

1. Quantities x and y are believed to be related by a law of the form $y = ax^n$, where a and n are constants. Experimental values of x and corresponding values of y are:

x	0.8	2.3	5.4	11.5	21.6	42.9
y	8	54	250	974	3028	10410

Show that the law is true and determine the values of 'a' and 'n'. Hence determine the value of y when x is 7.5 and the value of x when y is 5000.

2. Show from the following results of voltage V and admittance Y of an electrical circuit that the law connecting the quantities is of the form $V = kY^n$, and determine the values of k and n.

Voltage, V volts	2.88	2.05	1.60	1.22	0.96
Admittance, Y siemens	0.52	0.73	0.94	1.23	1.57

3. Quantities x and y are believed to be related by a law of the form $y = mx^n$. The values of x and corresponding values of y are:

x	0.5	1.0	1.5	2.0	2.5	3.0
y	0.53	3.0	8.27	16.97	29.69	46.77

Verify the law and find the values of m and n.

39.3 Graphs of the form $y = ab^x$

Taking logarithms to a base of 10 of both sides of $y = ab^x$

gives:
$$\lg y = \lg (ab^x) = \lg a + \lg b^x$$
$$= \lg a + x \lg b$$

i.e.
$$\lg y = (\lg b)x + \lg a$$

which compares with $Y = mX + c$

Thus, by plotting $\lg y$ vertically against x horizontally a straight line results, i.e. the graph $y = ab^x$ is reduced to linear form. In this case, graph paper having a linear horizontal scale and a logarithmic vertical scale may be used. This type of graph paper is called **log-linear graph paper**, and is specified by the number of cycles on the logarithmic scale. For example, graph paper having 3 cycles on the logarithmic scale is called 'log 3 cycle × linear' graph paper.

Problem 4. Experimental values of quantities x and y are believed to be related by a law of the form $y = ab^x$, where 'a' and 'b' are constants. The values of x and corresponding values of y are:

x	0.7	1.4	2.1	2.9	3.7	4.3
y	18.4	45.1	111	308	858	1850

Verify the law and determine the approximate values of 'a' and 'b'. Hence evaluate (i) the value of y when x is 2.5, and (ii) the value of x when y is 1200.

Since $y = ab^x$ then $\lg y = (\lg b)x + \lg\ a$ (from above), which is of the form $Y = mX + c$, showing that to produce a straight line graph $\lg y$ is plotted vertically against x horizontally. Using log-linear graph paper, values of x are marked on the horizontal scale to cover the range 0.7 to 4.3. Values of y range from 18.4 to 1850 and 3 cycles are needed (i.e. 10 to 100, 100 to 1000 and 1000 to 10000). Thus, using 'log 3 cycles × linear' graph paper the points are plotted as shown in Fig. 39.5. **A straight line is drawn through the co-ordinates, hence the law $y = ab^x$ is verified.**
Gradient of straight line, $\lg b = AB/BC$. Direct measurement (say in centimetres) is not made with log-linear graph paper since the vertical scale is logarithmic and the horizontal scale is linear.

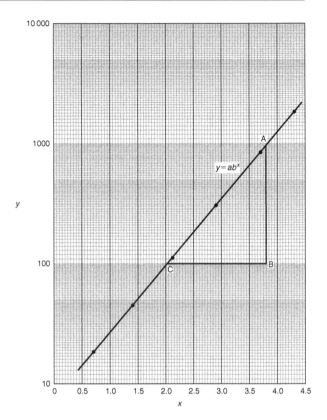

Figure 39.5

Hence,
$$\lg b = \frac{AB}{BC} = \frac{\lg 1000 - \lg 100}{3.82 - 2.02}$$
$$= \frac{3-2}{1.80} = \frac{1}{1.80} = 0.5556$$

Hence,
$$b = \text{antilog } 0.5556$$
$$= 10^{0.5556} = \textbf{3.6}, \text{ correct to 2 significant figures.}$$

Point A has co-ordinates $(3.82, 1000)$.
Substituting these values into $y = ab^x$

gives:
$$1000 = a(3.6)^{3.82}$$

i.e.
$$a = \frac{1000}{(3.6)^{3.82}} = \textbf{7.5}, \text{ correct to 2 significant figures.}$$

Hence, the law of the graph is: $y = 7.5(3.6)^x$

(i) When $x = 2.5$, $y = 7.5(3.6)^{2.5} = \textbf{184}$

(ii) When $y = 1200$, $1200 = 7.5(3.6)^x$,

hence $(3.6)^x = \dfrac{1200}{7.5} = 160$

Taking logarithms gives:

$$x \lg 3.6 = \lg 160$$

i.e.

$$x = \frac{\lg 160}{\lg 3.6} = \frac{2.2041}{0.5563} = \mathbf{3.96}$$

Now try the following Practice Exercise

Practice Exercise 160   Logarithmic graphs of the form $y = ab^x$ (answers on page 439)

1. Experimental values of p and corresponding values of q are shown below.

p	-13.2	-27.9	-62.2	-383.2	-1581	-2931
q	0.30	0.75	1.23	2.32	3.17	3.54

    Show that the law relating p and q is $p = ab^q$, where 'a' and 'b' are constants. Determine (i) values of 'a' and 'b', and state the law, (ii) the value of p when q is 2.0, and (iii) the value of q when p is -2000

39.4   Graphs of the form $y = ae^{kx}$

Taking logarithms to a base of e of both sides of $y = ae^{kx}$

gives:

$$\ln y = \ln(ae^{kx}) = \ln a + \ln e^{kx}$$

$$= \ln a + kx \ln e$$

i.e.

$$\ln y = kx + \ln a \quad \text{since} \quad \ln e = 1$$

which compares with $Y = mX + c$

Thus, by plotting $\ln y$ vertically against x horizontally, a straight line results, i.e. the equation $y = ae^{kx}$ is reduced to linear form. In this case, graph paper having a linear horizontal scale and a logarithmic vertical scale may be used.

Problem 5. The data given below is believed to be related by a law of the form $y = ae^{kx}$, where 'a' and 'k' are constants. Verify that the law is true and determine approximate values of 'a' and 'k'. Also, determine the value of y when x is 3.8 and the value of x when y is 85.

x	-1.2	0.38	1.2	2.5	3.4	4.2	5.3
y	9.3	22.2	34.8	71.2	117	181	332

Since $y = ae^{kx}$ then $\ln y = kx + \ln a$ (from above), which is of the form $Y = mX + c$, showing that to produce a straight line graph $\ln y$ is plotted vertically against x horizontally. The value of y ranges from 9.3 to 332 hence 'log 3 cycle × linear' graph paper is used. The plotted co-ordinates are shown in Fig. 39.6 and **since a straight line passes through the points the law $y = ae^{kx}$ is verified**.

Gradient of straight line,

$$k = \frac{AB}{BC} = \frac{\ln 100 - \ln 10}{3.12 - (-1.08)} = \frac{2.3026}{4.20}$$

$$= \mathbf{0.55}, \text{ correct to 2 significant figures.}$$

Since $\ln y = kx + \ln a$, when $x = 0, \ln y = \ln a$ i.e. $y = a$.

The vertical axis intercept value at $x = 0$ is 18, hence, $\boldsymbol{a = 18}$

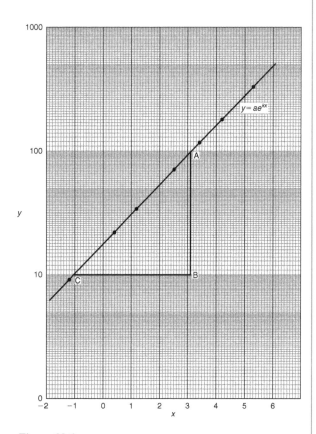

Figure 39.6

The law of the graph is thus: $y = 18e^{0.55x}$

When x is 3.8,

$$y = 18e^{0.55(3.8)} = 18e^{2.09} = 18(8.0849) = \mathbf{146}$$

When y is 85, $85 = 18e^{0.55x}$

Hence, $\qquad e^{0.55x} = \dfrac{85}{18} = 4.7222$

and $\qquad 0.55x = \ln 4.7222 = 1.5523$

Hence, $\qquad x = \dfrac{1.5523}{0.55} = \mathbf{2.82}$

Problem 6. The voltage, v volts, across an inductor is believed to be related to time, t ms, by the law $v = Ve^{t/T}$, where V and T are constants. Experimental results obtained are:

v volts	883	347	90	55.5	18.6	5.2
t ms		10.4	21.6	37.8	43.6	56.7	72.0

Show that the law relating voltage and time is as stated, and determine the approximate values of V and T. Find also the value of voltage after 25 ms and the time when the voltage is 30.0 V

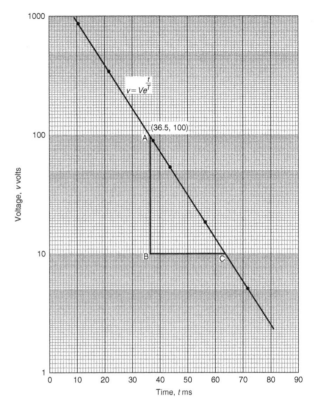

Figure 39.7

Since $v = Ve^{t/T}$ then: $\ln v = \ln(Ve^{t/T})$

$$= \ln V + \ln e^{t/T} = \ln V + \dfrac{t}{T} \ln e = \dfrac{t}{T} + \ln V$$

i.e. $\ln v = \dfrac{1}{T}t + \ln V$ which is of the form $Y = mX + c$

Using 'log 3 cycle × linear' graph paper, the points are plotted as shown in Fig. 39.7.
Since the points are joined by a straight line the law $v = Ve^{t/T}$ is verified.

Gradient of straight line,

$$\dfrac{1}{T} = \dfrac{AB}{BC} = \dfrac{\ln 100 - \ln 10}{36.5 - 64.2} = \dfrac{2.3026}{-27.7}$$

Hence, $\qquad T = \dfrac{-27.7}{2.3026} = \mathbf{-12.0}$, correct to 3 significant figures.

Since the straight line does not cross the vertical axis at $t = 0$ in Fig. 39.7, the value of V is determined by selecting any point, say A, having co-ordinates (36.5, 100) and substituting these values into $v = Ve^{t/T}$

Thus, $\qquad 100 = Ve^{36.5/-12.0}$

i.e. $\qquad V = \dfrac{100}{e^{-36.5/12.0}}$

$$= \mathbf{2090 \ volts}, \text{ correct to 3 significant}$$

figures.

Hence, the law of the graph is: $v = 2090e^{-t/12.0}$

When time $\quad t = 25$ ms, voltage $v = 2090e^{-25/12.0}$

$$= \mathbf{260 \, V}$$

When the voltage is 30.0 volts, $30.0 = 2090e^{-t/12.0}$

Hence, $\quad e^{-t/12.0} = \dfrac{30.0}{2090}$ and $e^{t/12.0} = \dfrac{2090}{30.0} = 69.67$

Taking Napierian logarithms gives:

$$\dfrac{t}{12.0} = \ln 69.67 = 4.2438$$

from which, time $t = (12.0)(4.2438) = \mathbf{50.9 \, ms}$

Now try the following Practice Exercise

Practice Exercise 161   Logarithmic graphs of the form $y = ae^{kx}$ **(answers on page 439)**

1. Atmospheric pressure p is measured at varying altitudes h and the results are as shown below:

Altitude, h m	500	1500	3000	5000	8000
pressure, p cm	73.39	68.42	61.60	53.56	43.41

Show that the quantities are related by the law $p = ae^{kh}$, where 'a' and 'k' are constants. Determine the values of 'a' and 'k' and state the law. Find also the atmospheric pressure at 10 000 m.

2. At particular times, t minutes, measurements are made of the temperature, $\theta°$C, of a cooling liquid and the following results are obtained:

Temperature $\theta°$C	92.2	55.9	33.9	20.6	12.5
Time t minutes	10	20	30	40	50

Prove that the quantities follow a law of the form $\theta = \theta_0 e^{kt}$, where θ_0 and k are constants, and determine the approximate value of θ_0 and k.

For fully worked solutions to each of the problems in Practice Exercises 159 to 161 in this chapter, go to the website:
www.routledge.com/cw/bird

Revision Test 15: Numbers, inequalities and logarithmic graphs

This assignment covers the material contained in Chapters 36 to 39. *The marks for each question are shown in brackets at the end of each question.*

1. Determine the 20th term of the series 15.6, 15, 14.4, 13.8, ... (3)

2. The sum of 13 terms of an arithmetic progression is 286 and the common difference is 3. Determine the first term of the series. (5)

3. An engineer earns £21 000 per annum and receives annual increments of £600. Determine the salary in the 9th year and calculate the total earnings in the first 11 years. (5)

4. Determine the 11th term of the series 1.5, 3, 6, 12, ... (2)

5. Find the sum of the first eight terms of the series $1, 2\frac{1}{2}, 6\frac{1}{4}, \ldots$, correct to 1 decimal place. (4)

6. Determine the sum to infinity of the series 5, 1, $\frac{1}{5}$, ... (3)

7. A machine is to have seven speeds ranging from 25 rev/min to 500 rev/min. If the speeds form a geometric progression, determine their value, each correct to the nearest whole number. (10)

8. Convert the following to decimal numbers:
   (a) 11010_2 (b) 101110_2 (6)

9. Convert the following decimal numbers into binary numbers:
   (a) 53   (b) 29 (8)

10. Determine the binary addition: $1011 + 11011$ (3)

11. Convert the hexadecimal number 3B into its binary equivalent. (2)

12. Convert 173_{10} into hexadecimal. (4)

13. Convert 1011011_2 into hexadecimal. (3)

14. Convert DF_{16} into its binary equivalent. (3)

15. Solve the following inequalities:

    (a) $2 - 5x \leq 9 + 2x$ (b) $|3 + 2t| \leq 6$

    (c) $\dfrac{x-1}{3x+5} > 0$ (d) $(3t+2)^2 > 16$

    (e) $2x^2 - x - 3 < 0$ (22)

16. State the minimum number of cycles on logarithmic graph paper needed to plot a set of values ranging from 0.065 to 480 (2)

17. The current i flowing in a discharging capacitor varies with time as shown below:

i (mA)	50.0	17.0	5.8	1.7	0.58	0.24
t (ms)	200	255	310	375	425	475

Using logarithmic graph paper, show that these results are connected by the law of the form $i = I e^{\frac{t}{T}}$ where I, the initial current flowing, and T are constants. Determine the approximate values of I and T. (15)

Multiple choice questions Test 7
Number Sequences, Numbering Systems, Inequalities and Logarithmic Graphs
This test covers the material in Chapters 36 to 39

All questions have only one correct answer (Answers on page 440).

1. In decimal, the binary number 1101101 is equal to:

   (a) 218   (b) 31   (c) 127   (d) 109

2. The 5th term of an arithmetic progression is 18 and the 12th term is 46. The 18th term is:

   (a) 70   (b) 74   (c) 68   (d) 72

3. The solution of the inequality $\dfrac{3t+2}{t+1} \leq 1$ is:

   (a) $t \geq -2\dfrac{1}{2}$ (b) $-1 < t \leq \dfrac{1}{2}$

   (c) $t < -1$ (d) $-\dfrac{1}{2} < t \leq 1$

Questions 4 to 7 relate to the following information. A straight line graph is plotted for the equation $y = ax^n$, where y and x are the variables and a and n are constants.

4. On the vertical axis is plotted:

   (a) y (b) x (c) $\lg y$ (d) a

5. On the horizontal axis is plotted:

   (a) $\lg x$ (b) x (c) x^n (d) a

6. The gradient of the graph is given by:

   (a) y (b) a (c) x (d) n

7. The vertical axis intercept is given by:

   (a) n (b) $\lg a$ (c) x (d) $\lg y$

8. In hexadecimal, the decimal number 237 is equivalent to:

   (a) 355   (b) ED   (c) 1010010011   (d) DE

9. To depict a set of values from 0.05 to 275, the minimum number of cycles required on logarithmic graph paper is:

   (a) 5   (b) 4   (c) 3   (d) 2

10. The first term of a geometric progression is 9 and the fourth term is 45. The eighth term is:

    (a) 225   (b) 150.5   (c) 384.7   (d) 657.9

11. The solution of the inequality $x^2 - x - 2 < 0$ is:

    (a) $1 < x < -2$ (b) $x > 2$

    (c) $-1 < x < 2$ (d) $x < -1$

Questions 12 to 15 relate to the following information. A straight line graph is plotted for the equation $y = ae^{bx}$, where y and x are the variables and a and b are constants.

12. On the vertical axis is plotted:

    (a) y (b) x (c) $\ln y$ (d) a

13. On the horizontal axis is plotted:

    (a) $\ln x$ (b) x (c) e^x (d) a

14. The gradient of the graph is given by:

    (a) y (b) a (c) x (d) b

15. The vertical axis intercept is given by:

    (a) b (b) $\ln a$ (c) x (d) $\ln y$

The companion website for this book contains the above multiple-choice test. If you prefer to attempt the test online then visit:

www.routledge.com/cw/bird

For a copy of this multiple choice test, go to:

www.routledge.com/cw/bird

List of formulae

Laws of indices:

$$a^m \times a^n = a^{m+n} \quad \frac{a^m}{a^n} = a^{m-n} \quad (a^m)^n = a^{mn}$$

$$a^{m/n} = \sqrt[n]{a^m} \quad a^{-n} = \frac{1}{a^n} \quad a^0 = 1$$

Quadratic formula:

If $ax^2 + bx + c = 0$ then $x = \dfrac{-b \pm \sqrt{b^2 - 4ac}}{2a}$

Equation of a straight line:

$$y = mx + c$$

Definition of a logarithm:

If $y = a^x$ then $x = \log_a y$

Laws of logarithms:

$$\log(A \times B) = \log A + \log B$$

$$\log\left(\frac{A}{B}\right) = \log A - \log B$$

$$\log A^n = n \times \log A$$

Exponential series:

$$e^x = 1 + x + \frac{x^2}{2!} + \frac{x^3}{3!} + \cdots \quad \text{(valid for all values of } x\text{)}$$

Theorem of Pythagoras:

$$b^2 = a^2 + c^2$$

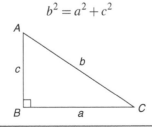

Areas of plane figures:

(i) **Rectangle** Area $= l \times b$

(ii) **Parallelogram** Area $= b \times h$

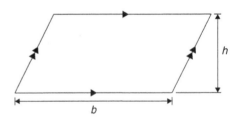

(iii) **Trapezium** Area $= \dfrac{1}{2}(a + b)h$

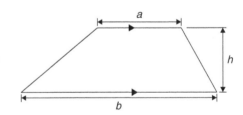

(iv) **Triangle** Area $= \dfrac{1}{2} \times b \times h$

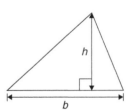

(v) **Circle** Area $= \pi r^2$ Circumference $= 2\pi r$

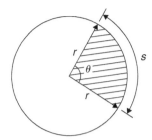

Radian measure: 2π radians $= 360$ degrees

For a sector of circle:

arc length, $s = \dfrac{\theta^\circ}{360}(2\pi r) = r\theta$ (θ in rad)

shaded area $= \dfrac{\theta^\circ}{360}(\pi r^2) = \dfrac{1}{2}r^2\theta$ (θ in rad)

Equation of a circle, centre at origin, radius r:

$$x^2 + y^2 = r^2$$

Equation of a circle, centre at (a, b), radius r:

$$(x - a)^2 + (y - b)^2 = r^2$$

Volumes and surface areas of regular solids:

(i) **Rectangular prism (or cuboid)**

$$\text{Volume} = l \times b \times h$$

$$\text{Surface area} = 2(bh + hl + lb)$$

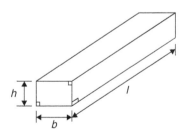

(ii) **Cylinder**

$$\text{Volume} = \pi r^2 h$$

$$\text{Total surface area} = 2\pi rh + 2\pi r^2$$

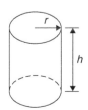

(iii) **Pyramid**

If area of base $= A$ and
perpendicular height $= h$ then:

$$\text{Volume} = \frac{1}{3} \times A \times h$$

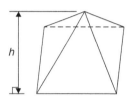

Total surface area $=$ sum of areas of triangles
forming sides $+$ area of base

(iv) **Cone**

$$\text{Volume} = \frac{1}{3}\pi r^2 h$$

$$\text{Curved surface area} = \pi rl$$

$$\text{Total surface area} = \pi rl + \pi r^2$$

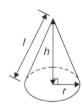

(v) **Sphere**

$$\text{Volume} = \frac{4}{3}\pi r^3$$

$$\text{Surface area} = 4\pi r^2$$

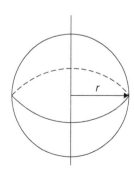

Areas of irregular figures by approximate methods:

Trapezoidal rule

$$\text{Area} \approx \binom{\text{width of}}{\text{interval}} \left[\frac{1}{2}\binom{\text{first + last}}{\text{ordinate}} + \text{sum of remaining ordinates} \right]$$

Mid-ordinate rule

$$\text{Area} \approx (\text{width of interval})(\text{sum of mid-ordinates})$$

Simpson's rule

$$\text{Area} \approx \frac{1}{3}\binom{\text{width of}}{\text{interval}} \left[\binom{\text{first + last}}{\text{ordinate}} + 4\binom{\text{sum of even}}{\text{ordinates}} + 2\binom{\text{sum of remaining}}{\text{odd ordinates}} \right]$$

Mean or average value of a waveform:

$$\text{mean value, } y = \frac{\text{area under curve}}{\text{length of base}}$$

$$= \frac{\text{sum of mid-ordinates}}{\text{number of mid-ordinates}}$$

Triangle formulae:

Sine rule: $\dfrac{a}{\sin A} = \dfrac{b}{\sin B} = \dfrac{c}{\sin C}$

Cosine rule: $a^2 = b^2 + c^2 - 2bc\cos A$

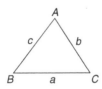

Area of any triangle

$$= \frac{1}{2} \times \text{base} \times \text{perpendicular height}$$

$$= \frac{1}{2}ab\sin C \quad \text{or} \quad \frac{1}{2}ac\sin B \quad \text{or} \quad \frac{1}{2}bc\sin A$$

$$= \sqrt{[s(s-a)(s-b)(s-c)]} \quad \text{where} \quad s = \frac{a+b+c}{2}$$

For a **general sinusoidal function** $y = A\sin(\omega t \pm \alpha)$, then

$$A = \text{amplitude}$$

$$\omega = \text{angular velocity} = 2\pi f \text{ rad/s}$$

$$\frac{\omega}{2\pi} = \text{frequency, } f \text{ hertz}$$

$$\frac{2\pi}{\omega} = \text{periodic time } T \text{ seconds}$$

$$\alpha = \text{angle of lead or lag (compared with } y = A\sin\omega t)$$

Cartesian and polar co-ordinates:

If co-ordinate $(x, y) = (r, \theta)$ then

$$r = \sqrt{x^2 + y^2} \quad \text{and} \quad \theta = \tan^{-1}\frac{y}{x}$$

If co-ordinate $(r, \theta) = (x, y)$ then

$$x = r\cos\theta \quad \text{and} \quad y = r\sin\theta$$

Arithmetic progression:

If a = first term and d = common difference, then the arithmetic progression is: $a, a+d, a+2d, \ldots$

The nth term is: $a + (n-1)d$

Sum of n terms, $S_n = \dfrac{n}{2}[2a + (n-1)d]$

Geometric progression:

If a = first term and r = common ratio, then the geometric progression is: $a, ar, ar^2, \ldots$

The nth term is: ar^{n-1}

Sum of n terms, $S_n = \dfrac{a(1-r^n)}{(1-r)} \quad \text{or} \quad \dfrac{a(r^n-1)}{(r-1)}$

If $-1 < r < 1$, $\quad S_\infty = \dfrac{a}{(1-r)}$

Statistics:

Discrete data:

$$\text{mean, } \bar{x} = \frac{\sum x}{n}$$

$$\text{standard deviation, } \sigma = \sqrt{\left[\frac{\sum(x-\bar{x})^2}{n}\right]}$$

Grouped data:

$$\text{mean, } \bar{x} = \frac{\sum f x}{\sum f}$$

$$\text{standard deviation, } \sigma = \sqrt{\left[\frac{\sum\{f(x - \bar{x})^2\}}{\sum f}\right]}$$

Standard derivatives

y or $f(x)$	$\dfrac{dy}{dx} =$ or $f'(x)$
ax^n	anx^{n-1}
$\sin ax$	$a\cos ax$
$\cos ax$	$-a\sin ax$
e^{ax}	ae^{ax}
$\ln ax$	$\dfrac{1}{x}$

Standard integrals

y	$\int y\,dx$
ax^n	$a\dfrac{x^{n+1}}{n+1} + c$ (except when $n = -1$)
$\cos ax$	$\dfrac{1}{a}\sin ax + c$
$\sin ax$	$-\dfrac{1}{a}\cos ax + c$
e^{ax}	$\dfrac{1}{a}e^{ax} + c$
$\dfrac{1}{x}$	$\ln x + c$

For a copy of List of formulae, go to website:
www.routledge.com/cw/bird

Answers to Practice Exercises

Chapter 1

Exercise 1 (page 3)

1. 19 kg 2. 16 m 3. 479 mm
4. −66 5. £565 6. −225
7. −2136 8. −36121 9. £107701
10. −4 11. 1487 12. 5914
13. 189 g 14. −70872 15. $15333
16. $d = 64$ mm, $A = 136$ mm, $B = 10$ mm

Exercise 2 (page 5)

1. (a) 468 (b) 868 2. (a) £1827 (b) £4158
3. (a) 8613 kg (b) 584 kg
4. (a) 351 mm (b) 924 mm
5. (a) 10304 (b) −4433 6. (a) 48 m (b) 89 m
7. (a) 259 (b) 56 8. (a) 1648 (b) −1060
9. (a) 8067 (b) 3347 10. 18 kg
11. 89.25 cm 12. 29

Exercise 3 (page 6)

1. (a) 4 (b) 24 2. (a) 12 (b) 360
3. (a) 10 (b) 350 4. (a) 90 (b) 2700
5. (a) 2 (b) 210 6. (a) 3 (b) 180
7. (a) 5 (b) 210 8. (a) 15 (b) 6300
9. (a) 14 (b) 420420 10. (a) 14 (b) 53900

Exercise 4 (page 8)

1. 59 2. 14 3. 88 4. 5 5. 33
6. 22 7. 68 8. 5 9. 2 10. 5
11. −1

Chapter 2

Exercise 5 (page 11)

1. $2\frac{1}{7}$ 2. $7\frac{2}{5}$ 3. $\frac{22}{9}$ 4. $\frac{71}{8}$
5. $\frac{4}{11}$ 6. $\frac{3}{7}, \frac{4}{9}, \frac{1}{2}, \frac{3}{5}, \frac{5}{8}$ 7. $\frac{8}{25}$ 8. $\frac{11}{15}$
9. $\frac{17}{30}$ 10. $\frac{9}{10}$ 11. $\frac{3}{16}$ 12. $\frac{43}{77}$
13. $\frac{47}{63}$ 14. $1\frac{1}{15}$ 15. $\frac{4}{27}$ 16. $8\frac{51}{52}$
17. $1\frac{9}{40}$ 18. $1\frac{16}{21}$ 19. $\frac{17}{60}$ 20. $\frac{17}{20}$

Exercise 6 (page 13)

1. $\frac{8}{35}$ 2. $2\frac{2}{9}$ 3. $\frac{6}{11}$ 4. $\frac{5}{12}$ 5. $\frac{3}{28}$
6. $\frac{3}{5}$ 7. 11 8. $\frac{1}{13}$ 9. $1\frac{1}{2}$ 10. $\frac{8}{15}$
11. $2\frac{2}{5}$ 12. $\frac{5}{12}$ 13. $3\frac{3}{4}$ 14. $\frac{12}{23}$ 15. 4
16. $\frac{3}{4}$ 17. $\frac{1}{9}$ 18. 13 19. 15 20. 400 litres
21. (a) £60, (b) P £36, Q £16 22. 2880 litres

Exercise 7 (page 15)

1. $2\frac{1}{18}$ 2. $-\frac{1}{9}$ 3. $1\frac{1}{6}$ 4. $4\frac{3}{4}$ 5. $\frac{13}{20}$
6. $\frac{7}{15}$ 7. $4\frac{19}{20}$ 8. 2 9. $7\frac{1}{3}$ 10. $\frac{1}{15}$
11. 4 12. $2\frac{17}{20}$

Chapter 3

Exercise 8 (page 19)

1. $\dfrac{13}{20}$ 2. $\dfrac{9}{250}$ 3. $\dfrac{7}{40}$ 4. $\dfrac{6}{125}$

5. (a) $\dfrac{33}{50}$ (b) $\dfrac{21}{25}$ (c) $\dfrac{1}{80}$ (d) $\dfrac{141}{500}$ (e) $\dfrac{3}{125}$

6. $4\dfrac{21}{40}$ 7. $23\dfrac{11}{25}$ 8. $10\dfrac{3}{200}$ 9. $6\dfrac{7}{16}$

10. (a) $1\dfrac{41}{50}$ (b) $4\dfrac{11}{40}$ (c) $14\dfrac{1}{8}$ (d) $15\dfrac{7}{20}$ (e) $16\dfrac{17}{80}$

11. 0.625   12. 6.6875   13. 0.21875   14. 11.1875

15. 0.28125

Exercise 9 (page 20)

1. 14.18   2. 2.785   3. 65.38   4. 43.27
5. 1.297   6. 0.000528

Exercise 10 (page 21)

1. 80.3   2. 329.3   3. 72.54   4. -124.83
5. 295.3   6. 18.3   7. 12.52 mm

Exercise 11 (page 22)

1. 4.998   2. 47.544   3. 385.02   4. 582.42
5. 456.9   6. 434.82   7. 626.1   8. 1591.6
9. 0.444   10. 0.62963   11. 1.563   12. 53.455
13. 13.84   14. 8.69   15. (a) 24.81 (b) 24.812
16. (a) 0.00639 (b) 0.0064   17. (a) $8.\dot{4}$ (b) $62.\dot{6}$
18. 2400

Chapter 4

Exercise 12 (page 24)

1. 30.797   2. 11927   3. 13.62   4. 53.832
5. 84.42   6. 1.0944   7. 50.330   8. 36.45
9. 10.59   10. 12.325

Exercise 13 (page 25)

1. 12.25   2. 0.0361   3. 46.923   4. 1.296×10^{-3}
5. 2.4430   6. 2.197   7. 30.96   8. 0.0549
9. 219.26   10. 5.832×10^{-6}

Exercise 14 (page 26)

1. 0.571   2. 40   3. 0.13459   4. 137.9
5. 14.96   6. 19.4481   7. 515.36×10^{-6}
8. 1.0871   9. 15.625×10^{-9} 10. 52.70

Exercise 15 (page 27)

1. 2.182   2. 11.122   3. 185.82
4. 0.8307   5. 0.1581   6. 2.571
7. 5.273   8. 1.2555   9. 0.30366
10. 1.068   11. 3.5×10^{6} 12. 37.5×10^{3}
13. 4.2×10^{-6} 14. 202.767×10^{-3} 15. 18.32×10^{6}

Exercise 16 (page 28)

1. 0.4667   2. $\dfrac{13}{14}$ 3. 4.458   4. 2.732

5. $\dfrac{1}{21}$ 6. 0.7083   7. $-\dfrac{9}{10}$ 8. $3\dfrac{1}{3}$

9. 2.567   10. 0.0776

Exercise 17 (page 29)

1. 0.9205   2. 0.7314   3. 2.9042   4. 0.2719
5. 0.4424   6. 0.0321   7. 0.4232   8. 0.1502
9. -0.6992 10. 5.8452

Exercise 18 (page 29)

1. 4.995   2. 5.782   3. 25.72   4. 69.42
5. 0.6977   6. 52.92   7. 591.0   8. 17.90
9. 3.520   10. 0.3770

Exercise 19 (page 31)

1. $A = 66.59\,\text{cm}^2$ 2. $C = 52.78\,\text{mm}$ 3. $R = 37.5$
4. 159 m/s   5. 0.407 A   6. 5.02 mm
7. 0.144 J   8. $628.8\,\text{m}^2$ 9. 224.5
10. $14\,230\,\text{kg}/\text{m}^3$ 11. 281.1 m/s   12. $2.526\,\Omega$

Exercise 20 (page 32)

1. £589.27   2. 508.1 W   3. $V = 2.61\,\text{V}$
4. $F = 854.5$ 5. $I = 3.81\,\text{A}$ 6. $t = 14.79\,\text{s}$
7. $E = 3.96\,\text{J}$ 8. $I = 12.77\,\text{A}$ 9. $s = 17.25\,\text{m}$
10. $A = 7.184\,\text{cm}^2$ 11. $v = 7.327$
12. (a) 12.53 h   (b) 1 h 40 min, 33 m.p.h.
     (c) 13.02 h   (d) 13.15 h

Chapter 5

Exercise 21 (page 37)

1. 0.32% 2. 173.4% 3. 5.7% 4. 37.4%
5. 128.5% 6. 0.20 7. 0.0125 8. 68.75%
9. 38.462% 10. (a) 21.2% (b) 79.2% (c) 169%
11. (b), (d), (c), (a) 12. $\dfrac{13}{20}$ 13. $\dfrac{5}{16}$ 14. $\dfrac{9}{16}$
15. $A = \dfrac{1}{2}$, $B = 50\%$, $C = 0.25$, $D = 25\%$, $E = 0.30$, $F = \dfrac{3}{10}$, $G = 0.60$, $H = 60\%$, $I = 0.85$, $J = \dfrac{17}{20}$
16. 779 Ω to 861 Ω

Exercise 22 (page 38)

1. 21.8 kg 2. 9.72 m
3. (a) 496.4 t (b) 8.657 g (c) 20.73 s 4. 2.25%
5. (a) 14% (b) 15.67% (c) 5.36% 6. 37.8 g
7. 14 minutes 57 seconds 8. 76 g 9. £611
10. 37.49% 11. 39.2% 12. 17% 13. 38.7%
14. 2.7% 15. 5.60 m 16. 3.5%
17. (a) (i) 544 Ω (ii) 816 Ω
    (b) (i) 44.65 kΩ (ii) 49.35 kΩ
18. 2592 rev/min

Exercise 23 (page 40)

1. 2.5% 2. 18% 3. £310 4. £175 000
5. £260 6. £20 000 7. £8778.05 8. £50.25
9. £39.60 10. £917.70 11. £185 000 12. 7.2%
13. A 0.6 kg, B 0.9 kg, C 0.5 kg
14. 54%, 31%, 15%, 0.3 t
15. 20 000 kg (or 20 tonnes)
16. 13.5 mm, 11.5 mm 17. 600 kW

Chapter 6

Exercise 24 (page 45)

1. 36 : 1 2. 3.5 : 1 or 7 : 2 3. 47 : 3
4. 96 cm, 240 cm. 5. $5\dfrac{1}{4}$ hours or 5 hours 15 minutes
6. £3680, £1840, £920 7. 12 cm 8. £2172

Exercise 25 (page 46)

1. 1 : 15 2. 76 ml 3. 25% 4. 12.6 kg
5. 14.3 kg 6. 25 000 kg

Exercise 26 (page 48)

1. £556 2. £66 3. 264 kg 4. 450 N 5. 14.56 kg
6. (a) 0.00025 (b) 48 MPa 7. (a) 440 K (b) 5.76 litre
8. 8960 bolts

Exercise 27 (page 49)

1. (a) 2 mA (b) 25 V 2. 434 fr 3. 685.8 mm
4. 83 lb 10 oz 5. (a) 159.1 litres (b) 16.5 gallons
6. 29.4 MPa 7. 584.2 mm 8. $1012

Exercise 28 (page 51)

1. 3.5 weeks 2. 20 days
3. (a) 9.18 (b) 6.12 (c) 0.3375 4. 50 minutes
5. (a) 300×10^3 (b) $0.375 \, \text{m}^2$ (c) 240×10^3 Pa
6. (a) 32 J (b) 0.5 m

Chapter 7

Exercise 29 (page 54)

1. 27 2. 128 3. 100 000 4. 96 5. 2^4
6. ±5 7. ±8 8. 100 9. 1 10. 64

Exercise 30 (page 56)

1. 128 2. 3^9 3. 16 4. $\dfrac{1}{9}$ 5. 1 6. 8
7. 100 8. 1000 9. $\dfrac{1}{100}$ or 0.01 10. 5 11. 7^6
12. 3^6 13. 3^6 or 729 14. 3^4 15. 1 16. 25
17. $\dfrac{1}{3^5}$ or $\dfrac{1}{243}$ 18. 49 19. $\dfrac{1}{2}$ or 0.5 20. 1

Exercise 31 (page 57)

1. $\dfrac{1}{3 \times 5^2}$ 2. $\dfrac{1}{7^3 \times 3^7}$ 3. $\dfrac{3^2}{2^5}$ 4. $\dfrac{1}{2^{10} \times 5^2}$
5. 9 6. 3 7. $\dfrac{1}{2}$ 8. $\pm\dfrac{2}{3}$
9. $\dfrac{147}{148}$ 10. $-1\dfrac{19}{56}$ 11. $-3\dfrac{13}{45}$ 12. $\dfrac{1}{9}$
13. $-\dfrac{17}{18}$ 14. 64 15. $4\dfrac{1}{2}$

Chapter 8

Exercise 32 (page 63)

1. cubic metres, m^3 2. farad
3. square metres, m^2 4. metres per second, m/s
5. kilogram per cubic metre, kg/m^3
6. joule   7. coulomb   8. watt
9. radian or degree   10. volt   11. mass
12. electrical resistance 13. frequency
14. acceleration   15. electric current
16. inductance   17. length
18. temperature   19. pressure
20. angular velocity   21. $\times 10^9$ 22. m, $\times 10^{-3}$
23. $\times 10^{-12}$ 24. M, $\times 10^6$

Exercise 33 (page 64)

1. (a) 7.39×10 (b) 2.84×10 (c) 1.9762×10^2
2. (a) 2.748×10^3 (b) 3.317×10^4 (c) 2.74218×10^5
3. (a) 2.401×10^{-1} (b) 1.74×10^{-2} (c) 9.23×10^{-3}
4. (a) 1.7023×10^3 (b) 1.004×10 (c) 1.09×10^{-2}
5. (a) 5×10^{-1} (b) 1.1875×10
   (c) 3.125×10^{-2} (d) 1.306×10^2
6. (a) 1010 (b) 932.7 (c) 54100 (d) 7
7. (a) 0.0389 (b) 0.6741 (c) 0.008
8. (a) 1.35×10^2 (b) 1.1×10^5
9. (a) 2×10^2 (b) 1.5×10^{-3}
10. (a) $2.71 \times 10^3 \, kg\,m^{-3}$ (b) 4.4×10^{-1}
    (c) $3.7673 \times 10^2 \, \Omega$ (d) $5.11 \times 10^{-1} \, MeV$
    (e) $9.57897 \times 10^7 \, C\,kg^{-1}$
    (f) $2.241 \times 10^{-2} \, m^3 \, mol^{-1}$

Exercise 34 (page 66)

1. 60 kPa   2. $150 \, \mu W$ or 0.15 mW
3. 50 MV   4. 55 nF
5. 100 kW   6. 0.54 mA or $540 \, \mu A$
7. $1.5 \, M\Omega$ 8. 22.5 mV
9. 35 GHz   10. 15 pF
11. $17 \, \mu A$ 12. $46.2 \, k\Omega$
13. $3 \, \mu A$ 14. 2.025 MHz
15. 50 kN   16. 0.3 nF
17. 62.50 m   18. 0.0346 kg
19. 13.5×10^{-3} 20. 4×10^3
21. $3.8 \times 10^5 \, km$ 22. 0.053 nm
23. 5.6 MPa   24. $4.3 \times 10^{-3} m$ or 4.3 mm

Chapter 9

Exercise 35 (page 70)

1. $-3a$ 2. $a + 2b + 4c$
3. $3x - 3x^2 - 3y - 2y^2$ 4. $6ab - 3a$
5. $6x - 5y + 8z$ 6. $1 + 2x$
7. $4x + 2y + 2$ 8. $3a + 5b$
9. $-2a - b + 2c$ 10. $3x^2 - y^2$

Exercise 36 (page 72)

1. $p^2 q^3 r$ 2. $8a^2$ 3. $6q^2$
4. 46   5. $5\frac{1}{3}$ 6. $-\frac{1}{2}$
7. 6   8. $\frac{1}{7y}$ 9. $5xz^2$
10. $3a^2 + 2ab - b^2$
11. $6a^2 - 13ab + 3ac - 5b^2 + bc$
12. $\frac{1}{3b}$ 13. $2ab$ 14. $3x$
15. $2x - y$ 16. $3p + 2q$ 17. $2a^2 + 2b^2$

Exercise 37 (page 74)

1. z^8 2. a^8 3. n^3
4. b^{11} 5. b^{-3} 6. c^4
7. m^4 8. x^{-3} or $\frac{1}{x^3}$ 9. x^{12}
10. y^{-6} or $\frac{1}{y^6}$ 11. t^8 12. c^{14}
13. a^{-9} or $\frac{1}{a^9}$ 14. b^{-12} or $\frac{1}{b^{12}}$ 15. b^{10} 16. s^{-9}
17. $p^6 q^7 r^5$ 18. $x^{-2} y z^{-2}$ or $\frac{y}{x^2 z^2}$
19. $x^5 y^4 z^3, 13\frac{1}{2}$ 20. $a^3 b^{-2} c$ or $\frac{a^3 c}{b^2}, 9$

Exercise 38 (page 75)

1. $a^2 b^{1/2} c^{-2}, \pm 4\frac{1}{2}$ 2. $\frac{1+a}{b}$ 3. $a\, b^6 \, c^{3/2}$
4. $a^{-4} b^5 c^{11}$ 5. $\frac{p^2 q}{q - p}$ 6. $x\, y^3 \sqrt[6]{z^{13}}$
7. $\frac{1}{ef^2}$
8. $a^{11/6} b^{1/3} c^{-3/2}$ or $\frac{\sqrt[6]{a^{11}} \sqrt[3]{b}}{\sqrt{c^3}}$

Chapter 10

Exercise 39 (page 77)

1. $x^2 + 5x + 6$
2. $2x^2 + 9x + 4$
3. $4x^2 + 12x + 9$
4. $2j^2 + 2j - 12$
5. $4x^2 + 22x + 30$
6. $2pqr + p^2q^2 + r^2$
7. $a^2 + 2ab + b^2$
8. $x^2 + 12x + 36$
9. $a^2 - 2ac + c^2$
10. $25x^2 + 30x + 9$
11. $4x^2 - 24x + 36$
12. $4x^2 - 9$
13. $64x^2 + 64x + 16$
14. $r^2s^2 + 2rst + t^2$
15. $3ab - 6a^2$
16. $2x^2 - 2xy$
17. $2a^2 - 3ab - 5b^2$
18. $13p - 7q$
19. $7x - y - 4z$
20. $4a^2 - 25b^2$
21. $x^2 - 4xy + 4y^2$
22. $9a^2 - 6ab + b^2$
23. 0
24. $4 - a$
25. $4ab - 8a^2$
26. $3xy + 9x^2y - 15x^2$
27. $2 + 5b^2$
28. $11q - 2p$

Exercise 40 (page 79)

1. $2(x + 2)$
2. $2x(y - 4z)$
3. $p(b + 2c)$
4. $2x(1 + 2y)$
5. $4d(d - 3f^5)$
6. $4x(1 + 2x)$
7. $2q(q + 4n)$
8. $r(s + p + t)$
9. $x(1 + 3x + 5x^2)$
10. $bc(a + b^2)$
11. $3xy(xy^3 - 5y + 6)$
12. $2pq^2(2p^2 - 5q)$
13. $7ab(3ab - 4)$
14. $2xy(y + 3x + 4x^2)$
15. $2xy(x - 2y^2 + 4x^2y^3)$
16. $7y(4 + y + 2x)$
17. $\dfrac{3x}{y}$
18. 0 19. $\dfrac{2r}{t}$
20. $(a + b)(y + 1)$
21. $(p + q)(x + y)$
22. $(x - y)(a + b)$
23. $(a - 2b)(2x + 3y)$
24. $\dfrac{A^2}{pg}\left(\dfrac{A}{pg^2} - \dfrac{1}{g} + A^3\right)$

Exercise 41 (page 81)

1. $2x + 8x^2$
2. $12y^2 - 3y$
3. $4b - 15b^2$
4. $4 + 3a$
5. $\dfrac{3}{2} - 4x$
6. 1
7. $10y^2 - 3y + \dfrac{1}{4}$
8. $9x^2 + \dfrac{1}{3} - 4x$
9. $6a^2 + 5a - \dfrac{1}{7}$
10. $-15t$
11. $\dfrac{1}{5} - x - x^2$
12. $10a^2 - 3a + 2$

Chapter 11

Exercise 42 (page 84)

1. 1   2. 2   3. 6   4. −4   5. 2
6. 1   7. 2   8. $\dfrac{1}{2}$ 9. 0   10. 3
11. 2   12. −10   13. 6   14. −2   15. 2.5
16. 2   17. 6   18. −3

Exercise 43 (page 86)

1. 5   2. −2   3. $-4\dfrac{1}{2}$ 4. 2   5. 12
6. 15   7. −4   8. $5\dfrac{1}{3}$ 9. 2   10. 13
11. −10   12. 2   13. 3   14. −11   15. −6
16. 9   17. $6\dfrac{1}{4}$ 18. 3   19. 4   20. 10
21. ±12   22. $-3\dfrac{1}{3}$ 23. ±3   24. ±4

Exercise 44 (page 88)

1. 10^{-7} 2. $8\,\text{m/s}^2$ 3. 3.472
4. (a) $1.8\,\Omega$ (b) $30\,\Omega$
5. digital camera battery £9, camcorder battery £14
6. $800\,\Omega$ 7. $30\,\text{m/s}^2$
8. 176 MPa

Exercise 45 (page 90)

1. $12\,\text{cm}, 240\,\text{cm}^2$ 2. 0.004   3. 30
4. 45°C   5. 50   6. £312, £240
7. 30 kg   8. 12 m, 8 m   9. 3.5 N

Chapter 12

Exercise 46 (page 96)

1. $d = c - e - a - b$ 2. $x = \dfrac{y}{7}$
3. $v = \dfrac{c}{p}$ 4. $a = \dfrac{v - u}{t}$
5. $R = \dfrac{V}{I}$ 6. $y = \dfrac{1}{3}(t - x)$
7. $r = \dfrac{c}{2\pi}$ 8. $x = \dfrac{y - c}{m}$

9. $T = \dfrac{I}{PR}$ **10.** $L = \dfrac{X_L}{2\pi f}$

11. $R = \dfrac{E}{I}$ **12.** $x = a(y - 3)$

13. $C = \dfrac{5}{9}(F - 32)$ **14.** $f = \dfrac{1}{2\pi\, CX_C}$

Exercise 47 (page 98)

1. $r = \dfrac{S - a}{S}$ or $1 - \dfrac{a}{S}$

2. $x = \dfrac{d}{\lambda}(y + \lambda)$ or $d + \dfrac{yd}{\lambda}$

3. $f = \dfrac{3F - AL}{3}$ or $f = F - \dfrac{AL}{3}$

4. $D = \dfrac{AB^2}{5Cy}$ **5.** $t = \dfrac{R - R_0}{R_0\alpha}$ **6.** $R_2 = \dfrac{RR_1}{R_1 - R}$

7. $R = \dfrac{E - e - Ir}{I}$ or $R = \dfrac{E - e}{I} - r$

8. $b = \sqrt{\left(\dfrac{y}{4ac^2}\right)}$ **9.** $x = \dfrac{ay}{\sqrt{(y^2 - b^2)}}$

10. $L = \dfrac{gt^2}{4\pi^2}$ **11.** $u = \sqrt{v^2 - 2as}$

12. $R = \sqrt{\left(\dfrac{360A}{\pi\theta}\right)}$ **13.** $a = N^2 y - x$

14. $L = \dfrac{\sqrt{Z^2 - R^2}}{2\pi f}, 0.080$ **15.** $v = \sqrt{\dfrac{2L}{\rho ac}}$

16. $V = \dfrac{k^2 H^2 L^2}{\theta^2}$

Exercise 48 (page 101)

1. $a = \sqrt{\left(\dfrac{xy}{m - n}\right)}$ **2.** $R = \sqrt[4]{\left(\dfrac{M}{\pi} + r^4\right)}$

3. $r = \dfrac{3(x + y)}{(1 - x - y)}$ **4.** $L = \dfrac{mrCR}{\mu - m}$

5. $b = \dfrac{c}{\sqrt{1 - a^2}}$ **6.** $r = \sqrt{\left(\dfrac{x - y}{x + y}\right)}$

7. $b = \dfrac{a(p^2 - q^2)}{2(p^2 + q^2)}$ **8.** $v = \dfrac{uf}{u - f}, 30$

9. $t_2 = t_1 + \dfrac{Q}{mc}, 55$ **10.** $v = \sqrt{\left(\dfrac{2dgh}{0.03L}\right)}, 0.965$

---

11. $l = \dfrac{8S^2}{3d} + d, 2.725$

12. $C = \dfrac{1}{\omega\left\{\omega L - \sqrt{Z^2 - R^2}\right\}}, 63.1 \times 10^{-6}$

13. 64 mm

14. $\lambda = \sqrt[5]{\left(\dfrac{a\mu}{\rho CZ^4 n}\right)^2}$

15. $w = \dfrac{2R - F}{L}$; 3 kN/m   **16.** $t_2 = t_1 - \dfrac{Qd}{kA}$

17. $r = \dfrac{v}{\omega}\left(1 - \dfrac{s}{100}\right)$

18. $F = EI\left(\dfrac{n\pi}{L}\right)^2$; 13.61 MN   **19.** $r = \sqrt[4]{\left(\dfrac{8\eta\ell V}{\pi p}\right)}$

20. $\ell = \sqrt[3]{\left(\dfrac{20g H^2}{I\rho^4 D^2}\right)^2}$

Chapter 13

Exercise 49 (page 105)

1. $x = 4, y = 2$ **2.** $x = 3, y = 4$
3. $x = 2, y = 1.5$ **4.** $x = 4, y = 1$
5. $p = 2, q = -1$ **6.** $x = 1, y = 2$
7. $x = 3, y = 2$ **8.** $a = 2, b = 3$
9. $a = 5, b = 2$ **10.** $x = 1, y = 1$
11. $s = 2, t = 3$ **12.** $x = 3, y = -2$
13. $m = 2.5, n = 0.5$ **14.** $a = 6, b = -1$
15. $x = 2, \ y = 5$ **16.** $c = 2, d = -3$

Exercise 50 (page 107)

1. $p = -1, \ q = -2$ **2.** $x = 4, y = 6$
3. $a = 2, \ b = 3$ **4.** $s = 4, t = -1$
5. $x = 3, \ y = 4$ **6.** $u = 12, v = 2$
7. $x = 10, y = 15$ **8.** $a = 0.30, b = 0.40$

Exercise 51 (page 109)

1. $x = \dfrac{1}{2}, y = \dfrac{1}{4}$ **2.** $a = \dfrac{1}{3}, b = -\dfrac{1}{2}$

3. $p = \dfrac{1}{4}, q = \dfrac{1}{5}$ **4.** $x = 10, y = 5$

5. $c = 3, d = 4$ **6.** $r = 3, s = \dfrac{1}{2}$

7. $x = 5, y = 1\dfrac{3}{4}$ **8.** 1

Exercise 52 (page 112)

1. $a = 0.2, b = 4$
2. $I_1 = 6.47, I_2 = 4.62$
3. $u = 12, a = 4, v = 26$ 4. £15 500, £12 800
5. $m = -0.5, c = 3$
6. $\alpha = 0.00426, R_0 = 22.56\,\Omega$ 7. $a = 12, b = 0.40$
8. $a = 4, b = 10$
9. $F_1 = 1.5, F_2 = -4.5$
10. $R_1 = 5.7$ kN, $R_2 = 6.3$ kN

Exercise 53 (page 114)

1. $x = 2, y = 1, z = 3$ 2. $x = 2, y = -2, z = 2$
3. $x = 5, y = -1, z = -2$ 4. $x = 4, y = 0, z = 3$
5. $x = 2, y = 4, z = 5$ 6. $x = 1, y = 6, z = 7$
7. $x = 5, y = 4, z = 2$ 8. $x = -4, y = 3, z = 2$
9. $x = 1.5, y = 2.5, z = 4.5$
10. $i_1 = -5, i_2 = -4, i_3 = 2$
11. $F_1 = 2, F_2 = -3\ F_3 = 4$

Chapter 14

Exercise 54 (page 119)

1. 4 or -4 2. 4 or -8 3. 2 or -6
4. -1.5 or 1.5 5. 0 or $-\dfrac{4}{3}$ 6. 2 or -2
7. 4   8. -5 9. 1
10. -2 or -3 11. -3 or -7 12. 2 or -1
13. 4 or -3 14. 2 or 7   15. -4
16. 2   17. -3 18. 3 or -3
19. -2 or $-\dfrac{2}{3}$ 20. -1.5 21. $\dfrac{1}{8}$ or $-\dfrac{1}{8}$
22. 4 or -7 23. -1 or 1.5 24. $\dfrac{1}{2}$ or $\dfrac{1}{3}$
25. $\dfrac{1}{2}$ or $-\dfrac{4}{5}$ 26. $1\dfrac{1}{3}$ or $-\dfrac{1}{7}$ 27. $\dfrac{3}{8}$ or -2
28. $\dfrac{2}{5}$ or -3 29. $\dfrac{4}{3}$ or $-\dfrac{1}{2}$ 30. $\dfrac{5}{4}$ or $-\dfrac{3}{2}$
31. $x^2 - 4x + 3 = 0$ 32. $x^2 + 3x - 10 = 0$
33. $x^2 + 5x + 4 = 0$ 34. $4x^2 - 8x - 5 = 0$
35. $x^2 - 36 = 0$ 36. $x^2 - 1.7x - 1.68 = 0$

Exercise 55 (page 121)

1. -3.732 or -0.268 2. -3.137 or 0.637
3. 1.468 or -1.135 4. 1.290 or 0.310
5. 2.443 or 0.307 6. -2.851 or 0.351

Exercise 56 (page 122)

1. 0.637 or -3.137 2. 0.296 or -0.792
3. 2.781 or 0.719   4. 0.443 or -1.693
5. 3.608 or -1.108 6. 1.434 or 0.232
7. 0.851 or -2.351 8. 2.086 or -0.086
9. 1.481 or -1.081 10. 4.176 or -1.676
11. 4 or 2.167   12. 7.141 or -3.641
13. 4.562 or 0.438

Exercise 57 (page 124)

1. 1.191 s   2. 0.345 A or 0.905 A 3. 7.84 cm
4. 0.619 m or 19.38 m   5. 0.0133
6. 1.066 m   7. 86.78 cm
8. 1.835 m or 18.165 m   9. 7 m
10. 12 ohms, 28 ohms
11. 0.52 s and 5.73 s   12. 400 rad/s

Exercise 58 (page 125)

1. $x = 1, y = 3$ and $x = -3, y = 7$
2. $x = \dfrac{2}{5}, y = -\dfrac{1}{5}$ and $x = -1\dfrac{2}{3}, y = -4\dfrac{1}{3}$
3. $x = 0, y = 4$ and $x = 3, y = 1$

Chapter 15

Exercise 59 (page 128)

1. 4   2. 4   3. 3   4. -3 5. $\dfrac{1}{3}$
6. 3   7. 2   8. -2 9. $1\dfrac{1}{2}$ 10. $\dfrac{1}{3}$
11. 2   12. 10000 13. 100000 14. 9   15. $\dfrac{1}{32}$
16. 0.01 17. $\dfrac{1}{16}$ 18. e^3

Exercise 60 (page 130)

1. $\log 6$ 2. $\log 15$ 3. $\log 2$ 4. $\log 3$
5. $\log 12$ 6. $\log 500$ 7. $\log 100$ 8. $\log 6$
9. $\log 10$ 10. $\log 1 = 0$ 11. $\log 2$
12. $\log 243$ or $\log 3^5$ or $5 \log 3$
13. $\log 16$ or $\log 2^4$ or $4 \log 2$

14. $\log 64$ or $\log 2^6$ or $6\log 2$

15. 0.5   **16.** 1.5   **17.** $x = 2.5$ **18.** $t = 8$

19. $b = 2$ **20.** $x = 2$ **21.** $a = 6$ **22.** $x = 5$

Exercise 61 (page 132)

1. 1.690 **2.** 3.170 **3.** 0.2696 **4.** 6.058 **5.** 2.251
6. 3.959 **7.** 2.542 **8.** -0.3272 **9.** 316.2

Chapter 16

Exercise 62 (page 134)

1. (a) 0.1653   (b) 0.4584   (c) 22030
2. (a) 5.0988   (b) 0.064037 (c) 40.446
3. (a) 4.55848   (b) 2.40444   (c) 8.05124
4. (a) 48.04106 (b) 4.07482   (c) -0.08286
5. 2.739   **6.** 120.7 m

Exercise 63 (page 136)

1. 2.0601   **2.** (a) 7.389 (b) 0.7408

3. $1 - 2x^2 - \dfrac{8}{3}x^3 - 2x^4$

4. $2x^{1/2} + 2x^{5/2} + x^{9/2} + \dfrac{1}{3}x^{13/2}$
$\qquad\qquad + \dfrac{1}{12}x^{17/2} + \dfrac{1}{60}x^{21/2}$

Exercise 64 (page 137)

1. $3.95, 2.05$ **2.** $1.65, -1.30$
3. (a) $28\,\text{cm}^3$ (b) $116\,\text{min}$ **4.** (a) $70°\text{C}$ (b) 5 minutes

Exercise 65 (page 140)

1. (a) 0.55547   (b) 0.91374   (c) 8.8941
2. (a) 2.2293   (b) -0.33154 (c) 0.13087
3. -0.4904 **4.** -0.5822 **5.** 2.197   **6.** 816.2
7. 0.8274 **8.** 11.02   **9.** 1.522   **10.** 1.485
11. 1.962 **12.** 3   **13.** 4
14. 147.9 **15.** 4.901 **16.** 3.095

17. $t = e^{b + a\ln D} = e^b e^{a\ln D} = e^b e^{\ln D^a}$ i.e. $t = e^b D^a$

18. 500   **19.** $W = PV\ln\left(\dfrac{U_2}{U_1}\right)$

20. 992 m/s **21.** 348.5 Pa

Exercise 66 (page 143)

1. (a) $150°\text{C}$ (b) $100.5°\text{C}$ **2.** 99.21 kPa

3. (a) 29.32 volts (b) $71.31 \times 10^{-6}\,\text{s}$

4. (a) 1.993 m (b) 2.293 m

5. (a) $50°\text{C}$ (b) 55.45 s

6. 30.37 N

7. (a) 3.04 A (b) 1.46 s

8. $2.45\,\text{mol/cm}^3$

9. (a) 7.07 A (b) 0.966 s

10. £2066

11. (a) 100% (b) 67.03% (c) 1.83%

12. 2.45 mA   **13.** 142 ms

14. 99.752%   **15.** 20 min 38 s

Chapter 17

Exercise 67 (page 153)

1. (a) Horizontal axis: $1\,\text{cm} = 4\,\text{V}$ (or $1\,\text{cm} = 5\,\text{V}$),
      vertical axis: $1\,\text{cm} = 10\,\Omega$
   (b) Horizontal axis: $1\,\text{cm} = 5\,\text{m}$, vertical axis:
      $1\,\text{cm} = 0.1\,\text{V}$
   (c) Horizontal axis: $1\,\text{cm} = 10\,\text{N}$, vertical axis:
      $1\,\text{cm} = 0.2\,\text{mm}$
2. (a) -1 (b) -8 (c) -1.5 (d) 4   **3.** 14.5
4. (a) -1.1 (b) -1.4
5. The 1010 rev/min reading should be 1070 rev/min;
   (a) 1000 rev/min (b) 167 V

Exercise 68 (page 159)

1. Missing values: $-0.75, 0.25, 0.75, 1.75, 2.25, 2.75$;
   Gradient $= \dfrac{1}{2}$

2. (a) $4, -2$ (b) $-1, 0$ (c) $-3, -4$ (d) $0, 4$

3. (a) $2, \dfrac{1}{2}$ (b) $3, -2\dfrac{1}{2}$ (c) $\dfrac{1}{24}, \dfrac{1}{2}$

4. (a) $6, -3$ (b) $-2, 4$ (c) $3, 0$ (d) $0, 7$

5. (a) $2, -\dfrac{1}{2}$ (b) $-\dfrac{2}{3}, -1\dfrac{2}{3}$ (c) $\dfrac{1}{18}, 2$ (d) $10, -4\dfrac{2}{3}$

6. (a) $\dfrac{3}{5}$ (b) -4 (c) $-1\dfrac{5}{6}$

7. (a) and (c), (b) and (e)

8. $(2, 1)$ **9.** $(1.5, 6)$ **10.** $(1, 2)$

11. (a) 89 cm (b) 11 N (c) 2.4 (d) $l = 2.4W + 48$

12. $P = 0.15W + 3.5$ **13.** $a = -20, b = 412$

Exercise 69 (page 164)

1. (a) $40°C$ (b) $128\,\Omega$

2. (a) $850\,\text{rev/min}$ (b) $77.5\,V$

3. (a) 0.25 (b) 12 (c) $F = 0.25L + 12$

   (d) $89.5\,N$ (e) $592\,N$ (f) $212\,N$

4. $-0.003, 8.73\,N/\,cm^2$

5. (a) $22.5\,m/s$ (b) $6.5\,s$ (c) $v = 0.7t + 15.5$

6. $m = 26.8L$

7. (a) $1.25t$ (b) 21.6% (c) $F = -0.095\,W + 2.2$

8. (a) $96 \times 10^9\,\text{Pa}$ (b) 0.00022 (c) $29 \times 10^6\,\text{Pa}$

9. (a) $\dfrac{1}{5}$ (b) 6 (c) $E = \dfrac{1}{5}L + 6$ (d) $12\,N$ (e) $65\,N$

10. $a = 0.85, b = 12, 254.3\,\text{kPa}, 275.5\,\text{kPa}, 280\,K$

Chapter 18

Exercise 70 (page 170)

1. (a) y (b) x^2 (c) c (d) d 2. (a) y (b) $\sqrt{x}$ (c) b (d) a

3. (a) y (b) $\dfrac{1}{x}$ (c) f (d) e 4. (a) $\dfrac{y}{x}$ (b) x (c) b (d) c

5. (a) $\dfrac{y}{x}$ (b) $\dfrac{1}{x^2}$ (c) a (d) b

6. $a = 1.5, b = 0.4, 11.78\,\text{mm}^2$ 7. $y = 2x^2 + 7, 5.15$

8. (a) 950 (b) $317\,\text{kN}$

9. $a = 0.4, b = 8.6$ (a) 94.4 (b) 11.2

Exercise 71 (page 174)

1. (a) $\lg y$ (b) x (c) $\lg a$ (d) $\lg b$
2. (a) $\lg y$ (b) $\lg x$ (c) L (d) $\lg k$
3. (a) $\ln y$ (b) x (c) n (d) $\ln m$
4. $I = 0.0012\,V^2, 6.75$ candelas
5. $a = 3.0, b = 0.5$
6. $a = 5.6, b = 2.6, 37.86, 3.0$
7. $R_0 = 25.1, c = 1.42$ 8. $y = 0.08e^{0.24x}$
9. $T_0 = 35.3\,N, \mu = 0.27, 64.8\,N, 1.29$ radians

Chapter 19

Exercise 72 (page 177)

1. $x = 2, y = 4$ 2. $x = 1, y = 1$

3. $x = 3.5, y = 1.5$ 4. $x = -1, y = 2$
5. $x = 2.3, y = -1.2$ 6. $x = -2, y = -3$
7. $a = 0.4, b = 1.6$

Exercise 73 (page 181)

1. (a) Minimum $(0, 0)$ (b) Minimum $(0, -1)$
   (c) Maximum $(0, 3)$ (d) Maximum $(0, -1)$
2. -0.4 or 0.6 3. -3.9 or 6.9
4. -1.1 or 4.1 5. -1.8 or 2.2
6. $x = -1.5$ or -2, Minimum at $(-1.75, -0.1)$
7. $x = -0.7$ or 1.6 8. (a) ± 1.63 (b) 1 or -0.3
9. $(-2.6, 13.2), (0.6, 0.8); x = -2.6$ or 0.6
10. $x = -1.2$ or 2.5 (a) -30 (b) 2.75 and -1.50
    (c) 2.3 or -0.8

Exercise 74 (page 182)

1. $x = 4, y = 8$ and $x = -0.5, y = -5.5$
2. (a) $x = -1.5$ or 3.5 (b) $x = -1.24$ or 3.24
   (c) $x = -1.5$ or 3.0

Exercise 75 (page 183)

1. $x = -2.0, -0.5$ or 1.5
2. $x = -2, 1$ or 3, Minimum at $(2.1, -4.1)$,
   Maximum at $(-0.8, 8.2)$
3. $x = 1$ 4. $x = -2.0, 0.4$ or 2.6
5. $x = -1.2, 0.70$ or 2.5
6. $x = -2.3, 1.0$ or 1.8 7. $x = -1.5$

Chapter 20

Exercise 76 (page 190)

1. $82°27'$ 2. $27°54'$ 3. $51°11'$ 4. $100°6'52''$
5. $15°44'17''$ 6. $86°49'1''$ 7. $72.55°$ 8. $27.754°$
9. $37°57'$ 10. $58°22'52''$

Exercise 77 (page 192)

1. reflex 2. obtuse 3. acute   4. right angle
5. (a) $21°$ (b) $62°23'$ (c) $48°56'17''$
6. (a) $102°$ (b) $165°$ (c) $10°18'49''$
7. (a) $60°$ (b) $110°$ (c) $75°$ (d) $143°$ (e) $140°$
   (f) $20°$ (g) $129.3°$ (h) $79°$ (i) $54°$

8. Transversal (a) 1 & 3, 2 & 4, 5 & 7, 6 & 8
(b) 1 & 2, 2 & 3, 3 & 4, 4 & 1, 5 & 6, 6 & 7,
7 & 8, 8 & 5, 3 & 8, 1 & 6, 4 & 7 or 2 & 5
(c) 1 & 5, 2 & 6, 4 & 8, 3 & 7 (d) 3 & 5 or 2 & 8

9. $59°20'$ **10.** $a = 69°, b = 21°, c = 82°$ **11.** $51°$

12. 1.326 rad   **13.** 0.605 rad   **14.** $40°55'$

Exercise 78 (page 196)

1. (a) acute-angled scalene triangle
(b) isosceles triangle (c) right-angled isosceles triangle
(d) obtuse-angled scalene triangle
(e) equilateral triangle (f) right-angled triangle

2. $a = 40°, b = 82°, c = 66°,$
$d = 75°, e = 30°, f = 75°$

3. DF, DE   **4.** $52°$ **5.** $122.5°$

6. $\phi = 51°, x = 161°$

7. $40°, 70°, 70°, 125°$, isosceles

8. $a = 18°50', b = 71°10', c = 68°, d = 90°,$
$e = 22°, f = 49°, g = 41°$

9. $a = 103°, b = 55°, c = 77°, d = 125°,$
$e = 55°, f = 22°, g = 103°, h = 77°,$
$i = 103°, j = 77°, k = 81°$

10. $17°$ **11.** $A = 37°, B = 60°, E = 83°$

Exercise 79 (page 198)

1. (a) congruent *BAC, DAC* (SAS)
(b) congruent *FGE, JHI* (SSS)
(c) not necessarily congruent
(d) congruent *QRT, SRT* (RHS)
(e) congruent *UVW, XZY* (ASA)

2. proof

Exercise 80 (page 201)

1. $x = 16.54$ mm, $y = 4.18$ mm   **2.** 9 cm, 7.79 cm

3. (a) 2.25 cm (b) 4 cm   **4.** 3 m

Exercise 81 (page 203)

1–5. Constructions – see similar constructions in
worked Problems 30 to 33 on pages 201–203.

Chapter 21

Exercise 82 (page 206)

1. 9 cm   **2.** 24 m   **3.** 9.54 mm

4. 20.81 cm   **5.** 7.21 m   **6.** 11.18 cm

7. 24.11 mm   **8.** $8^2 + 15^2 = 17^2$

9. (a) 27.20 cm each   (b) $45°$ **10.** 20.81 km

11. 3.35 m, 10 cm   **12.** 132.7 nautical miles

13. 2.94 mm   **14.** 24 mm

Exercise 83 (page 208)

1. $\sin Z = \dfrac{9}{41}, \cos Z = \dfrac{40}{41}, \tan X = \dfrac{40}{9}, \cos X = \dfrac{9}{41}$

2. $\sin A = \dfrac{3}{5}, \cos A = \dfrac{4}{5}, \tan A = \dfrac{3}{4}, \sin B = \dfrac{4}{5},$
$\cos B = \dfrac{3}{5}, \tan B = \dfrac{4}{3}$

3. $\sin A = \dfrac{8}{17}, \tan A = \dfrac{8}{15}$

4. $\sin X = \dfrac{15}{113}, \cos X = \dfrac{112}{113}$

5. (a) $\dfrac{15}{17}$ (b) $\dfrac{15}{17}$ (c) $\dfrac{8}{15}$

6. (a) $\sin\theta = \dfrac{7}{25}$ (b) $\cos\theta = \dfrac{24}{25}$

7. (a) 9.434 (b) -0.625

Exercise 84 (page 211)

1. 2.7550   **2.** 4.846   **3.** 36.52

4. (a) 0.8660 (b) -0.1010 (c) 0.5865

5. $42.33°$ **6.** $15.25°$ **7.** $73.78°$ **8.** $7°56'$

9. $31°22'$ **10.** $41°54'$ **11.** $29.05°$ **12.** $20°21'$

13. 0.3586   **14.** 1.803

15. (a) $40°$ (b) 6.79 m

Exercise 85 (page 213)

1. (a) 12.22 (b) 5.619 (c) 14.87 (d) 8.349
(e) 5.595 (f) 5.275

2. (a) $AC = 5.831$ cm, $\angle A = 59.04°, \angle C = 30.96°$
(b) $DE = 6.928$ cm, $\angle D = 30°, \angle F = 60°$
(c) $\angle J = 62°, HJ = 5.634$ cm, $GH = 10.60$ cm
(d) $\angle L = 63°, LM = 6.810$ cm, $KM = 13.37$ cm
(e) $\angle N = 26°, ON = 9.124$ cm, $NP = 8.201$ cm
(f) $\angle S = 49°, RS = 4.346$ cm, $QS = 6.625$ cm

3. 6.54 m   **4.** 9.40 mm   **5.** 5.63 m

Exercise 86 (page 216)

1. 36.15 m   **2.** 48 m   **3.** 249.5 m   **4.** 110.1 m
5. 53.0 m   **6.** 9.50 m   **7.** 107.8 m
8. 9.43 m, 10.56 m   **9.** 60 m

Chapter 22

Exercise 87 (page 223)

1. (a) $42.78°$ and $137.22°$ (b) $188.53°$ and $351.47°$
2. (a) $29.08°$ and $330.92°$ (b) $123.86°$ and $236.14°$
3. (a) $44.21°$ and $224.21°$ (b) $113.12°$ and $293.12°$
4. $t = 122°7'$ and $237°53'$
5. $\alpha = 218°41'$ and $321°19'$
6. $\theta = 39°44'$ and $219°44'$

Exercise 88 (page 227)

1. 5   **2.** $180°$ **3.** 30   **4.** $120°$
5. $1, 120°$ **6.** $2, 144°$ **7.** $3, 90°$ **8.** $5, 720°$
9. $3.5, 960°$ **10.** $6, 360°$ **11.** $4, 180°$ **12.** 5 ms
13. 40 Hz   **14.** $100 \,\mu$s or 0.1 ms
15. 625 Hz   **16.** leading   **17.** leading

Exercise 89 (page 229)

1. (a) 40 mA (b) 25 Hz (c) 0.04 s or 40 ms
(d) 0.29 rad (or $16.62°$) leading $40 \sin 50\pi t$

2. (a) 75 cm (b) 6.37 Hz (c) 0.157 s
(d) 0.54 rad (or $30.94°$) lagging $75 \sin 40t$

3. (a) 300 V (b) 100 Hz (c) 0.01 s or 10 ms
(d) 0.412 rad (or $23.61°$) lagging $300 \sin 200\pi t$

4. (a) $v = 120 \sin 100\pi t$ volts
(b) $v = 120 \sin (100\pi t + 0.43)$ volts

5. $i = 20 \sin \left(80\pi t - \dfrac{\pi}{6}\right)$ A or
$i = 20 \sin(80\pi t - 0.524)$ A

6. $3.2 \sin(100\pi t + 0.488)$ m

7. (a) 5 A, 50 Hz, 20 ms, $24.75°$ lagging
(b) -2.093 A (c) 4.363 A (d) 6.375 ms (e) 3.423 ms

Chapter 23

Exercise 90 (page 232)

1. $C = 83°, a = 14.1$ mm, $c = 28.9$ mm,
area = 189 mm^2

2. $A = 52°2', c = 7.568$ cm, $a = 7.152$ cm,
area = 25.65 cm^2

3. $D = 19°48', E = 134°12', e = 36.0$ cm,
area = 134 cm^2

4. $E = 49°0', F = 26°38', f = 15.09$ mm,
area = 185.6 mm^2

5. $J = 44°29', L = 99°31', l = 5.420$ cm,
area = 6.133 cm^2, or, $J = 135°31', L = 8°29',$
$l = 0.811$ cm, area = 0.917 cm^2

6. $K = 47°8', J = 97°52', j = 62.2$ mm,
area = 820.2 mm^2 or $K = 132°52', J = 12°8',$
$j = 13.19$ mm, area = 174.0 mm^2

Exercise 91 (page 234)

1. $p = 13.2$ cm, $Q = 47.34°, R = 78.66°,$
area = 77.7 cm^2

2. $p = 6.127$ m, $Q = 30.83°, R = 44.17°,$
area = 6.938 m^2

3. $X = 83.33°, Y = 52.62°, Z = 44.05°,$
area = 27.8 cm^2

4. $X = 29.77°, Y = 53.50°, Z = 96.73°,$
area = 355 mm^2

Exercise 92 (page 236)

1. 193 km   **2.** (a) 122.6 m (b) $94.80°, 40.66°, 44.54°$
3. (a) 11.4 m (b) $17.55°$ **4.** 163.4 m
5. $BF = 3.9$ m, $EB = 4.0$ m   **6.** 6.35 m, 5.37 m
7. 32.48 A, $14.31°$

Exercise 93 (page 238)

1. $80.42°, 59.38°, 40.20°$ **2.** (a) 15.23 m (b) $38.07°$
3. 40.25 cm, $126.05°$ **4.** 19.8 cm   **5.** 36.2 m
6. $x = 69.3$ mm, $y = 142$ mm   **7.** $130°$ **8.** 13.66 mm

Chapter 24

Exercise 94 (page 241)

1. $(5.83, 59.04°)$ or $(5.83, 1.03$ rad$)$
2. $(6.61, 20.82°)$ or $(6.61, 0.36$ rad$)$
3. $(4.47, 116.57°)$ or $(4.47, 2.03$ rad$)$
4. $(6.55, 145.58°)$ or $(6.55, 2.54$ rad$)$
5. $(7.62, 203.20°)$ or $(7.62, 3.55$ rad$)$
6. $(4.33, 236.31°)$ or $(4.33, 4.12$ rad$)$
7. $(5.83, 329.04°)$ or $(5.83, 5.74$ rad$)$
8. $(15.68, 307.75°)$ or $(15.68, 5.37$ rad$)$

Exercise 95 (page 242)

1. $(1.294, 4.830)$ 2. $(1.917, 3.960)$
3. $(-5.362, 4.500)$ 4. $(-2.884, 2.154)$
5. $(-9.353, -5.400)$ 6. $(-2.615, -3.027)$
7. $(0.750, -1.299)$ 8. $(4.252, -4.233)$
9. (a) $40\angle18°, 40\angle90°, 40\angle162°, 40\angle234°, 40\angle306°$
   (b) $(38.04, 12.36), (0, 40), (-38.04, 12.36),$
   $(-23.51, -32.36), (23.51, -32.36)$
10. $47.0\,$mm

Chapter 25

Exercise 96 (page 251)

1. $p = 105°, q = 35°$ 2. $r = 142°, s = 95°$
3. $t = 146°$

Exercise 97 (page 255)

1. (i) rhombus (a) $14\,$cm^2 (b) $16\,$cm (ii) parallelogram
   (a) $180\,$mm^2 (b) $80\,$mm (iii) rectangle (a) $3600\,$mm^2
   (b) $300\,$mm (iv) trapezium (a) $190\,$cm^2 (b) $62.91\,$cm
2. $35.7\,$cm^2 3. (a) $80\,$m (b) $170\,$m   4. $27.2\,$cm^2
5. $18\,$cm   6. $1200\,$mm
7. (a) $29\,$cm^2 (b) $650\,$mm^2 8. $560\,$m^2
9. $3.4\,$cm   10. $6750\,$mm^2 11. $43.30\,$cm^2
12. 32 13. $230, 400$

Exercise 98 (page 257)

1. $482\,$m^2
2. (a) $50.27\,$cm^2 (b) $706.9\,$mm^2 (c) $3183\,$mm^2
3. $2513\,$mm^2 4. (a) $20.19\,$mm (b) $63.41\,$mm
5. (a) $53.01\,$cm^2 (b) $129.9\,$mm^2 6. $5773\,$mm^2
7. $1.89\,$m^2

Exercise 99 (page 259)

1. $1932\,$mm^2 2. $1624\,$mm^2 3. (a) $0.918\,$ha (b) $456\,$m

Exercise 100 (page 260)

1. $80\,$ha   2. $80\,$m^2 3. $3.14\,$ha

Chapter 26

Exercise 101 (page 263)

1. $45.24\,$cm   2. $259.5\,$mm   3. $2.629\,$cm   4. $47.68\,$cm
5. $38.73\,$cm   6. $12730\,$km   7. $97.13\,$mm

Exercise 102 (page 264)

1. (a) $\dfrac{\pi}{6}$ (b) $\dfrac{5\pi}{12}$ (c) $\dfrac{5\pi}{4}$
2. (a) 0.838 (b) 1.481 (c) 4.054
3. (a) $210°$ (b) $80°$ (c) $105°$
4. (a) $0°43'$ (b) $154°8'$ (c) $414°53'$ 5. $104.7\,$rad/s

Exercise 103 (page 266)

1. $113\,$cm^2 2. $2376\,$mm^2 3. $1790\,$mm^2
4. $802\,$mm^2 5. $1709\,$mm^2 6. $1269\,$m^2
7. $1548\,$m^2 8. (a) $106.0\,$cm (b) $783.9\,$cm^2
9. $21.46\,$m^2 10. $17.80\,$cm, $74.07\,$cm^2
11. (a) $59.86\,$mm (b) $197.8\,$mm   12. $26.2\,$cm
13. $202\,$mm^2 14. $8.67\,$cm, $54.48\,$cm 15. $82.5°$
16. 748 17. (a) $0.698\,$rad (b) $804.2\,$m^2
18. $10.47\,$m^2
19. (a) $396\,$mm^2 (b) 42.24% 20. $701.8\,$mm
21. $7.74\,$mm

Exercise 104 (page 269)

1. (a) 2 (b) $(3, -4)$ 2. Centre at $(3, -2)$, radius 4
3. Circle, centre $(0, 1)$, radius 5
4. Circle, centre $(0, 0)$, radius 6

Chapter 27

Exercise 105 (page 275)

1. $1.2\,$m^3 2. $5\,$cm^3 3. $8\,$cm^3
4. (a) $3840\,$mm^3 (b) $1792\,$mm^2
5. $972\,$litres   6. $15\,$cm^3, $135\,$g   7. $500\,$litres
8. $1.44\,$m^3 9. (a) $35.3\,$cm^3 (b) $61.3\,$cm^2
10. (a) $2400\,$cm^3 (b) $2460\,$cm^2 11. $37.04\,$m
12. $1.63\,$cm   13. $8796\,$cm^3
14. $4.709\,$cm, $153.9\,$cm^2
15. $2.99\,$cm   16. $28060\,$cm^3, $1.099\,$m^2
17. $8.22\,$m by $8.22\,$m   18. $62.5\,$min
19. $4\,$cm   20. $4.08\,$m^3

Exercise 106 (page 279)

1. $201.1\,$cm^3, $159.0\,$cm^2 2. $7.68\,$cm^3, $25.81\,$cm^2
3. $113.1\,$cm^3, $113.1\,$cm^2 4. $5.131\,$cm   5. $3\,$cm
6. $2681\,$mm^3 7. (a) $268083\,$mm^3 or $268.083\,$cm^3
   (b) $20106\,$mm^2 or $201.06\,$cm^2
8. $8.53\,$cm
9. (a) $512 \times 10^6\,$km^2 (b) $1.09 \times 10^{12}\,$km^3 10. 664
11. $92\,$m^3, $92,000\,$litres

Exercise 107 (page 284)

1. $5890\,\text{mm}^2$ or $58.90\,\text{cm}^2$
2. (a) $56.55\,\text{cm}^3$ (b) $84.82\,\text{cm}^2$ 3. $13.57\,\text{kg}$
4. $29.32\,\text{cm}^3$ 5. $393.4\,\text{m}^2$
6. (i) (a) $670\,\text{cm}^3$ (b) $523\,\text{cm}^2$ (ii) (a) $180\,\text{cm}^3$
   (b) $154\,\text{cm}^2$ (iii) (a) $56.5\,\text{cm}^3$ (b) $84.8\,\text{cm}^2$
   (iv) (a) $10.4\,\text{cm}^3$ (b) $32.0\,\text{cm}^2$ (v) (a) $96.0\,\text{cm}^3$
   (b) $146\,\text{cm}^2$ (vi) (a) $86.5\,\text{cm}^3$ (b) $142\,\text{cm}^2$
   (vii) (a) $805\,\text{cm}^3$ (b) $539\,\text{cm}^2$
7. (a) $17.9\,\text{cm}$ (b) $38.0\,\text{cm}$ 8. $125\,\text{cm}^3$
9. $10.3\,\text{m}^3, 25.5\,\text{m}^2$ 10. $6560\,\text{litres}$
11. $657.1\,\text{cm}^3, 1027\,\text{cm}^2$ 12. $220.7\,\text{cm}^3$
13. (a) $1458\,\text{litres}$ (b) $9.77\,\text{m}^2$ (c) £140.45

Exercise 108 (page 288)

1. $147\,\text{cm}^3, 164\,\text{cm}^2$ 2. $403\,\text{cm}^3, 337\,\text{cm}^2$
3. $10480\,\text{m}^3, 1852\,\text{m}^2$ 4. $1707\,\text{cm}^2$
5. $10.69\,\text{cm}$ 6. $55910\,\text{cm}^3, 6051\,\text{cm}^2$
7. $5.14\,\text{m}$

Exercise 109 (page 289)

1. $8:125$ 2. $137.2\,\text{g}$

Chapter 28

Exercise 110 (page 292)

1. 4.5 square units   2. 54.7 square units   3. $63.33\,\text{m}$
4. $4.70\,\text{ha}$ 5. $143\,\text{m}^2$

Exercise 111 (page 293)

1. $42.59\,\text{m}^3$ 2. $147\,\text{m}^3$ 3. $20.42\,\text{m}^3$

Exercise 112 (page 297)

1. (a) $2\,\text{A}$ (b) $50\,\text{V}$ (c) $2.5\,\text{A}$ 2. (a) $2.5\,\text{mV}$ (b) $3\,\text{A}$
3. $0.093\,\text{As}, 3.1\,\text{A}$ 4. (a) $31.83\,\text{V}$ (b) 0
5. $49.13\,\text{cm}^2, 368.5\,\text{kPa}$

Chapter 29

Exercise 113 (page 301)

1. A scalar quantity has magnitude only; a vector
   quantity has both magnitude and direction.
2. scalar   3. scalar   4. vector   5. scalar
6. scalar   7. vector   8. scalar   9. vector

Exercise 114 (page 307)

1. $17.35\,\text{N}$ at $18.00°$ to the $12\,\text{N}$ force
2. $13\,\text{m/s}$ at $22.62°$ to the $12\,\text{m/s}$ velocity
3. $16.40\,\text{N}$ at $37.57°$ to the $13\,\text{N}$ force
4. $28.43\,\text{N}$ at $129.29°$ to the $18\,\text{N}$ force
5. $32.31\,\text{m}$ at $21.80°$ to the $30\,\text{m}$ displacement
6. $14.72\,\text{N}$ at $-14.72°$ to the $5\,\text{N}$ force
7. $29.15\,\text{m/s}$ at $29.04°$ to the horizontal
8. $9.28\,\text{N}$ at $16.70°$ 9. $6.89\,\text{m/s}$ at $159.56°$
10. $15.62\,\text{N}$ at $26.33°$ to the $10\,\text{N}$ force
11. $21.07\,\text{knots}$, E $9.22°$S

Exercise 115 (page 310)

1. (a) $54.0\,\text{N}$ at $78.16°$ (b) $45.64\,\text{N}$ at $4.66°$
2. (a) $31.71\,\text{m/s}$ at $121.81°$ (b) $19.55\,\text{m/s}$ at $8.63°$

Exercise 116 (page 311)

1. $83.5\,\text{km/h}$ at $71.6°$ to the vertical
2. 4 minutes 55 seconds, $60°$
3. $22.79\,\text{km/h}$, E $9.78°$N

Exercise 117 (page 312)

1. $i - j - 4k$ 2. $4i + j - 6k$
3. $-i + 7j - k$ 4. $5i - 10k$
5. $-3i + 27j - 8k$ 6. $-5i + 10k$
7. $i + 7.5j - 4k$ 8. $20.5j - 10k$
9. $3.6i + 4.4j - 6.9k$ 10. $2i + 40j - 43k$

Chapter 30

Exercise 118 (page 315)

1. $4.5\sin(A + 63.5°)$
2. (a) $20.9\sin(\omega t + 0.63)$ volts
   (b) $12.5\sin(\omega t - 1.36)$ volts
3. $13\sin(\omega t + 0.393)$ volts

Exercise 119 (page 316)

1. $4.5\sin(A + 63.5°)$
2. (a) $20.9\sin(\omega t + 0.62)$ volts
   (b) $12.5\sin(\omega t - 1.33)$ volts
3. $13\sin(\omega t + 0.40)$

Exercise 120 (page 318)

1. $4.472\sin(\omega t + 63.44°)$
2. (a) $20.88\sin(\omega t + 0.62)$ volts
   (b) $12.50\sin(\omega t - 1.33)$ volts
3. $13\sin(\omega t + 0.395)$ **4.** $11.11\sin(\omega t + 0.324)$
5. $8.73\sin(\omega t - 0.173)$ **6.** $1.01\sin(\omega t - 0.698)$A

Exercise 121 (page 320)

1. $11.11\sin(\omega t + 0.324)$
2. $8.73\sin(\omega t - 0.173)$
3. $i = 21.79\sin(\omega t - 0.639)$
4. $v = 5.695\sin(\omega t + 0.695)$
5. $x = 14.38\sin(\omega t + 1.444)$
6. (a) $305.3\sin(314.2t - 0.233)$ V (b) $50\,$Hz
7. (a) $10.21\sin(628.3t + 0.818)$ V (b) $100\,$Hz
   (c) $10\,$ms
8. (a) $79.83\sin(300\pi t + 0.352)$V (b) $150\,$Hz
   (c) $6.667\,$ms
9. $150.6\sin(\omega t - 0.247)$ volts

Chapter 31

Exercise 122 (page 327)

1. (a) continuous (b) continuous (c) discrete
   (d) continuous
2. (a) discrete (b) continuous (c) discrete (d) discrete

Exercise 123 (page 330)

1. If one symbol is used to represent 10 vehicles, working correct to the nearest 5 vehicles, gives 3.5, 4.5, 6, 7, 5 and 4 symbols respectively.
2. If one symbol represents 200 components, working correct to the nearest 100 components gives: Mon 8, Tues 11, Wed 9, Thurs 12 and Fri 6.5.
3. Six equally spaced horizontal rectangles, whose lengths are proportional to 35, 44, 62, 68, 49 and 41, respectively.
4. Five equally spaced horizontal rectangles, whose lengths are proportional to 1580, 2190, 1840, 2385 and 1280 units, respectively.
5. Six equally spaced vertical rectangles, whose heights are proportional to 35, 44, 62, 68, 49 and 41 units, respectively.
6. Five equally spaced vertical rectangles, whose heights are proportional to 1580, 2190, 1840, 2385 and 1280 units, respectively.

7. Three rectangles of equal height, subdivided in the percentages shown in the columns of the question. P increases by 20% at the expense of Q and R.
8. Four rectangles of equal height, subdivided as follows: week 1: 18%, 7%, 35%, 12%, 28%; week 2: 20%, 8%, 32%, 13%, 27%; week 3: 22%, 10%, 29%, 14%, 25%; week 4: 20%, 9%, 27%, 19%, 25%. Little change in centres A and B, a reduction of about 8% in C, an increase of about 7% in D and a reduction of about 3% in E.
9. A circle of any radius, subdivided into sectors having angles of $7.5°, 22.5°, 52.5°, 167.5°$ and $110°$, respectively.
10. A circle of any radius, subdivided into sectors having angles of $107°, 156°, 29°$ and $68°$, respectively.
11. (a) £495 (b) 88   **12.** (a) £16 450 (b) 138

Exercise 124 (page 336)

1. There is no unique solution, but one solution is:
39.3–39.4   1; 39.5–39.6   5; 39.7–39.8   9;
39.9–40.0 17; 40.1–40.2 15; 40.3–40.4   7;
40.5–40.6   4; 40.7–40.8   2.
2. Rectangles, touching one another, having midpoints of $39.35, 39.55, 39.75, 39.95, \ldots$ and heights of $1, 5, 9, 17, \ldots$
3. There is no unique solution, but one solution is:
20.5–20.9   3; 21.0–21.4 10; 21.5–21.9 11;
22.0–22.4 13; 22.5–22.9   9; 23.0–23.4   2.
4. There is no unique solution, but one solution is:
1–10   3; 11–19   7; 20–22   12; 23–25   11;
26–28 10; 29–38   5; 39–48   2.
5. 20.95 3; 21.45 13; 21.95 24; 22.45 37; 22.95 46; 23.45 48
6. Rectangles, touching one another, having midpoints of 5.5, 15, 21, 24, 27, 33.5 and 43.5. The heights of the rectangles (frequency per unit class range) are 0.3, 0.78, 4, 4.67, 2.33, 0.5 and 0.2.
7. (10.95 2), (11.45 9), (11.95 19), (12.45 31), (12.95 42), (13.45 50)
8. A graph of cumulative frequency against upper class boundary having co-ordinates given in the answer to Problem 7.
9. (a)   There is no unique solution, but one solution is:
2.05–2.09   3; 2.10–2.14 10; 2.15–2.19 11;
2.20–2.24 13; 2.25–2.29   9; 2.30–2.34   2.

   (b)   Rectangles, touching one another, having midpoints of $2.07, 2.12, \ldots$ and heights of $3, 10, \ldots$

(c) Using the frequency distribution given in the solution to part (a) gives 2.095 3; 2.145 13; 2.195 24; 2.245 37; 2.295 46; 2.345 48

(d) A graph of cumulative frequency against upper class boundary having the co-ordinates given in part (c).

Chapter 32

Exercise 125 (page 340)

1. Mean 7.33, median 8, mode 8
2. Mean 27.25, median 27, mode 26
3. Mean 4.7225, median 4.72, mode 4.72
4. Mean 115.2, median 126.4, no mode

Exercise 126 (page 341)

1. $23.85\,\text{kg}$ 2. $171.7\,\text{cm}$
3. Mean 89.5, median 89, mode 88.2
4. Mean $2.02158\,\text{cm}$, median $2.02152\,\text{cm}$, mode $2.02167\,\text{cm}$

Exercise 127 (page 343)

1. 4.60   2. $2.83\,\mu\text{F}$
3. Mean $34.53\,\text{MPa}$, standard deviation $0.07474\,\text{MPa}$
4. $0.296\,\text{kg}$ 5. $9.394\,\text{cm}$ 6. $0.00544\,\text{cm}$

Exercise 128 (page 344)

1. 30, 27.5, 33.5 days   2. 27, 26, 33 faults
3. $Q_1 = 164.5\,\text{cm}$, $Q_2 = 172.5\,\text{cm}$, $Q_3 = 179\,\text{cm}$, $7.25\,\text{cm}$
4. 37 and 38; 40 and 41   5. 40, 40, 41; 50, 51, 51

Chapter 33

Exercise 129 (page 348)

1. (a) $\dfrac{2}{9}$ or 0.2222 (b) $\dfrac{7}{9}$ or 0.7778

2. (a) $\dfrac{23}{139}$ or 0.1655 (b) $\dfrac{47}{139}$ or 0.3381
   (c) $\dfrac{69}{139}$ or 0.4964

3. (a) $\dfrac{1}{6}$ (b) $\dfrac{1}{6}$ (c) $\dfrac{1}{36}$ 4. $\dfrac{5}{36}$

5. (a) $\dfrac{2}{5}$ (b) $\dfrac{1}{5}$ (c) $\dfrac{4}{15}$ (d) $\dfrac{13}{15}$

6. (a) $\dfrac{1}{250}$ (b) $\dfrac{1}{200}$ (c) $\dfrac{9}{1000}$ (d) $\dfrac{1}{50000}$

Exercise 130 (page 350)

1. (a) 0.6 (b) 0.2 (c) 0.15   2. (a) 0.64 (b) 0.32
3. 0.0768   4. (a) 0.4912 (b) 0.4211
5. (a) 89.38% (b) 10.25%
6. (a) 0.0227 (b) 0.0234 (c) 0.0169

Chapter 34

Exercise 131 (page 356)

1. 1, 5, 21, 9, 61   2. 0, 11, −10, 21   3. proof

Exercise 132 (page 357)

1. 16, 8

Exercise 133 (page 360)

1. $28x^3$ 2. 2   3. $2x - 1$

4. $6x^2 - 5$ 5. $-\dfrac{1}{x^2}$ 6. 0

7. $1 + \dfrac{2}{x^3}$ 8. $15x^4 - 8x^3 + 15x^2 + 2x$

9. $-\dfrac{6}{x^4}$ 10. $4 - 8x$ 11. $\dfrac{1}{2\sqrt{x}}$

12. $\dfrac{3}{2}\sqrt{t}$ 13. $-\dfrac{3}{x^4}$

14. $3 + \dfrac{1}{2\sqrt{x^3}} - \dfrac{1}{x^2}$ 15. $2x + 2$

16. $1 + \dfrac{3}{2\sqrt{x}}$ 17. $2x - 2$

18. $-\dfrac{10}{x^3} + \dfrac{7}{2\sqrt{x^9}}$ 19. $6t - 12$ 20. $1 - \dfrac{4}{x^2}$

21. (a) 6 (b) $\dfrac{1}{6}$ (c) 3 (d) $-\dfrac{1}{16}$ (e) $-\dfrac{1}{4}$ (f) -7

22. $12x - 3$ (a) -15 (b) 21
23. $6x^2 + 6x - 4, 32$ 24. $-6x^2 + 4, -9.5$

Exercise 134 (page 362)

1. (a) $12\cos 3x$ (b) $-12\sin 6x$
2. $6\cos 3\theta + 10\sin 2\theta$ 3. -0.707 4. -3
5. $270.2\,\text{A/s}$ 6. $1393.4\,\text{V/s}$
7. $12\cos(4t + 0.12) + 6\sin(3t - 0.72)$

Exercise 135 (page 364)

1. (a) $15e^{3x}$ (b) $-\dfrac{4}{7e^{2x}}$ **2.** $\dfrac{5}{\theta} - \dfrac{4}{\theta} = \dfrac{1}{\theta}$ **3.** 16

4. 2.80 **5.** 664

Exercise 136 (page 364)

1. (a) -1 (b) 16

2. $-\dfrac{4}{x^3} + \dfrac{2}{x} + 10\sin 5x - 12\cos 2x + \dfrac{6}{e^{3x}}$

Exercise 137 (page 365)

1. (a) $36x^2 + 12x$ (b) $72x + 12$ **2.** $8 + \dfrac{2}{x^3}$

3. (a) $\dfrac{4}{5} - \dfrac{12}{t^5} + \dfrac{6}{t^3} + \dfrac{1}{4\sqrt{t^3}}$ (b) -4.95

4. $-12\sin 2t - \cos t$ **5.** Proof

Exercise 138 (page 366)

1. $-2542\,\text{A/s}$ **2.** (a) $0.16\,\text{cd/V}$ (b) $312.5\,\text{V}$

3. (a) $-1000\,\text{V/s}$ (b) $-367.9\,\text{V/s}$

4. $-1.635\,\text{Pa/m}$

Chapter 35

Exercise 139 (page 371)

1. (a) $4x + c$ (b) $\dfrac{7x^2}{2} + c$

2. (a) $\dfrac{5}{4}x^4 + c$ (b) $\dfrac{3}{8}t^8 + c$

3. (a) $\dfrac{2}{15}x^3 + c$ (b) $\dfrac{5}{24}x^4 + c$

4. (a) $\dfrac{2}{5}x^5 - \dfrac{3}{2}x^2 + c$ (b) $2t - \dfrac{3}{4}t^4 + c$

5. (a) $\dfrac{3x^2}{2} - 5x + c$

   (b) $4\theta + 2\theta^2 + \dfrac{\theta^3}{3} + c$

6. (a) $\dfrac{5}{2}\theta^2 - 2\theta + \theta^3 + c$

   (b) $\dfrac{3}{4}x^4 - \dfrac{2}{3}x^3 + \dfrac{3}{2}x^2 - 2x + c$

7. (a) $-\dfrac{4}{3x} + c$ (b) $-\dfrac{1}{4x^3} + c$

8. (a) $\dfrac{4}{5}\sqrt{x^5} + c$ (b) $\dfrac{1}{9}\sqrt[4]{x^9} + c$

9. (a) $\dfrac{10}{\sqrt{t}} + c$ (b) $\dfrac{15}{7}\sqrt[5]{x} + c$

10. (a) $\dfrac{3}{2}\sin 2x + c$ (b) $-\dfrac{7}{3}\cos 3\theta + c$

11. (a) $-6\cos\dfrac{1}{2}x + c$ (b) $18\sin\dfrac{1}{3}x + c$

12. (a) $\dfrac{3}{8}e^{2x} + c$ (b) $\dfrac{-2}{15e^{5x}} + c$

13. (a) $\dfrac{2}{3}\ln x + c$ (b) $\dfrac{u^2}{2} - \ln u + c$

14. (a) $8\sqrt{x} + 8\sqrt{x^3} + \dfrac{18}{5}\sqrt{x^5} + c$

   (b) $-\dfrac{1}{t} + 4t + \dfrac{4t^3}{3} + c$

Exercise 140 (page 373)

1. (a) 1.5 (b) 0.5 **2.** (a) 105 (b) -0.5

3. (a) 6 (b) -1.333 **4.** (a) -0.75 (b) 0.8333

5. (a) 10.67 (b) 0.1667 **6.** (a) 0 (b) 4

7. (a) 1 (b) 4.248 **8.** (a) 0.2352 (b) 2.638

9. (a) 19.09 (b) 2.457 **10.** (a) 0.2703 (b) 9.099

11. $77.7\,\text{m}^3$

Exercise 141 (page 377)

1. proof **2.** proof **3.** 32 **4.** $29.33\,\text{N m}$

5. 37.5 **6.** 7.5 **7.** 1

8. 1.67 **9.** 2.67 **10.** $140\,\text{m}$

Chapter 36

Exercise 142 (page 383)

1. 21, 25 **2.** 48, 96 **3.** 14, 7

4. $-3, -8$ **5.** 50, 65 **6.** 0.001, 0.0001

7. 54, 79

Exercise 143 (page 384)

1. 1, 3, 5, 7, .. **2.** 7, 10, 13, 16, 19, ..

3. 6, 11, 16, 21, .. **4.** $5n$

5. $6n - 2$ **6.** $2n + 1$

7. $4n - 2$ **8.** $3n + 6$

9. $6^3 (= 216), 7^3 (= 343)$

Exercise 144 (page 385)

1. 68   2. 6.2   3. 85.25
4. 23.5   5. 11th   6. 209
7. 346.5

Exercise 145 (page 386)

1. −0.5   2. 1.5, 3, 4.5
3. 7808   4. 25
5. 8.5, 12, 15.5, 19
6. (a) 120 (b) 26070 (c) 250.5
7. £19,840, £223,680   8. £8720

Exercise 146 (page 388)

1. 2560   2. 273.25   3. 512, 4096
4. 812.5   5. 8   6. $1\frac{2}{3}$

Exercise 147 (page 390)

1. (a) 3 (b) 2 (c) 59022
2. 10th
3. £1566, 11 years
4. 56.68 M
5. 71.53 g
6. (a) £599.14 (b) 19 years
7. 100, 139, 193, 268, 373, 518, 720, 1000 rev/min.

Chapter 37

Exercise 148 (page 392)

1. (a) 6_{10} (b) 11_{10} (c)14_{10} (d) 9_{10}
2. (a) 21_{10} (b) 25_{10} (c) 45_{10} (d) 51_{10}
3. (a) 42_{10} (b) 56_{10} (c) 65_{10} (d) 184_{10}
4. (a) 0.8125_{10} (b) 0.78125_{10} (c) 0.21875_{10}
   (d) 0.34375_{10}
5. (a) 26.75_{10} (b) 23.375_{10} (c) 53.4375_{10}
   (d) 213.71875_{10}

Exercise 149 (page 394)

1. (a) 101_2 (b) 1111_2 (c) 10011_2 (d) 11101_2
2. (a) 11111_2 (b) 101010_2 (c) 111001_2 (d) 111111_2
3. (a) 101111_2 (b) 111100_2 (c) 1001001_2
   (d) 1010100_2
4. (a) 0.01_2 (b) 0.00111_2 (c) 0.01001_2 (d) 0.10011_2
5. (a) 101111.01101_2 (b) 11110.1101_2
   (c) 110101.11101_2 (d) 111101.10101_2

Exercise 150 (page 395)

1. 101   2. 1011
3. 10100   4. 101100
5. 1001000   6. 100001010
7. 1010110111   8. 1001111101
9. 111100   10. 110111
11. 110011   12. 1101110

Exercise 151 (page 397)

1. (a) 101010111_2 (b) 1000111100_2 (c) 10011110001_2
2. (a) 0.01111_2 (b) 0.1011_2 (c) 0.10111_2
3. (a) 11110111.00011_2 (b) 1000000010.0111_2
   (c) 11010110100.11001_2
4. (a) 7.4375_{10} (b) 41.25_{10} (c) 7386.1875_{10}

Exercise 152 (page 399)

1. 231_{10} 2. 44_{10} 3. 152_{10}
4. 753_{10} 5. 36_{16} 6. $C8_{16}$
7. $5B_{16}$ 8. EE_{16}

Exercise 153 (page 400)

1. $D7_{16}$ 2. EA_{16} 3. $8B_{16}$
4. $A5_{16}$ 5. 110111_2 6. 111011101_2
7. 10011111_2 8. 1010001000001_2

Chapter 38

Exercise 154 (page 402)

1. (a) $t > 2$ (b) $x < 5$
2. (a) $x > 3$ (b) $x \geq 3$
3. (a) $t \leq 1$ (b) $x \leq 6$
4. (a) $k \geq \frac{3}{2}$ (b) $z > \frac{1}{2}$
5. (a) $y \geq -\frac{4}{3}$ (b) $x \geq -\frac{1}{2}$

Exercise 155 (page 403)

1. $-5 < t < 3$
2. $-5 \leq y \leq -1$
3. $-\frac{3}{2} < x < \frac{5}{2}$
4. $t > 3$ and $t < \frac{1}{3}$
5. $k \geq 4$ and $k \leq -2$

Exercise 156 (page 404)

1. $-4 \leq x \leq 3$ **2.** $t > 5$ or $t < -9$
3. $-5 < z \leq 14$ **4.** $-3 < x \leq -2$

Exercise 157 (page 405)

1. $z > 4$ or $z < -4$ **2.** $-4 < z < 4$
3. $x \geq \sqrt{3}$ or $x \leq -\sqrt{3}$ **4.** $-2 \leq k \leq 2$
5. $-5 \leq t \leq 7$ **6.** $t \geq 7$ or $t \leq -5$
7. $y \geq 2$ or $y \leq -2$ **8.** $k > -\dfrac{1}{2}$ or $k < -2$

Exercise 158 (page 406)

1. $x > 3$ or $x < -2$
2. $-4 \leq t \leq 2$
3. $-2 < x < \dfrac{1}{2}$
4. $y \geq 5$ or $y \leq -4$
5. $-4 \leq z \leq 0$
6. $\left(-\sqrt{3}-3\right) \leq x \leq \left(\sqrt{3}-3\right)$
7. $t \geq \left(\sqrt{11}+2\right)$ or $t \leq \left(2-\sqrt{11}\right)$

8. $k \geq \left(\sqrt{\dfrac{13}{4}}+\dfrac{1}{2}\right)$ or $k \leq \left(-\sqrt{\dfrac{13}{4}}+\dfrac{1}{2}\right)$

Chapter 39

Exercise 159 (page 411)

1. $a = 12, n = 1.8, 451, 28.5$

2. $k = 1.5, n = -1$ **3.** $m = 3, n = 2.5$

Exercise 160 (page 413)

1. (i) $a = -8, b = 5.3, p = -8(5.3)^{q}$ (ii) -224.7
(iii) 3.31

Exercise 161 (page 415)

1. $a = 76, k = -7 \times 10^{-5}, p = 76\, e^{-7 \times 10^{-5} h}$,
37.74 cm

2. $\theta_0 = 152, k = -0.05$

Answers to multiple choice questions

Test 1 (Page 92)

1. (a)	**2.** (c)	**3.** (c)	**4.** (c)	**5.** (b)
6. (a)	**7.** (d)	**8.** (d)	**9.** (b)	**10.** (b)
11. (b)	**12.** (b)	**13.** (c)	**14.** (d)	**15.** (a)
16. (b)	**17.** (d)	**18.** (d)	**19.** (a)	**20.** (c)
21. (a)	**22.** (c)	**23.** (c)	**24.** (a)	**25.** (d)

Test 2 (Page 146)

1. (b)	**2.** (b)	**3.** (c)	**4.** (b)	**5.** (a)
6. (b)	**7.** (c)	**8.** (a)	**9.** (a)	**10.** (a)
11. (d)	**12.** (b)	**13.** (d)	**14.** (c)	**15.** (b)
16. (b)	**17.** (c)	**18.** (d)	**19.** (a)	**20.** (a)
21. (d)	**22.** (d)	**23.** (c)	**24.** (c)	**25.** (b)
26. (c)	**27.** (c)	**28.** (a)	**29.** (d)	**30.** (d)

Test 3 (page 245)

1. (d)	**2.** (a)	**3.** (b)	**4.** (d)	**5.** (a)
6. (b)	**7.** (d)	**8.** (a)	**9.** (c)	**10.** (d)
11. (c)	**12.** (a)	**13.** (b)	**14.** (a)	**15.** (c)
16. (c)	**17.** (b)	**18.** (d)	**19.** (c)	**20.** (b)
21. (a)	**22.** (d)	**23.** (a)	**24.** (b)	**25.** (c)
26. (a)	**27.** (b)	**28.** (c)	**29.** (d)	**30.** (b)
31. (d)	**32.** (a)	**33.** (b)	**34.** (c)	**35.** (b)
36. (c)	**37.** (c)	**38.** (a)	**39.** (d)	**40.** (a)

Test 4 (page 323)

1. (c)	**2.** (c)	**3.** (d)	**4.** (a)	**5.** (d)
6. (c)	**7.** (b)	**8.** (d)	**9.** (a)	**10.** (b)
11. (a)	**12.** (c)	**13.** (d)	**14.** (a)	**15.** (d)
16. (c)	**17.** (b)	**18.** (c)	**19.** (b)	**20.** (b)
21. (a)	**22.** (a)	**23.** (b)	**24.** (d)	**25.** (b)

Test 5 (page 353)

1. (c)	**2.** (c)	**3.** (d)	**4.** (b)	**5.** (b)
6. (d)	**7.** (a)	**8.** (b)	**9.** (c)	**10.** (b)
11. (a)	**12.** (d)	**13.** (b)	**14.** (c)	**15.** (a)

Test 6 (page 380)

1. (b)	**2.** (d)	**3.** (a)	**4.** (a)	**5.** (a)
6. (b)	**7.** (a)	**8.** (d)	**9.** (d)	**10.** (c)
11. (b)	**12.** (a)	**13.** (c)	**14.** (b)	**15.** (c)
16. (c)	**17.** (b)	**18.** (d)	**19.** (c)	**20.** (c)
21. (a)	**22.** (b)	**23.** (d)	**24.** (b)	**25.** (d)

Test 7 (page 417)

1. (d)	**2.** (a)	**3.** (b)	**4.** (c)	**5.** (a)
6. (d)	**7.** (b)	**8.** (d)	**9.** (a)	**10.** (c)
11. (c)	**12.** (c)	**13.** (b)	**14.** (d)	**15.** (b)

Index

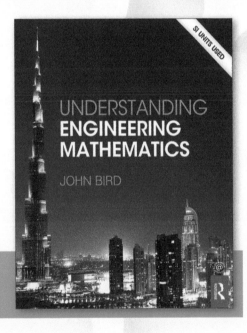